T0306120

HEAT TRANSFER AND FLUID FLOW IN BIOLOGICAL PROCESSES

HEAT TRANSFER AND FLUID FLOW IN BIOLOGICAL PROCESSES

Edited by

SID M. BECKER
*University of Canterbury,
Christchurch, New Zealand*

ANDREY V. KUZNETSOV
*North Carolina State University,
Raleigh, NC, USA*

AMSTERDAM • BOSTON • HEIDELBERG • LONDON
NEW YORK • OXFORD • PARIS • SAN DIEGO
SAN FRANCISCO • SINGAPORE • SYDNEY • TOKYO
Academic Press is an imprint of Elsevier

Academic Press is an imprint of Elsevier
32 Jamestown Road, London NW1 7BY, UK
525 B Street, Suite 1800, San Diego, CA 92101-4495, USA
225 Wyman Street, Waltham, MA 02451, USA
The Boulevard, Langford Lane, Kidlington, Oxford OX5 1GB, UK

Notices
Knowledge and best practice in this field are constantly changing. As new research
and experience broaden our understanding, changes in research methods,
professional practices, or medical treatment may become necessary.

Practitioners and researchers must always rely on their own experience and
knowledge in evaluating and using any information, methods, compounds, or
experiments described herein. In using such information or methods they should
be mindful of their own safety and the safety of others, including parties for whom
they have a professional responsibility.

To the fullest extent of the law, neither the Publisher nor the authors, contributors,
or editors, assume any liability for any injury and/or damage to persons or
property as a matter of products liability, negligence or otherwise, or from any use
or operation of any methods, products, instructions, or ideas contained in the
material herein.

Library of Congress Cataloging-in-Publication Data
A catalog record for this book is available from the Library of Congress

British Library Cataloguing-in-Publication Data
A catalogue record for this book is available from the British Library

For information on all Academic Press publications
visit our website at http://store.elsevier.com/

Printed and bound in the USA

ISBN: 978-0-12-408077-5

ELSEVIER Book Aid International Working together to grow libraries in developing countries

www.elsevier.com • www.bookaid.org

Contents

Contributors

Terpsichori S. Alexiou University of Patras, Patras, Greece

Stavros Pavlou University of Patras, Patras, Greece

C. Pozrikidis University of Massachusetts, Amherst, MA, USA

Sid M. Becker University of Canterbury, Christchurch, New Zealand

Claudio Chiastra Politecnico di Milano, Milano, Italy

R.J. Clarke University of Auckland, Auckland, New Zealand

Filippo de Monte University of L'Aquila, L'Aquila, Italy

Katharine Fraser Imperial College London, London, UK

Mohit Ganguly Florida Institute of Technology, Melbourne, FL, USA

B.P. Hierck Leiden University Medical Center, Leiden, The Netherlands

Tzyy-Leng Horng Feng Chia University, Taichung, Taiwan

Huang-Wen Huang Tamkang University, Taipei, Taiwan

George E. Kapellos University of Patras, Patras, Greece

Jong Won Kim Central Michigan University, Mount Pleasant, MI, USA

A.V. Kuznetsov North Carolina State University, Raleigh, NC, USA

I.A. Kuznetsov Johns Hopkins University, Baltimore, MD, USA

Francesco Migliavacca Politecnico di Milano, Milano, Italy

Kunal Mitra Florida Institute of Technology, Melbourne, FL, USA

Ryan O'Flaherty Florida Institute of Technology, Melbourne, FL, USA

C. Poelma Delft University of Technology, Delft, The Netherlands

Giuseppe Pontrelli Institute for Applied Mathematics—CNR, Rome, Italy

Amir Sajjadi Massachusetts General Hospital, Boston, MA, USA

Xiuhua A. Si Calvin College, Grand Rapids, MI, USA

Neil T. Wright Michigan State University, East Lansing, MI, USA

Jinxiang Xi Central Michigan University, Mount Pleasant, MI, USA

Preface

This collection of 13 chapters provides a compendium of the most recent advances in the theoretical and computational modeling of transport phenomena associated with current biomedical therapies. This compendium is distinguished by its focus on the applications of fluid flow and heat transfer that are relevant to diseases, pathologies, and biomedical technology. Although this field is growing at a rapid rate, strict mechanistic interpretations of the involved transport processes are still required.

Each of this book's contributing authors is internationally recognized for contributions to the field of computational transport modeling and its applications, and the book has been written for expert researchers. Yet, each chapter also provides a detailed introduction so that a reader new to the field is not alienated by the technical jargon. In fact, we anticipate that both the advanced academic clinician and the emerging researcher will find the book's contents to be laid out in a clear and interesting manner.

Chapters 1-4 concern the modeling of heat transfer in biological media. Chapter 1 reviews the models of heat transfer in perfuse tissue, providing a technical review of the methods used to capture the effect of blood flow in the context of hyperthermia treatments.

Chapter 2 details the computational approaches for considering the thermal response of living tissue to short pulse laser irradiation, and the modeling presented in the chapter has been developed with commercially available software in mind,

providing comparisons between the models and experimental results.

Chapter 3 offers an extraordinary review of the quantification of thermal damage. It also presents clinical and theoretical considerations of heat response at the molecular, cellular, and tissue levels in a thorough and categorical manner.

Chapter 4 then introduces the development of the exact solutions to the Pennes' bioheat equation, and it lists solutions addressing the handling of steady and transient nonhomogenous terms, along with the handling of composite systems.

Microscale and continuum transport modeling are covered in Chapters 5-10. Chapter 5 presents recent advances in the modeling of the thermohumidity phenomena in human upper airways, before conducting a rigorous examination of hygroscopic factors that influence the growth and deposition of inhaled particles.

Chapter 6 provides an exciting and insightful look at the physics involved in the transport mechanisms employed by microorganisms, and it discusses physiological elements in a captivating narrative that engages the reader in a theoretical description of an unseen world.

Chapter 7 offers a critical review of experimental and numerical work on specific flow fields associated with the leftward flow in a mouse ventral node. A new method for modeling flow and morphogen transport in the node, which is based on assuming a given vorticity distribution at the edge of the ciliated surface, is also discussed.

Chapter 8 extensively reviews the current understanding of transport in biofilms, which includes transport in the extracellular matrix, and a multiscale perspective is provided that includes advanced continuum model development.

Chapter 9 presents the mathematical modeling of fluid flow through a tube with a permeable wall, which is discussed with reference to physiological and biomedical applications. Exact, approximate, and numerical solutions are presented and compared, the physics of the flow is delineated, and the significance of the relevant flow parameters is explained.

Chapter 10 introduces a combined physiological-mathematical perspective for the development of theoretical models predicting the passive transient delivery of drugs through the skin. The current understanding of models is reviewed, and the approach for the development of a multilayered model is provided.

Topics related to hemodynamic modeling are discussed in Chapters 11-13. Chapter 11 provides important physiological considerations of mechanical stresses experienced by blood in different cardiovascular devices. These aspects are then considered in a comprehensive description of state-of-the-art numerical models for analyzing blood trauma.

Chapter 12 reviews the modeling of the hemodynamics of stented coronary arteries, with a focus on recent computational fluid dynamics models, and the underlying physics is considered in both idealized situations as well as from actual image-based geometries.

Chapter 13 concludes the book by considering the effect that hemodynamic forces have on the developing human cardiovascular system. This consideration includes a discussion of conjugate fluid-solid physics and provides a comprehensive description of the experimental and theoretical models used to investigate the behavior in such a complex environment.

This book presents some well-developed paths through the physics, physiology, and mathematics involved in modeling heat transport and fluid behavior within biological media. We believe that the reader will find this book to be a great help in bridging the gap between transport modeling and biology.

Sid M.Becker
University of Canterbury,
Christchurch, New Zealand

Andrey V.Kuznetsov
North Carolina State University,
Raleigh, NC, USA

1

Bioheat Transfer and Thermal Heating for Tumor Treatment

Huang-Wen Huang[a], Tzyy-Leng Horng[b]

[a]Tamkang University, Taipei, Taiwan
[b]Feng Chia University, Taichung, Taiwan

1.1 PENNES' AND OTHER BIOHEAT TRANSFER EQUATIONS

1.1.1 Introduction

The investigation of heat transfer and fluid flow in biological processes requires accurate mathematical models. Biological processes basically involve two phases—solid and liquid (fluid). During the past 50 years, through development of thermal modeling in biological processes, heat transfer processes have been established that include the impact of fluid flow which is due to blood. Table 1.1 shows the significance of thermal transport modes in typical components of biothermal systems, as our subject of discussion refers to cancer treatments using heat. For example, thermal diffusion plays a dominant transport mode in tissues, and convection is less significant as blood perfuses in solid tissues at capillary level vessels (which are small in size and slow in blood motion).

TABLE 1.1   Significance of Thermal Transport Modes in Typical Components of Biothermal Systems

	Conduction	**Convection**	**Radiation**
Tissues	Significant	Less significant	Insignificant
Bones	Significant	Insignificant	Insignificant
Blood vessels	Less significant	Significant	Insignificant
Skins	Insignificant	Significant	Significant

TABLE 1.2   Temperature Ranges with Their Tissue Interactions in Biological Processes

Temperature range (°C)	**Interaction and terminology with tissues**
35-40	Normothermia
42-46	Hyperthermia
46-48	Irreversible cellular damage at 45 min
50-52	Coagulation necrosis, 4-6 min
60-100	Near instantaneous coagulation necrosis
> 110	Tissue vaporization

Thermal ablation therapy is an application of heat transfer and fluid flow in biological processes. Temperature plays a significant role with tissue interactions (e.g., coagulation necrosis). To give readers a picture of temperature treatments with tissue (and terminology), Table 1.2 shows temperature ranges with their tissue interactions in biological processes. A thermal model that satisfied the following three criteria was needed to predict temperatures in a perfused tissue: (1) the model satisfied conservation of energy; (2) the heat transfer rate from blood vessels to tissue was modeled without following a vessel path; and (3) the model applied to any unheated and heated tissue. To meet these criteria, many research groups around the world have proposed mathematical models in an attempt to properly describe the heat transfer and fluid flow in biological processes in a heated, vascularized, finite tissue by making a few simplifying assumptions. We will highlight some of the key models and some models considering the impact of large blood vessel(s) by starting with Pennes' model.

1.1.2 Pennes' Bioheat Transfer Equation

The Pennes' [1] bioheat transfer equation (PBHTE) has been a standard model for predicting temperature distributions in living tissues for more than a half century. The equation was established by conducting a sequence of experiments measuring temperatures of tissue and arterial blood in the resting human forearm. The equation includes a special term that describes the heat exchange between blood flow and solid tissues. The blood temperature is assumed to be constant arterial blood temperature.

In 1948, Pennes [1] performed a series of experiments that measured temperatures on human forearms of volunteers and derived a thermal energy conservation equation: the well-known bioheat transfer equation (BHTE) or the traditional BHTE. Tissue matrix thermal equations can be explained most succinctly by considering the PBHTE as the most general formulation. It is written as:

$$\nabla \cdot k \nabla T + q_p + q_m - W c_b (T - T_a) = \rho c_p \frac{\partial T}{\partial t}, \tag{1.1}$$

where $T(°C)$ is the local tissue temperature, $T_a(°C)$ is the arterial temperature, $c_b(J/kg/°C)$ is the blood specific heat, $c_p(J/kg/°C)$ is the tissue specific heat, $W(kg/m^3/s)$ is the local tissue-blood perfusion rate, $k(w/m/°C)$ is the tissue thermal conductivity, $\rho(kg/m^3)$ is the tissue density, $q_p(w/m^3)$ is the energy deposition rate, and $q_m(w/m^3)$ is the metabolism, which is usually very small compared to the external power deposition term q_p [2]. The term $W c_b (T - T_a)$, which accounts for the effects of blood perfusion, can be the dominant form of energy removal when considering heating processes. It assumes that the blood enters the control volume at some arterial temperature T_a, and then comes to equilibrium at the tissue temperature. Thus, as the blood leaves the control volume it carries away the energy, and hence acts as an energy sink in hyperthermia treatment.

Because Pennes' equation is an approximation equation and does not have a physically consistent theoretical basis, it is surprising that this simple mathematical formulation predicted temperature fields well in many applications. The reasons why PBHTE has been widely used in the hyperthermia modeling field are twofold: (1) its mathematical simplicity; and (2) its ability to predict the temperature field reasonably well in application.

Nevertheless, the equation does have some limitations. It does not, nor was it ever intended to, handle several physical effects. The most significant problem is that it does not consider the effect of the directionality of blood flow, and hence does not describe any convective heat transfer mechanism.

1.1.3 The Chen and Holmes Model

Several investigators have developed alternative formulations to predict temperatures in living tissues. In 1980, Chen and Holmes (CH) [3] derived one with a very strong physical and physiological basis. The equation can be written as:

$$\nabla \cdot (k + k_p) \nabla T + q_p + q_m - W c_b (T - T_a) - \rho_b c_b u \cdot \nabla T = \rho c_p \frac{\partial T}{\partial t}. \tag{1.2}$$

Comparing this equation with Pennes' equation, two extra terms have been added. The term $-\rho_b c_b u \cdot \nabla T$ is the convective heat transfer term, which accounts for the thermal interactions between blood vessels and tissues. The term $\nabla \cdot k_p \nabla T$ accounts for the enhanced tissue conductive heat transfer due to blood perfusion term in tissues, where k_p is called the perfusion conductivity, and is a function of the blood perfusion rate. The blood perfusion term $-W c_b (T - T_a)$, shown in the CH model, accounts for the effects of the large number of capillary structures whose individual dimensions are small relative to the macroscopic phenomenon under their study. Relatively, the CH model has a more solid physical basis than Pennes' model. However, it requires knowledge of the details of the vascular anatomy and flow pattern to solve it, and while this does increase the accuracy, it adds a great deal of complication to the solution.

1.1.4 The Weinbaum and Jiji Model

In 1985, Weinbaum and Jiji (WJ) [4] proposed an alternative mathematical formulation of the BHTE. Their formulation is based on their observations from the vascular network of rabbit thighs that blood vessels that are significant for heat transfer in tissues always occur in countercurrent pairs. Hence, the major heat transfer mechanism between blood and tissues is the "incomplete countercurrent heat exchanger" between thermally significant arteries and veins (with diameters about 50-500 μm). Their formulation uses tensor notation and it can be written as:

$$\rho c \frac{\partial \theta}{\partial t} - \frac{\partial}{\partial x_i}\left[\left(k_{ij}\right)_{\text{eff}} \frac{\partial \theta}{\partial x_j}\right] = -\frac{\pi^2 n a^2 k_b^2}{4\sigma k} Pe^2 l_j \frac{\partial l_i}{\partial k_j} \frac{\partial \theta}{\partial x_j} + Q_m, \tag{1.3}$$

where θ is the local temperature, ρc is the volume average tissue density and specific heat product, a is the local blood vessel radius, σ is a shape factor for the thermal conduction resistance between adjacent countercurrent vessels, n is the number density of blood vessels of size a, k_b is the blood thermal conductivity, Pe is the local Peclet number ($=2\ \rho_b$ cnau$/k_b$), u is average blood flow velocity in the vessels, and l_i is the direction cosine of the ith pair of countercurrent vessels (i.e., ϕ is the angle of the ith pair of countercurrent vessels' axes relative to the temperature gradient and l_i is expressed as cos ϕ). The effective conductivity tensor element, $(k_{ij})_{\text{eff}}$, is given by:

$$\left(k_{ij}\right)_{\text{eff}} = k\left(\delta_{ij} + \frac{\pi^2 n a^2 k_b^2}{4\sigma k^2} Pe^2 l_i l_j\right), \tag{1.4}$$

where δ_{ij} is the kronecker delta function, and k is the tissue thermal conductivity. Clearly, this equation represents one of the most significant contributions to the bioheat transfer formulation. However, in practical situations, this equation needs detailed knowledge of the sizes, orientations, and blood flow velocities in the countercurrent vessels to solve it and that presents a formidable task. Furthermore, there are several issues related to the WJ model. First, thorough comparisons for both predicted temperatures and macroscopic experiments are required. Second, the formulation was developed for superficial normal tissues in which countercurrent heat transfer occurs. In tumors, the vascular anatomy is different from the superficial normal tissues, and therefore a new model should be derived for tumors. Wissler [5,6] has questioned the two basic assumptions of the WJ model: first, that the arithmetic mean of the arteriole and venule blood temperature can be approximated by the mean tissue temperature; and second, that there is negligible heat transfer between the thermally significant arteriole-venule pairs and surrounding tissue.

1.1.5 The Weinbaum, Jiji, and Lemons Model

The Weinbaum, Jiji, and Lemons (WJL) model [7] attempted to describe the blood flow effect in the heat transfer process when limited to small blood vessels. Keller and Seiler [8] used the effective conductivity of the nonisothermal region, which is determined under various blood flow conditions. The WJL model's approach resembles that of Keller and Seiler [8] mathematically in its use of three equations, but the WJL model is based on completely different vascular generations—the WJL equations apply to thermally significant small vessels and not to major supply blood vessels.

1.1.6 Baish et al

According to Baish et al. [9,10], one of the underlying assumptions in deriving the WJ model was that, due to the proximity of the vessels in a countercurrent pair, almost all of the heat conducted through the arterial wall reaches the venous wall—a process in which the temperature of the tissue between the vessels remains unaffected. They [9,10] criticized this hypothesis, postulating that part of the heat leaving the wall of a small arteriole will remain within the tissue. They suggested that the heat transfer between countercurrent vessels depends not only on $T_a - T_v$, but also on the difference between the tissue temperature, T, and the average blood temperature, $(T_a + T_v)/2$. T_a is the arterial temperature and T_v is the venous temperature.

1.1.7 Others

Efforts have been directed, for the most part, toward the WJL and WJ models. Following the publication of the CH, WJL, and WJ models, several studies were performed to evaluate the validity of these new approaches. Here are some arguments and approximations to examine the blood flow impacts on biothermal modeling.

In 1987, Wissler [5,6] strongly criticized the WJ model because of the assumption made on the blood temperature at arterial and venous vessels as well as the nearby tissue temperature. Wissler proposed a new model that described tissue-blood vessels heat exchange that differs from the respective equations in WJL by virtue of an additional perfusion term. For example, temperature profiles along an artery-vein pair is approximated as $T \approx \frac{(T_a + T_v)}{2}$ in the WJ model. He rejected the hypothesis that blood and tissue temperatures are closely coupled which was basically used for the derivation of the thermal conductivity tensor defined in the WJL equation. The thermal conductivity tensor form in the WJL equation is to remove the blood flow term.

During 1989-1990, Charny et al. [11,12] introduced a "modified" WJL model for the blood vessels by changing the governing equation with a tissue energy conservation equation. Based on their analysis of both steady state and transient temperature fields in the limb under hyperthermic and normothermic conditions, the tissue temperature profiles predicted by that model were very similar to those predicted by Pennes' model in the tissue regions with large vessels ($d > 0.4$ mm).

1.2 BLOOD FLOW IMPACTS ON THERMAL LESIONS WITH PULSATION AND DIFFERENT VELOCITY PROFILES

In this section, we will develop a solution to PBHTE coupled with an energy transport equation of pulsatile blood flow in a thermally significant blood vessel surrounded by a tumor tissue. The purpose of this design is to study the cooling effect of pulsatile blood flow in large blood vessels.

1.2.1 Introduction

Though PBHTE, shown in Equation (1.1), is the most popular model in hyperthermia modeling, a fundamental criticism of this model by Nelson [13] is that the treatment of blood flow term as a distributed heat source (or sink) mistakenly presumes that the capillary vasculature is the major site of heat exchange. In other words, the blood flow term is a scalar property. In fact, the blood flow in a tissue usually has a direction from artery to vein passing through the capillary bed. Furthermore, the blood and its surrounding tissues are not in thermal equilibrium when the blood vessel diameter is larger than 500 µm [14–22]. This means the energy equations for tissue and blood in significantly large vessels must be treated individually.

One of the key issues of thermal treatments is blood flow. Blood flow usually drains the delivered heat from the heating region, which causes insufficient thermal dose in the targeted volume. This is an important factor needing to be considered carefully in thermal treatments [23–26]. In fact, the differential therapeutic effect of thermal treatments between malignant and normal tissue may primarily depend on the vascular characteristics of the tumor [27]. Craciunescu and Clegg [28] solved the fully coupled Navier-Stokes and energy equations to obtain the temperature distribution of pulsatile blood flow within a rigid blood vessel. They found that the reversed flow enhances as the Womersley number becomes larger, which results in a smaller temperature difference between forward and reverse flows. Nevertheless, in their model they only focused on the temperature distribution in blood vessels without considering the surrounding tissue. Khanafer et al. [26] and Horng et al. [29] further studied the effects of pulsatile blood flow on temperature distributions during hyperthermia by considering both the pulsatile blood flow in a blood vessel and its surrounding tissue. Here we focus on the results of Horng et al. [29] to discuss how the blood flow velocity profile (including pulsation frequency), size of blood vessel (including blood flow rate), and heating rate affect the thermal dose distribution of the surrounding tissue during heat treatment.

1.2.2 Mathematical Model and Numerical Method

1.2.2.1 Velocity Profile of Pulsatile Blood Flow in a Circular Blood Vessel

It is of interest to not only consider simple steady uniform or parabolic blood velocity profile, but also the pulsatile blood flow in thermally significant blood vessels (i.e., larger than 200 µm in diameter) [23,25,29], with the assumptions that the blood vessel segment is straight, the vessel wall is rigid and impermeable, and the flow is incompressible and Newtonian. Considering the steady blood flow passing through a rigid vessel of inner radius r_0, the axial Hagen-Poiseuille velocity profile can be expressed as:

$$w(r) = -\frac{1}{4\mu}(r_0^2 - r^2)\frac{\mathrm{d}p}{\mathrm{d}z}, \tag{1.5}$$

where μ is the dynamic viscosity and $\frac{\mathrm{d}p}{\mathrm{d}z}$ the constant pressure gradient along the axial (z) direction. Because the blood flow in the cardiovascular system is periodic, the pressure gradient cannot remain a constant. Here, it is modified to have an additional sinusoidal component in time, shown as follows:

$$\frac{\partial p}{\partial z} = c_0 + c_1 \mathrm{e}^{i\omega t},$$

where ω is the angular frequency and the associated period of time is denoted as $\widetilde{T} = \frac{2\pi}{\omega}$. Then the corresponding axial velocity profile, $W(r,t)$, can be expressed as:

$$W(r,t) = w(r) + w_1(r)e^{i\omega t}. \tag{1.6}$$

Here, c_0 can be related to the average volume flow rate over the time period $\widetilde{T}$ as follows:

$$\dot{Q}_{avg} = \frac{1}{\widetilde{T}} \int_0^{\widetilde{T}} \int_0^{r_0} 2\pi W r\, dr\, dt = -\frac{\pi r_0^4}{8\mu} c_0. \tag{1.7}$$

The average velocity can be further deduced from above:

$$\overline{w} = \frac{\dot{Q}_{avg}}{\pi r_0^2} = -\frac{c_0 r_0^2}{8\mu}, \tag{1.8}$$

and $w(r)$ can then be alternatively expressed as:

$$w(r) = 2\overline{w}\left(1 - \frac{r^2}{r_0^2}\right). \tag{1.9}$$

The term w_1 can be derived from the Navier-Stokes equations [30]. Together with $w(r)$ in Equation (1.9), $W(r,t)$ in Equation (1.6) can be expressed as follows:

$$W(r,t) = 2\overline{w}\left(1 - \frac{r^2}{r_0^2}\right) + \frac{ic_1 r_0^2}{\mu\alpha^2}\left[1 - \frac{J_0\left(\alpha\frac{r}{r_0}i^{\frac{3}{2}}\right)}{J_0\left(\alpha i^{\frac{3}{2}}\right)}\right]e^{i\omega t}, \tag{1.10}$$

where $\alpha = \frac{r_0}{\sqrt{\nu/\omega}}$ denotes the Womersley number describing the competition between the inertia and viscous forces; ν denotes the kinematic viscosity of blood; and J_0 is the Bessel function of the first kind of order zero. If oscillatory driving pressure gradient is $c_1 \cos(\omega t)$, the corresponding velocity will then be the real part of Equation (1.10). If the oscillatory driving pressure gradient is $c_1 \sin(\omega t)$, the corresponding velocity will then be the imaginary part of Equation (1.10). Here we also define

$$\text{fac} = c_1/c_0 = c_1 \bigg/ \left(-\frac{8\mu\overline{w}}{r_0^2}\right), \tag{1.11}$$

and use it to characterize the relative intensity of pulsation in the blood flow. Reasonable value of fac ranging from 0.2 to 1 is considered here. When the Womersley number, α, is large, the effect of viscosity cannot propagate far from the vessel wall, and the blood flow in the central part of a vessel acts like an inviscid flow and can be chiefly determined by the balance between the inertia force and the pressure gradient. Under this situation, the velocity profile of an oscillatory component has a rather flattop shape at certain phases compared with a parabolic profile of Poiseuille flow. When the Womersley number, α, is large enough, the velocity profile of an oscillatory component may even display two peaks at certain phases [29,31]. Some examples of the diameters of thermally significant blood vessels and their associated average velocities are listed in Table 1.3 [29]. Taking the largest blood vessel considered in

TABLE 1.3   List of the Blood Vessel Parameters Used in Current Study

Diameter (mm)	Average blood velocity in tumor ($\overline{w}$) (mm/s)
0.2	3.4
0.6	6
1.0	8
1.4	10.5
2.0	20

Table 1.3 (diameter$=2$ mm), and varying heart beat frequency from 1 to 3 Hz as suggested by Huo and Kasab [32], the velocity profiles of an oscillatory component are respectively shown in Figure 1.1 at selected time phases. It can be observed that, as the Womersley number increases with increasing beating frequency, the oscillatory velocity component exhibits flat-top and even two-peak behaviors.

1.2.2.2 Governing Equations and Numerical Method

Here we assume that the absorbed power density in blood and tissue is equal to the heating power density. Although this is a very limiting assumption and generally not true, it still serves its purpose because we focus our study on the cooling effect of large blood vessels here. The axis-symmetric geometric configuration considered here is a cylindrical perfused tissue, including tumor and normal tissues, with a coaxial rigid blood vessel inside and throughout the tissue as shown in Figure 1.2. The whole computational domain is bounded by $r=r_{max}$, $z=0$, and $z=z_{max}$; the blood vessel is surrounded by $r=r_0, z=0$, and $z=z_{max}$; and the heating target (tumor and a part of blood vessel inside the tumor) is bounded by $r=r_1, z=z_1$, and $z=z_2$. The diameters of blood vessels and their associated average flow velocities considered in the study of this section are presented in Table 1.3.

Under the axis-symmetric geometric configuration mentioned above, the governing equations for the temperature evolution are PBHTE in Equation (1.12) for tissue and energy transport equation in Equation (1.13) for blood vessels:

$$\rho_t c_t \frac{\partial T_t}{\partial t} = k_t \left[\frac{1}{r}\frac{\partial}{\partial r}\left(r\frac{\partial T_t}{\partial r}\right) + \frac{\partial^2 T_t}{\partial z^2}\right] - W_b c_b (T_t - T_a) + Q_t(r,z,t), \tag{1.12}$$

$$\rho_b c_b \left(\frac{\partial T_b}{\partial t} + w\frac{\partial T_b}{\partial z}\right) = k_b \left[\frac{1}{r}\frac{\partial}{\partial r}\left(r\frac{\partial T_b}{\partial r}\right) + \frac{\partial^2 T_b}{\partial z^2}\right] + Q_b(r,z,t), \tag{1.13}$$

where $T(r,z,t)$ denotes the temperature that is distributed axis-symmetrically; ρ, k, c are density, thermal conductivity, and specific heat, respectively, that are all assumed to be constant; $Q(r,z,t)$ is the power of heat added axis-symmetrically; W_b is the perfusion mass flow rate; T_a is the ambient temperature that is usually set to be 37 °C; $w(r,t)$ is that axial velocity of blood flow; and subscripts t and b represent tissue and blood, respectively. Notice that in the study of this section the tissue metabolic heat production Q_m is neglected compared with heating power as mentioned in Section 1.1.2. The heat sink $-W_b c_b (T_t - T_a)$ in Equation (1.12) is used to describe the perfusion effect by the microvascular network of blood flow (i.e., blood vessels with diameter generally less than 200 μm), while the heat transfer due to the thermally significant large

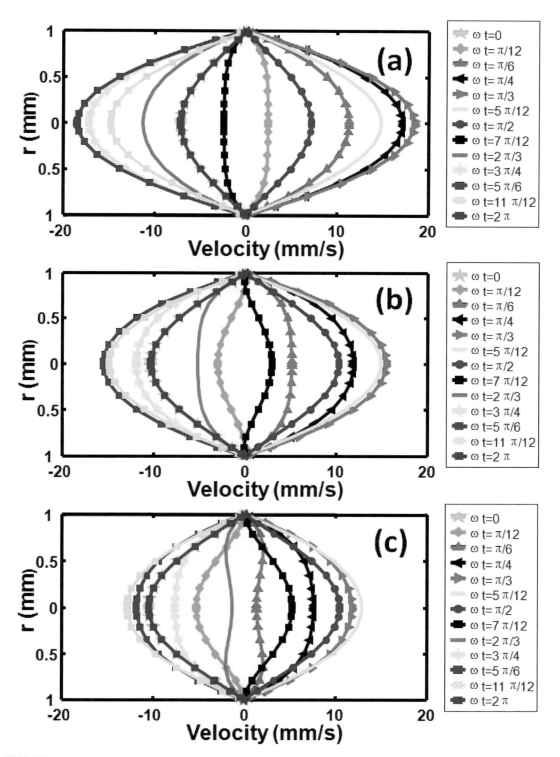

FIGURE 1.1 Effect of Womersley number, α, on the oscillatory component of the velocity profile for blood flow with blood vessel diameter being 2 mm. The velocity profile is shown at several selected phases between 0 and 2π for (a) $f=1$ Hz, $\alpha=1.2843$, (b) $f=2$ Hz, $\alpha=1.8162$, and (c) $f=3$ Hz, $\alpha=2.2244$. Flattop and two peak features can be observed when α is large as shown in (c). The Womersley number is calculated based on the density of blood $\rho_b=1050$ kg/m^3, and dynamic viscosity of blood $\mu_b=4\times10^{-3}$ Pa s.

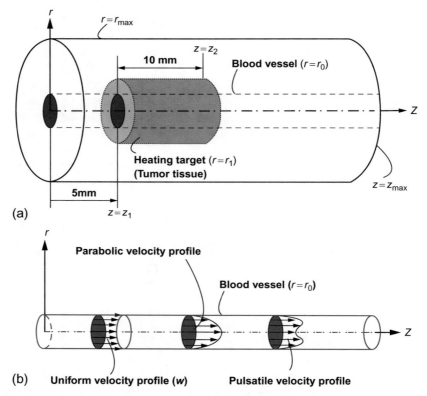

FIGURE 1.2   Geometric configuration in current simulations. (a) The treatment target (heating target) is specified as $z_1 \leq z \leq z_2, 0 \leq r \leq r_1$, with $z_1 = 5$ mm, $z_2 = 15$ mm, $r_1 = 5$ mm considered here. (b) Schematic illustration of the three kinds of velocity profile of blood flow in blood vessels. Left: steady uniform velocity profile; Middle: steady parabolic velocity profile; Right: pulsatile velocity profile.

blood vessel has to be separately described by Equation (1.13). Together with initial, boundary, and interface conditions (see details in Ref. [29]), Equations (1.12) and (1.13) were numerically solved by the highly accurate multiblock Chebyshev pseudospectral method under the framework of method of lines (MOL). Notice that the computational domain is decomposed to nine blocks and is shown in Figure 1.3 with numerical meshes. Blocks 1, 2, and 3 are the blood vessel; block 5 is the tumor; the others are normal tissue. Heating zones are blocks 2 and 5. For further numerical details, see Horng et al. [29] and Shih et al. [33].

1.2.2.3 Calculation of Thermal Dose

The accumulated thermal dose to tissue is a function of heating duration and the temperature level. The estimate of tissue damage is based on the thermal dose the formula for which was proposed by Sapareto and Dewey [34]. The thermal dose or equivalent minutes at 43 °C (EM_{43}) is shown as follows:

$$EM_{43}(\text{in min}) = \int_0^{t_f} R^{T-43}\,\mathrm{d}t, \tag{1.14}$$

FIGURE 1.3 The overall computational domain is decomposed into nine rectangular blocks in $r-z$ coordinates. Notice that blocks 1-3 are for the blood vessel, and blocks 4-9 are for the tissue with block 5 being the tumor and the others being the normal tissue. Heating zone indicated in Figure 1.2(a) would be blocks 2 and 5.

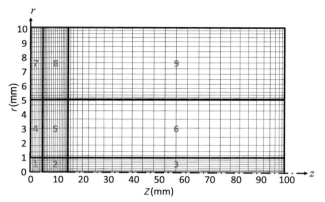

TABLE 1.4 Parameters of Six Different Heating Schemes Used in Current Study

Heating case	I	II	III	IV	V	VI
Heating power density Q (W/cm^3)	100	50	25	10	5	2
Heating duration t_h (s)	1	2	4	10	20	50
Total heated energy density (J/cm^3)	100	100	100	100	100	100

where $R=2$ for $T \geq 43\,°C$, $R=4$ for $37\,°C < T < 43\,°C$, and $t_f = 60$ s in the study of this section. The threshold dose for necrosis is $EM_{43}=240$ min for tumor muscle tissue, and the region encircled by the level curve $EM_{43}=240$ min is taken as the thermal lesion region. Covering tumor tissue but not normal tissue by thermal lesion region as fully as possible is most desired in the thermal treatment, though normal tissue may still survive at such a thermal dose because of the fact that tumor tissue is much less heat-bearable than normal tissue. If we define the deficit region as the tumor region excluding a thermal lesion region, it would serve as an evaluation of the effectiveness of the treatment. This region is greatly influenced by the size of the blood vessel [34] and the heating rate [35]. Six different heating schemes characterizing different heating rates under the same amount of heat added (100 J/cm^3 from preliminary energy analysis in lump [36]) are depicted in Table 1.4.

1.2.3 Results and Discussions

Figure 1.4 compares the effects of steady uniform and parabolic velocity profiles for blood flow on a thermal lesion region with heating scheme II in Table 1.4. Likewise, Figure 1.5 compares the effects of pulsatile blood flow with various pulsatile frequencies (1, 1.5, and 2 Hz) with the relative intensity of pulsation fac$=0.2$ on a thermal lesion region with the same heating scheme. Figures 1.4 and 1.5 generally show that there is almost no difference in a thermal lesion region among all these velocity profiles under the same size of blood vessel. Only a

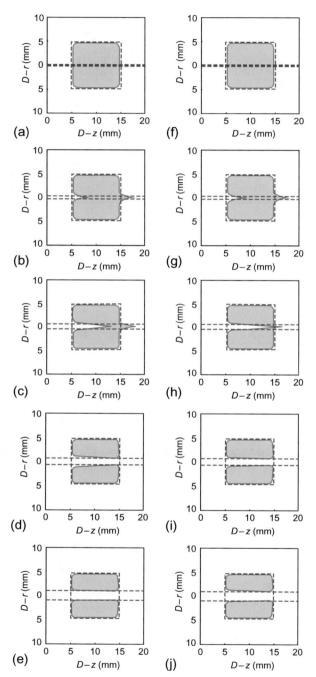

FIGURE 1.4   Effect of the steady velocity profiles of blood flow on the thermal lesion region (shaded region) for the blood vessels (a and f) 0.2 mm, (b and g) 0.6 mm, (c and h) 1 mm, (d and i) 1.4 mm, (e and j) 2 mm in diameter. (a-e) are results of a uniform velocity profile, and (f-j) are results of a parabolic one. The blood vessel boundaries are denoted with the horizontal dashed lines. The heated target region (tumor) is denoted by a square with dashed lines. Here $r_1 = 5$ mm, $W_b = 2$ kg/m^3/s^1, $\rho_b = \rho_t = 1050$ kg/m^3, $c_b = c_t = 3770$ J/kg^1/$^\circ$C^1, $k_b = k_t = 0.5$ W/m^3/$^\circ$C^1. The target is heated in the way of heating case II (i.e., $Q_t = Q_b = 50$ W/cm^3, and the heating duration $= 2$ s) in Table 1.4.

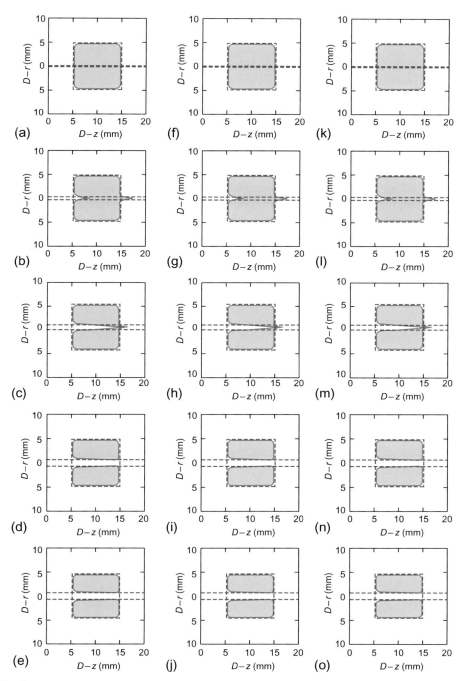

FIGURE 1.5 Effect of the frequency of the pulsatile blood flow on the thermal lesion region (shaded region) for the blood vessels (a, f, and k) 0.2 mm, (b, g, and l) 0.6 mm, (c, h, and m) 1 mm, (d, i, and n) 1.4 mm, (e, j, and o) 2 mm in diameter. (a-e) have the frequency 1 Hz, (f-j) have the frequency 1.5 Hz, and (k-o) have the frequency 2 Hz. The blood vessel boundaries are denoted with the horizontal dashed lines. The rest of the parameters and conditions are the same as Figure 1.4.

minor difference of the thermal lesion region in the blood vessel is observed in the middle-sized blood vessels (see details in Figure 1.6 of Horng et al. [29]).

Although the thermal lesion region is rather insensitive to the velocity profile of blood flow, it is deeply influenced by the size of the blood vessel because the heat convection by the blood flow in a blood vessel usually serves as a stronger heat sink than the blood perfusion in tissue. That means the temperature would drop faster in a blood vessel than in its surrounding tissue. This may cause a deficit in the thermal lesion region in the blood vessel and the surrounding tissue, which can be easily observed from Figures 1.4 and 1.5. Generally, the deficit of the thermal lesion region is less for smaller vessels. In the case of the smallest vessel (here with a diameter 0.2 mm), the thermal lesion region covers almost the entire blood vessel that is inside the tumor and the deficit is naturally the least. For the middle-sized vessels (here with diameters of 0.6 and 1 mm), the thermal lesion region in the blood vessel becomes smaller and shifts downstream. For large vessels (here with diameters of 1.4 and 2 mm), there is a total deficit of the thermal lesion region in the blood vessel, and this would cause a deficit in the tumor tissue near the blood vessel and especially in the upstream area.

Besides having a dependence on the blood vessel size, the thermal lesion region is also very sensitive to the heating rate. Figure 1.6 generally shows a larger thermal lesion region for

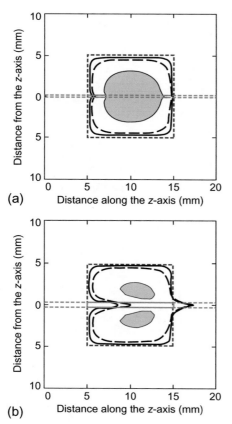

(a)

(b)

FIGURE 1.6 Comparison of the effect of different heating schemes and blood vessel diameters on the thermal lesion region with a steady uniform velocity profile. The solid and dashed lines represent heating cases II and IV, and the shaded area represents the heating case VI in Table 1.4. (a) The diameter of the blood vessel is 0.2 mm. (b) The diameter of the blood vessel is 1 mm.

faster heating, and there is a more pronounced effect on the heating rate when the blood vessel is larger. As shown in Figure 1.6b, there exists an obvious shift of thermal lesion region to the downstream of the blood vessel with diameter 1 mm when heating is fast, and this may cause unwanted thermal injury in normal tissue nearby. The effect of pulsation amplitude, in terms of relative intensity fac, of pulsatile blood flow on the thermal lesion region generally has little difference among various facs, except a minor difference for middle-sized blood vessels. This is further noted in Horng et al. [29] in the discussions of Figures 1.9–1.11. Generally, when fac increases, the blood flows more in a stick-slip fashion, and this may considerably influence the heat convection when incorporated with pulsation frequency. With large pulsation amplitudes like fac = 0.8 and 1, it even shows two-peak behavior in a thermal dose contour at the downstream of the blood vessel.

1.2.4 Conclusion

The current investigation shows that the effect of velocity profiles of blood flow, ranging from uniform, to parabolic, to pusatile, has almost no difference in the thermal lesion region on the tumor region and only a minor difference on the blood vessel when the blood vessel is of middle size. This result suggests that we might just as well use the simplest steady uniform or parabolic velocity profile to do the simulation. In fact, the thermal lesion region is much more sensitive to the heating rate and the size of the blood vessel. Faster heating would form a much better thermal lesion region, and it works best on small blood vessels with a better covering of both the tumor and the blood vessel by the thermal lesion region because the heat convection by the blood flow is least in the blood vessel. For large vessels, it has a total deficit in the blood vessel and some deficit in the tumor near the upstream of blood vessel. As to middle-sized vessels, a shift of a partially deficient thermal lesion region to the downstream of the blood vessel may cause unwanted thermal injury to the normal tissue nearby.

1.3 THERMAL RELAXATION TIME FACTOR IN BLOOD FLOW DURING THERMAL THERAPY

Non-Fourier heat conduction has been observed in biotissues, which implies PBHTE has to be modified by considering the thermal relaxation time factor. Here we study this effect under the same geometric configuration for tissue and blood vessel as in Section 1.2 to see how it affects hyperthermia generally.

1.3.1 Introduction

PBHTE, shown in Equation (1.1), is based on the classical Fourier law, which assumes that a temperature disturbance in any part of the materials leads to an instantaneous perturbation at each point of the whole. This implies that the propagation speed of thermal perturbation is infinite even when the intervening distance is very large, and causes some doubts and discussions [37–39]. Actually, non-Fourier heat conduction behavior has been observed in biomaterials with inhomogeneous inner structures [40], in biological tissues [41,42], in canine thigh muscles [43], and in processed meats [44,45]. Considering the finite propagation speed for the thermal disturbance, Cattaneo [46] and Vernotte [47] formulated a modified heat flux equation, as shown in Equation (1.16) with the Fourier law shown in Equation (1.15) for comparison

$$q\left(\vec{r},t\right)=-k\nabla T\left(\vec{r},t\right),\tag{1.15}$$

$$q\left(\vec{r},t\right)+\tau\frac{\partial q\left(\vec{r},t\right)}{\partial t}=-k\nabla T\left(\vec{r},t\right),\tag{1.16}$$

where $T,q,k,$ and τ are temperature, heat flux, thermal conductivity, and thermal relaxation time, respectively. If we formally treat $-k\nabla T$ in Equation (1.16) as a constant A, the solution of Equation (1.16) would be simply:

$$q=A+Be^{-\frac{t}{\tau}}.\tag{1.17}$$

We can then see how q is relaxed to A in the time scale of τ from Equation (1.17). The thermal relaxation time for biological tissues has typically been found to be large, leading to significant non-Fourier thermal behavior. Mitra et al. [44] conducted an experiment in which they measured the thermal relaxation time in processed meat and reported that τ could be as large as 16 s. Kaminski [40] reported that τ ranges from 10 to 50 s in his experiment with materials with inhomogeneous inner structures. Roetzel et al. [48] also confirmed the hyperbolic behavior of thermal propagation with τ about 1.77 s in a similar experiment. Using the thermal properties of tissue and blood from some literatures, Zhang [49] computed and argued that reasonable τ should range from 0.464 to 6.825 s. He further found that the dual-phase lag phenomenon in temperature and its gradient due to the wave feature is more pronounced when the blood vessel is large. Shih et al. [33] continued to explore the heat wave caused by thermal relaxation time and further investigated the coupled effect of blood flow in large blood vessels and thermal relaxation time on the heating of tumor tissues. Based on the results of Shih et al. [33], we discuss the related heat wave behavior caused by non-zero thermal relaxation time.

1.3.2 Mathematical Model and Numerical Method

1.3.2.1 Features of the Hyperbolic Heat Equation

Consider the 1D transient heat equation without any heat source as follows:

$$\rho c_{\text{p}}\frac{\partial T(x,t)}{\partial t}+\frac{\partial q(x,t)}{\partial x}=0,\tag{1.18}$$

with ρ being the density and c_{p} being the specific heat. Differentiating Equation (1.18) with respect to t, and multiplying it by τ, we can obtain a new equation. Adding this new equation to Equation (1.18), and applying Equation (1.16), we can obtain the following hyperbolic heat equation:

$$\rho c_{\text{p}}\tau\frac{\partial^2 T}{\partial t^2}+\rho c_{\text{p}}\frac{\partial T}{\partial t}=k\frac{\partial^2 T}{\partial x^2},\tag{1.19}$$

or

$$\frac{\partial^2 T}{\partial t^2} + \frac{1}{\tau}\frac{\partial T}{\partial t} = c^2\frac{\partial^2 T}{\partial x^2}, \tag{1.20}$$

with the wave speed $c = \sqrt{\frac{k}{\rho c_p \tau}}$. Here we conduct a normal mode analysis of Equation (1.20) by studying its solution form in traveling wave $T(x,t) = e^{i(\xi x - \omega t)}$. Substituting this traveling wave solution into Equation (1.20), we can obtain its dispersion relationship:

$$\omega = \frac{-i \pm \sqrt{4\tau^2 c^2 \xi^2 - 1}}{2\tau}, \tag{1.21}$$

and then

$$T(x,t) = e^{-\frac{t}{2\tau}} e^{i\xi(x - c_{\text{eff}}t)}, \quad c_{\text{eff}} = \pm\frac{\sqrt{4\tau^2 c^2 \xi^2 - 1}}{2\xi\tau}, \quad \text{when } 4\tau^2 c^2 \xi^2 \geq 1, \tag{1.22}$$

$$T(x,t) = e^{\left(\frac{-1 \pm \sqrt{1 - 4\tau^2 c^2 \xi^2}}{2\tau}\right)t} e^{i\xi x}, \quad \text{when } 4\tau^2 c^2 \xi^2 < 1. \tag{1.23}$$

From Equation (1.22), high-frequency modes travel in two directions and damp at the same time, while low-frequency modes simply decay without propagation from Equation (1.23). The attenuation rate decays with increasing τ for both high-frequency and low-frequency modes. All these mean that while temperature is decaying as a whole, we can only observe high-frequency waves traveling.

1.3.2.2 Thermal Governing Equations and the Numerical Method

Here we consider the same geometric configuration and blood velocity profile (Equation 1.10) as in Section 1.2. Usually, the governing equations for the temperature evolution are the PBHTE shown in Equation (1.24) for solid tissue and energy transport equation shown in Equation (1.25) for blood flow in terms of the cylindrical coordinate under an axis-symmetric situation:

$$\rho_t c_t \frac{\partial T_t}{\partial t} = k_t \left[\frac{1}{r}\frac{\partial}{\partial r}\left(r\frac{\partial T_t}{\partial r}\right) + \frac{\partial^2 T_t}{\partial z^2}\right] - W_b c_b(T_t - T_a) + Q_t(r,z,t), \tag{1.24}$$

$$\rho_b c_b \left(\frac{\partial T_b}{\partial t} + W\frac{\partial T_b}{\partial z}\right) = k_b \left[\frac{1}{r}\frac{\partial}{\partial r}\left(r\frac{\partial T_b}{\partial r}\right) + \frac{\partial^2 T_b}{\partial z^2}\right] + Q_b(r,z,t), \tag{1.25}$$

with symbol of notations and meaning of each term the same as in Section 1.2.

Taking into account the finite thermal propagation speed in living solid tissues, we modified Equation (1.24) by the heat flux formula in Equation (1.16) following the same procedure used to arrive at Equation (1.19), and obtained a hyperbolic bioheat transfer equation (HBTE) as shown in Equation (1.26) to replace Equation (1.24):

$$\rho_t c_t \left(\tau_t \frac{\partial^2 T_t}{\partial t^2} + \frac{\partial T_t}{\partial t}\right) = k_t \left[\frac{1}{r}\frac{\partial}{\partial r}\left(r\frac{\partial T_t}{\partial r}\right) + \frac{\partial^2 T_t}{\partial z^2}\right] + W_b c_b(T_a - T_t) + Q_t + \tau_t\left(-W_b c_b \frac{\partial T_t}{\partial t} + \frac{\partial Q_t}{\partial t}\right), \tag{1.26}$$

in which the terms on the left side represent heat wave and heat diffusion, respectively. They are competing with each other with the thermal relaxation time τ_t characterizing the strength of wave. When $\tau_t = 0$, HBTE (Equation 1.26) will totally reduce to a parabolic-type PBHTE (Equation 1.24), and heat wave reduces to heat diffusion. The external heating rate Q_t in Equation (1.26) and Q_b in Equation (1.25) are designated as follows:

$$Q_t(r, z, t) = \begin{cases} \tilde{Q}_t \dfrac{\pi}{2} \sin\left(\dfrac{\pi t}{t_h}\right), & r_0 \leq r \leq r_1, z_1 \leq z \leq z_2, 0 \leq t \leq t_h, \\ 0, & (r, z, t) \text{ otherwise,} \end{cases}$$

$$Q_b(r, z, t) = \begin{cases} \tilde{Q}_b \dfrac{\pi}{2} \sin\left(\dfrac{\pi t}{t_h}\right), & r \leq r_0, z_1 \leq z \leq z_2, 0 \leq t \leq t_h, \\ 0, & (r, z, t) \text{ otherwise,} \end{cases} \tag{1.27}$$

where $\tilde{Q}_t$ and $\tilde{Q}_b$ are the time averaged values of Q_t and Q_b, respectively, and t_h is the duration of time of heating. In the study of this section, we let $\tilde{Q}_t = \tilde{Q}_b = Q$. Six heating schemes consisting of various combinations of Q and t_h are shown in Table 1.4, and schemes I-V are particularly employed here to study the effect of heating rate.

The initial conditions for the blood vessel and the tissue are

$$T_t(r, z, 0) = T_b(r, z, 0) = 37, \quad \text{and} \quad \frac{\partial T_t}{\partial t}(r, z, 0) = 0°C/s.$$

At the interface $\Gamma(r = r_0, 0 \leq z \leq z_{max})$ between the blood vessel and tissue, temperature and heat flux continuity conditions are imposed:

$$T_t = T_b, \quad \text{and} \quad k_t \frac{\partial T_t}{\partial n} = k_b \frac{\partial T_b}{\partial n} \quad \text{at } \Gamma,$$

where n denotes the direction normal to Γ. At $r = 0$, the pole condition was applied for the blood vessel:

$$\frac{\partial T_b}{\partial r} = 0.$$

The boundary conditions at $r = r_{max}, z = 0$, and $z = z_{max}$ are all set to:

$$T_t = T_b = 37°C,$$

except that the convective boundary condition is employed for the blood vessel part at $z = z_{max}$:

$$\frac{\partial T_b}{\partial t} + W \frac{\partial T_b}{\partial z} = 0, \quad \text{at } z = z_{max}.$$

Together with these initial boundary and interface conditions, Equations (1.25) and (1.26) were numerically solved by the highly accurate, multiblock Chebyshev pseudospectral method under the framework of MOL. Again, the computational domain is decomposed to nine blocks, and designation of heating zones is the same as Section 1.2. For further numerical details, see Horng et al. [29] and Shih et al. [33].

1.3.3 Results and Discussions

Here we discuss how the thermal relaxation time τ_t affects the thermal treatment. Numerical experiments were conducted under exhaustive combinations of heating rate Q (first five schemes listed in Table 1.4), and thermal relaxation time ($\tau_t = 0, 0.464, 1.756,$ and 6.825 s as suggested in [49]). Here we only consider the case of the blood vessel diameter being 2 mm in Table 1.3, because larger blood vessels have been proved to be more thermally significant with sensitivity to heating rate [29]. The time evolution of maximum temperature and the thermal dose are particularly chosen here to demonstrate the effect of the thermal relaxation time. Though we did consider the blood flow to be pulsatile as periodically driven by the heart as in Section 1.2 in the beginning, blood flow pulsation was found to make negligible difference when just considering blood flow to be simply steady axial Hagen-Poiseuille flow, both in the time evolution of maximum temperature and thermal dose for all heating rates and τ_t considered here. We can, therefore, conclude that the thermal behavior is actually quite insensitive to the pulsation of blood flow [33].

The wave feature in HBTE (Equation 1.26) with large τ_t is found to be most pronounced when the heating rate is fast. In Figures 1.7 and 1.8, we compare the time evolution of the temperature distribution for $\tau_t = 0$ and 6.825 s under the case of the fastest heating rate (heating scheme I in Table 1.4). A non-smooth temperature distribution in space is clearly observed during time evolution in Figure 1.8 for $\tau_t = 6.825$ s, featuring high-frequency wave propagating, while only smooth temperature distribution is observed all the time in Figure 1.7 for $\tau_t = 0$ s, featuring the parabolic tendency of PBHTE in Equation (1.24). Notice that only high-frequency waves are pronounced in Figure 1.8. This is because high-frequency modes propagate and at the same time attenuate, while low-frequency modes attenuate only. This can be well explained by the analysis in Section 1.3.2.1.

The time evolution of the maximum temperature in space, $\max_{(r,z)\in\Omega}T$, can be particularly useful to demonstrate the effect of thermal relaxation time under different heating rates as shown in Figure 1.9. Generally, this maximum temperature happens near the center of zone 5 (tumor tissue). For $\tau_t = 0$, we can see $\max_{(r,z)\in\Omega}T$ always reaches its maximum in time right at the end of heating as expected, while $\max_{(r,z)\in\Omega}T$ generally exhibits a plateau in time after the end of heating for non-zero τ_t's, which is more pronounced as τ_t is large and heating rate is fast by comparing sub-figures in Figure 1.9. This can be easily understood by the larger lagging of heating and smaller attenuation rate from Equation (1.23) for low-frequency modes when τ_t is large. Obviously, high temperature also tends to accumulate and preserve in time at the tumor zone when the heating is fast (less time for tumor to respond and relax thermally). Notice that, for slow heating in Figure 1.9d and e, the maximum temperature plateau is far less pronounced compared with fast heating in Figure 1.9a-c and this is because the heating is so slow that the tumor has enough time to respond and relax thermally. In Figure 1.9d and e, we can also observe non-zero τ_t's even lead $\tau_t = 0$ in time to reach their maxima, and the larger τ_t's are, the larger maxima are reached. Again, this can be explained by the larger heating lagging and less attenuation as τ_t gets large.

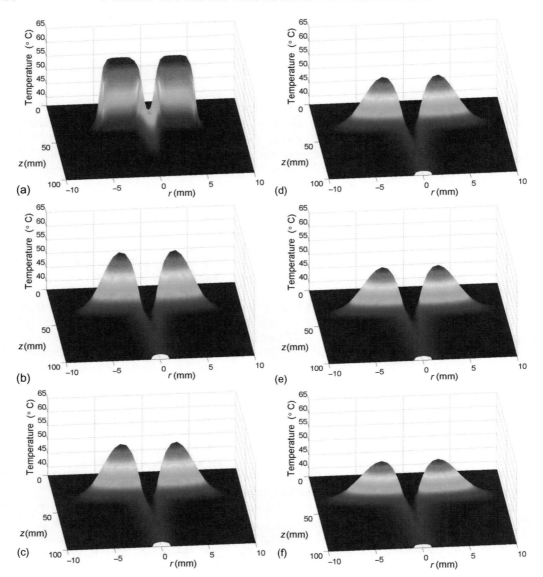

FIGURE 1.7   Time development of temperature distribution in space for $\tau_t = 0$ s under heating scheme I in Table 1.4 of Section 1.2 is shown at (a) $t = 1$ s, (b) $t = 7.5$ s, (c) $t = 10$ s, (d) $t = 12$ s, (e) $t = 14.5$ s, (f) $t = 19.5$ s. The blood vessel diameter is 2 mm.

From Figure 1.10, we can observe that the tumor tissue is generally covered better by a thermal lesion region based on $EM_{43} = 240$ min level curve when heating is fast, and that this effect is further enhanced when τ_t is large. Some of these regions even cover a small part of the normal tissue near the downstream junction of the blood vessel, tumor, and normal tissue as shown in Figure 1.10a and b, which is actually not desired in thermal treatment.

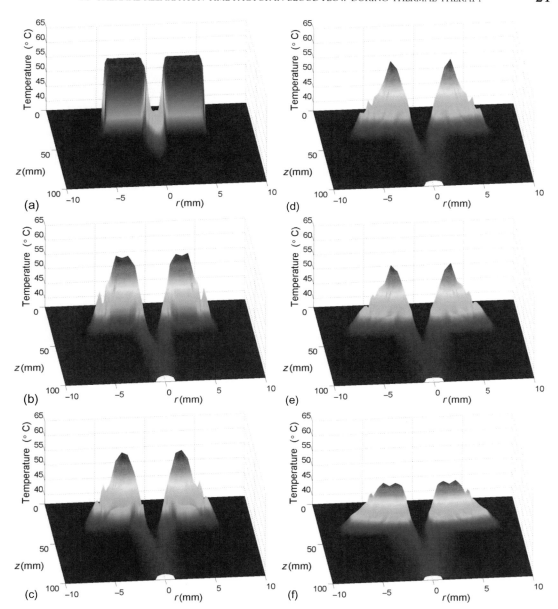

FIGURE 1.8 Time development of temperature distribution in space for $\tau_t = 6.825$ s under heating scheme I in Table 1.4 of Section 1.2 is shown at (a) $t = 1$ s, (b) $t = 7.5$ s, (c) $t = 10$ s, (d) $t = 12$ s, (e) $t = 14.5$ s, (f) $t = 19.5$ s. The blood vessel diameter is 2 mm.

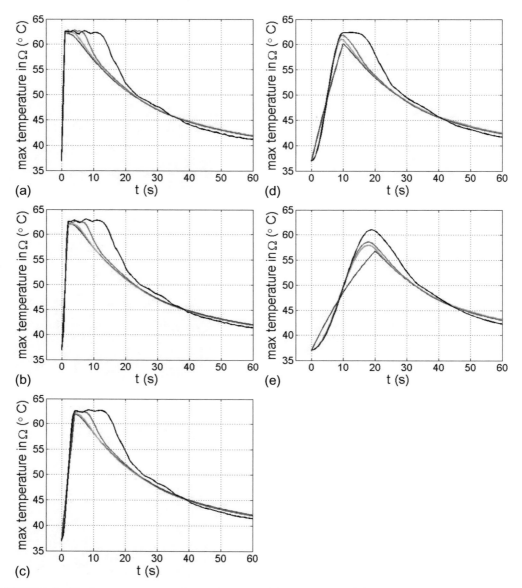

FIGURE 1.9 Maximum temperature in space $\max_{(r,z)\in\Omega} T$ versus time with heating schemes I-V in Table 1.4 in Section 1.2 shown in (a-e), respectively. $\tau_t = 0, 0.464, 1.756,$ and 6.825 s are represented by blue, green, red, and black curves, respectively. The blood vessel diameter is 2 mm.

The phenomenon above is comprehensible from the fact that the high temperature is preserved for longer durations and the heat is drained more slowly when heating is fast and τ_t is large (as demonstrated in Figure 1.9). Generally speaking, the traditional simulations based on PBHTE may underestimate the thermal lesion region by neglecting nontrivial thermal relaxation time in living tissues.

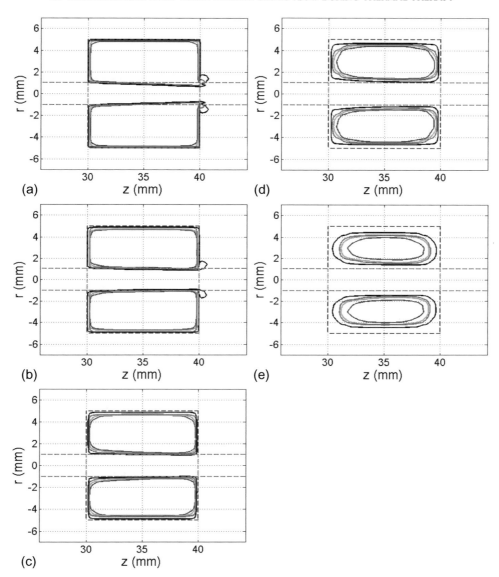

FIGURE 1.10   Thermal lesion region represented by $EM_{43} = 240$ min contour with heating schemes I-V in Table 1.4 of Section 1.2 shown in (a-e), respectively. $\tau_t = 0, 0.464, 1.756,$ and 6.825 s are represented by blue, green, red, and black curves, respectively. The blood vessel diameter is 2 mm.

1.3.4 Conclusion

From the coupling model of thermal hyperbolic bioheat transfer in solid tissues and energy transport of blood flow in thermal significant blood vessels developed above, the thermal behavior is found to be very sensitive to the heating rate and the thermal relaxation time. Heat leaves the target region more slowly and tumor tissue preserves high temperature longer when heating is fast and thermal relaxation time is large. This is all due to the larger lagging of heating and less attenuation for low-frequency modes as τ_t gets large. The wave feature characterized by large thermal relaxation time also causes the thermal lesion region based on $EM_{43} = 240$ min level curve to cover the tumor region better. It implies that the traditional simulations based on PBHTE may underestimate thermal dose because thermal relaxation time is actually nontrivial in living tissues.

1.4 PBHTE WITH THE VASCULAR COOLING NETWORK MODEL

As stated in the previous section, blood flow accounts for up to 90% of heat removal [50]. Thus, attempting to model complicated vascular geometry in which blood circulates along those different sizes of vessels is an important approach to analyze precisely the heat transfer processing in a biothermal system. To account for the impact of large blood vessels on the heat transfer processes, thermal models have to take conductive and convective heat transport into account. And many researchers have attempted to describe the impacts of the blood vessels [18]. Some of the vascular models have been designed not for application in hyperthermia treatment planning, but to obtain basic insight into the heat transfer between large blood vessels and tissue. Therefore, these models are relatively simple with straight vessels represented by a tube with a specified diameter [16,51–53]. These basic models are also very useful in calculating the temperature distributions induced by thermal ablation, where it is necessary to account for a single large blood vessel passing through a target region. The heat transfer modes—conduction and convection—are seen during biological processes. Investigation of the basics of heat transfer using basic discrete vessel models and a selection of these models are discussed below.

1.4.1 Thermally Significant Blood Vessel Model

During the late 1980s, many investigators [9–11], following the rationale that was similar to that which initiated the CH and WJ models presented in Section 1.1, began to question the handling of the blood perfusion term and how to better approximate the blood temperature and the local tissue temperatures where blood vessels (countercurrent vessels) are involved. Because arterial and venous capillary vessels are small, their thermal contributions to local tissue temperatures are insignificant when compared with large blood vessels. However, for vessel sizes larger than the capillaries, there are noticeable, thermally significant impacts on the local tissue temperatures during either the cooling or heating processes. Several investigators [16,51] examined the effect of large blood vessels on the temperature distribution using theoretical studies. Huang et al. [54], in 1996, presented a more fundamental approach to model temperatures in tissues than do the generally used approximate equations, such as the PBHTE or effective thermal conductivity equations. As such, this type of model can be used to study many important questions at a more basic level. For example, in the particular hyperthermia application studied [54], a simple vessel network model predicts that the role of countercurrent veins is minimal and that their presence does not significantly affect the tissue temperature profiles. The arteries, however, removed a significant fraction of the power deposited in the tissue. The Huang model used a simple convective energy balance equation to calculate the blood temperature as a function of position:

$$\dot{M}_i c_b \frac{dT_b}{dx_i} = \dot{Q}_{ap} - h_i A_i (T_b - T_w). \tag{1.28}$$

Here, $\dot{M}_i$ is the mass flow rate of blood in artery i, c_b is the specific heat of blood, $T_b(x_i)$ is the average blood temperature at position x_i, x_i indicates the direction along the vessel i (either x, y, or z depending on the vessel level). $\dot{Q}_{ap}$ is the applied power deposition, x_i is the position x along blood vessel i, h_i is the heat transfer coefficient between the blood and the tissue, A_i is the perimeter of blood vessel i, and $T_w(x_i)$ is the temperature of the tissue at the vessel wall. For the smallest, terminal arterial vessels, a decreasing blood flow rate is present, resulting in the energy balance equation:

$$\dot{M}_i c_b \frac{dT_b}{dx_i} = \dot{Q}_{ap} - h_i A_i (T_b - T_w) - \frac{d\dot{M}_i}{dx_i} c_b T_b. \tag{1.29}$$

The blood leaving these terminal arterial vessels at any cross-section is assumed to perfuse throughout the tissue at a constant rate. A detailed description is given by Huang et al. [54]. As to the venous thermal model, for all veins except the smallest terminal veins, Equation (1.28) holds. For the smallest veins, the blood temperature, T_b, is replaced by the venous return temperature, $T_{vr}(x_i)$. In the presented study, this temperature is taken to be the average temperature of four tissue nodes adjacent to the terminal vein in the plane perpendicular to that vein:

$$T_{vr} = \frac{1}{4} \sum_{i=1}^{4} T_{i,adj}, \tag{1.30}$$

$T_{i,adj}$ is the tissue temperatures adjacent to the venous vessel. As a terminal vessel runs in any x, y, or z straight direction, there are four neighboring tissue nodes considered in terms of the computational scheme (i.e., finite difference method). In order to graphically illustrate the models and assumptions above, Figure 1.11 shows models depicting Equations (1.28)–(1.30).

1.4.2 Vessel Network Geometry and Fully Conjugated Blood Vessel Network Model

The Fully Conjugated Blood Vessel Network Model (FCBVNM) is a model formulation which describes the solid tissue matrix having thermally significant vessel generations (seven levels) by Huang et al. [54]. The effects of all vessels smaller than the terminal (level 7) vessels are not explicitly modeled in FCBVNM. Thus, those smaller vessels (connected to the terminal arteries and the terminal veins in the network) are implicitly assumed to be thermally insignificant in the FCBVNM.

The tissue geometry used in this study [54] consists of a regular, branching vessel network as partially shown (only the arterial vessels are shown) in Figure 1.12 that is embedded in a

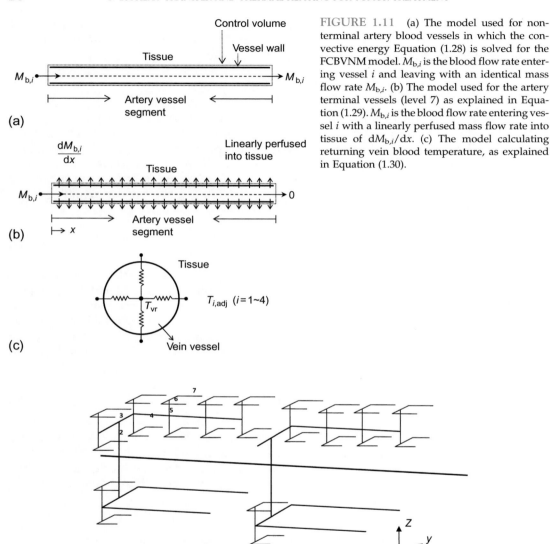

FIGURE 1.11   (a) The model used for non-terminal artery blood vessels in which the convective energy Equation (1.28) is solved for the FCBVNM model. $M_{b,i}$ is the blood flow rate entering vessel i and leaving with an identical mass flow rate $M_{b,i}$. (b) The model used for the artery terminal vessels (level 7) as explained in Equation (1.29). $M_{b,i}$ is the blood flow rate entering vessel i with a linearly perfused mass flow rate into tissue of $dM_{b,i}/dx$. (c) The model calculating returning vein blood temperature, as explained in Equation (1.30).

FIGURE 1.12   Schematic diagram to show a portion of the arterial vessel network used in this study. All seven vessel levels (level 1-7) for the arterial network are shown, and the venous network, which is not shown, is parallel to the arterial network, with a grid size in the x, y, and z dimensions away from the arterial network. '1' is not shown in Fig 1.12. It represents the main artery (level 1) and largest artery of 7-level blood vessel model. The level 1 artery lies along the long central, lengthwise (x) axis in Fig.1.12.

control volume, which is an (approximate) cube of dimensions $L=8.2$ cm and $W=H=8$ cm in the x, y, and z directions, respectively. All vessels are straight-line segments and are parallel to one of the three Cartesian axes. There are up to seven levels of arteries, beginning with the main artery (level 1) which lies along the central, lengthwise (x) axis of the cube. Table 1.5

TABLE 1.5   Vascular Parameters for Each Vessel Level in the Vascular Network

Vessel level	Total number of vessels	Individual vessel length	Vessel diameter	Vessel spacing (x, y, z)	Maximum mass flow rate in vessel	Total vessel surface area
1	1	$L/2^a$	D	NA	$128\,PV_{\text{tsv}} + M_{\text{TA}}{}^b$	$\pi LD/2$
2	4	$H/4$	γD	NA, NA, $L/2$	$32\,PV_{\text{tsv}}$	$\gamma \pi HD$
3	8	$W/4$	$\gamma^2 D$	NA, $H/2$, $L/2$	$16\,PV_{\text{tsv}}$	$2\gamma^2 \pi WD$
4	8	$3L/8^c$	$\gamma^3 D$	$W/2$, $H/2$, NA	$12\,PV_{\text{tsv}}$	$3\gamma^3 \pi LD$
5	64	$H/8$	$\gamma^4 D$	$W/2$, NA, $L/8$	$2\,PV_{\text{tsv}}$	$8\gamma^4 \pi HD$
6	128	$W/8$	$\gamma^5 D$	$W/4$, $H/4$, $L/8$	PV_{tsv}	$16\gamma^5 \pi WD$
7	128	$L/8 - \Delta^d$	$\gamma^5 D$	$W/4$, $H/4$, NA	PV_{tsv}	$16\gamma^5 \pi (L - 8\Delta)D$

NA indicates that the particular parameter in not applicable for that vessel level.
[a] *The level 1 vessel terminates at the center of the control for all cases i in this paper.*
[b] *The value at the inlet to the control volume.*
[c] *Each level 4 vessel has three segments.*
[d] *The terminal ends of one set of level 7 vessels are separated from the beginning of the next set of level 7 vessels by a gap of one finite difference nodal space, $\Delta = 2$ mm.*

lists the basic vessel network properties used in this section (vascular parameters for each vessel level in the vascular network model in simulation). The diameters of the arteries decrease by a constant ratio, γ, between successive levels of branching vessels (the ratio of diameters of successive vessel generations), in other words:

$$\gamma = \frac{D_{i+1}}{D_i},\qquad (1.31)$$

where D_i and D_{i+1} are the diameters of two successive levels of branching arteries. When two successive levels of numbered vessels do not branch but only change direction (i.e., levels 6 and 7 in this model), the vessel diameter does not change. In this study, we used $\gamma = 0.9$ in the presented results.

The desired treated tumor region is a cube described in Figure 1.13a with 20 mm in x, y, and z dimensions. The locations and paths of arterial vessels inside the treated region are described in Figure 1.13b and c. The geometric arrangement of the countercurrent veins is essentially identical to that of the arteries, with all of the veins offset from the arteries by one finite difference node in x, y, and z dimensions as appropriate to avoid intersections of vessels.

1.4.3 Discrete Vessel Modeling with Semicurved Vessel Network and Real 3D Vasculature Network

A vasculature model with straight lines provides a tool for a better analysis in a heat transfer process that is influenced by convection from different sizes of vessels. A more flexible algorithm in which the model geometry is subdivided into a vessel space and a tissue space has been developed by Mooibroek and Lagendijk [55] to obtain a more realistic, semicurved representation of vessel networks for use in hyperthermia treatment planning. This semicurved

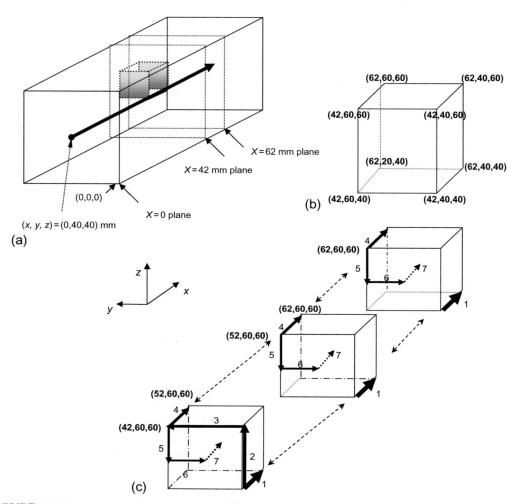

FIGURE 1.13   (a) A transparent view of parallelepiped showing the desired heated tumor region, which is a cube of 20 mm in each dimension. The level 1 blood vessel (largest) runs through the desired heated cube's edge from (42, 40, 40) to (62, 40, 40). The inlet temperature of the level 1 vessel starting at (0, 40, 40) is set at 37 °C. (b) Shows the coordinates of eight corners of the desired treated cube (unit: mm). (c) Is a dissected transparent view showing all associated arterial blood vessel paths (or segments) in the cubic volume, and venous vessels do not appear in the figure. There are two branches of level 5-7 blood vessels and one of level 5 and 6 on the back boundary as the dissected view indicates.

model uses a finite difference method, and the vessel segments are represented by connected strings of vessel nodes. The nodes with their centers closest to the vessel axis are considered vessel nodes. The description of a 3D vessel segment is used as a building block, which allows modeling of realistic complex vessel networks. A vessel-specific second discretization step in time is performed to describe the convective heat transfer within the vessel space. This makes it possible to incorporate vessels with different flows and diameters. Furthermore, the predicted thermal equilibration lengths were compared to theoretical values, and a reasonable agreement was observed. Comparison with phantom measurements showed accuracy within the range of the experimental error. This result shows that cylindrically shaped vessels can be modeled accurately using a square grid, which is a very important simplification for numerical techniques used by practitioners. However, because of the cubic subdivision of blood vessels, a one-to-one description of a true 3D vessel network cannot be realized with this model.

To address the above-mentioned problems, a more sophisticated thermal model that allows thermal modeling with real detailed 3D discrete vasculature networks (DIVA) has been developed by Kotte et al. [56,57]. Similar to Mooibroek and Lagendijk [55], the model geometry is subdivided into a vessel space and a tissue space, but in the DIVA model, the vasculature is described by 3D curves with a specified diameter. This approach allows the consideration of all relevant blood vessels independent of the voxel size. Besides the clear advantage of the tissue-voxel resolution independence, modeling blood vessels as 3D curves makes the model compatible with MR/CT (Magnetic Resonance/Computerized Tomography) angiography vessel reconstruction software, which is essential for routine use in treatment planning. As to the heat transfer calculation, the exact heat flow between a vessel segment and its surrounding tissue is difficult to calculate for a realistic situation with a heterogeneous vessel network that includes power and temperature distributions. To solve this problem, a method to estimate the heat flow using a simplified situation has been developed [56]. For a vessel segment embedded in a tissue cylinder, an analytical expression for the heat flow can be derived when cylindrically symmetric boundary conditions and a thermally fully developed flow are assumed. The latter is justified because the entrance length is considerably shorter than the equilibration length [18].

1.4.4 Conclusion

There has been substantial progress in thermal modeling with discrete vasculature. Many basic models have been developed to provide insight into the cooling effects of vasculature and temperature gradients around large blood vessels. Modeling of straight and semicurved vessel networks has led to improved characterization of heat transfer between vasculature and tissue. However accuracy and time efficiency will be great concerns when complicated vasculature thermal models are developed and used in the real time clinic treatments.

1.5 HYPERTHERMIA TREATMENT PLANNING

A treatment planning for hyperthermia in biological processes is essential for adequate treatment control. Reliable temperature information during clinical hyperthermia and thermal ablation must comprise a thermal model, but conventional temperature measurements do not provide 3D temperature information. The model must take conductive and convective heat transport into account, as blood flow plays a significant role in hyperthermia [50]. Hyperthermia cancer treatment requires precise thermal absorbed power deposition to raise tumor tissue temperature up to the therapeutic range with a sufficient amount of time duration to prevent overheating the normal tissues. Many researchers [58–61] have investigated a noninvasive heating modality for exploring power deposition with fine spatial resolution and/or optimization within the tumor region. Hyperthermia applicator technology is currently one of the most important things that can improve temperature homogeneity in the treated region as well as help to reach an optimal applied power field. This section investigates the significance of blood vessels in the absorbed power and temperature distributions when optimization was employed during hyperthermia. The treated tumor region is simulated using a three-dimensional (3D) tissue model embedded with a countercurrent blood vessel network [54]. 3D absorbed power depositions are obtained by using optimization to reach a uniform temperature of 43 °C for the desired treated region. The results show that the absorbed power deposition for optimization with fine spatial resolution produces a uniform temperature distribution maintained at 43 °C in the desired treated tumor region except for some cold spots and/or small cold strips caused by thermally significant large vessels. The amount of total absorbed power suggests that a region with thermally significant vasculature requires much more power deposited than one without vasculature. In addition, optimization with coarse spatial resolution results in a highly inhomogeneous temperature distribution in the treated region due to the strong cooling effect of blood vessels. Therefore, prior to hyperthermia treatments, thermally significant blood vessels should be identified and handled carefully to effectively reduce their strong cooling effect, particularly those vessels flowing into the treated region.

1.5.1 Optimization with Fine Spatial Power Deposition: Based on Local Temperature Response in the Treated Region

Figure 1.14 is a flow chart describing continuously adjusting absorbed power deposition in the desired treated tumor region in order to reach ideal temperature (uniform temperature throughout the treated tumor region with a temperature of 43 °C). The evaluation criterion of absorbed power deposition is shown in Equation (1.32). It states that the root mean square of the difference between the ideal temperature (43 °C) and the calculated temperature of all heated target nodes which is normalized by the difference, 43-37 °C, reaches less than the criterion value (set to be 10% of the temperature difference of 43-37 °C). If this criterion is achieved, we obtain the optimization of absorbed power deposition such that the heating temperature distribution is close to the ideal temperature distribution. Otherwise, the absorbed power deposition will be adjusted according to the local temperature. The readjusted power deposition (P_{n+1}) is described in Equation (1.33):

$$\text{Evaluation criterion:} \quad \frac{\sqrt{\dfrac{\sum_{\text{all target nodes}}(\Delta T(x,y,z))^2}{\text{Total number of target nodes}}}}{43-37 \ ^\circ\text{C}} \leq 0.1, \tag{1.32}$$

$$P_{n+1}(x,y,z) = P_n(x,y,z) + \Delta P(x,y,z), \tag{1.33}$$

with $\Delta P(x,y,z) = C \cdot \Delta T(x,y,z)$, C is 10,000, n is the iteration number, and $\Delta T(x, y, z)$ is the difference of ideal temperature (43 °C) and calculated temperature.

To investigate the significance of blood vessels in the temperature distribution for optimal hyperthermia treatment, an optimization scheme as described in Equations (1.32) and (1.33).

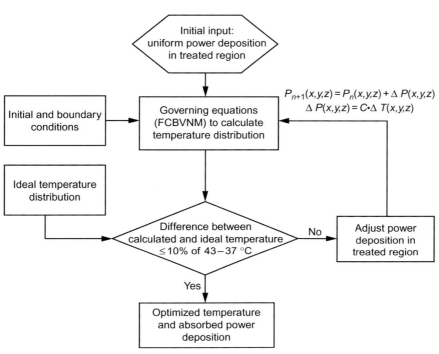

FIGURE 1.14   The flow chart of optimization used in this study. The absorbed power deposition in the desired treated cube (20 mm in each dimension) is adjusted locally in order to achieve an ideal therapeutic temperature of 43 °C uniformly for the entire cube.

The computational flow chart, shown in Figure 1.15, determines the absorbed power deposition, which includes the heating of blood vessels to achieve an optimal treatment. The flow chart to find the optimal solutions is used in this section. Figure 1.16a-e is the optimal temperature distributions on the planes 4 mm away from the front boundary, the middle, the back boundary, and 4 mm away from the back boundary of the treated region, respectively, and Figure 1.16f-h is the absorbed power depositions on the planes of the front boundary, the middle, and the back boundary of the treated region, respectively, for a blood perfusion of 0.5 kg/m^3/s and a blood flow velocity of 320 mm/s in the level 1 vessel. Figure 1.16a shows that the temperature is approximately 40.0 °C near the treated region and displays a cold spot at the center which is due to a level 1 artery blood vessel running perpendicular inwards to the plane. At its southeastern diagonal direction about 2.8 mm away from the level 1 artery, a level 1 vein is running in an opposite direction outwards to the plane. The vein appears to be collecting some thermal energy by convection through the treated region. Figure 1.16b shows that the temperature on the boundary of the treated region is close to ideal temperature (43 °C), and there are steep thermal gradients near the level 2 artery running upwards from the center point. As seen in Figure 1.16f, large amounts of thermal power were deposited on level 2, 3, and 5 arteries. The maximum thermal power deposition is approximately 3.7×10^6 W/m^3. Figure 1.16c shows that the temperature in the treated region is close to ideal

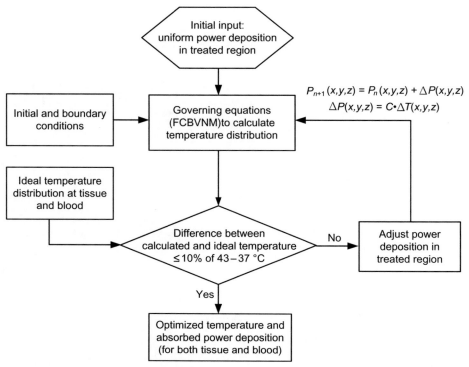

FIGURE 1.15   The computational flow chart to find the optimal power and temperature distributions during hyperthermia.

temperature, while high temperature appears outside of the treated region due to a level 6 artery carrying convective thermal energy leftwards. It illustrates that thermally significant blood vessels within the treated region have effectively been heated and carried the convective thermal energy out of the region. Figure 1.16c also shows maximum temperature located on (or near) the left-side artery branch node of level 5 and 6 arteries in the treated region. Figure 1.16g shows that large thermal power is deposited on the corner of the treated region, which is an area with dense blood vessels (level 3, 4, and 5 arteries and veins). In Figure 1.16d, the temperature distribution at the back boundary of the treated region shows that a cold spot is found near the northwestern corner of the heated region. The spot is 1.7 °C below the ideal temperature, and it is caused by the level 4 vein flowing into the heated region. As expected, Figure 1.16h shows a large amount of thermal power deposited on the same corner as shown in Figure 1.16g to compensate the heat loss caused by vessels. Dense blood vessels act as energy sinks, and a large amount of thermal power deposition is required in that area in order to maintain the local temperature at the desired level. Figure 1.16e, temperature on the plane 4 mm away from the back boundary of the heated region, shows some hot spots, and these spots are approximately 42 °C. One spot, located in the northeastern direction more than 4 mm away from the heated region, has a temperature of about 38.3 °C. Those hot spots are caused by arteries carrying hot blood flow.

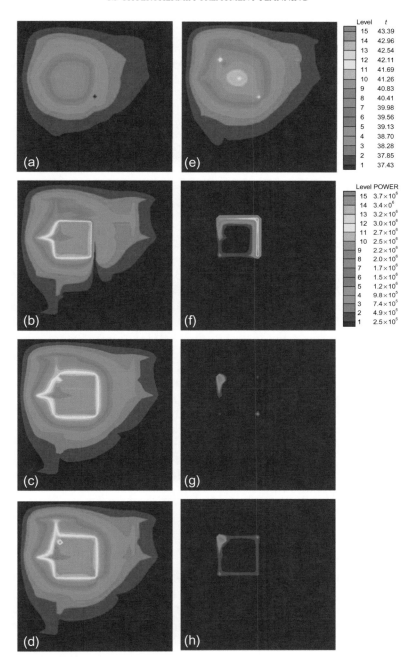

FIGURE 1.16 Temperatures and absorbed power depositions for a blood perfusion rate of 0.5 kg/m³/s after optimization with fine spatial power deposition. The ideal temperature is set to be 43 °C, and the blood flow velocity is about 320 mm/s in level 1 vessel. (a-e) are the temperature distributions at $x = 38$ mm (4 mm away from the front boundary), $x = 42$ mm (the front boundary), $x = 52$ mm (the middle of the treated region), $x = 62$ mm (the back boundary), and $x = 66$ mm (4 mm away from the back boundary) planes, respectively. (f-h) are the absorbed power depositions at $x = 42$, 52, and 62 mm planes, respectively, after optimization (units in figure, t: °C and power: W/m³).

1.5.2 Optimization with Lumped Power Deposition: Uniform Absorbed Power Deposition in the Treated Tumor Region

To investigate the spatial resolution of absorbed power deposition on the temperature distribution, a uniform power deposition in the entire desired treated tumor region is applied. Two important parameters need to be introduced for this optimization. One is cost function and the other is the power coefficient. Cost function at nth iteration is set to be $C_n = \sqrt{\sum_{\text{All target nodes}} (T(x,y,z) - 43\,^\circ\text{C})^2}$, with $\Delta C_{n+1} = C_{n+1} - C_n$, and the absorbed power is $P_{n+1} = P_n + h_{\text{coef}} \cdot \Delta C_n$ with the coefficient h_{coef}. h_{coef} is the updated power coefficient, with a constant value of 3050. It is chosen based on a smoothly converging (i.e., no oscillating) search and with less computational time required during optimization. The optimization process will be terminated when ΔC_n is smaller than 10^{-4}. Optimization of uniform power deposition allows the attention to be focused on the effect of the vasculature on the temperature distribution.

To investigate the effect of lumped power deposition on the resulting temperature distribution, a uniform power deposition for the entire treated region is used. This represents the limitation of the heating system to tune its power spatially fine enough to meet the treatment requirement. Figure 1.17a-e is the optimized temperature distributions on the planes of 4 mm away from the front boundary, the front boundary, the middle, the back boundary, and 4 mm away from the back boundary of the treated region, respectively, and Figure 1.17f-h is absorbed power depositions on the planes of the front boundary, the middle, and the back boundary, respectively, for a blood perfusion of 0.5 kg/m^3/s and a blood flow velocity of 320 mm/s in the level 1 vessel. The initial guess of the uniform power deposition in the treated region was 10^5 W/m^3, and the optimized absorbed power deposition was obtained as the difference (ΔC_n) between two successive cost function values was smaller than 10^{-4}. With this optimization of lumped power deposition, the temperatures shown in Figure 1.17b-d indicates that the temperatures in the treated region are highly inhomogeneous, with a temperature about 3 °C below the desired therapeutic temperature in the places near the boundary planes. These temperature distributions show that blood flow of vessels results in significantly lower temperature strips along the vessels in the treated region, particularly a large vessel located at the boundary of the treated region.

1.5.3 Effect of Blood Perfusion and Blood Flow Rates on the Optimization

As blood perfusion increases, the flow rates in vessels get higher due to the conservation of blood mass, and the higher flow rate will produce a stronger thermal impact on the treated region. Figure 1.18 shows temperature and power depositions for a blood perfusion of 2.0 kg/m^3/s and a blood flow velocity about 1280 mm/s in level 1 blood vessel. Figure 1.19 shows temperature and power depositions for a blood perfusion of 0.123 kg/m^3/s and a blood flow velocity of about 80 mm/s in a level 1 blood vessel, which in size and blood flow velocity of vessel is identical to dog data from CH [3]. The optimized power deposition pattern is similar to, but with a higher or lower value than, that shown in Figure 1.15, and the temperature distribution shows that uniform temperature close to the ideal value can be obtained in the

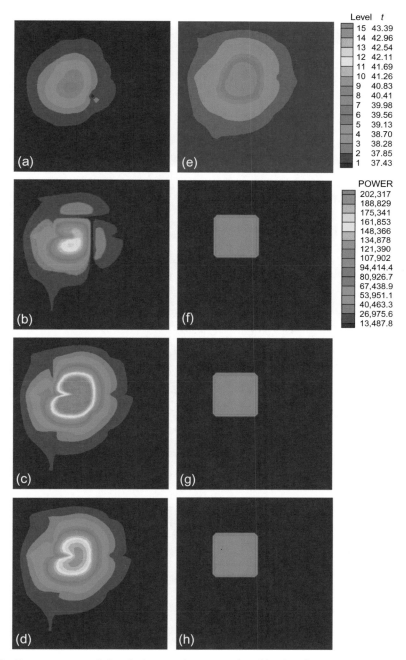

FIGURE 1.17 Temperatures and absorbed power depositions for a blood perfusion rate of 0.5 kg/m³/s after optimization with lumped power deposition. The ideal temperature is set to be 43 °C, and the blood flow velocity is about 320 mm/s in level 1 vessel. (a-e) are the temperature distributions at $x = 38$ mm (4 mm away from the front boundary), $x = 42$ mm (the front boundary), $x = 52$ mm (the middle of the treated region), $x = 62$ mm (the back boundary), and $x = 66$ mm (4 mm away from the back boundary) planes, respectively. (f-h) are the absorbed power depositions at $x = 42$, 52, and 62 mm planes, respectively, after optimization (units in figure, t: °C and power: W/m³).

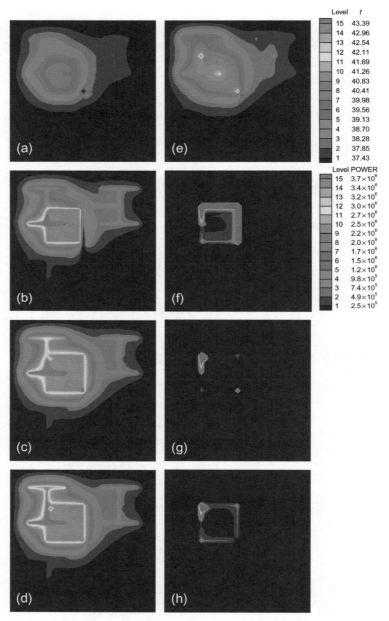

FIGURE 1.18 Temperatures and absorbed power depositions for a blood perfusion rate of 2.0 kg/m^3/s after optimization with fine spatial power deposition. The ideal temperature is set to be 43 °C, and the blood flow velocity is about 1280 mm/s in level 1 vessel. (a-e) are the temperature distributions at $x=38$ mm (4 mm away from the front boundary), $x=42$ mm (the front boundary), $x=52$ mm (middle of the treated region), $x=62$ mm (the back boundary), and $x=66$ mm (4 mm away from the back boundary) planes, respectively, after optimization. (f-h) are the absorbed power depositions at $x=42$, 52, and 62 mm planes, respectively, after optimization (units in figure, t: °C and power: W/m^3).

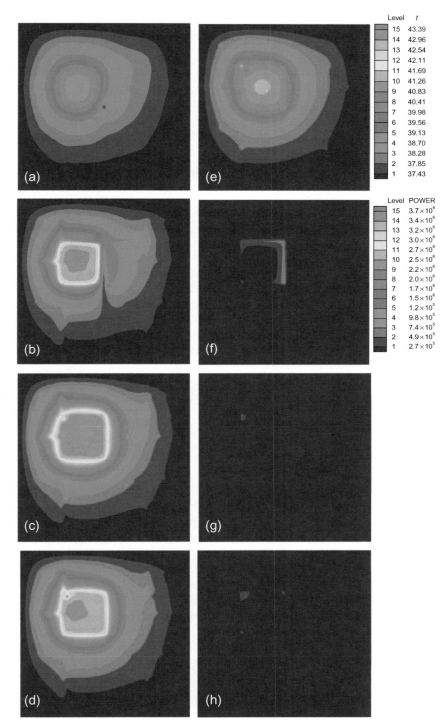

FIGURE 1.19 Temperatures and absorbed power depositions for a blood perfusion rate of 0.123 kg/m³/s after optimization with fine spatial power deposition. The ideal temperature is set to be 43 °C, and the blood flow velocity is about 80 mm/s in level 1 vessel. (a-e) are the temperature distributions at $x=38$ mm (4 mm away from the front boundary), $x=42$ mm (the front boundary), $x=52$ mm (middle of the treated region), $x=62$ mm (the back boundary), and $x=66$ mm (4 mm away from the back boundary) planes, respectively, after optimization. (f-h) are the absorbed power depositions at $x=42, 52$, and 62 mm planes, respectively, after optimization (units in figure, t: °C and power: W/m³).

treated region except for some cold spots which are produced by the arteries, the same as in the case of 0.5 kg/m^3/s.

Smooth and homogeneous temperatures in treated tumor volume could be achieved easily with less power deposition by using preheating [62]. Furthermore, for a better homogeneous temperature distribution in the treated region, modification of an adaptive optimization scheme is required [63].

1.5.4 Optimization Without Thermally Significant Blood Vessels in the Tissues

The PBHTE is used to investigate temperature and absorbed power deposition in the treated region for the condition without thermally significant blood vessels. A uniform blood perfusion rate of 0.5 kg/m^3/s in the entire tissue was studied using the optimization. Figures 1.20a and e show two unheated parallel temperature distribution planes next to the front boundary plane at 1 and 2 discretization steps. Figures 1.20f-h show the optimized power deposition on the front boundary, the middle, and the back boundary planes, respectively. Most of the power is deposited on the corners and edges of the treated region to compensate for thermal energy loss through conduction due to the strong conductive effects near corners and edges. The deposited power pattern of Figure 1.20h is identical to that shown in Figure 1.20f, and Figure 1.20g (the middle plane of the treated region) shows that there is less power deposited on corners and center area as compared to Figure 1.20f (front boundary plane). It indicates that the thermal diffusion rate is much smaller in the middle region. Figures 1.20b-d show a very uniform therapeutic temperature distribution on the front boundary, the middle, and the back boundary planes, respectively, in the treated region. They can be achieved as there are no thermally significant blood vessels present.

1.5.5 Conclusion

To produce a uniform therapeutic temperature distribution in the desired treated region while minimizing the overheating of the surrounding normal tissue is desirable for hyperthermia treatment. To reach this goal requires a powerful heating system that is able to deposit power in the treated region to raise the temperature of the entire treated region up to the desired value and overcome the loss of energy by blood perfusion, boundary conduction, and blood flow from the vasculature. The temperature results after optimization show the cold spots and/or cold strips along the blood vessels. These temperatures display the tremendous effects of blood vessels on the resulting heating temperature and the limitation of heating systems. A powerful heating system with fine spatial resolution for power deposition has a better ability to deliver suitable power to locally overcome the convective effect caused by the thermally significant vessels. Although a heating system with fine power deposition is a very important factor during treatment, the complexity of existing thermally significant blood vessels plays a crucial role in successful hyperthermia treatments. This complexity is related to mass flow rates, inlet temperatures, and directions of vessels. Therefore, prior to hyperthermia treatments, thermally significant blood vessels should be identified and handled carefully in order to reduce their cooling effects on the treated region, particularly to those vessels flowing into the treated region.

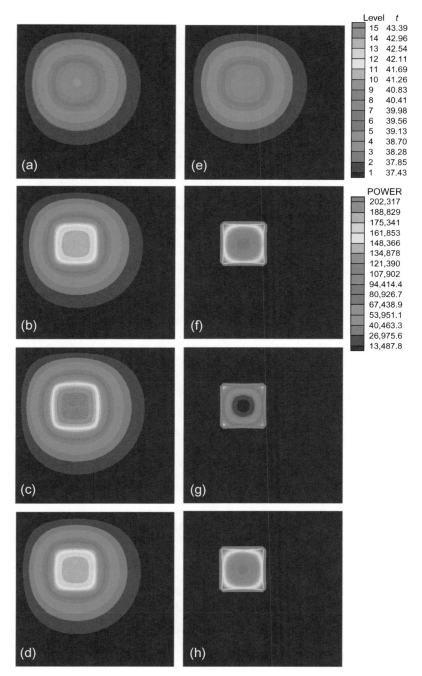

FIGURE 1.20   Temperatures and absorbed power depositions for a blood perfusion rate of 0.5 kg/m³/s with no vasculature present and after optimization with fine spatial power deposition. The ideal temperature is set to be 43 °C. (a-e) are the temperature distributions at $x=38$ mm (4 mm away from the front boundary), $x=42$ mm (the front boundary), $x=52$ mm (middle of the treated region), $x=62$ mm (the back boundary), and $x=66$ mm (4 mm away from the back boundary) planes, respectively, after optimization. (f-h) are the absorbed power depositions at $x=42$, 52 and 62 mm planes, respectively, after optimization (units in figure, t: °C and power: W/m³).

References

[1] Pennes HH. Analysis of tissue and arterial blood temperature in the resting human forearm. J Appl Phys 1948;1:93–122.

[2] Roemer RB, Forsyth K, Oleson JR, Clegg ST, Sim DA. The effect of hydralazine dose on blood perfusion changes during hyperthermia. Int J Hyperthermia 1988;4(4):401–15.

[3] Chen MM, Holmes KR. Micro vascular contributions in tissue heat transfer. Ann N Y Acad Sci 1980;335:137–50.

[4] Weinbaum S, Jiji LM. A new simplified bioheat equation for the effect of blood flow on local average tissue temperature. J Biomech Eng 1985;107:131–9.

[5] Wissler EH. Comments on the new bioheat equation proposed by Weinbaum and Jiji. J Biomech Eng 1987;109:131–9.

[6] Wissler EH. Comments on Weinbaum and Jiji discussion of their proposed bioheat equation. J Biomech Eng 1987;109:355–6.

[7] Weinbaum S, Jiji LM, Lemons DE. Theory and experiment for the effect of vascular temperature on surface tissue heat transfer—part 11: model formulation and solution. J Biomech Eng 1984;106:331–41.

[8] Keller KH, Seiler L. An analysis of peripheral heat transfer in man. J Appl Physiol 1971;30:779–86.

[9] Baish JW, Ayyaswamy PS, Foster KU. Small scale temperature fluctuations in perfused tissue during local hyperthermia. J Biomech Eng 1986;108:246–51.

[10] Baish JW, Ayyaswamy PS, Foster KR. Heat transport mechanisms in vascular tissues: a model comparison. J Biomech Eng 1986;108:324–31.

[11] Charny CK, Weinbaum S, Levin UL. An evaluation of the Weinbaum-Jiji bioheat equation for normal and hyperthermic conditions. J Biomech Eng 1990;I(12):80–7.

[12] Charny CK, Levin RL. Bioheat transfer in a branching countercurrent network during hyperthermia. J Biomech Eng 1989;111:263–70.

[13] Nelson DA. Invited editorial on "Pennes' 1948 paper revisited". J Appl Physiol 1998;85:2–3.

[14] Kolios MC, Sherar MD, Hunt JW. Blood flow cooling and ultrasonic lesion formation. Med Phys 1996;23:1287–98.

[15] Kotte A, van Leeuwen G, de Bree J, van der Koijk J, Crezee H, Lagendijk J. A description of discrete vessel segments in thermal modeling of tissues. Phys Med Biol 1996;41:865–84.

[16] Chato JC. Heat transfer to blood vessels. J Biomech Eng 1980;102:110–8.

[17] Lagendijk JJW. The influence of blood flow in large vessels on the temperature distribution in hyperthermia. Phys Med Biol 1982;27:17–23.

[18] Crezee J, Lagendijk JJW. Temperature uniformity during hyperthermia: the impact of large vessels. Phys Med Biol 1992;37:1321–37.

[19] Hariharan P, Myers MR, Banerjee RK. HIFU procedures at moderate intensities-effect of large blood vessels. Phys Med Biol 2007;52:3493–513.

[20] Consiglieri L, dos Santos I, Haemmerich D. Theoretical analysis of the heat convection coefficient in large vessels and the significance for thermal ablative therapies. Phys Med Biol 2003;48:4125–34.

[21] Shrivastava D, Roemer RB. Readdressing the issue of thermally significant blood vessels using a countercurrent vessel network. J Biomech Eng 2006;128:210–6.

[22] He Q, Zhu L, Lemonds DE, Weinbaum S. Experimental measurements of the temperature variation along artery-vein pairs from 200 to 1000 μm diameter in rat hind limb. J Biomech Eng 2002;124:656–61.

[23] Shih TC, Kou HS, Lin WL. Cooling effect of thermally significant blood vessels in perfused tumor tissue during thermal therapy. Int Commun Heat Mass 2006;33:135–41.

[24] Kolios MC, Sherar MD, Hunt JW. Large blood vessel cooling in heated tissues: a numerical study. Phys Med Biol 1995;40:477–94.

[25] Shih TC, Kou HS, Lin WL. The impact of thermally significant blood vessels in perfused tumor tissue on thermal distributions during thermal therapies. Int Commun Heat Mass 2003;30:975–85.

[26] Khanafer K, Bull JL, Pop I, Berguer R. Influence of pulsatile blood flow and heating scheme on the temperature distribution during hyperthermia treatment. Int J Heat Mass Tran 2007;50:4883–90.

[27] Song CW, Lokshina A, Rhee JG, Paten M, Levitt SH. Implication of blood flow in hyperthermia treatments of tumors. IEEE Trans Biomed Eng 1984;31:9–16.

[28] Craciunescu OI, Clegg ST. Pulsatile blood flow effects on temperature distribution and heat transfer in rigid vessels. J Biomech Eng 2001;123:500–5.

[29] Horng TL, Lin WL, Liauh CT, Shih TC. Effects of pulsatile blood flow in large vessels on thermal dose distribution during thermal therapy. Med Phys 2007;34:1312–20.

[30] Fung YC. Biomechanics: motion, flow stress, and growth. New York: Springer-Verlag; 1996.

[31] Nichols WW, O'Rourke MF. McDonald's blood flow in arteries: theoretic, experimental and clinical principles. Philadelphia: Lea & Febiger; 1990.

[32] Huo Y, Kasab GS. Pulsatile blood flow in the entire coronary arterial tree: theory and experiment. Am J Physiol Heart Circ Physiol 2006;291:H1074–87.

[33] Shih T-C, Horng T-L, Huang H-W, Ju K-C, Huang T-C, Chen P-Y, et al. Numerical analysis of coupled effects of pulsatile blood flow and thermal relaxation time during thermal therapy. Int J Heat Mass Tran 2012;55:3763–73.

[34] Sapareto SA, Dewey WC. Thermal dose determination in cancer therapy. Int J Radiat Oncol Biol Phys 1984;10:787–800.

[35] Damianou C, Hynynen K. Focal spacing and near-field heating during pulsed high temperature ultrasound therapy. Ultrasound Med Biol 1993;19:777–87.

[36] Kou HS, Shih TC, Lin WL. Effect of directional blood flow on thermal dose distribution during thermal therapy: an application of Green's function based on the porous model. Phys Med Biol 2003;48:1577–89.

[37] Joseph DD, Preziosi L. Heat waves. Rev Mod Phys 1989;61:41–73.

[38] Özisik MN, Tzou DY. On the wave theory in heat conduction. J Heat Trans-T ASME 1994;116:525–6.

[39] Tzou DY. Marco- to micro-scale heat transfer: the lagging behavior. Washington, DC: Taylor and Francis; 1997.

[40] Kaminski K. Hyperbolic heat conduction equation for materials with a nonhomogeneous inner structure. J Heat Trans-T ASME 1990;112:555–60.

[41] Zhou J, Zhang Y, Chen JK. Non-Fourier heat conduction effect on laser-induced thermal damage in biological tissues. Numer Heat Transfer 2008;54:1–19.

[42] Zhou J, Chen JK, Zhang Y. Dual-phase effects on thermal damage to biological tissues caused by laser irradiation. Comput Biol Med 2009;39:286–93.

[43] Roemer RB, Oleson JR, Cetas TC. Oscillatory temperature response to constant power applied to canine muscle. Am J Physiol 1985;249:R153–8.

[44] Mitra K, Kumar S, Vedavarz A, Moallemin MK. Experimental evidence of hyperbolic heat conduction in processed meat. J Heat Trans-T ASME 1995;117:568–73.

[45] Antaki PJ. New interpretation of non-Fourier heat conduction in processed meat. J Heat Trans-T ASME 2005;127:189–93.

[46] Cattaneo MC. Sur une forme de l'équation de la chaleur éliminant le paradoxe d'une propagation instanteé. Comptes Rendus de L'Academie des Sciences: Series I-Mathematics 1958;247:431–3.

[47] Vernotte P. Les paradoxes de la théorie continue de l'équation de la chaleur. Comptes Rendus 1958;246:3154–5.

[48] Roetzel W, Putra N, Das SK. Experiment and analysis for non-Fourier conduction in materials with non-homogeneous inner structure. Int J Therm Sci 2003;42:541–52.

[49] Zhang Y. Generalized dual-phase lag bioheat equations based on non-equilibrium heat transfer in living biological tissues. Int J Heat Mass Tran 2009;52:4829–34.

[50] Lagendijk JJ, Hofman P, Schipper J. Perfusion analyses in advanced breast carcinoma during hyperthermia. Int J Hyperthermia 1988;4:479–95.

[51] Huang H, Chan C, Roemer RB. Analytical solutions of Pennes bio-heat transfer equation with a blood vessel. J Biomech Eng 1994;116:208–12.

[52] Crezee J, Lagendijk JJW. Experimental verification of bio-heat transfer theories: measurement of temperature profiles around large artificial vessels in perfused tissue. Phys Med Biol 1990;35:905–23.

[53] Chen ZP, Roemer RB. The effects of large blood vessels on temperature distributions during simulated hyperthermia. J Biomech Eng 1992;114:473–81.

[54] Huang HW, Chen ZP, Roemer RB. A countercurrent vascular network model of heat transfer in tissues. J Biomech Eng 1996;118:120–9.

[55] Mooibroek J, Lagendijk JJ. A fast and simple algorithm for the calculation of convective heat transfer by large vessels in three dimensional inhomogeneous tissues. IEEE Trans Biomed Eng 1991;38:490–501.

[56] Kotte ANTJ, van Leeuwen GMJ, de Bree J, van der Koijk JF, Crezee J, Lagendijk JJW. A description of discrete vessel segments in thermal modelling of tissues. Phys Med Biol 1996;41:865–84.

[57] Kotte AN, van Leeuwen GM, Lagendijk JJ. Modelling the thermal impact of a discrete vessel tree. Phys Med Biol 1999;44:57–74.

[58] Tharp HS, Roemer RB. Optimal power deposition with finite-sized, planar hyperthermia applicator arrays. IEEE Trans Biomed Eng 1992;39(6):569–79.

[59] Lin WL, Liang TC, Yen JY, Liu HL, Chen YY. Optimization of power deposition and a heating strategy for external ultrasound thermal therapy. Med Phys 2001;28(10):2172–81.

[60] Kumaradas JC, Sherar MD. Optimization of a beam shaping bolus for superficial microwave hyperthermia waveguide applicators using a finite element method. Phys Med Biol 2003;48(1):1–18.

[61] Cheng TY, Ju KC, Ho CS, Chen YY, Chang H, Lin WL. Split-focused ultrasound transducer with multidirectional heating for breast tumor thermal surgery. Med Phys 2008;35(4):1387–97.

[62] Huang H-W, Liauh C-T, Horng T-L, Shih T-C, Chiang C-F, Lin W-L. Effective heating for tumors with thermally significant blood vessels during hyperthermia treatment. Appl Therm Eng 2013;50:837–47.

[63] Huang H-W, Liauh C-T, Chou C-Y, Shih T-C, Lin W-L. A fast adaptive power scheme based on temperature distribution and convergence value for optimal hyperthermia treatment. Appl Therm Eng 2012;37:103–11.

Tissue Response to Short Pulse Laser Irradiation

Mohit Ganguly[a], Ryan O'Flaherty[a], Amir Sajjadi[b],
Kunal Mitra[a]

[a]Florida Institute of Technology, Melbourne, FL, USA
[b]Massachusetts General Hospital, Boston, MA, USA

2.1 INTRODUCTION

The use of lasers has become common in many medical applications for the diagnosis and treatment of cancerous tumors. Laser ablation is a minimally invasive tumor treatment modality that utilizes photo-thermal, photo-chemical, photo-mechanical, or plasma-mediated mechanisms.

Photo-thermal interactions are caused by the absorption of laser energy in the tissue, and the subsequent conversion of that energy into heat. The effect in the tissue depends on the deposited heat which can vary in amount and rate. If the tissue temperature is slightly elevated for a long duration ($\sim10\,°C$ for seconds to minutes), cell death can occur without causing structural damage to the tissue. This is known as photo-heating. If the tissue temperature is more dramatically elevated for a brief period of time (~20-$30\,°C$ for 2-3 s), photo-coagulation occurs, where cell death is accompanied by structural damage due to the denaturing of tissue proteins. In the most extreme case, where drastic temperature elevation occurs over a time frame smaller than one second, photo-vaporization occurs.

In photo-vaporization, thermally mediated ablation occurs as the laser delivers enough energy to boil the water in the tissue, causing explosive vaporization [1].

Photo-chemical effects are due to the breakdown of molecular bonds. Photons are absorbed in the tissue and excite molecules to higher electronic states. If the incident photon energy is high enough, the molecule can disassociate and form a free radical or possibly break down. This interaction is more relevant when laser light has a wavelength in the visible spectrum. Another mechanism by which photo-chemical ablation occurs is via the use of photo-sensitizing agents which, when injected into the targeted sites in the body, absorb laser radiation and trigger the production of toxic biochemical agents that cause tissue necrosis [2].

Photo-mechanical interactions are generally due to the generation of pressure waves which propagate through the tissue. The pressure waves are generated through one of three mechanisms. The first mechanism is the thermo-elastic expansion of tissue due to laser heating. The stress produced due to the expansion causes physical cellular damage or spallation, which is the ejection of material fragments due to stress that removes surface layers of tissue—the second mechanism by which pressure waves are generated. The third pressure wave generating mechanism is the breakdown of vapor cavities [3]. The pressure waves can travel at speeds upward of 1500 m s^{-1} through the tissue and thus can cause mechanical tissue damage far away from the original ablation zone.

At very high power densities ($\sim 10^{11}$ W cm^{-2}) [3], matter is transformed to plasma through a process called laser-induced optical breakdown. The plasma state consists of a high density of ions and free electrons. Once this occurs, the plasma will absorb subsequent laser energy and cause the vaporization of surrounding tissue.

The photo-thermal mechanism of tissue ablation is the predominant mechanism for laser-based therapeutic applications which primarily use continuous wave (CW) lasers. Techniques such as laser-induced hyperthermia (LIH), laser-induced interstitial thermotherapy (LITT) [4], and laser immunotherapy [5] have been developed for the ablation of subsurface tumors. In LIH and LITT, the laser heats both normal tissue and tumor tissue indiscriminately. This often results in the unwanted damage and death of the surrounding healthy tissues. The temperature increase in the case of LITT leads to the phase change of the tissue material. However, laser immunotherapy utilizes a highly selective absorption of light energy in the tumor tissue by injecting photo-absorptive dyes at the tumor location. Continuous wave (CW) or long pulse laser sources often produce heat-affected zones that are larger than the boundaries of the tumor. This in turn leads to the potential of collateral damage of healthy tissue.

Lasers with pulse durations in the range of nanoseconds (short pulse) to femtoseconds (ultra-short) are being increasingly used for the treatment of subsurface tumors [6]. Short and ultra-short pulse laser can achieve high peak temperatures inside the irradiation zone resulting in localized material removal with minimal surrounding tissue damage.

Numerical studies have been performed to analyze bioheat transfer using various approaches. The finite difference method (FDM) and the finite element method (FEM) are two such numerical approaches. In the study by He et al. [7], FDM is used to simulate bioheat transfer in irregular tissues for applications in hyperthermia and cryosurgery. Zhao et al. [8] used the finite difference scheme for the 1D Pennes' bioheat equation and further validated it using numerical experiments for a skin-heating model. Dai et al. [9] developed a fourth-ordered finite difference scheme for solving a 1-D Pennes' bioheat transfer in a triple-layered skin structure. Scott [10] developed a finite element model of heat transport of the human eye based on the Pennes' bioheat transfer. Gonzalez et al. [11] used FEM to simulate laser-induced heating of gold nanoparticles for specific ablation treatments. Torvi and Dale [12] developed a finite element model to predict skin temperature and times to second and third degree burns under simulated flash fire conditions. FEM is often used for analyzing complex three-dimensional tissue geometries where multiple physical processes must be considered. FEM is favored over FDM and FVM (finite volume method) as it is better equipped to handle unstructured meshes, extensive groups of element choices, and easy handling of boundary conditions. The use of FDM is generally restricted to simple geometries. Studying irregular geometries and boundary conditions with FDM is computationally cumbersome. While FVM can handle irregular domains using unstructured grids (stemming from the FEM), the required averaging over the volume limits the method to second-order spatial accuracy.

The tissue geometries used for studying bioheat transfer have been developed based on either vascular or continuum models. Stanczyk and Telega [13] were two of the first to review the limitations of modeling of heat exchange in perfused tissues. They proposed the division of models into two classes: continuum and vascular. The key distinction between these two approaches is how the influence of blood flow on the temperature distribution of the soft tissue is considered. Continuum models account for the effects of blood flow in terms of an effective thermal conductivity of the tissue. This includes a bulk perfusion term to describe the net blood flow through the region. Thus, continuum models use a single equation to calculate the temperature distribution in the entire tissue domain. Vascular models, on the other hand, use one or more coupled equations to account for the variations in temperature distribution within tissues and blood vessels. This approach requires knowledge of the blood vessel geometry within the tissue. When examining the effects of vasculature on heat transfer in perfused tissues, two sub volumes should be considered. The first sub volume is the interior of the vessels, which consists of liquid blood moving due to the pressure created by contraction of the heart muscles. Another sub volume is the surrounding tissues through which the vessel passes. It is considered a solid with no blood flow. These two sub volumes share a common internal boundary which is the vessel wall. Calculating the velocity of the blood inside each vessel and incorporating it into the model is computationally challenging because the blood flow inside individual vessels is highly nonuniform and may change from one vessel to the next. Also, inside a given section of tissue of size one cubic centimeter, there may be around 10,000 different vessels which may be specific to an organ [14]. Therefore, assigning an average velocity to the blood flow reduces the computational time and complexity of the problem.

Several vascular models have been studied in the literature. Weinbaum and Jiji [15] were two of the first to study effective conductivity models to describe energy transport in tissue in terms of heat conduction. They demonstrated that the vascularization of tissue causes it to behave as an anisotropic heat transfer medium. Brink and Werner analyzed the temperature distribution in human soft tissue during exercise or rest periods [16]. In that study, the authors analyzed tissue temperature and arterial/venous blood temperatures in individual vessels in a small tissue volume. The study concluded that the analysis was limited to a small tissue volume due to the complexity of the vessel geometry in the tissue.

In addition to the continuum and vascular models, porous models have been proposed in which the tissue is treated as a porous media and a porosity term is included to capture the ratio of blood to tissue [17]. Another approach to consider effects of blood flow in the heat transfer of soft tissues is the hybrid model, which uses a combination of continuum and vascular models [18]. Vascular models make the study of bioheat transfer realistic but are computationally more complex than the continuum model.

The objective of this chapter is to analyze the influence of embedded blood vessels on the temperature distribution in soft tissue during short pulse laser irradiation of tissues using both an experimental approach and numerical simulation. Two different skin tissue geometries, with and without embedded vasculature, are studied using the FEM solver COMSOL. The embedded blood vessel geometry is considered using the vascular approach, by coupling two equations for the temperature in tissue and blood vessel domains. Blood flow is assumed to be fully developed with a parabolic velocity profile which is characterized with an average velocity to reduce the complexity of the computation. Parametric study is done to determine the change in temperature rise with the change in average velocity. The geometry with no embedded blood vessels is analyzed using a continuum approach. The resulting laser induced temperature distributions for both models are compared with the experimentally measured temperature distribution during pulsed laser irradiation of live anesthetized mice. To our knowledge of existing literature, an FEM model of laser irradiation of tissues with embedded vasculature has not been validated with experimental data and is the focus of the paper.

2.2 MATHEMATICAL FORMULATION

2.2.1 Numerical Modeling of Laser-Tissue Interactions

To calculate the source term Q_L, which is the heat generated due to the irradiation, the laser beam is assumed to be Gaussian in both the spatial and temporal domains and it is expressed as follows:

$$Q_L = \frac{\mu_a}{4\pi} I_c (\mu^c \mu + \eta^c \eta + \xi^c \xi) \tag{2.1}$$

where the unit vector of μ^c, η^c, ξ^c represents the collimated laser-incident direction and μ_a is the absorption coefficient (m^{-1}). The collimated intensity is given by

$$I_c = L_0 \exp\left\{ -4\ln 2 \times \left[\frac{(t - \frac{z}{c})}{t_p} - 1.5 \right]^2 \right\} \times \exp\left(\frac{-2r^2}{\sigma^2} \right) \exp(-z\mu_e) \tag{2.2}$$

where t is the time, t_p is the laser pulse width, c is the velocity of light in the tissue medium, r is the spatial variable denoting radial distance from the center of the laser beam (m), z is the spatial variable denoting depth from the surface of the tissue (m), μ_e is the extinction coefficient (m^{-1}), L_0 is the peak intensity of the laser beam at the sample surface, $\sigma(z)$ is the beam radius varying with z for which the peak intensity drops to the $1/e^2$ value. L_0 is given as:

$$L_0 = \frac{P}{\pi r_0^2} \tag{2.3}$$

where P is the power of the laser (W) and r_0 is the laser beam radius at the tissue surface (m). For the case of the focused laser beam, $\sigma(z)$ is given by [19]:

$$\sigma(z) = \sigma(0) \left(\frac{-(r_0 - r_d)}{r_0} \frac{z}{F_d} + 1 \right) \quad 0 \le z \le F_d \tag{2.4}$$

$$\sigma(z) = \sigma(0) \left(\frac{-(r_0 - r_d)}{r_0} \frac{z}{F_d} - \frac{(r_0 - 2r_d)}{r_0} \right) \quad z > F_d \tag{2.5}$$

where F_d is the focal depth (m), r_d is the beam radius at the focal depth (m), and $\sigma(0)$ is the beam radius at the tissue surface for which the peak intensity drops to the $1/e^2$ value (m).

The various laser parameters used in the model are summarized in Table 2.1.

2.2.2 Continuum Model Development

The axis-symmetrical geometry of the continuum model is shown in Figure 2.1. The geometry simulates the three layers of the skin: the epidermis, dermis, and hypodermis. The tissue block is 6 mm square, with a thickness of 6.5 mm. The epidermis is 0.5 mm thick, while the dermis and hypodermis are each 3 mm thick. The thermo-physical properties and the perfusion rates of the three tissue layers [20] and blood are obtained from literature and are presented in Table 2.2.

The geometry mesh consists of 82,985 elements with maximum and minimum element sizes of 6.5×10^{-4} m and 1.17×10^{-4} m, respectively. Figure 2.2 shows the meshed geometry.

TABLE 2.1   Laser Beam Geometric Parameters

Parameter	F_d (m)	r_d (m)	$\sigma(0)$ (m)	r_0 (m)
Value	0.002	100×10^{-6}	76×10^{-6}	20×10^{-6}

FIGURE 2.1   Continuum model geometry.

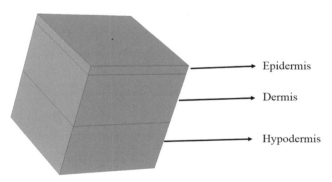

Epidermis

Dermis

Hypodermis

TABLE 2.2   Thermophysical Properties and Perfusion Rates of Tissue Layers

Components	k (W m^{-1} K^{-1})	C (J kg^{-1} K^{-1})	ρ (kg m^{-3})	w_{bl} (s^{-1})
Epidermis	0.25	3600	1200	0
Dermis	0.45	3300	1200	0.0004
Hypodermis	0.2	3000	1000	0.0026
Blood	1.3	3600	1000	–

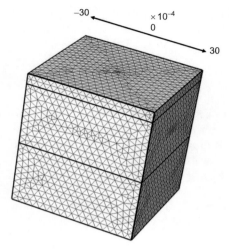

−30
× 10⁻⁴
0
30

FIGURE 2.2   Meshed geometry of continuum model.

In the entire domain, the temperature distribution is obtained by solving the bioheat transfer equation (BHTE) given by [21]:

$$\rho_{\text{tissue}} c_{\text{tissue}} \frac{\partial T_{\text{tissue}}}{\partial t} = \nabla \cdot (k_{\text{tissue}} \nabla T_{\text{tissue}}) + Q_{\text{bl}} + Q_{\text{met}} + Q_{\text{L}} \tag{2.6}$$

The heat sink term due to blood perfusion is given by

$$Q_{\text{bl}} = \rho_{\text{bl}} w_{\text{bl}} c_{\text{bl}} (T_{\text{a}} - T_{\text{tissue}}) \tag{2.7}$$

where k is the thermal conductivity (W m^{-1} K^{-1}), ρ is density (kg m^{-3}), c is specific heat (J kg^{-1} K^{-1}), Q_{met} is the volumetric metabolic heat generation (W m^{-3}), w_{bl} (m^3 blood/s/m^3 tissue) is the blood perfusion rate, and T_{tissue} is temperature of the tissue (K). The subscript bl is for blood and a for arterial. Due to the relatively high magnitudes of the induced laser energy and the relatively short time constants associated with the thermal energy, it is assumed that metabolic heat generation Q_{met} may be safely neglected. The arterial blood temperature, T_{a}, is set to the physiological temperature of 310.15 K (37 °C). The specific heat of blood C_{bl} is obtained from Sassaroli et al. [22] as 4000 J kg^{-1} K^{-1}. w_{bl}, the blood perfusion term, is given in units of m^3 blood/s/m^3 tissue and multiplied by the blood density (1000 kg m^{-3}) to obtain the proper units of kg m^{-3} s^{-1}. The optical properties, μ_{s} (scattering coefficient) and μ_{a} (absorption coefficient), are detailed for each tissue layer and blood in Table 2.3, and the values are derived from Jacques [23].

For the boundary conditions, the top layer of the epidermis is subjected to convective cooling. The convective heat transfer coefficient (h) is set to 10 W (m^{2}K)$^{-1}$ and the ambient temperature (T_{amb}) is assumed to be 298.15 K (25 °C). The contribution of convective cooling to the thermal load is given by:

$$q''_{\text{conv}} = h(T_{\text{amb}} - T_{\text{tissue}}) \tag{2.8}$$

All other external boundaries of the geometry are maintained at the physiological temperature of 310.15 K (37 °C).

TABLE 2.3   Optical Properties of Tissue Layers

Components	μ_s (1/m)	μ_a (1/m)
Epidermis	8000	355
Dermis	8000	49
Hypodermis	7000	50
Blood	1000	100

2.2.3 Vascular Model Development

The vascular model is an alternate approach which has been developed to account for the localized cooling effect of vessels at the target tissue site.

In this model, it is assumed that the vessels exist in artery–vein pairs and heat transfer is due to temperature difference between countercurrent vessel pairs and the surrounding tissue. The geometry of the vascular model having three layers is shown in Figure 2.3.

The arteries and veins in the dermal layer have a diameter of 150 μm and those in the hypodermis have diameters of 300 μm. The geometry mesh consists of 323,896 tetrahedral elements with maximum and minimum element sizes of 6.5×10^{-4} m and 1.17×10^{-4} m, respectively. The mesh is refined in the regions near the countercurrent vessel pairs. Figure 2.4 shows the meshed geometry which is generated in COMSOL for solving the BHTE in the tissue geometry by the FEM.

The heat equation is solved simultaneously in two domains: (1) in the individual blood vessels of the countercurrent pairs; and (2) in the tissue. In the blood vessels, the equation is of the form:

$$\rho_{bl}c_{bl}\frac{\partial T_{bl}}{\partial t} + \rho_{bl}c_{bl}u\nabla T_{bl} = \nabla(k_{bl}\nabla T_{bl}) + Q_L \tag{2.9}$$

where u (m s^{-1}) is the velocity of flow field of the blood flow. In this case, an average uniform velocity is assumed for the fully developed parabolic laminar flow of the blood flowing in the

FIGURE 2.3   Vascular model geometry.

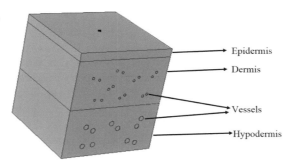

$\times 10^{-4}$

-30   0   30

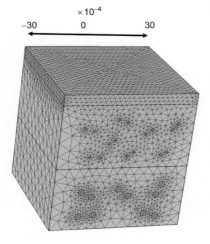

FIGURE 2.4   Meshed geometry of vascular model (dimensions in m).

vessels, which is influenced by vessel diameter. The heat equation in the tissue domain is given by:

$$\rho_{\text{tissue}} C_{\text{tissue}} \frac{\partial T_{\text{tissue}}}{\partial t} = \nabla \cdot (k_{\text{tissue}} \nabla T_{\text{tissue}}) + Q_{\text{L}} \qquad (2.10)$$

The blood velocity is assumed to act entirely in the flow direction of the vessels. Blood flowing in 300 µm vessels was set to 60 mm s^{-1}, while for 150 µm vessels it was set to 30 mm s^{-1} [12]. The interface between the tissue domain and the blood vessel domain is satisfied by the FEM. Therefore, the heat transfer can occur locally across these boundaries and the influence of each individual blood vessel on the temperature distribution can be studied. The thermo-physical properties of the tissue layers are the same as those used in the continuum model, as presented in Table 2.2. The optical properties of the blood flowing in the vessel pairs of the vascular model are given in Table 2.3. The initial conditions are the same as in the continuum model, with the additional condition that the blood flow is initially at a physiological temperature which is equal to 37 °C.

2.3 EXPERIMENTAL METHODS

The schematic of the experimental setup developed by the author's group [17] is shown in Figure 2.5. A Q-switched short pulse Nd:YAG (neodymium-doped yttrium aluminum garnet) laser having a pulse width of 200 ns and a wavelength of 1064 nm is used to study the laser-induced thermal effects in tissues. A converging lens is used to focus the laser beam at a depth of 2 mm beneath the skin surface. A thermal imaging camera (Agema Thermovision 450) is utilized to measure surface temperature of samples during irradiation testing. During experimentation, the camera captured images at one second intervals. The images are recorded using a data acquisition system and are then processed with National Instruments IMAQ Vision Builder image-processing software. The camera provided a

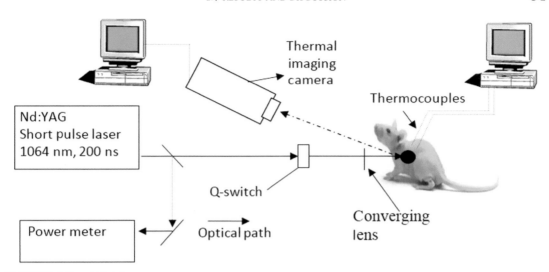

FIGURE 2.5   Nd:YAG mouse experiment setup.

measurement range of −4 to 932 °F with a sensitivity of ±0.18 °F and accuracy of ±3.6 °F (2%). The camera has a spectral response rated at 2-5 μm. To record temperatures at subsurface locations, T-type thermocouples are inserted to a specified depth within the sample. Experiments are performed on live mice which are anesthetized with 90 mg sodium pentobarbital per kilogram of mouse body weight. All animal experimental protocols are approved by Florida Tech's Institutional Animal Care and Use Committee (IACUC).

2.4 RESULTS AND DISCUSSION

In this chapter, the influence of embedded vessels on the temperature distribution in soft tissues during short pulse laser irradiation of tissues has been studied. Two tissue geometries, with and without embedded vasculature, have been studied using the FEM. The temperature distributions for both geometries are compared to experimentally measured temperature distribution.

Figure 2.6 shows the temperature distribution on the tissue surface for the case of the continuum model tissue block after 10 s of irradiation exposure with an average laser power of 0.36 W.

The temperature distribution indicates an elevated temperature at the center of the tissue surface. This point corresponds to the central $r=0$ mm axis and is the point where the laser intensity is maximum in that plane. The surface temperature decreases uniformly as the radial distance from the center increases.

In a vascular model, all the laser parameters are the same as that of the continuum model. The temperature distribution at the surface of the tissue for the vascular model is shown in Figure 2.7.

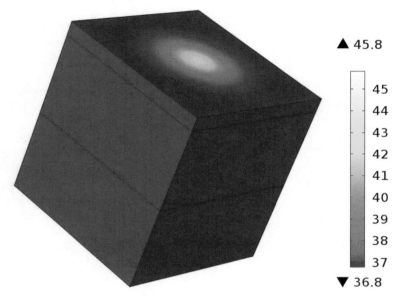

FIGURE 2.6 Surface temperature distribution of continuum model, 10 s exposure time and 0.36 W average power.

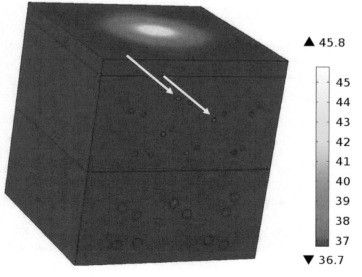

FIGURE 2.7 Surface temperature distribution of vascular model, 10 s exposure time and 0.36 W average power (spatial dimensions in m, temperature measurements in °C).

As expected, the maximum temperature occurs at the center of the laser beam which coincides with the center of the surface. The white arrows point to individual blood vessels with elevated surface temperatures as they exit the tissue domain in the geometry. These vessels exhibit higher temperature than the surrounding vessels because of their proximity to the laser beam. The maximum surface temperature rises are 8.97 and 8.38 °C for the continuum and vascular models, respectively. The results obtained through numerical modeling in COMSOL are compared to the live anesthetized mouse tissue irradiation measurements in Figure 2.8.

The surface temperature rise for the vascular model is 9.01 °C, whereas for the continuum model, the temperature rise is about 8.82 °C. Thus, both models are in very close agreement. However, the experimental measurements in live anesthetized mice show that the surface temperature rise is 10.25 ± 0.56 and 10.54 ± 0.57 °C as reported for two sample experimental trials.

Figure 2.9 displays the temperature distribution of the mid-plane slice in the continuum model.

The maximum temperature is observed at the focal point (depth $= 2$ mm, $r = 0$ mm) of the laser beam. The localized heating at the focal depth of 2 mm shows the efficacy of the focused laser beam to perform targeted subcutaneous heating. The tissue temperature increases with increase in depth from the surface to the focal plane where the laser beam converges. With further increase in depth, the temperature starts to decrease due to the impact of the scattering coefficient of the tissue. This is evident from the decreased temperature rise beyond the focal depth. Figure 2.10 displays the temperature distribution of the mid-plane cross section perpendicular to the blood flow direction in the vascular model.

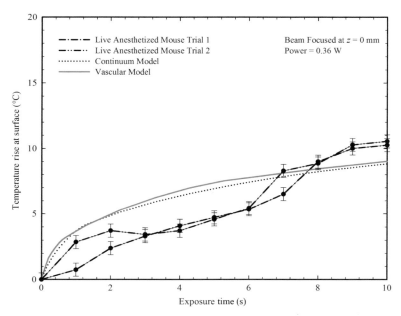

FIGURE 2.8 Temperature rise at surface, $r = 0$ mm, 10 s exposure time, and 0.36 W average power.

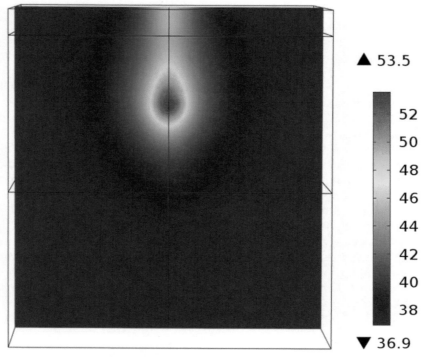

FIGURE 2.9   Temperature distribution of mid-plane slice in continuum model, 10 s exposure time and 0.36 W average power (spatial dimensions in m, temperature measurements in °C).

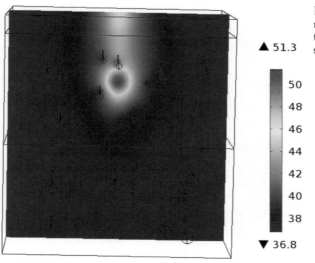

FIGURE 2.10   Temperature distribution of mid-plane slice in vascular model, 10 s exposure time and 0.36 W average power (spatial dimensions in m, temperature measurements in °C).

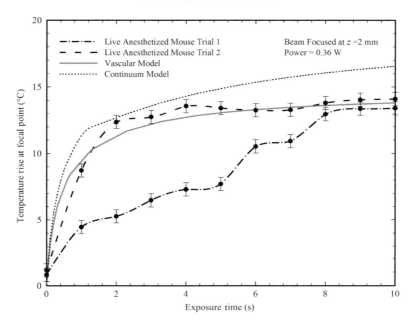

FIGURE 2.11 Temperature rise at focal point, 10 s exposure time and 0.36 W average power.

The maximum temperature for this case is observed at the focal point of the laser beam; however, the temperature distribution is influenced by the embedded vessel pairs. The presence of vessel pairs results in relatively lower temperature compared to the surrounding non-vascularized region. Additionally, the maximum temperature rise in the vascular model is 2.75 °C less than the continuum model maximum temperature rise. Figure 2.11 compares the focal point temperature rise results from both models with those observed in live anesthetized mice experiments.

For the anesthetized mice trials, the final temperature rise after 10 s at the focal point is in good agreement, increasing by 14.1 ± 0.76 and 13.4 ± 0.72 °C in two sample trials. The continuum model predicts a temperature rise after 10 s of 16.5 °C, whereas the vascular model predicts a temperature rise of 13.8 °C. The closer agreement between the final temperature rise of the vascular model and the live anesthetized mice experiments suggests that the local cooling effects of blood flowing near the focal point of the laser beam is best modeled via discrete embedded vasculature, as opposed to the bulk perfusion term based on the continuum model. The blood perfusion present in the vessels dissipates the heat away from the focal point, leading to a slightly lesser temperature in the vascular model than in the continuum model where this blood flow effect is not considered.

To further analyze the effects of embedded vasculature pairs, a parametric study is conducted where the average blood velocity in each vessel is varied according to parameters presented in Table 2.4.

Figure 2.12 depicts the focal point temperature rise in the vascular model for the different parametric cases considered in Table 2.4. A parametric study of four blood velocities is

TABLE 2.4   Blood Velocity Parametric Study Cases

Parametric Case	Blood Velocity in 150 μm Vessels (mm s^{-1})	Blood Velocity in 300 μm Vessels (mm s^{-1})
1	15	30
2	30	60
3	45	90
4	60	120

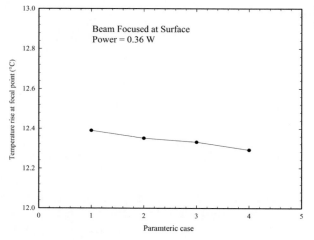

FIGURE 2.12   Variation of temperature rise with blood velocity.

conducted. The temperature rise in the focal point is plotted against change in blood velocity in the dermal vessels which are 150 μm in diameter.

The parametric variation depicted in Figure 2.12 shows that blood velocity has little effect on magnitude of temperature rise at the focal point, where the laser beam is converged. This may be due to the extremely small size of the vessels. Change of blood flow velocity in such small vessels does not have any significant effect on the temperature rise due to laser irradiation. A small decrease in temperature rise is observed as the blood flow increases. This can be attributed to the increased rate of heat loss through convection as the blood flow increases.

2.5 CONCLUSION

Two different computational models, a nonvascular continuum model and a vascular hybrid model, have been developed to analyze the experimental temperature measurements during laser irradiation of live anesthetized mice. It is demonstrated, using the FEM solver COMSOL, that the laser-induced temperature distribution predicted by the vascular models are in good agreement with experimental measurements. The implications of such embedded vascularity on the efficacy of modern clinical applications are discussed.

The potential applications of this model might be to predict the ablation depth and the thermal damage to the tissues more accurately during the laser therapy of tumors. The thermal damage to the tissues can be calculated by evaluating the effective thermal dosage using the method published by Sapareto and Dewey [24]. The probability of the tissues undergoing necrosis after damage due to laser irradiation can be determined using the Arrhenius damage integral [25]. The vessel geometry used in this paper is a simple representation of a very complex vessel structure in the body. Future studies should focus on improving the existing model by incorporating branched vessels and also take into account the rheology of blood flow. The finite element model developed in this chapter provides a good foundation on which to model the various processes associated with short pulse laser irradiation of tissues, such as generation of pressure waves leading to tissue material ejection and subsequent change of phase leading to ablation [26].

References

[1] Waynant W. Lasers in medicine, CRC in medicine. Florida: Boca Raton; 2002.
[2] Walsh JT, van Leeuwen TG, Jansen D. Pulsed laser tissue interaction. In: Welch AJ, van Germet MJC, editors. Optical-thermal response of laser-irradiated tissue. 2nd ed. The Netherlands: Springer Media; 2011.
[3] Panjehpour M, Wilke A, Frazier DL, Overholt BF. Hyperthermia treatment using a computer controlled Nd:YAG laser system in combination with surface cooling. Proc SPIE 1991;1427:307–15.
[4] Manns F, Milne PJ, Gonzales-Cirre X, Denam DB, Parel JM, Robinson DS. In-site temperature measurements with thermocouple probes during laser interstitial thermometry (LITT): quantification and correction of a measurement artifact. Lasers Surg Med 1998;23:94–103.
[5] O'Neal DP, Hirsch LR, Halas NJ, Payne JD, West JL. Photo-thermal tumor ablation in mice using near infrared-absorbing nanoparticles. Cancer Lett 2004;209:201–6.
[6] Yousef Sajjadi A, Mitra K, Grace M. Ablation of subsurface tumors using an ultra-short pulse laser. Opt Lasers Eng 2001;49(3):451–6.
[7] Zhu He Z, Xue X, Liu J. An effective finite difference method for simulation of bio-heat transfer in irregular tissues. J Heat Transf 2013;135(7):071003–8.
[8] Zhao JJ, Zhang J, Kang N, Yang F. A two level finite difference scheme for one dimensional Pennes' bio-heat equation. Appl Math Comput 2005;171(1):320–31.
[9] Dai W, Tzou DY. A fourth-order compact finite difference scheme for solving an N-carrier system with Neumann boundary conditions. Numer Methods Partial Differential Equations 2010;25:274–89.
[10] Scott JA. A finite element model of heat transport in the human eye. Phys Med Biol 1988;33:227–41.
[11] Gonzalez M, Krishnana KM. Synthesis of magnetoliposomes with monodisperse iron oxide nanocrystal cores for hyperthermia. J Magnetism Magn Mater 2005;293:265–70.
[12] Torvi DA, Dale JD. A finite element model of skin subjected to a flash fire. J Biomech Eng 1994;116(3):250–5.
[13] Stanczyk M, Telega JJ. Modeling of heat transfer in biomechanics—a review. Acta Bioeng Biomech 2002;4 (1):31–61.
[14] Baish JW, Mukundakrishnan K, Ayyaswamy PS. Numerical models of blood flow effects in biological tissues in advances in numerical heat transfer, vol. 3. UK: Taylor and Francis; 2009.
[15] Weinbaum S, Jiji LM. A new simplified bio-heat equation for the effect of blood flow on local average tissue temperature. J Biomech Eng 1985;107:131–9.
[16] Brinck H, Werner J. Estimation of the thermal effect of blood flow in a branching countercurrent network using a three-dimensional vascular model. J Biomech Eng 1994;116:324–30.
[17] Khaled ARA, Vafai K. The role of porous media in modeling flow and heat transfer in biological tissues. Int J Heat Mass Transf 2003;46:4989–5003.

[18] Wren J, Karlsson M, Loyd D. A hybrid equation for simulation of perfused tissue during thermal treatment. J Hypothermia 2001;17(6):483–98.

[19] Jaunich M, Raje S, Kim K, Mitra K, Guo Z. Bio-heat transfer analysis during short pulse laser irradiation of tissues. Int J Heat Mass Transf 2008;51:5511–21.

[20] Jain RK. Temperature distributions in normal and neoplastic tissues during normothermia and hyperthermia. Annals NY Acad Sci 1980;335:48–66.

[21] Pennes HH. Analysis of tissue and arterial blood temperatures in the resting human forearm. J Appl Physiol 1948;1:93–122.

[22] Sassaroli E., Li K.C.P., & O'Neill B.E., Modeling of the impact of blood vessel flow on the temperature distribution during focused ultrasound exposure, Proceedings of the COMSOL Conference 2010, Boston, U.S.

[23] Jacques SL. Laser-tissue interactions. Photochemical, photothermal and photomechanical. Surg Clin North Am 1992;72:531–58.

[24] Sapareto SA, Dewey WC. Thermal dose determination in cancer therapy. Int J Radiat Oncol 1984;10(6):787–800.

[25] Moritz AR, Henriques FC. Studies of thermal injury. Am J Pathol 1947;23(5):695–720.

[26] Tungjitkusolmun S, Tyler Staelin S, Haemmerich D, Tsai JZ, Cao H, Webster JG, et al. Three-dimensional finite element analyses for radio-frequency hepatic tumor ablation. IEEE Trans Biomed Eng 2002;49(1):3–9.

Quantitative Models of Thermal Damage to Cells and Tissues

Neil T. Wright

Michigan State University, East Lansing, MI, USA

3.1 INTRODUCTION

The human core body temperature is typically about 37 °C. This temperature may vary somewhat with time of day, illness, environmental conditions, or exertion, but the body has a highly developed thermoregulatory system to maintain the core body temperature in a narrow range. A core body temperature that is in excess of 40 °C is clinically defined as heat stroke although patients have recovered from warmer core body temperatures [1]. Local tissue temperatures, however, may exceed 40 °C, either accidentally (e.g., thermal burns) or deliberately during some clinical procedures (e.g., laser or microwave irradiation, drilling of hard tissue, curing of bone cement). The heating of cells, proteins, and tissues produces a wide variety of changes depending on the specific sample, temperature level, and duration of heating. Supraphysiological temperatures can destabilize lipid membranes, denature protein, arrest the cell cycle, and induce heat shock proteins (HSP) 27 and 72 [2]. Hyperthermia in oncology can be defined as treatments occurring in the range of temperatures from ~38 to 50 °C [3]. Tumors heated at these temperatures experience increased susceptibility to chemo- or radiotherapies, or their cells may undergo apoptosis or necrosis due to the heating itself. Temperatures greater than 50 °C, whether occurring accidentally in a burn or clinically as a treatment, are often termed ablative and can lead to necrosis in cells. These high temperatures also have the potential to alter structural proteins in the extracellular matrix, such as the shrinkage of collagen. These responses occur more quickly as temperature increases and increase in severity as heating is prolonged.

Quantitative models of cell death and tissue damage could be combined with bioheat transfer models to help develop future thermotherapies. An early analysis of burns used such an approach using the Henriques damage integral and a one-dimensional conduction analysis, albeit ignoring blood perfusion [4,5]. Such models might also be useful in the optimal design of experiments to improve the understanding of the pathophysiology of cell death and tissue damage due to heating. Quantitative models of these changes are increasingly necessary as researchers and clinicians seek to improve the design of therapies, whether developing, for example, a new thermotherapy or a treatment for accidental burns.

Mammalian cell cultures for experiments or tissue engineering are typically held at 37 °C, with occasional exceptions [6]. If, however, the cells and other constituents of tissue are heated at, say, 42 °C or warmer, then they will likely undergo some detrimental response [7]. Initial changes may include expression of HSPs which can protect cells from further heating. In addition to variations in temperature due to normal metabolic and environmental influences, clinical applications of heating include adjuvants to radio- and chemotherapy, as well as ablation of tissue.

3.2 HEAT TRANSFER IN TISSUE

Modeling the temperature distribution within tissue is an important starting point in predicting the changes that elevated temperatures cause to tissues, cells, and proteins. The most commonly used model of heat transfer in tissue is the Pennes' bioheat transfer equation [8], which may be written as

$$\rho_t c_t \frac{\partial T}{\partial t} = \nabla(k_t \nabla T) + \omega \rho_b c_b (T_a - T) + \dot{q}_m + \dot{q}_e \tag{3.1}$$

where T is the temperature of the tissue, T_a is the temperature of the arterial blood, ρ_t and ρ_b are mass density of the tissue and blood, respectively, c_t and c_b are the specific heat of the tissue and blood, respectively, k_t is the thermal conductivity of the tissue, ω the blood perfusion rate, $\dot{q}_m$ is the rate of metabolic heat generation, and $\dot{q}_e$ is the rate of heat generation due to external sources, such as electromagnetic radiation or ultrasound. Blood perfusion ω is the ratio of the volume flow rate of blood to the volume of tissue in the control volume of interest and it has units of s^{-1}. A typical value of $\dot{q}_m$ is 145 W m^{-3} [9]. A representative value of ω in the liver for fasting males is 0.285 s^{-1} [10]. The limitations of Equation (3.1) have led to the development of a number of other models [11]. While there were some questionable assumptions made in analyzing the data used in developing the Pennes' model, it has still been successfully applied in many applications. For example, Pennes used incorrect values for thermophysical property data of the tissue—the values used for both the blood perfusion and the thermal conductivity were small, as compared with accepted values [12]. Fortunately, the errors were counterbalancing. He failed to scale the data, specifically the radius of the forearm, so that some of the temperatures would appear to be outside of the median arm. Wissler [12] demonstrated that this does not have a significant effect on many results. Furthermore, Pennes assumed that the capillary bed is the principal site of heat exchange, whereas others have shown that the blood temperature is in equilibrium with the tissue at this point [13]. Keller and Seiler [14] proposed a countercurrent heat exchange between vessel pairs. Chen and Holmes expanded this concept by adding two terms, one accounting for the blood perfusion velocity and another for an effective thermal conductivity [13]. Zhu et al. [15] developed a model to predict the axial temperature variation along artery-vein pairs of vessels with diameters in the range of 300-1000 μm and the contribution of these vessels to the local tissue temperature. Weinbaum et al. [16] added incomplete countercurrent heat exchange, in addition to a capillary bleed-off term. Yet, for its broad, and sometimes questionable, assumptions, the Pennes' bioheat transfer equation provides a reasonable starting point for analysis of the temperature in metabolically active tissues [12].

The thermoregulatory response of changing blood perfusion has been studied by Romer who examined four perfusion regimes in response to different levels of electromagnetic heating [17]. Xu et al. measured the response of blood perfusion to transurethral microwave heating of a canine prostate [18]. Their results showed that blood perfusion in the prostate changed in response to temperature level. Shitzer [19] demonstrated the temperature response of cold-stressed fingers in a 4°C environment. The skin temperature of the gloved middle finger first cooled to about 10 °C before undergoing an oscillatory cold-induced vasodilatation response, with the tip temperature rising to about 15 °C before cooling to about 7 °C.

Pennes [8] developed one-dimensional steady-state solutions of Equation (3.1) to analyze his measurements. Others have developed more complete exact and numerical solutions [20]. Typically, T_a is taken as the core body temperature, that is, 37 °C, although more sophisticated models of whole body temperature distribution take local values based on energy balances [21]. For suitable simplifications in geometry, tissue properties, and boundary conditions, exact solutions of the Pennes' equation have been expressed using separation of variables [9,22] or Green's functions [23,24].

3.3 REACTION RATES AND TEMPERATURE

Many of the responses of proteins, cells, and tissues are described by chemical kinetics. These responses are functions of both the temperature level and the duration for which the specimen is heated. The importance of chemical kinetics is widely noted, as illustrated by the review article entitled "Arrhenius relationships from the molecule and cell to the clinic" [25].

Consider an nth-order reaction

$$\frac{dc_i}{dt} = -k(T)c_i^n,\qquad(3.2)$$

where c_i denotes the concentration of chemical species i, $k(T)$ is the rate parameter, T is the absolute temperature, t is time, and n is the order of the reaction. Equation (3.2) may represent the change in concentration of a protein in its native structure due to heating. Similar descriptions have been used to describe the more complex responses of cells and tissues. Many models of the rate of change in c_i are available depending on the specific species (c.f. [26,27]), but nth-order reactions, and specifically first-order reactions, are often cited in biological contexts.

The Arrhenius equation cited by Dewey [25] relates $k(T)$ in terms of T as

$$k(T) = Ae^{-E_a/RT} \tag{3.3}$$

where A is the frequency factor, E_a is the activation energy, and R is the gas constant. The ranges of values for the parameters E_a and especially A are quite large for biological materials. For example, Wright [28] listed values from various studies ranging from the shrinkage of the joint capsule over temperatures of $44 \leq T \leq 60\,°C$, with $A = 4.0 \times 10^5\,s^{-1}$ and $E_a = 0.034\,MJ\,mol^{-1}$ [29] to a calorimetric study of rat tail tendon (predominantly type 1 collagen) in acetic acid over the range $35 \leq T \leq 37\,°C$ with $A = 3.81 \times 10^{218}\,s^{-1}$ and $E_a = 1.31\,MJ\,mol^{-1}$ [30]. Others (e.g., [31]) have noted similarly large ranges of A and E_a for wide arrays of biological samples.

Another description of the temperature dependence of the kinetics parameter is the Eyring equation [32]

$$k(T) = \kappa \frac{k_b T}{h} e^{s_a/R} e^{-H_a/RT} \tag{3.4}$$

where κ is a coefficient usually taken as unity, $k_b = 1.381 \times 10^{-23}\,J\,K^{-1}$ is the Boltzmann constant, $h = 6.62608 \times 10^{-34}\,J\,s$ is the Planck constant, S_a is the activation entropy, and H_a is the activation enthalpy. In condensed systems, $H_a \approx E_a$ [27]. Comparing Equations (3.3) and (3.4) suggests that $\ln A \sim S_a$, and thus it might be more advantageous to compare $12.9 \leq \ln A \leq 42.7$ than to compare $4.0 \times 10^5 \leq A \leq 3.81 \times 10^{218}$.

In the context of thermal damage to cells and protein denaturation, several have noted that the values of E_a and $\ln A$ appear to be linearly correlated for a variety of samples [28,31,33,34]. This appears to be a manifestation of the entropy-enthalpy compensation effect, which has been the topic of considerable debate [35–37]. While a number cite evidence of the existence of entropy-enthalpy compensation [36,38], there are statistical [35] and thermodynamic [37] arguments against it. In the context of protein denaturation, Harrington and Wright [39] analyzed the scaled sensitivity coefficients [40] X for A and E_a. Comparing the X for various parameters in a given model will reveal if the parameters can be independently determined for a given set of experiments. The X are defined for a first-order reaction, Equation (3.2) with $n = 1$ and using the Arrhenius model for $k(T)$, as

$$X_A = A\frac{\partial c_i(\tau)}{\partial A} = A\frac{\partial c_i(\tau)}{\partial k}\frac{\partial k}{\partial A}$$

$$= Ae^{-E_a/RT}\frac{\partial c_i(\tau)}{\partial k} = k\frac{\partial c_i(\tau)}{\partial k} \tag{3.5}$$

and

$$X_{E_a} = E_a\frac{\partial c_i(\tau)}{\partial E_a} = E_a\frac{\partial c_i(\tau)}{\partial k}\frac{\partial k}{\partial E_a} = -\frac{E_a}{RT}k\frac{\partial c_i(\tau)}{\partial k}. \tag{3.6}$$

The ratio of these is $X_A/X_{E_a} = -RT/E_a$, which varies by only 18% over the range of $35\,°C < T < 95\,°C$ (i.e., $308\,K < T < 388\,K$). Few studies of thermal damage to biological materials examine such a wide temperature range, thus, A and E_a are difficult to distinguish by variations in temperature only.

A means to improve the confidence region is available by using a reference temperature T_{ref} [41]. Recently, Schwaab and Pinto [42] demonstrated the utility of an optimal reference temperature for reactions involving a single rate constant. In this method, Equation (3.3) is reparameterized using a reference temperature, such as

$$k(T) = k_{T_{ref}} \exp\left[-\frac{E_a}{R}\left(\frac{1}{T} - \frac{1}{T_{ref}}\right)\right], \tag{3.7}$$

where $k_{T_{ref}}$ is the reaction rate at T_{ref}.

A related model that is commonly used in microbiology and food science to describe the number of organisms (or the number of colony forming units [CFU]) surviving following a heat treatment is the Bigelow model [43]. It is formulated in terms of base-10 logarithms as

$$\log_{10}\frac{N_0}{N(t)} = \frac{1}{D}\int_0^t 10^{(T(t)-T_{ref})/z}\,dt \tag{3.8}$$

where $N(t)$ is the number of CFUs after a treatment of duration t at temperature T, N_0 is the number of CFUs in an untreated sample, D is the time required to reduce the CFUs to 10% of N_0 for treatment at T_{ref}, and z is the increase in temperature required for the same reduction in CFUs for a constant D. A relatively thermally stable material may have a value of $z = 10\,°C$. Miles [30] tabulated relationships between the parameters of the Arrhenius relation, the Eyring equation, and the Bigelow model (also known as the D-z model). The kinetics parameter of Equation (3.3) may be written in terms of the Bigelow model as

$$k(T) = \frac{2.303}{D}10^{(T-T_{ref})/z} = \frac{2.303}{D}e^{(2.303(T-T_{ref}))/z}. \tag{3.9}$$

The reparameterization using a reference temperature has been used to enhance confidence in parameters estimated from limited ranges of temperature in changes due to thermal treatment of *Salmonella* in food samples [44].

3.4 THERMAL DENATURATION OF PROTEINS

Proteins can be grouped as globular, fibrous (scleroproteins), and membrane proteins. The globular proteins, as the name implies, typically form three-dimensional structures and have many binding points. The scleroproteins are structural proteins, such as collagen and elastin, and tend to form fibrils. Type I collagen, for example, has molecules composed of three polypeptide chains of about 1300-1700 amino acid residues, with the central portion (~1000 residues) forming a triple helix [45]. For consideration of thermal stability, it might be convenient to classify proteins as those that form a cooperative unit (i.e., globular proteins) [46] and those that do not (i.e., fibrous proteins) [47]. Many studies characterize the thermal stability of proteins by a "melting temperature" [48]. The melting temperature may be more appropriate for proteins forming a single cooperative unit but is insufficient to describe the denaturation of fibrous proteins [49].

The seminal rate model of thermal denaturation is the Lumry-Eyring model [50]

$$N \underset{k_2}{\overset{k_1}{\rightleftharpoons}} U \overset{k_3}{\rightarrow} D, \tag{3.10}$$

where N represents the native protein, U is the reversibly unfolded protein, D is the irreversibly denatured protein, and k_1, k_2, and k_3 are the reaction rates. Type I collagen, for example, is functional in its native state N. The triple-helix structure of type I collagen is stabilized by hydrogen-bonded water-bridges, often at hydroxyproline residues [51]. With moderate heating, the collagen may unfold reversibly to U, returning to its native state N when cooled. The unfolding appears to be due to the breaking of a small number of consecutive hydrogen bonds between amino acid residues 877 and 936 [30]. More severe heating overcomes a larger free energy barrier, and this results in a time-dependent irreversible transformation of the native triple helical structure into a more random (coiled) structure, the denatured state D. Differential scanning calorimetry (DSC) has been commonly used to measure enthalpy changes during thermal denaturation of proteins. In DSC, a test sample is heated at a constant rate of temperature increase, the energy required to maintain this temperature rise is compared to a reference sample, and the enthalpy increase may thereby be determined. Two analyses revealed that the kinetics parameters of the irreversible step of the reaction (3.10) limit the rate of temperature increase used during DSC testing [52,53]. Increasing the sample temperature at faster rates can give misleading values of the A and E_a in Equation (3.3). Several other techniques are also used to measure changes in protein structure during thermal denaturation. Fourier transform infrared spectroscopy (FTIR) and circular dichroism, for example, can measure changes in the secondary structure of proteins as they denature, where as DSC measures the energetic effects [54].

Thermal damage of collagen can be measured on a gross scale by its shrinkage during heating. Some tendons are convenient specimens to measure shrinkage because they are nearly cylindrical, composed predominantly of type I collagen. Weir [55] measured the uniaxial shrinkage of kangaroo tendon. He developed a first-order reaction kinetics model of the denaturation of collagen based on the shrinkage and modeled the overall trend as

$$\xi_w = e^{-k(T)t}, \tag{3.11}$$

where $k(T)$ was described as in Equation (3.3). Weir [55] also determined that uniaxial shrinkage is slower if the sample is dehydrated, stretched, or chemically cross-linking, such as in tanning of leather.

Chen et al. [56] measured the uniaxial free shrinkage of bovine chordae tendineae. The chordae are tendons connecting papillary muscles to the tricuspid valves of the heart, and these specimens were approximately 0.7 mm in diameter and 10 mm long. Measuring the shrinkage, defined as $\xi(t, T) = (L_0 - l(t))/L_0$, where L_0 is the original length, and l is the current length, revealed the existence of a characteristic time τ_2 that was related to temperature by an Arrhenius-type relationship. Dividing the time of heating by τ_2 collapsed the shrinkage curves at various temperatures to a single curve. The characteristic time τ_2 was defined as the second inflection point in the shrinkage curve, which was relatively easy to measure. Wright et al. [57] showed that other points on the shrinkage curves, such as the point at which 50% of the shrinkage had occurred (τ_{50}), could be used in a similar fashion. Because tissue *in vivo* is often in a state of mechanical load, Chen et al. [58] measured the uniaxial shrinkage of mechanically loaded chordae tendineae and described the load and temperature dependence of τ_2 as

$$\tau_2 = e^{(\alpha + \beta P)} e^{m/T}, \tag{3.12}$$

where α, β, and m are material parameters and P is the first Piola-Kirchhoff stress. Thus, Equation (3.12) shows that the effect of tensile loading on the tendon is to delay shrinkage at a given temperature. This effect may be related to the polymer-in-a-box model of Miles and Ghelashvili [59].

Chen et al. [60] sought a model to predict the shrinkage of a specimen that was undergoing a time-varying temperature change. They broke the shrinkage into three parts: a short-term linear response, a long-term linear response, and a sigmoidal transition between the two linear responses. Shrinkage during the initial response was modeled as

$$\xi(v) = A_0 + A_1 v, \tag{3.13}$$

where $v = t/\tau_2$ and A_0 and A_1 are material constants. The late response was modeled as

$$\xi(v) = a_0 + a_1, v, \tag{3.14}$$

where a_0 and a_1 are also material constants. Shrinkage in the transition regime was extrapolated from the short-term response and the long-term response as

$$\xi(v) = [1 - f(v)](A_0 + A_1 v) + [f(v)](a_0 + a_1 v). \tag{3.15}$$

Modeling the shrinkage as a combination of the denatured and native proteins, an empirically developed expression for $f(v)$ was given as

$$f(v) = \frac{e^{a(v - v_m)}}{1 + e^{a(v - v_m)}}. \tag{3.16}$$

The specimens exhibited some recovery in length after heating had ended, and the specimen temperature was returned to 37 °C. It was assumed that this partial recovery, described by

$$\xi(t) = B_0 - B_1 \left(1 - e^{-t/t_e}\right), \quad \forall t \tag{3.17}$$

was the result of rehydration and the recovery of some of the hydrogen bonds that bind the tropocollagen molecules.

Wall et al. [61] measured the $\xi(t, T)$ of bovine extensor tendons for temperatures of 59.5-65.5 °C. They expressed the shrinkage using a 7-parameter logistic equation

$$\xi(t, T) = 35.35 + \frac{(0.53T - 70.09)}{\left[1 + (t/4.0 e^{[29.76 - 0.48T]})^{(0.38T - 21.31)}\right]} \tag{3.18}$$

where T in Equation (3.18) is the relative temperature (i.e., °C). They also found, in agreement with Chen et al. [60], that the tendon became more compliant with increased shrinkage and that the degree of compliance was a function of the extent of shrinkage and not the specific parameters during heating to get that level of shrinkage.

3.5 CELLS

Cells may respond to hyperthermia in several ways depending on the temperature level and the duration of heating. The precise cellular target of damage due to heating is unknown, although proteins are the most likely target, with about 5% denaturation needed for cell death [62]. Mild heating resulting in temperatures of about $T \leq 42$ °C can lead to the expression of HSPs, other modes of thermal tolerance, or can lead to cell death [25]. The transient expression of HSPs can be controlled by regulating the protocol of stress on the cells [2,63]. Cell death can be necrosis, apoptosis, or necroptosis, depending on the temperature and the duration of heating. Necrosis is characterized by loss of membrane integrity and the release of the cell contents [64]. This tends to occur at higher temperatures. Apoptosis is a programmed cell death that results from a signal that triggers the caspase pathway. Temperatures in the 42-50 °C range often lead to apoptotic cell death. In most experiments in the apoptotic range, the cells are returned to the incubator at 37 °C after the specified heating time. Assays of cell survival are then made at some later time. Rylander et al. [65], for example, found that for cancerous and normal prostate cells, the mechanism of injury made a transition at 54 °C. Necroptosis, a more recently identified form of cell death, may also be activated by heating. Necroptosis is driven by molecular pathways, as with apoptosis, but results in the loss of cell membrane integrity and the loss of the cell contents, as with necrosis [66]. Some have suggested a critical temperature for cell death, such as the 45-48 °C in osteoblasts for 10 min of heating [67]. These references are typically noted for specific experimental or clinical circumstances (i.e., for a specific cell line and heating protocol) and are not meant as a general model of cell death. For reviews of the cellular targets of heating, see [62,68,69].

Models of cell survival following heating have usually been developed based on a hypothetical chemical model, either an analogy to models of ionizing radiation damage, a stochastic model of multiple chemical processes, or a model based on statistical thermodynamic arguments [70,71]. Some of these models have been based on first principle arguments, while others have been empirically motivated. Some models have been inspired by analogs of damage to cells by ionizing radiation. Each of these methods often leads to a model which includes descriptions using chemical kinetics.

Wright compared six of the models of cell death following heating [72]. A standard model against which some researchers compare is a first-order reaction, which can be written as

$$S(\tau) = e^{-k_1 \tau}, \tag{3.19}$$

where S is the fraction of cells surviving after heating, k_1 is the temperature-dependent rate parameter for the first-order model, and τ is the duration of the heat treatment of the cells. The rate parameter k_1 may be related to temperature using the Arrhenius relation Equation (3.3). The surviving fraction S is the ratio of the number of surviving cells to the number of cells before heating, or to the number in a control group that has been maintained at 37 °C. Ideally, $S = 1$ if there is no heat treatment (i.e., $\tau = 0$) and has a limiting value of $S = 0$ for long heating times. The value of τ for which $S \approx 0$ decreases as the treatment temperature increases. Westra and Dewey [73] used this model to describe cell survival in the exponential range for Chinese hamster ovary cells. Equation (3.19) can capture the behavior of the exponential range of cell survival curves [74] but is insufficient to capture the shoulder region that is often apparent

[72,75]. If one considers A and E_a as separate parameters, then Equation (3.19) may be considered a two-parameter model.

Models adding one or more parameters capture the shoulder region of the cell survival curves after shorter durations of heating. Johnson and Pavelec [76], for example, proposed a model where n_j independent molecular events are required for fatal cell injury, each being a first-order reaction with the rate parameter k_j. The cells survive according to

$$S(\tau) = 1 - \left(1 - e^{-k_j \tau}\right)^{n_j},\qquad(3.20)$$

where k_j is related to temperature via Equation (3.4). Similar models have been developed by Dewey et al. [77], Hahn [78], and Roti Roti and Henle [79]. Jung [80] assumed that temperature-dependent random events produce irreversible nonlethal lesions with a rate parameter p. A subsequent set of events that convert the nonlethal lesions to lethal ones are assumed to occur with rate parameter c. Cells may remain viable with the hypothetical nonlethal lesions, but a single lethal lesion causes cell death. The resulting equation can be written as

$$S(\tau) = \exp\left\{\frac{p}{c}[1 - c\tau - e^{-c\tau}]\right\}.\qquad(3.21)$$

Jung subsequently developed a model to accommodate thermotolerence [81]. This model considered an initial heating at T_1 for time τ_1 followed immediately by heating at T for duration τ. The surviving cell population is then described as

$$S(\tau_1, \tau) = \exp\left\{\frac{p_1}{c_1}\exp(-c\tau)[1 - c_1\tau_1 - e^{c\tau} - e^{c_1\tau_1}]\frac{p}{c}[1 - c\tau - c^{-c\tau}]\right\}.\qquad(3.22)$$

where p_1 and c_1 are analogous to p and c, respectively, in Equation (3.21) but for the heating at T_1. The immediate heating at T after the preheating at T_1 may not be optimal for expression of thermotolerance by cells [2,63,65]. Nevertheless, results for Chinese hamster ovary (CHO) cells showed significant thermotolerance after preheating at 40 °C for between 15 min and 16 h [81].

Mackey and Roti Roti [82] took a somewhat different approach by postulating a temperature-dependent parameter ϵ assumed to be normally distributed, with a unit variance, within a population of cells. The parameter ϵ may be modeled as

$$f(\epsilon) = \frac{1}{\sqrt{2\pi}}\exp\left[-(\epsilon-\bar{\epsilon})^2/2\right],\qquad(3.23)$$

where

$$\bar{\epsilon} = \bar{\epsilon}_f\left(1 - e^{-k_m * \tau}\right) + \bar{\epsilon}_0 e^{-k_m * \tau},\qquad(3.24)$$

with a mean initial value of $\bar{\epsilon}_0$ (which can be assumed to equal zero), a temperature-dependent final value of $\bar{\epsilon}_f$, and a temperature independent k_m that is a constant for each cell line. The fraction of cells with $\epsilon \geq \epsilon_{min}$ are clonogenic. The resulting description of survival can be written, in integrated form, as

$$S = \frac{1}{2}\text{erfc}\left[\frac{-\left(\epsilon_{min} - \bar{\epsilon}_f[1 - e^{-k_m * \tau}]\right)}{\sqrt{2}}\right].\qquad(3.25)$$

Mackey and Roti Roti took $\epsilon_{min} = -3$ giving an asymptote of $S \approx 1$ at $\tau = 0$. Some sets of data were better described by relaxing the restriction on k_m and allowing it to vary with temperature [72].

Feng et al. [70] started from concepts in statistical thermodynamics to predict the probability density of a cell surviving. The model can be written as

$$S(T) = \frac{1}{1 + e^{(\gamma/T + \alpha\tau + \beta)}} \tag{3.26}$$

where α, β, and γ are constants that depend on the cells studied and the temperature of treatment. Feng et al. found parameters for Equation (3.26) using two sets of data, one for PC3 cells and the other for RWPE-1 cells, each set of cells heated in the temperature range of 44-58 °C.

O'Neill et al. [71] proposed a three-state model that may be written in terms of the normalized populations of undamaged cells S, dead cells D, and vulnerable cells V such that

$$\frac{dS}{d\tau} = -k_f S + k_b V \tag{3.27}$$

$$\frac{dD}{d\tau} = k_f V \tag{3.28}$$

$$V = 1 - S - D \tag{3.29}$$

where k_f is the forward reaction rate for the transition from S to V and from V to D, and k_b is a backward reaction rate for transition from V to S. To account for the small rate of damage at 37 °C and the assumption that damaged cells may affect the rate of further damage, k_f is defined as

$$k_f = \bar{k}_f e^{t/t_k}(1 - S). \tag{3.30}$$

where $\bar{k}_f$ and t_k are fitting parameters. Note that t and t_k have units of °C; ratios of relative temperatures are unusual in thermodynamics. Because the system is nonlinear with the dependence of k_f on S, a numerical solution was sought. The model was fit to survival curves of Hep G2 and MSRC-5 cells heated in the range of $55 \leq T \leq 100$ °C, which is hotter than is typically considered in cell survival studies. The heating times were from 300 to 900 s (5 to 15 min). Although it might be characterized as a three-parameter model, the solution also depends on the choice of the initial value of S, suggesting that four parameters must be chosen.

Each of the preceding models contains at least one parameter that needs to be determined experimentally, and often a model contains several parameters. The multiple parameters often improve the comparisons of the model predictions to the experimental results. The physical justification for the various parameters is greater in some models than others. Even in the most fundamental of models, the parameters may be correlated, suggesting that caution should be exercised in ascribing too much insight into their values. Examining the sensitivity coefficients of the parameters in these models can provide insight into designing experiments to determine these parameters [40]. If the sensitivity coefficients are correlated over the range of experimental conditions used, then they cannot be determined independently. This suggests that while the agreement between a model and a given set of data may be good, the set of parameters may not be unique, and this may limit the predictive capability of the model.

3.6 TISSUE-LEVEL DESCRIPTIONS

Models intended for clinical applications must characterize changes to tissues *in vivo*. For many therapies, these models must address the composite response of the constituent cells and the extracellular matrix, as well as potential interactions due to cell-to-cell signaling and the thermoregulatory system, in the case of relatively long-term heating. The models of cell survival *in vitro* suggest that the complexity of cellular responses *in vivo* and the extracellular matrix itself is composed of structural proteins, proteoglycans, and extracellular fluid.

Two prominent models used to quantify composite responses in clinical applications are the Henriques burn integral [5], developed to quantify the extent of thermal burns, and cumulative equivalent minutes at 43 °C (CEM_{43}) [77], which gives a common basis to compare hyperthermia dosage for the typically nonuniform and nonsteady temperature fields during treatment. While these two descriptors have different intended uses, each incorporates an Arrhenius-type relationship between temperature and heating time. The result is that they allow for the quantification of thermal damage when temperatures vary with space and time, as is the most likely case for accidental burns or clinical hyperthermia. Temperature variations in living tissues are expected given the range of tissue thermophysical properties, variations in blood perfusion, and irregular geometries.

3.6.1 Burns

At the same time that Pennes was publishing his work involving the bioheat transfer equation [8], Henriques and Moritz were publishing their work involving the progression of burns [4,5,83,84]. Henriques and Moritz wrote a series of papers that culminated in an integral representation of the evolution of a burn injury [5]. This integral model of burn injury may be written

$$\Omega = \int_{\tau=0}^{t} A e^{-E_a/RT(\tau)} d\tau. \tag{3.31}$$

where Ω is the damage integral and t is the time during which the supraphysiological temperature is experienced. Henriques et al. reported the values of $A = 3.1 \times 10^{98}$ s^{-1} and $E_a = 6.27 \times 10^8$ J k mol^{-1} based on measurements of porcine and human skin. Based on their results of heating the skin of human forearms using direct contact with water [83], $\Omega = 0.53$ was identified as a first-degree burn (onset of erythema) and $\Omega = 1.0$ as a second-degree burn (partial dermal thickness). Later, Takata [85] extrapolated results of other studies and suggested that $\Omega = 10^4$ represents a third-degree burn (full dermal thickness). Other values of A and E_a are reported for other tissues. Pearce et al. [86], for example, examined damage to heart muscle and found $A = 3.5 \times 10^{22}$ s^{-1} and $E_a = 1.64 \times 10^5$ J k mol^{-1}. Several reviews have included lists of A and E_a for different tissue types (c.f., [54]).

The integrand in Equation (3.31) is recognized as the Arrhenius relation Equation (3.3), and since Ω is independent of concentration, it may be thought of as zeroeth-order kinetics, that is $n = 0$ in Equation (3.2). Assuming that the burn injury is likely due to damage to proteins, Henriques suggested that Q could be considered to describe the first-order denaturation of proteins, such that $\ln(\Omega) = [N]_0/[N]$, where $[N]_0$ is the initial concentration of native protein and $[N]$ is the current concentration of native protein. Equation (3.31) then equals a first-order kinetics description of the change in concentration of native molecules. Pearce et al. [87] extended this idea that Ω describes changes in native protein concentration to studies in birefringence of thermally coagulated collagen. As collagen denatures, its birefringence also changes leading to a reduction in the intensity of light $I(t)$ transmitted through, for example, rat skin specimens. Pearce et al. modeled this change using Equation (3.31) such that

$$\Omega_p(t) = \ln\left(\frac{I_0}{I(t)}\right),\tag{3.32}$$

where $\Omega_p(t)$ is the damage integral of Henriques modified in terms of light intensity and I_0 is the initial intensity of the transmitted light. As a further extension of the damage integral for comparison of thermal damage, Thomsen and Pearce [88] defined a critical temperature of thermal damage T_{crit}. Taking the point where $d\Omega/dt = 1$, T_{crit} was defined as

$$T_{crit} = \frac{E_a}{R \ln(A)}.\tag{3.33}$$

For representative values of $A = 1.0 \times 10^{75}$ s^{-1} and $E_a = 5 \times 10^5$ J mol^{-1} K^{-1}, T_{crit} is calculated to be 74.8 °C [88]. They carefully noted that this temperature is a guide, and not a "melting temperature."

Xu and Qian [89] took a different approach by assuming that the rate of burn injury equals the rate of inactivation of a hypothetical enzyme. They then used their model to fit the burn data of the Henriques et al. [4]. Specifically, based on enzyme deactivation kinetics, the model may be written as

$$X(t) = \int_{\tau=0}^{t} \frac{Be^{-\alpha z}}{1 + Ce^{-\beta z}} dt\tag{3.34}$$

where $X(t)$ is the fraction of inactivated enzyme (which is assumed to equal $\Omega(t)$, $z = 1 - T_0/T$, T_0 is a reference temperature, and B, C, α, and β, are found by regression fits of data. The parameter β is related to the difference in Gibbs energy functions for the formation and decomposition of the substrate-enzyme complex, and α is related to the activation free energy of enzyme deactivation. With $T_0 = 32.5$ °C, the curve fit of $X(t)$ agrees well with the $\Omega(t)$ measured by Henriques et al. [4], except at the highest temperatures.

3.6.2 Normalizing Hyperthermia to Time at 43 °C

Hyperthermia, either focal or whole-body, has been used to increase the effectiveness of some chemo- and radiotherapies in cancer treatment. In clinical applications, superficial lesions may be heated by contact, but deep tissue tumors require absorption of electromagnetic or acoustic energy that is focused on the target tissue. For the deep tissue lesions especially, it is difficult to raise the temperature of the target tissue quickly and uniformly. Therefore, an easily implemented means is necessary to assess the thermal dose. Dewey and co-workers [77,90] developed the concept of cumulative equivalent minutes at 43 °C (CEM$_{43}$), which may be written as

$$CEM_{43} = \int_0^t \alpha^{(43 - T(t))} dt\tag{3.35}$$

where $T(t)$ is the time-dependent temperature in °C. Sapareto and Dewey [90] define α as $\alpha = \exp[-H_a/(R(T_{43})(T_{43}+1))]$, where $T_{43} = (273 + 43)$ K and H_a is the activation enthalpy, which in condensed matter is approximately equal to the activation energy E_a. Equation (3.35) assumes that there is a first-order kinetics description of thermal dose and that the temperature dependent rate parameter follows an Arrhenius-type dependence on temperature. This is effectively a non-dimensionalization of time so that the heating for any known temperature history can be related to a common time. Values of α are often taken as $\alpha = 0.25$ for $T(t) \leq 43$ °C (with $H_a = 1.53$ MJ/mol) and $\alpha = 0.5$ for $T(t) \geq 43$ °C (with $H_a = 0.590$ MJ/mol), which suggests that one activation enthalpy is sufficient for every cell line treated at temperatures less than 43 °C, and another one for every cell line treated at temperatures higher than 43 °C. Other values for α, and even for the breakpoint temperature of 43 °C, are sometimes cited [91,92]. Foster et al. [93] state that while CEM_{43} of less than 10 min was the threshold for thermal damage to mouse testis and brain, a CEM_{43} of more than 150 min was required for porcine skin, fat, and muscle.

3.6.3 Other Studies Addressing Clinical Response

Some researchers have modeled thermal damage in transient temperature fields. These models usually combine finite difference or finite element modeling of the Pennes' bioheat equation with a metric of thermal damage [94]. For example, Lim et al. [95] used finite element modeling to examine the roles of the energy waveform and the existence of large arteries on the effectiveness of RF ablation on liver tumors. They used a simple temperature metric of 47 °C to signal the initiation of tumor cell death and 64 °C to indicate 63% cell death based on the earlier work of Chang and Nguyen [96]. Similarly, Brinton et al. [97] modeled the temperature rise in the use of ultrasound to reduce hyperplasia due to expanded polytetrafluoroethylene vascular grafts. The finite difference model of the temperature field included a simple temperature threshold of 50 °C to signify cell death. Ng and Chua, in contrast, used a finite difference solution for temperature combined with the Henriques model [4,5] of tissue damage [98,99].

Breen et al. [100] developed a model predicting cell death during laser thermal ablation that incorporates transient local temperatures. A cell is predicted to be dead if the accumulated damage Ω_B is greater than a critical value $\Omega_{B,c}$. The accumulated damage, in turn, relies on a critical temperature T_c such that

$$\Omega(t)_B = 1 - \exp\left[-\int_0^t \beta(T(\tau))d\tau\right], \tag{3.36}$$

where

$$\beta(T(t)) = \begin{cases} 0, & T(t) < T_c \\ A(T(t) - T_c)^N, & T(t) \geq T_c \end{cases} \tag{3.37}$$

which includes three experimentally determined parameters (A, N, T_c). When $T > T_c$, cell death occurs, and when $T < T_c$, damage stops. The rationale for this critical temperature comes from studies having shown cellular homeostasis at temperatures slightly warmer than the basal temperature for several hours. Though a departure from the chemical kinetics models often proposed (i.e., the Arrhenius relation), β is expected to be a monotonically increasing function with respect to temperature. Breen et al. found the four parameters (A, N, T_c, and $\Omega_{B,c}$) by minimizing an objective function of histological pictures of dead and normal cells postablation. The results depended somewhat on the value of the weighting function W between 0.5 and 0.95, with $0.063 \leq A \leq 0.72$, $0.8974 \leq N \leq 1.0731$, $45.07 \leq T_c \leq 47.66$, and $0.6570 \leq \Omega \leq 0.7015$. Chen and Saidel [101] used this model in a study of MRI-based temperature measurement of tissue with $\beta = \beta_0 (T - T_c)$, where $\beta_0 = 1 \times 20^{-4}\,°\mathrm{C}^{-1}$ and $T_c = 43\,°\mathrm{C}$.

Yung et al. [102] compared Henriques's damage integral (using Henriques's parameter values of $A = 3.1 \times 10^{98}\,\mathrm{s}^{-1}$ and $E_a = 6.28 \times 10^5\,\mathrm{J\,mol}^{-1}$), CEM_{43}, and a threshold temperature for evaluation of laser ablation of canine brain tissue. Temperatures were measured using magnetic resonance temperature imaging. The measured temperatures were used in the three models, and the prediction of thermal damage was then compared using the Dice similarity coefficient [103]. The authors found that the region damaged immediately after treatment was equally well predicted for $\Omega = 0.65$, $\mathrm{CEM}_{43} = 690$ min, and $T_{th} = 61\,°\mathrm{C}$. Fanjul-Velez et al. [104] compared CEM_{43} with Henriques's damage integral of Equation (3.31), which they refer to as the Arrhenius method. They found good agreement between these two methods, although they used the CEM_{43} for constant temperatures only, rather than using an integral or summation with binning. Liljemalm and Nyberg [105] used a finite element model to solve the bioheat transfer equation during laser irradiation and calculated damage using the damage integral. They used this model to analyze data for damage to astrocytes due to laser heating in a study of astrocyte migration due to temperature gradients. Liljemalm and Nyberg determined the kinetics constants $E_a = 321.4\,\mathrm{kJ\,mol}^{-1}$ and $A = 9.47 \times 10^{48}\,\mathrm{s}^{-1}$, for peak temperatures of $59.5\,°\mathrm{C}$.

3.7 DISCUSSION

Most models of thermal damage account for multiple responses to heating of proteins and cells. In most cases, it is temperature level and heating duration that determine the outcome. Structural proteins, type I collagen in particular, have been measured to undergo reversible alteration, but if sufficient time-dependent temperature level is maintained, irreversible denaturation occurs. Cellular responses are even more complex as cells may undergo apoptosis, necrosis, or necroptosis. Furthermore, mild heating, in the range of 37-42 °C, can produce thermotolerance, usually in the form of HSP expression, which offers some protection to cells from subsequent heating at higher temperatures.

It is clear from the number of models proposed for the protein denaturation and cell death that a first-order kinetics model, such as described by Equation (3.19), is insufficient to account for the damage across the range of temperatures that may occur naturally, accidentally, or clinically. As experiments become more refined and reveal changes with increasing spatial and temporal resolution, new models will be developed that provide improved predictive capabilities to a wider range of samples and conditions. As Pearce has noted [75], the biochemical networks, such as the caspase network model of apoptosis [106,107], may at some point offer the most refined understanding of the prediction of cell death due to hyperthermia. Currently, the precise target molecules of heating are unknown and the kinetics parameters of the caspase pathway are unknown to great precision. Thus, other models may offer more reliable predictions of the changes due to heating.

References

[1] Yaqub B, Al Deeb S. Heat strokes: aetiopathogenesis, neurological characteristics, treatment and outcome. J Neurol Sci 1998;156:144–51.

[2] Wang S, Diller KR, Aggarwal SJ. The kinetics study of endogenous heat shock protein 70 expression. J Biomech Eng 2003;125:794–7.

[3] Hildebrandt B, Wust P, Ahlers O, Dieing A, Sreenivasa G, Kerner T, et al. The cellular and molecular basis of hyperthermia. Crit Rev Oncol 2002;43:33–56.

[4] Henriques Jr FC, Moritz AR. Studies of thermal injury. I. The conduction of heat to and through skin and the temperatures attained therein. Am J Pathol 1947;23:531–49.

[5] Henriques Jr FC. Studies of thermal injury V. The predictability and the significance of thermally induced rate processes leading to irreversible epidermal injury. Arch Pathol 1947;43:489–502.

[6] Bal-Price A, Coecke S. Guidance on Good Cell Culture Practice (GCCP). In: Aschner M, editor. Cell culture techniques, vol. 56, LLC: Springer Science+Business Media; 2011. p. 1–25.

[7] Wismeth C, Dudel C, Pascher C, Ramm P, Pietsch T, Hirschmann B, et al. Transcranial electro-hyperthermia combined with alkylating chemotherapy in patients with relapsed high-grade gliomas: phase i clinical results. J Neuroomcol 2009;98:395–405.

[8] Pennes HH. Analysis of tissue and arterial blood temperature in the resting human forearm. J Appl Physiol 1948;1:93–122.

[9] Durkee Jr JW, Antich PP, Lee CE. Exact solutions to the multiregion, time-dependent bioheat equation. I: solution development. Phys Med Biol 1990;35:847–67.

[10] Dobson EL, Warner GF, Finney CR, Johnston ME. The measurement of liver circulation by means of the colloid disappearance rate: I. Liver blood flow in normal young men. Circulation 1953;7:690–5.

[11] Charny CK. Mathematical models of bioheat transfer. In: Cho YI, editor. Advances in heat transfer, 22. Boston: Academic Press; 1992. p. 19–156.

[12] Wissler EH. Pennes' 1948 paper revisited. J Appl Physiol 1998;85:35–41.

[13] Chen MM, Holmes KR. Microvascular contributions in tissue heat transfer. Ann N Y Acad Sci 1980;335:137–50.

[14] Keller KH, Seiler Jr L. An analysis of peripheral heat transfer in man. J Appl Physiol 1971;30:779–86.

[15] Zhu L, Xu LX, He Q, Weinbaum S. A new fundamental bioheat equation for muscle tissue—part II: temperature of SAV vessels. J Biomech Eng 2002;124:121–32.

[16] Weinbaum S, Jiji LM, Lemons DE. Theory and experiments for the effects of vascular microstructure on surface tissue heat transfer—part I: anatomical foundation and model conceptualization. J Biomech Eng 1984;106:321–30.

[17] Roemer RB, Oleson JR, Cetas TC. Oscillatory temperature response to constant power applied to canine muscle. Am J Physiol 1985;249:153–8.

[18] Xu LX, Zhu L, Holmes KR. Blood perfusion measurements in the canine prostate during transurethral hyperthermia. Ann N Y Acad Sci 1998;858:21–9.

[19] Shitzer A. On the thermal efficiency of cold-stressed fingers. Ann N Y Acad Sci 1998;858:74–87.

[20] Baish JW, Mukundakrishnan K, Ayyaswamy PS. Numerical models of blood flow effects in biological tissues. In: Minkowycz WJ, Sparrow EM, editors. Advances in numerical heat transfer volume 3: numerical implementation of bioheat transfer models and equations. New York: CRC Press; 2009. p. 29–74.

[21] Hensley DW, Mark AE, Abella JR, Netscher GM, Wissler EH, Diller KR. 50 years of computer simulation of the human thermoregulatory system. J Biomech Eng 2013;135, 021006–1.

[22] Durkee Jr JW, Antich PP. Characterization of bioheat transport using an exact solution of the cylindrical geometry, multi-region, time-dependent bioheat equation. Phys Med Biol 1991;36:1377–406.

[23] Klinger HG. Green's function formulation of the bioheat transfer problem. In: Shitzer A, Eberhart RC, editors. Heat transfer in medicine and biology. New York: Plenum Press; 1985. p. 245–60.

[24] Deng ZS, Liu J. Analytical study on bioheat transfer problems with spatial or transient heating on skin surface or inside biological bodies. J Biomech Eng 2002;124:638–49.

[25] Dewey WC. Arrhenius relationships from the molecule and cell to the clinic. Int J Hyperthermia 2009;25:3–20.

[26] Laidler KJ. Chemical kinetics. New York: Harper & Row; 1987.

[27] Atkins PW. Physical chemistry. 6th ed. New York: W.H. Freeman; 1998.

[28] Wright NT. On a relationship between the Arrhenius parameters from thermal damage studies. J Biomech Eng 2003;125:300–4.

[29] Moran K, Anderson P, Hutcheson J, Flock S. Thermally induced shrinkage of joint capsule. Clin Orthop Relat Res 2000;381:248–55.

[30] Miles CA, Burjanadze TV, Bailey AJ. The kinetics of thermal de-naturation of collagen in unrestrained rat tail tendon determined by differential scanning calorimetry. J Mol Biol 1995;245:437–46.

[31] He X, Bischof JC. Quantification of temperature and injury response in thermal therapy and cryosurgery. Crit Rev Biomed Eng 2003;31(5):355–422.

[32] Eyring H. The activated complex in chemical reactions. J Chem Phys 1935;3:107–15.

[33] Rosenberg B, Kemeny G, Switzer RC, Hamilton TC. Quantitative evidence for protein denaturation as the cause of thermal death. Can J Forest Res 1971;232:471–3.

[34] Jacques SL. Ratio of entropy to enthalpy in thermal transitions in biological tissues. J Biomed Opt 2006;11, 041108.

[35] Krug RR, Hunter WG, Grieger RA. Statistical interpretation of enthalpy-entropy compensation. Nature 1976;261:566–7.

[36] Olsson TSG, Ladbury JE, Pitt WR, Williams MA. Extent of enthalpy–entropy compensation in protein–ligand interactions. Protein Sci 2011;20:1607–18.

[37] Sharp K. Entropy-enthalpy compensation: Fact or artifact? Protein Sci 2001;10:661–7.

[38] Fenley AT, Muddana HS, Gilson MK. Entropy-enthalpy transduction caused by conformational shifts can obscure the forces driving protein-ligand binding. Proc Natl Acad Sci USA 2012;109:20006–12.

[39] Harrington PL, Wright NT. Sensitivity analysis of Arrhenius parameters for denaturation of collagen, Proceedings of 2005 ASME Summer Bioengineering Conference, June 2005. pp. 1434–1435.

[40] Beck JV, Arnold K. Parameter estimation in engineering and science. New York: John Wiley and Sons; 1977.

[41] Himmelblau DM. Process analysis by statistical methods. New York: John Wiley and Sons; 1970.

[42] Schwaab M, Pinto JC. Optimum reference temperature for reparameterization of the Arrhenius equation. Part 1: problems involving one kinetic constant. Chem Eng Sci 2007;62:2750–64.

[43] Van Loey A, Guiavarc'h Y, Claeys W, Hendrickx M. The use of time- temperature integrators (TTIs) to validate thermal processes. In: Richardson PS, editor. Improving thermal processing of foods. London: Woodhead Publishing Limited; 2002.

[44] Dolan KD, Mishra DK. Parameter estimation in food science. Annu Rev Food Sci Technol 2013;4:401–22.

[45] Ayad s, Boot-Handford RP, Humphries MJ, Kadler KE, Shuttleworth CA. The extracellular matrix facts book. New York: Academic Press; 1994.

[46] Privalov PL. Stability of proteins: small globular proteins. In: Richards FM, Anfinsen CB, Edsall JT, editors. Protein chemistry, 33. New York: Academic Press; 1979. p. 167–241.

[47] Privalov PL. Stability of proteins: proteins which do not present a single cooperative system. In: Richards FM, Anfinsen CB, Edsall JT, editors. Advances in protein chemistry, vol. 36. New York. Academic Press; 1982. p. 1–104.

[48] Bull HB, Breese K. Thermal stability of proteins. Arch Biochem Biophys 1973;158:681–6.

[49] Sanchez-Ruiz JM. Protein kinetic stability. Biophys Chem 2010;148:1–15.

[50] Lumry R, Eyring H. Conformation changes of proteins. J Phys Chem 1954;58:110–20.

[51] Miles CA, Bailey AJ. Thermal denaturation of collagen revisited. Proc Indian Acad Sci Chem Sci 1999;111:71–80.

[52] Sanchez-Ruiz JM. Theoretical analysis of Lumry-Eyring models in differential scanning calorimetry. Biophys J 1992;61:921–35.

[53] Lepock JR, Ritchie KP, Kolios MC, Rodahl AM, Heinz KA, Kruuv J. Influence of transition rates and scan rate on kinetic simulations of differential scanning calorimetry profiles of reversible and irreversible protein denaturation. Biochemistry 1992;31:12706–12.

[54] Bischof JC, He X. Thermal stability of proteins. Ann N Y Acad Sci 2005;1066:12–33.

[55] Weir CE. Rate of shrinkage of tendon collagen—heat, entropy, and free energy of activation of the shrinkage of untreated tendon: effect of acid, salt, pickle, and tannage on the activation of tendon collagen. J Am Leather Chem Assoc 1949;44:108–40.

[56] Chen SS, Wright NT, Humphrey JD. Heat-induced changes in the mechanics of a collagenous tissue: free shrinkage. J Biomech Eng 1997;119:372–8.

[57] Wright NT, Chen SS, Humphrey JD. Time-temperature equivalence of heat-induced chances in cells and proteins. J Biomech Eng 1998;120:22–6.

[58] Chen SS, Wright NT, Humphrey JD. Heat-induced changes in the mechanics of a collageneous tissue: isothermal isotonic shrinkage. J Biomech Eng 1998;120:382–8.

[59] Miles CA, Ghelashvili M. Polymer-in-a-box mechanism for the thermal stabilization of collagen molecules in fibers. Biophys J 1999;76:3243–52.

[60] Chen SS, Wright NT, Humphrey JD. Phenomenological evolution equations for heat-induced shrinkage of a collagenous tissue. IEEE Trans Biomed Eng 1998;45:1234–40.

[61] Wall MS, Deng X-H, Torzilli PA, Doty SB, OBrien SJ, Warren RF. Thermal modification of collagen. J Shoulder Elbow Surg 1999;8:339–44.

[62] Lepock JR. Cellular effects of hyperthermia: relevance to the minimum dose for thermal damage. Int J Hyperthermia 2003;19(3):252–66.

[63] Brown F, Diller KR. Calculating the optimum temperature for serving hot beverages. Burns 2008;34:648–54.

[64] Basu S, Binder RJ, Suto R, Anderson KM, Srivastava PK. Necrotic but not apoptotic cell death releases heat shock proteins, which deliver a partial maturation signal to dendritic cells and activate the NF-κB pathway. Int Immunol 2000;12:1539–46.

[65] Rylander MN, Feng Y, Zimmermann K, Diller KR. Measurement and mathematical modeling of thermally induced injury and heat shock protein expression kinetics in normal and cancerous prostate cells. Int J Hyperthermia 2010;26:748–64.

[66] Linkermann A, Green DR. Necroptosis. N Engl J Med 2014;370(5):455–65, PMID: 24476434.

[67] Li S, Chien S, Branemark P-I. Heat shock-induced necrosis and apoptosis in osteoblasts. J Orthop Res 1999;17:891–9.

[68] Roti Roti JL. Cellular responses to hyperthermia (40-46 °C): cell killing and molecular events. Int J Hyperthermia 2008;24(1):3–15.

[69] He X, Bhowmick S, Bischof J. Thermal therapy in urologic systems: a comparison of Arrhenius and thermal isoeffective dose models in predicting hyperthermic injury. J Biomech Eng 2009;131, 074507–1–12.

[70] Feng Y, Oden JT, Rylander MN. A two-state cell damage model under hyperthermic conditions: theory and in vitro experiments. J Biomech Eng 2008;130, 041016–1–10.

[71] O'Neill DP, Peng T, Stiegler P, Mayrhauser U, Koestenbauer S, Tscheiliessnigg K, et al. A three-state mathematical model of hyperthermic cell death. Ann Biomed Eng 2011;39(1):570–9.

[72] Wright NT. Comparison of models of post-hyperthermia cell survival. J Biomech Eng 2013;135:051001–1.

[73] Westra A, Dewey WC. Variation in sensitivity to heat shock during the cell cycle of Chinese hamster cells in vitro. Int J Rad Biol 1971;19:467–77.

[74] Bauer KD, Henle KJ. Arrhenius analysis of heat survival curves from normal and thermotolerant cho cells. Radiat Res 1979;78:251–63.

[75] Pearce JA. Hypothesis for thermal activation of the caspase cascade in apoptotic cell death at elevated temperatures. In: Ryan TP, editors. Proc. SPIE 8584, Energy-based treatment of tissue and assessment VII, vol. 5484, SPIE; 2013. p. 561–606.

[76] Johnson HA, Pavelec M. Thermal injury due to normal body temperature. Am J Pathol 1972;66:557–64.

[77] Dewey WC, Hopwood LE, Sapareto SA, Gerweck LE. Cellular responses to combinations of hyperthermia and radiation. Radiology 1977;123:463–79.

[78] Hahn GM. Hyperthermia and cancer. New York: Plenum Press; 1982.

[79] Roti Roti JL, Henle KJ. Comparison of two mathematical models for describing heat-induced cell killing. Rad Res 1980;81:374–83.

[80] Jung H. A generalized concept for cell killing by heat. Rad Res 1986;106:56–72.

[81] Jung H. A generalized concept for cell killing by heat: effect of chronically induced thermotolerance. Rad Res 1991;127:235–42.

[82] Mackey MA, Roti Roti JL. A model of heat-induced clonogenic cell death. J Theor Biol 1992;156(1):133–46.

[83] Moritz AR, Henriques Jr FC. Studies of thermal injury. II. The relative importance of time and surface temperature in the causation of cutaneous burns American. J Pathol 1947;23:695–720.

[84] Moritz AR. Studies of thermal injury. III. The pathology and pathogenesis of cutaneous burns, an experimental study. Am J Pathol 1947;23:915–41.

[85] Takata AN. Development of criterion for skin burns. Aerosp Med 1974;45:634–7.

[86] Pearce JA, Raghavan K, Thomsen S. Arrhenius model thermal damage coefficients for birefringence loss in rabbit myocardium. Int Mech Eng Congress Exposition 2003;4:421–3, ASME.

[87] Pearce JA, Thomsen SL, Vijverberg H, McMurray TJ. Kinetics for birefringence changes in thermally coagulated rat skin collagen. Proc SPIE 1876, Lasers in Otolaryngology, Dermatology, and Tissue Welding; July 1, 1993. p. 180.

[88] Pearce JA, Thomsen S. Rate process analysis of thermal damage. In: Welch AJ, van Germert MJC, editors. Optical and thermal response of laser-irradiated tissue. New York: Plenum Press; 1995. p. 561–606.
[89] Xu YS, Qian RZ. Analysis of thermal injury process based on enzyme deactivation mechanisms. J Biomech Eng 1995;117:462–5.
[90] Sapareto SA, Dewey WC. Thermal dose determination in cancer therapy. Int J Rad Oncol Biol Phys 1984;10:787–800.
[91] Dewhirst MW, Viglianti BL, Lora-Michiels M, Hanson M, Hoopes PJ. Basic principles of thermal dosimetry and thermal thresholds for tissue damage from hyperthermia. Int J Hyperthermia 2003;19:267–94.
[92] Jones E, Thrall D, Dewhirst MW, Vujaskovic Z. Prospective thermal dosimetry: the key to hyperthermia's future. Int J Hyperthermia 2006;22:247–53.
[93] Foster KR, Morrissey JJ. Thermal aspects of exposure to radiofrequency energy: report of a workshop. Int J Hyperthermia 2011;27:307–19.
[94] Diller KR. Modeling of bioheat transfer processes at high and low temperatures. In: Cho YI, editor. Advances in heat transfer, 22. Boston: Academic Press; 1992. p. 157–358.
[95] Lim D, Namgung B, Woo DG, Choi JS, Kim HS, Tack GR. Effect of input waveform pattern and large blood vessel existence on destruction of liver tumor using radiofrequency ablation: finite element analysis. J Biomech Eng 2010;132, 061003–1.
[96] Chang IA, Nguyen UD. Thermal modeling of lesion growth with radiofrequency ablation devices. Biomed Eng Online 2004;3:1–19.
[97] Brinton MR, Stewart RJ, Cheung AK, Christensen DA, Shiu Y-TE. Modelling ultrasound-induced mild hyperthermia of hyperplasia in vascular grafts. Theor Biol Med Model 2011;8:1–15.
[98] Ng EY-K, Chua LT. Prediction of skin burn injury. Part 1: numerical modeling. J Eng Med 2002;216:157–70.
[99] Prediction of skin burn injury. Part 2: parametric and sensitivity analysis. J Eng Med 2002;216:171–83.
[100] Breen MS, Breen M, Butts K, Chen L, Saidel GM, Wilson DL. MRI-guided thermal ablation therapy: model and parameter estimates to predict cell death from mr thermometry images. Ann Biomed Eng 2007;35:1391–403.
[101] Chen X, Saidel GM. Modeling of laser coagulation of tissue with MRI temperature monitoring. J Biomech Eng 2010;132, 064503–1.
[102] Yung JP, Shetty A, Elliott A, Weinberg JS, McNichols RJ, Gowda A, et al. Quantitative comparison of thermal dose models in normal canine brain. Med Phys 2010;37(10):5313–21.
[103] Dice LR. Measures of the amount of ecologic association between species. Ecology 1945;26:297–302.
[104] Fanjul-Vlez F, Ortega-Quijano N, Ramn Solana-Quirs J, Arce-Diego JL. Thermal damage analysis in biological tissues under optical irradiation: application to the skin. Int J Thermophys 2009;30:1423–37.
[105] Liljemaln R, Nyberg T. Quantification of a thermal damage threshold for astrocytes using infrared laser generated heat gradients. Ann Biomed Eng 2014;42:822–32.
[106] Eifiing T, Conzelmann H, Gilles ED, Allgöwer F, Bullinger E, Scheurich P. Bistability analyses of a caspase activation model for receptor-induced apoptosis. J Biol Chem 2004;279:36892–7.
[107] Eifiing T, Waldherr S, Allgöwer F, Scheurich P, Bullinger E. Steady state and (bi-) stability evaluation of simple protease signalling networks. BioSystems 2007;90:591–601.

Analytical Bioheat Transfer: Solution Development of the Pennes' Model

Sid M. Becker

Department of Mechanical Engineering, University of Canterbury, Christchurch, New Zealand

Nomenclature

a, b, c Constants of integration
c_b Specific heat of blood
c_p Specific heat of tissue
C_n Fourier coefficient
f Green's function boundary term
F Initial temperature
g Volumetric heat generation
G Heat conduction Green's function
G^P Bioheat transfer Green's function

h Convection heat transfer coefficient
k Thermal conductivity
L Layer thickness
N_n Norm
M Number of layers
q_O'' Surface heat flux
$\hat{r}$ Coordinate position vector
r Radial coordinate
r' Dummy variable
R Radial thickness
t Time coordinate
T Tissue temperature
T^* Unshifted tissue temperature
T^{HC} Solution of the heat conduction problem
T^{IH} Solution of the nonhomogenous bioheat problem
Ta Arterial blood temperature
T_O Tissue surface temperature
T_∞ Ambient temperature
w_P Rate of blood perfusion
x Cartesian coordinate
x' Dummy variable
$X_n(x)$ Eigenfunction

Greek Symbols

α Thermal diffusivity
$\Gamma_n(t)$ Transient solution component
λ_n Eigenvalue
ρ_p Density of tissue
ρ_b Density of blood
τ Dummy variable in time
$\Psi(x)$ Nonhomogenous steady solution component
ω Perfusion coefficient

Subscripted Index Notation

i,j Composite layer number
n,m Eigenvalue number
O Reference layer

4.1 PENNES' BIOHEAT EQUATION IN LIVING TISSUE ANALOGY

Passive transient heat transfer within living biological tissue is characterized by the effects of the passage of blood through the vascular circulatory system. As the blood moves through the microcapillary system, it has a stabilizing effect that returns the perfuse tissue to the natural equilibrium reference state (the state at which the tissue is found in the absence of externally applied fluxes). This is perfusion. In an exchange of thermal energy between the tissue and the microcapillary network, perfusion acts to bring the local tissue temperature closer to the body's core temperature. For example, when a tissue's thermal energy is increased by an externally applied heat flux, perfusion cools the tissue by delivering blood nearer the lower core body temperature and removing blood at the elevated temperature.

Perfusion may be viewed from a theoretical perspective as a volumetric phenomenon when the distribution of microcapillaries is uniform and the relative size scale of the capillary diameter is small compared to the length scale characterizing transport. The idea is that the network's blood flow, when viewed from a scale much larger than the diameter of an individual capillary, has no distinguishable direction and permeates the tissue in a generally homogenous manner. At such a scale, any local fluctuations in the blood temperature resulting from thermal exchange between the tissue and an individual capillary cannot be distinguished.

Mathematically, conduction of heat into perfuse tissue may be represented by the Pennes' bioheat equation [1]:

$$\rho c_p \frac{\partial T^*}{\partial t} = \nabla \cdot (k \nabla T^*) - \rho_b c_b w_P (T^* - Ta) + g \tag{4.1}$$

The symbols ρ, c_p, and k are respectively the tissue's density, specific heat, and thermal conductivity. The term g refers to volumetric heating, which can vary with time and with position. Heat generation can result from either low levels of metabolic sources or from some higher intensity externally induced source. The variable w_P is the blood perfusion rate. It is representative of the volume rate at which blood passes through the tissue per unit volume of tissue $(m^3 \text{blood} \cdot s^{-1})/(m^3 \text{tissue})$. The terms ρ_b and c_b refer respectively to the density and the specific heat of blood. The symbol Ta is the temperature of the blood at the body's core; thus the perfusion term has the effect of bringing the local tissue temperature closer to the body's core temperature. However, in some instances, this representation has an inherent weakness: implicit in the physical interpretation of the perfusion term of Equation (4.1) is that the assumption has been made that the temperature of the blood within the capillaries, Ta, is at the body's core temperature. This means that all of the blood's heat losses or gains are restricted to occur entirely within the capillary bed and that there are no losses or gains in the thermal energy of the blood in the higher order systemic components (large and intermediate sized vessels or veins). That is the primary criticism of Equation (4.1) and has been the motivation of alternate model representations of the perfusion term. This chapter will not address the relative strengths and shortcomings of the Pennes' bioheat equation. A detailed review of the field's view on this is provided in Chapter 1 of this book, and the inquisitive reader may find references [2–5] of interest.

The purpose of this chapter is to help the reader to develop analytic solutions to the transient Pennes' bioheat equation. Consider that in the absence of the perfusion term, the bioheat equation is identical to the heat conduction equation.

$$\rho c \frac{\partial T^*}{\partial t} = \nabla \cdot (k \nabla T^*) + g \tag{4.2}$$

The transient heat conduction equation has been well studied. The highly referenced textbooks *Conduction of Heat in Solids* by Carslaw and Jaeger [6], *Heat Conduction Using Green's Functions* by Beck et al. [8], *Heat Conduction* by Özisik [7], and its more recent rendition (available as an e-book) by Hahn and Özisik [9], all develop the solutions to the heat conduction equation for various boundary conditions and coordinate systems. Such books act as inventories of solutions to the heat equation. Because the solution of the heat conduction equation is readily available, this chapter will show how the existing solutions of the heat conduction equation may be altered in order to provide the corresponding solution to the bioheat equation.

4.1.1 Representation of the Governing Equations

In many applications of the bioheat equation, it is safe to assume that the material properties of the tissue are homogenous. In such cases the bioheat equation may be more compactly represented by

$$\frac{1}{\alpha}\frac{\partial T^*}{\partial t} = \nabla^2 T^* - \omega^2(T^* - Ta) + \frac{g}{k} \tag{4.3}$$

where the thermal diffusivity is $\alpha = k/\rho c$ and the perfusion rate is represented by $\omega^2 = \rho_b c_b w_P/k$. Consider momentarily the placement of the thermal diffusivity, α, with the transient derivative. This has been done because the thermal diffusivity is physically descriptive of transient behavior—it characterizes the ratio of the *rate* of thermal energy conducted to the *rate* of thermal energy stored.

For cases in which there are distinct spatial discontinuities in the thermophysical parameter values (k, α, and ω), a composite representation of the bioheat equation may be used. This is discussed in Section **4.5**.

4.1.2 Boundary Condition Types

In this chapter we will consider the three standard types of boundary conditions that might occur at the outer surface which is furthest from the body's core. In the following, the three types of boundary conditions are presented in the Cartesian coordinate system as depicted in Figure 4.1. In this figure, a section of perfuse tissue represents the outer region of a living body so that the outer surface corresponds to the position $x=0$ and the region that is closest to the body's core corresponds to the position $x=L$.

The tissue is exposed to a boundary condition at the position $x=0$ denoted BC1. This boundary condition may take on the form of any of the following three boundary condition types.

Type I boundary condition at $x=0$ (BC1 Type I): Specified Surface Temperature

$$T^*(x=0,t) = T_O^*(t) \tag{4.4}$$

Type II boundary condition at $x=0$ (BC1 Type II): Specified Heat Flux

$$-k\frac{\partial T^*}{\partial x}\bigg|_{x=0} = q_O''(t) \tag{4.5}$$

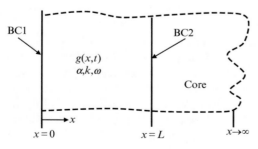

FIGURE 4.1   One dimensional bioheat transfer in the Cartesian coordinate system in which a section of tissue is exposed to the surroundings at $x=0$ and is exposed to the body's core at $x=L$.

Type III boundary condition at $x=0$ (BC1 Type III): Convection Heat Transfer

$$k\frac{\partial T^*}{\partial x}\bigg|_{x=0} = h(T^* - T^*_\infty(t)) \tag{4.6}$$

Far away from any externally imposed boundary condition (at $x=L$), the tissue may be considered to be in thermal equilibrium with the body's core. While this is not an explicit requirement for realistic applications of the bioheat equation, in the following discussion it seems most applicable. For this reason, we do not consider a Type III boundary condition at the body's core. The thermal equilibrium with the body's core may be interpreted as occurring either when the local temperature is equal to the arterial blood temperature or when there is a negligible heat flux into the core. This may be represented either through a Type I boundary condition.

Type I boundary condition at $x=L$ (BC2 Type I): Equilibrium in Temperature

$$T^*(x=L,t) = Ta \tag{4.7}$$

Type II boundary condition at $x=L$ (BC2 Type II): Equilibrium in Flux

$$q''_{x=L} = -k\frac{\partial T^*}{\partial x}\bigg|_{x=L} = 0 \tag{4.8}$$

In practice, the initial temperature distribution is often assumed to be homogenous (and equal to the arterial blood temperature) or some steady state distribution resulting from heat exchange between the body and the environment. In the discussion that follows, the initial condition (IC) is not restricted and the tissue's initial state is represented arbitrarily by some spatially dependent thermal profile.

$$T^*(x,t=0) = F^*(x) \tag{4.9}$$

4.1.3 Temperature Shift

The perfusion term of the governing Equation (4.3) is not homogenous. The problem can be slightly simplified by homogenizing this term. This can be accomplished through a simple linear translation of the temperature:

$$T = T^* - Ta \tag{4.10}$$

Such a temperature shift is described in the context of convection heat transfer in Ref. [9] on page 81 and in the context of heat transfer from extended surfaces in Ref. [8] on page 61 in Section 3.5.1. This shift requires that the arterial blood temperature, Ta, is constant. The result of applying this transformation on the bioheat equation is to homogenize the perfusion term so that the governing equation is represented by

$$\frac{1}{\alpha}\frac{\partial T}{\partial t} = \frac{\partial^2 T}{\partial x^2} - \omega^2 T + \frac{g}{k} \tag{4.11}$$

The boundary conditions and the initial condition must also account for this temperature shift. At the surface, $x = 0$, the shifted BC Types I–III are now represented by

$$\text{BC1 Type I:} \quad T(x = 0, t) = T_O(t)$$

$$\text{BC1 Type II:} \quad -k \frac{\partial T}{\partial x}\bigg|_{x=0} = q_O''(t)$$

(4.12)

$$\text{BC1 Type III:} \quad k \frac{\partial T}{\partial x}\bigg|_{x=0} = h(T - T_\infty(t))$$

where $T_O(t) = T_O^*(t) - Ta$ and $T_\infty(t) = T_\infty^*(t) - Ta$.

At the boundary core ($x = L$), this temperature shift has an added bonus which is to homogenize the Type I BC, while the Type II BC will remain unchanged:

$$\text{BC2 Type I:} \quad T(x = L, t) = 0$$

$$\text{BC2 Type II:} \quad k \frac{\partial T}{\partial x}\bigg|_{x=L} = 0$$

(4.13)

The shifted state of the initial temperature distribution is

$$\text{IC:} \quad T(x, t = 0) = F(x)$$

(4.14)

where $F(x) = F^*(x) - Ta$.

In Sections **4.2** and **4.3**, the temperature shift of relation (4.9) is employed so that the related governing equations are represented by the governing Equation (4.11), the applicable form of the boundary conditions of Equations (4.12) and (4.13), and the initial condition (4.14).

4.2 SOLUTIONS TO THE TRANSIENT HOMOGENOUS BIOHEAT EQUATION

In this chapter, the solution to the transient homogenous bioheat equation is developed using the separation of variables (SOV) method. This is done primarily to show how the existing corresponding solution of the heat conduction equation may be manipulated in order to arrive at the bioheat solution. This will involve special handling of the perfusion term. By applying the SOV method to the bioheat equation and comparing the result to the solution of pure conduction it will be shown that:

> The solution of the transient homogenous bioheat problem may be adopted from the solution of the corresponding heat conduction problem with identical initial and boundary conditions by multiplying the heat conduction solution by the term: $e^{-\alpha\omega^2 t}$.

This may be expressed symbolically as

$$T(\hat{r}, t) = T^{HC}(\hat{r}, t)\exp\left(-\alpha\omega^2 t\right)$$

(4.15)

where, T^{HC} is the solution to the heat conduction problem and $\hat{r}$ represents the coordinate set associated with the spatial domain. This effectively states that for transient homogenous problems, perfusion acts on the conduction solution in order to (uniformly in space) exponentially accelerate the process of reaching steady state.

The SOV method is frequently used in the evaluation of the solution to linear heat conduction problems. The details and underlying theory of the SOV method are well described in any of the editions of the book *Heat Conduction* by Özisik [7,9]. The reader unfamiliar with the SOV method is encouraged to read the excellent descriptions provided in Refs. [8,9]. In the following sections, the SOV method is applied to the homogenous bioheat equation.

The SOV method for transient or steady problems has the following restrictions:

1. The governing equation is homogenous
2. The governing equation and the boundary condition are linear
3. For transient problems, the boundary conditions are homogenous and the initial condition is nonhomogenous
4. For steady multidimensional problems, all boundaries except one are homogenous
5. The boundaries are orthogonal to one another

4.2.1 Bioheat Solution in the Cartesian Coordinate System

In the following example, the SOV method is illustrated in the Cartesian coordinate system. The results are equally applicable to the cylindrical and spherical coordinate systems. The solution derivation presented in Example 4.1 closely follows and references are described in Ref. [9].

EXAMPLE 4.1

Consider the example of the homogenous transient bioheat problem in the absence of energy generation with homogenous boundary conditions of the first kind at $x=0$ and of the second kind at $x=L$ and an initial condition that is represented by an arbitrary function in space:

$$\frac{1}{\alpha}\frac{\partial T}{\partial t} = \frac{\partial^2 T}{\partial x^2} - \omega^2 T \quad \text{in} [0 < x < L], \quad t > 0$$

$$\text{BC1:} \quad T(x=0,t) = 0 \quad \text{BC2:} \quad \left.\frac{\partial T}{\partial x}\right|_{x=L} = 0 \tag{4.16}$$

$$\text{IC:} \quad T(x,t=0) = F(x)$$

When the SOV method is applied to the bioheat problem, it is preferable to write the governing equation so that the diffusive term is isolated and the transient term is coupled with the perfusion term. In this example, the boundary conditions are homogenous, so that the heat conduction term (the Laplacian operator) is isolated. Because the initial condition is nonhomogenous, the transient derivative is grouped with the perfusion term. This results in the arrangement:

$$\frac{\partial^2 T}{\partial x^2} = \frac{1}{\alpha}\frac{\partial T}{\partial t} + \omega^2 T \quad \text{in} [0 < x < L], \quad t > 0 \tag{4.17}$$

While the difference in arrangement seems slight, the simplifications that follow are important.

The SOV method next assumes that the solution may be represented as the product of two independent variables: one that is purely space dependent, $X(x)$, and one that is time dependent, $\Gamma(t)$:

$$T(x,t) = X(x) \cdot \Gamma(t) \tag{4.18}$$

Substituting this into the governing equation results in the expression:

$$\frac{1}{\alpha}X(x)\cdot\frac{\partial\Gamma(t)}{\partial t} = \Gamma(t)\cdot\frac{\partial^2 X(x)}{\partial x^2} - \omega^2 X(x)\cdot\Gamma(t) \tag{4.19}$$

Note that since $X(x)$ is dependent only on position, x, and that $\Gamma(t)$ is dependent only on the time, t, the derivatives in Equation (4.19) are ordinary. The above expression may be rearranged so that the variables $X(x)$ and $\Gamma(t)$ are separated from one another and the governing equation may be equivalently expressed as equivalent ordinary differential equations (ODEs):

$$\frac{1}{\alpha\Gamma}\frac{d\Gamma}{dt} + \omega^2 = \frac{1}{X}\frac{d^2 X}{dx^2} \tag{4.20}$$

Note that the left side is purely a function of time and the right side is a function of position. This can only be true if the two sides of the equation are equal to a constant value, $-\lambda^2$, so that:

$$\frac{1}{\alpha\Gamma}\frac{d\Gamma}{dt} + \omega^2 = \frac{1}{X}\frac{d^2 X}{dx^2} = -\lambda^2 \tag{4.21}$$

At this point, special notice is made of the result of the choice to isolate the spatial component and to group the perfusion term with the time dependent component. Because, in Equation (4.21), the perfusion term is explicitly tied to the transient term $\Gamma(t)$, the perfusion term will only appear in the solution of the transient ODE. This is important because it implies that any subsequent solution of the spatial solution component $X(x)$ of Equation (4.21) will be independent of the existence of the perfusion term. This implies that the solution of $X(x)$ that has been developed for the heat conduction problem is identical to the solution of $X(x)$ in the corresponding bioheat problem.

It can be shown that that the sign of the constant, λ^2, is negative by considering the ODE governing the spatial coordinate $X(x)$ and its associated boundary conditions. By substituting relation (4.18) into boundary condition (4.16) and simplifying, the ODE and its boundary conditions are able to be represented independently of the transient component:

$$\frac{d^2 X}{dx^2} = -X(x)\lambda^2$$
$$\text{BC1: } X(0) = 0 \quad \text{BC2: } \left.\frac{\partial X}{\partial x}\right|_{x=L} = 0 \tag{4.22}$$

If the constant is equal to zero ($\lambda^2 = 0$), the solution of Equation (4.22) is linear: $X(x) = a + bx$, and when the boundary conditions are considered, the solution is trivial. If the square of the constant is negative ($\lambda^2 < 0$), the general solution of Equation (4.22) is represented by $X(x) = a\cosh(\lambda x) + b\sinh(\lambda x)$. And when the boundary conditions are considered, the solution is again trivial.

The nontrivial solution for Equation (4.22) may be found only for $\lambda^2 > 0$ and the solution for this ODE is the *eigenfunction*:

$$X(x) = a\sin(\lambda x) + b\cos(\lambda x) \tag{4.23}$$

Implementing BC1 at $x = 0$ shows that $b = 0$. Implementing BC2 at $x = L$ results in the expression:

$$a\cos(\lambda L) = 0 \tag{4.24}$$

To avoid the trivial solution, it is evident that Equation (4.24) is satisfied by an infinite number of *eigenvalues*, λ, whose values are

$$\lambda_n = \frac{(2n-1)\pi}{2L} \quad \text{for } n = 0,1,2,\ldots \tag{4.25}$$

For each eigenvalue, there exists a corresponding eigenfunction, which in this case is

$$X_n(x) = a_n \sin(\lambda_n x). \tag{4.26}$$

Note that in this particular case, the eigenvalue corresponding to $n=0$ provides no contribution to the eigenfunction in Equation (4.26) and may be safely omitted.

Again it is stressed that the arrival of the eigenvalues of Equation (4.25) and of the eigenfunction of Equation (4.26) were determined without any knowledge of the perfusion term so that the identical result would have resulted in the absence of perfusion. This means that the eigenvalues and eigenfunctions, which are documented for problems of pure heat conduction, are directly applicable to the spatial components of the corresponding transient bioheat equation.

Next the transient component of Equation (4.21), which is influenced by the perfusion term, is considered. For each eigenvalue, there is a corresponding transient component which obeys the ODE:

$$\frac{d\Gamma_n}{dt} = -\alpha(\lambda_n^2 + \omega^2)\Gamma_n \tag{4.27}$$

And the general solution of this ODE is

$$\Gamma_n(t) = c_n \exp\left(-\alpha(\lambda_n^2 + \omega^2)t\right). \tag{4.28}$$

The result of multiplying the transient component (4.28) by the spatial component (4.26) as in Equation (4.18) is that for each eigenvalue, λ_n, there exists a corresponding solution that satisfies the homogenous problem. Thus, for a particular eigenvalue, n, the corresponding solution may be represented:

$$T_n(x,t) = C_n \underbrace{\sin(\lambda_n x)}_{X_n(x)} \underbrace{\exp\left(-\alpha(\lambda_n^2 + \omega^2)t\right)}_{\Gamma_n(t)} \tag{4.29}$$

The solution can now be expressed as the linear sum of the series of orthogonal functions:

$$T(x,t) = \sum_{n=1}^{\infty} C_n \sin(\lambda_n x)\exp\left(-\alpha(\lambda_n^2 + \omega^2)t\right) \tag{4.30}$$

Here, a simplification in the representation of the integration constants $C_n = a_n \cdot c_n$ has been introduced. Finally, it is necessary to determine the value of C_n. This is done by substituting Equation (4.30) into the initial condition of Equation (4.16) yielding:

$$\sum_{n=1}^{\infty} C_n \sin(\lambda_n x) = F(x) \tag{4.31}$$

Multiplying both sides of Equation (4.31) by $\sin(\lambda_m x)$ and integrating over yields:

$$\int_{x=0}^{L} \sum_{n=1}^{\infty} C_n \sin(\lambda_n x)\sin(\lambda_m x)dx = \int_{x=0}^{L} F(x)\sin(\lambda_m x)dx \tag{4.32}$$

The eigenfunction (in this case represented by the sine function) has a special orthogonality property. This property can be illustrated by considering the result of the following integral:

$$\int_{x=0}^{L} \sin(\lambda_n x)\sin(\lambda_m x)dx = \frac{\sin((\lambda_n - \lambda_m)L)}{2(\lambda_n - \lambda_m)} - \frac{\sin((\lambda_n + \lambda_m)L)}{2(\lambda_n + \lambda_m)} \tag{4.33}$$

The result may be rewritten using the sum difference formula for sine functions so that

$$\int_{x=0}^{L} \sin(\lambda_n x)\sin(\lambda_m x)dx = \frac{\sin(\lambda_n L)\cos(\lambda_m L) - \sin(\lambda_m L)\cos(\lambda_n L)}{2(\lambda_n - \lambda_m)}$$
$$- \frac{\sin(\lambda_n L)\cos(\lambda_m L) + \sin(\lambda_m L)\cos(\lambda_n L)}{2(\lambda_n + \lambda_m)}. \tag{4.34}$$

Considering the boundary condition resulting in Equation (4.24), it is obvious that all cosine terms are zero. Thus, when the integral containing the summation on the LHS of Equation (4.32) is evaluated, a nonzero value is only returned for the term when $\lambda_n = \lambda_m$ and all other terms in the summation may be neglected. When this result is considered, Equation (4.32) may be equivalently expressed as:

$$C_m \int_{x=0}^{L} \sin(\lambda_m x)\sin(\lambda_m x)dx = \int_{x=0}^{L} F(x)\sin(\lambda_m x)dx \tag{4.35}$$

Replacing the index m with the index n, (these are arbitrary now that the summation of Equation (4.32) has been reduced to a single term) and then solving for the constant of integration, C_n, results in the expression:

$$C_n = \frac{\displaystyle\int_{x=0}^{L} F(x)\sin(\lambda_n x)dx}{\displaystyle\int_{x=0}^{L} \sin(\lambda_n x)\sin(\lambda_n x)dx} \tag{4.36}$$

Each value of C_n corresponds to the Fourier coefficient of the Fourier expansion of the initial condition $F(x)$ and the Fourier series expansions (see Chapter 2, pages 40–62 of the book [9] for further detail). The evaluation of the integral in the denominator is determined by relying on the orthogonal properties of the eigenfunctions that state:

$$\int_{x=0}^{L} \sin(\lambda_n x)\sin(\lambda_m x)dx = \begin{cases} 0 & \text{if } n \neq m \\ N_n(\lambda_n) & \text{if } n = m \end{cases} \tag{4.37}$$

The norm (denoted by the symbol N_n) is in this case:

$$N_n(\lambda_n) = \frac{L}{2} \quad \text{for } n \geq 1 \tag{4.38}$$

Thus the constant may now be represented:

$$C_n = \frac{2}{L}\int_{x=0}^{L} F(x)\sin(\lambda_n x)dx \tag{4.39}$$

So that the solution of the temperature is represented:

$$T(x,t) = \sum_{n=1}^{\infty} \underbrace{\frac{2}{L} \int_{x'=0}^{L} F(x')\sin(\lambda_n x')dx'}_{C_n} \cdot \underbrace{\sin(\lambda_n x)}_{X_n} \cdot \underbrace{\exp(-\alpha(\lambda_n^2 + \omega^2)t)}_{\Gamma_n} \tag{4.40}$$

Here, the symbol x' is a dummy variable that has been introduced in order to distinguish the space variables within the integral from those outside the integral. Again note that the eigenfunction, the eigenvalues, and the norm are independent of the perfusion coefficient. Thus, the solution to the bioheat problem provided in Equation (4.40) is identical to that of the corresponding homogenous heat conduction problem multiplied by the exponential: $\exp(-\alpha\omega^2 t)$.

For the special case in which the initial condition is equal to some constant, $F(x) = F_O$, the evaluation of the integral Equation (4.39) results in a temperature distribution of

$$C_n = \frac{2}{L} F_O \int_{x=0}^{L} \sin(\lambda_n x)dx = \frac{2}{L} \frac{(1 - \cos(\lambda_n L))}{\lambda_n}. \tag{4.41}$$

This may be simplified so that the solution is represented:

$$T(x,t) = \frac{2}{L} F_O \sum_{n=1}^{\infty} \frac{\sin(\lambda_n x)}{\lambda_n} \cdot \exp(-\alpha(\lambda_n^2 + \omega^2)t)$$

$$\text{for } \lambda_n = \frac{\pi(2n-1)}{2L} \tag{4.42}$$

4.2.1.1 Solutions to Transient Bioheat Transfer in a Slab

The SOV method for the homogenous transient bioheat equation in a slab results in an expression of the temperature as:

$$T(x,t) = \frac{C_O}{L} \cdot \exp(-\alpha\omega^2 t) + \sum_{n=1}^{\infty} C_n X_n(x)\exp(-\alpha(\lambda_n^2 + \omega^2)t) \tag{4.43}$$

The constant C_O/L only appears when both boundary conditions are of Type II and this is the only case in which an eigenvalue of zero value ($\lambda_0 = 0$) contributes to the solution. The eigenfunction $X_n(x)$ is governed by the ODE of Equation (4.22) and may be determined from the boundary condition at $x=0$. The eigenvalues, λ_n, are determined from the boundary condition at $x=L$. And finally the constant of integration, C_n, is developed from the application of the initial condition and for each term in the series, and is determined from the integral:

$$C_n = \frac{1}{N(\lambda_n)} \int_{x'=0}^{L} F(x')X_n(x')dx' \tag{4.44}$$

where x' is an integration variable and the norm is the result of the integral:

$$N(\lambda_n) = \int_{x=0}^{L} X_n^2(x)dx \tag{4.45}$$

TABLE 4.1 The Eigenfunctions, Norms, and Eigenvalues of $\frac{d^2 X_n}{dx^2} = -X_n(x)\lambda_n^2$

BC at $x=0$	BC at $x=L$	Eigenfunction: $X_n(x)$	Norm: $N(\lambda_n)$	Eigenvalues $\lambda_n, n=1,2,3\ldots$	C_O
Type I	Type I	$\sin(\lambda_n x)$	$L/2$	$n\pi/L$	0
	Type II	$\sin(\lambda_n x)$	$L/2$	$(2n-1)\pi/2L$	0
Type II	Type I	$\cos(\lambda_n x)$	$L/2$	$(2n-1)\pi/2L$	0
	Type II	$\cos(\lambda_n x)$	$L/2$	$n\pi/L$	1
Type III $H=\dfrac{h}{k}$	Type I	$\sin(\lambda_n(L-x))$	$\dfrac{L(\lambda_n^2+H^2)+H}{2(\lambda_n^2+H^2)}$	Roots of $\lambda_n\cot(\lambda_n L)=-H$	0
	Type II	$\cos(\lambda_n(L-x))$	$\dfrac{L(\lambda_n^2+H^2)+H}{2(\lambda_n^2+H^2)}$	Roots of $\lambda_n\tan(\lambda_n L)=H$	0

Adapted from Ref. [9].

In order to expedite the evaluation of the homogenous solution to the transient slab problem the functions: $X_n(x), N_n, C_O,$ and λ_n are provided in Table 4.1. These have been adapted from the existing solutions to pure heat conduction provided in Table 2.1 of Chapter 2 of the book [9] and may be directly used to determine the solutions to the homogenous problem of Equations (4.43–4.45).

Clearly, in order to complete the solution of Equation (4.43), the integral in Equation (4.44) of the product of the initial condition and the eigenfunction must be evaluated. In many cases, the initial condition defined by the function $F(x)$ will also satisfy the steady bioheat equation:

$$\frac{d^2 F(x)}{dx^2} - \omega_F^2 F(x) + \frac{g_F}{k} = 0 \tag{4.46}$$

where the subscript F is used to denote parameter values associated with the initial temperature distribution. For example, it is possible that the perfusion rate associated with the initial condition ω_F at $t=0$ is not equivalent to the perfusion rate of the transient system ω at $t>0$. When the eigenfunction is governed by Equation (4.21) and when the initial condition is governed by Equation (4.46), the evaluation of the integral that appears in Equation (4.44) may be expedited by the following relation:

$$\int_{x=0}^{x=L} (X_n \cdot F)\,dx = \frac{1}{(\omega_F^2 + \lambda_n^2)} \left(X \cdot \frac{dF}{dx} - F \cdot \frac{dX_n}{dx} - \left(\frac{1}{\lambda_n^2}\frac{g_F}{k}\right)\frac{dX_n}{dx} \right)\Bigg|_{x=0}^{x=L} \tag{4.47}$$

This relation is also valid for the special cases when $\omega_F=0$ and $g_F=0$. A proof of this relation is provided in Appendix III of Ref. [10].

4.2.1.2 Solutions of the Bioheat in the Infinite and Semi-infinite Domains

It is reasonable to consider that the flow of heat at locations very far from the surface (far away from $x=0$), are completely uninfluenced by the effects of the surface boundary conditions at $x=0$. In such cases, the physical domain is sometimes modeled as semi-infinite.

In general, problems existing in the semi-infinite domain are well suited to be handled using the Laplace transform technique. However, in some instances, the inverse transformations are not easily found. For this reason the SOV approach is used here to solve the bioheat equation in the semi-infinite domain.

It was shown in Example 4.1 that for the bioheat problem in a slab, the perfusion term may be explicitly linked to the transient solution component so that the spatial solution components of the bioheat problem are identical to those of the heat conduction problem. This is also true in the infinite and semi-infinite domains. Therefore, multiplying the homogenous solution of the heat conduction problem by $e^{-\alpha\omega^2 t}$ will result in the solution to the corresponding bioheat problem in the semi-infinite and infinite domains. This statement is reinforced through an example of the bioheat equation set in the semi-infinite domain. Although this example is set in the Cartesian coordinate system, the results are also valid in the cylindrical and spherical coordinate systems.

EXAMPLE 4.2 SEMI-INFINITE DOMAIN BC1 TYPE I

Consider the homogenous problem of the bioheat equation in the absence of energy generation with homogenous boundary conditions of the first kind at $x=0$ and an initial condition that is, for this example, not specified and is represented by a spatially arbitrary function. The governing equation is arranged in such a way that the spatial term is isolated and the transient derivative is grouped with the perfusion term:

$$\frac{\partial^2 T}{\partial x^2} = \frac{1}{\alpha}\frac{\partial T}{\partial t} + \omega^2 T \quad \text{in } [0<x<\infty], \quad t>0$$

$$\text{BC1:} \quad T(x=0,t)=0 \tag{4.48}$$

$$\text{IC:} \quad T(x,t=0)=F(x)$$

While not explicitly stated as a boundary condition, it should be noted that the temperature and the heat flux are finite as $x \to \infty$.

The solution to the bioheat problem described by Equation (4.48) may be found using the SOV method. The solution development shown here will mirror that of the corresponding heat conduction solution developed in Example 6.1 of the book [9] on pages 240–244. Again, it will be shown that the solution to the transient homogenous bioheat equation is equal to that of the corresponding heat conduction problem multiplied by the transient exponential decay factor: $e^{-\alpha\omega^2 t}$.

Following the SOV method, the temperature is represented by the product of a spatially dependent variable and a time dependent variable:

$$T(x,t) = X(x) \cdot \Gamma(t) \tag{4.49}$$

so that the governing equation may be represented as:

$$\frac{1}{\alpha\Gamma}\frac{d\Gamma}{dt} + \omega^2 = \frac{1}{X}\frac{d^2 X}{dx^2} = -\lambda^2 \tag{4.50}$$

Again, the transient component is represented by the ODE:

$$\frac{d\Gamma}{dt} = -\Gamma\alpha(\lambda^2 + \omega^2) \quad \text{for } t<0 \tag{4.51}$$

which has the general solution:

$$\Gamma(t) = c \exp\left(-\alpha(\lambda^2 + \omega^2)t\right) \tag{4.52}$$

where the constant of integration c depends on λ.

The spatially dependent component is governed by:

$$\frac{d^2 X}{dx^2} = -X\lambda^2 \tag{4.53}$$

$$\text{BC1:} \quad X(0) = 0 \quad \text{and} \quad x \to \infty, X \text{ and } \frac{dX}{dx} \text{ are finite} \tag{4.54}$$

This has the general solution:

$$X(x) = a \sin(\lambda x) + b \cos(\lambda x) \tag{4.55}$$

where the constants a and b might depend on λ.

BC1 shows that $b = 0$, so that:

$$X(x) = a \sin(\lambda x) \tag{4.56}$$

The temperature may be found by integrating over all λ:

$$T(x, t) = \int_{\lambda=0}^{\infty} C(\lambda) \sin(\lambda x) \exp\left(-\alpha(\lambda^2 + \omega^2)t\right) d\lambda \tag{4.57}$$

where the constant is represented by $C(\lambda) = a \cdot c$.

Applying the initial condition results in:

$$F(x) = \int_{\lambda=0}^{\infty} C(\lambda) \sin(\lambda x) d\lambda \tag{4.58}$$

This is a Fourier sine integral representation of the initial condition. In such a representation, the constant is identified by the relation:

$$C(\lambda) = \frac{2}{\pi} \int_{x'=0}^{\infty} F(x') \sin(\lambda x') dx' \tag{4.59}$$

Substituting this into Equation (4.58), the temperature may be represented by

$$T(x, t) = \frac{2}{\pi} \int_{x'=0}^{\infty} \left[F(x') \cdot \int_{\lambda=0}^{\infty} \sin(\lambda x') \sin(\lambda x) \exp\left(-\alpha(\lambda^2 + \omega^2)t\right) d\lambda dx' \right]. \tag{4.60}$$

The integral over λ may be evaluated following the description provided in Example 6.1 of the book [9] which employs the relation:

$$2\sin(\lambda x)\sin(\lambda x') = \cos(\lambda(x - x')) - \cos(\lambda(x + x')) \tag{4.61}$$

and the evaluation of the integral:

$$\int_{\lambda=0}^{\infty} \exp\left(-\alpha\lambda^2 t\right) \cos(\lambda(x \pm x')) d\lambda = \sqrt{\frac{\pi}{4\alpha t}} \exp\left(-\frac{(x \pm x')^2}{4\alpha t}\right) \tag{4.62}$$

The temperature may be represented by

$$T(x, t) = \frac{\exp(-\alpha\omega^2 t)}{\sqrt{4\pi\alpha t}} \int_{x'=0}^{\infty} \left[F(x') \cdot \left(\exp\left(-\frac{(x - x')^2}{4\alpha t}\right) - \exp\left(-\frac{(x + x')^2}{4\alpha t}\right) \right) \right] dx'. \tag{4.63}$$

This is identical to the product of the exponential decay term $e^{-\alpha\omega^2 t}$ and the corresponding solution of pure conduction provided in Equation (6.35) on page 241 of Ref. [9].

Without further derivation, we provide the solutions to the bioheat equation in the infinite and the semi-infinite domains using the existing referenced solutions of the heat conduction problem.

SEMI-INFINITE DOMAIN BC1 TYPE II

$$\frac{\partial^2 T}{\partial x^2} = \frac{1}{\alpha}\frac{\partial T}{\partial t} + \omega^2 T \quad \text{in } [0 < x < \infty], \quad t > 0$$

$$\text{BC1: } \frac{\partial T}{\partial x}\bigg|_{x=0} = 0 \quad \text{IC: } T(x, t = 0) = F(x) \tag{4.64}$$

$$T(x, t) = \frac{\exp(-\alpha\omega^2 t)}{\sqrt{4\pi\alpha t}} \int_{x'=0}^{\infty} \left[F(x') \cdot \left(\exp\left(-\frac{(x-x')^2}{4\alpha t}\right) + \exp\left(-\frac{(x+x')^2}{4\alpha t}\right) \right) \right] dx'$$

Which has been modified from the corresponding solution of pure heat conduction provided in Equation (6.71) on page 247 of Ref. [9].

SEMI-INFINITE DOMAIN BC1 TYPE III

$$\frac{\partial^2 T}{\partial x^2} = \frac{1}{\alpha}\frac{\partial T}{\partial t} + \omega^2 T \quad \text{in } [0 < x < \infty], \quad t > 0$$

$$\text{BC1: } -k\frac{\partial T}{\partial x}\bigg|_{x=0} = -hT \quad \text{IC: } T(x, t = 0) = F(x)$$

$$T(x, t) = \exp(-\alpha\omega^2 t) \times \left(\begin{array}{c} \dfrac{1}{\sqrt{4\pi\alpha t}} \displaystyle\int_{x'=0}^{\infty} \left[F(x') \cdot \left(\exp\left(-\dfrac{(x-x')^2}{4\alpha t}\right) + \exp\left(-\dfrac{(x+x')^2}{4\alpha t}\right) \right) \right] dx' \\[3ex] -\dfrac{h}{k} \displaystyle\int_{x'=0}^{\infty} \left[F(x') \cdot \exp\left(\dfrac{h}{k}(x+x') + \alpha\dfrac{h^2}{k^2}t\right) \cdot \text{erfc}\left(\dfrac{(x+x')}{\sqrt{4\pi\alpha t}} + \dfrac{h}{k}\sqrt{\alpha t}\right) \right] dx' \end{array} \right)$$

$$\tag{4.65}$$

Which has been modified from the corresponding Green's function of pure heat conduction provided in Table 6.1 on page 143 of Ref. [8].

INFINITE DOMAIN

$$\frac{\partial^2 T}{\partial x^2} = \frac{1}{\alpha}\frac{\partial T}{\partial t} + \omega^2 T \quad \text{in } [-\infty < x < \infty], \quad t > 0$$

$$\text{IC: } T(x, t = 0) = F(x) \tag{4.66}$$

$$T(x, t) = \frac{\exp(-\alpha\omega^2 t)}{\sqrt{4\pi\alpha t}} \cdot \int_{x'=-\infty}^{\infty} F(x') \cdot \exp\left(-\frac{(x-x')}{4\alpha t}\right) dx'$$

Which has been amended from the corresponding solution of pure heat conduction provided in Equation (6-132) on page 256 of Ref. [9].

4.2.2 Bioheat Equation in the Cylindrical and Spherical Coordinate Systems

The adaptation of existing transient solutions of the homogenous heat conduction equation to the bioheat equation is equally valid in the cylindrical and spherical coordinate systems. Again, one must simply multiply the corresponding heat conduction solution by the perfusion related transient term: $e^{-\alpha\omega^2 t}$. In the following discussion, some such solutions to the one dimensional transient bioheat problems are provided by altering the published solutions of the corresponding heat conduction problems.

Consider that, in general, one could anticipate an initial condition that has some arbitrary radial dependency: $F(r)$. In order to keep the discussion as general as possible, this dependency is not specified here. In either the spherical or the cylindrical coordinate systems it is reasonable to assume that the temperature at the center of the system is finite: that is $r \to 0$, $T(r,t) \neq \pm\infty$. In this chapter, the unrealistic scenarios involving either a cylindrical shell or spherical shell with two boundaries are not considered. Instead we consider a solid cylindrical region or a solid spherical region that is set within the perfuse tissue and centered about a heated region. Such a geometry is depicted in Figure 4.2 and is applicable in cases for which a region of tissue within the body is targeted with a heating source such as in the radio frequency ablation of tumors. At some location far away from the center, it is reasonable to expect the tissue to be at thermal equilibrium with the core. For discrete problems discussed in the previous section, this thermal equilibrium is represented as either a homogenous Type I BC or a homogenous Type II BC. In the infinite and semi-infinite domains, it is the initial state of the system that determines this equilibrium. Following are solutions to the bioheat equation in cylindrical and spherical coordinates which have been adapted from the published forms of the corresponding heat conduction equation.

4.2.2.1 Bioheat in Cylindrical Coordinates

Consider the bioheat problem in cylindrical coordinates subjected to the following constraints:

$$\frac{1}{\alpha}\frac{\partial T}{\partial t} = \frac{1}{r}\frac{\partial}{\partial r}\left(r\frac{\partial T}{\partial r}\right) - \omega^2 T \quad \text{in} [0 < r < R], \quad t > 0$$

$$T(r = 0, t) \Rightarrow \text{finite} \quad \text{IC:} \quad T(r,0) = F(r)$$

(4.67)

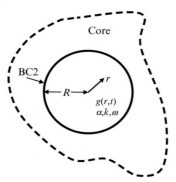

FIGURE 4.2   Radial bioheat transfer in either the cylindrical or spherical coordinate systems in which the system is in contact with the body's core at $r = R$.

At some location far away from the center, thermal equilibrium with the body's core may appear as either a homogenous Type I or Type II BC. The corresponding solutions follow.

4.2.2.1.1 CYLINDER WITH A HOMOGENOUS TYPE I BC2

The solution of the bioheat problem (4.67) with a Type I representation of thermal equilibrium at the outer boundary may be determined by modifying the corresponding heat conduction solution provided in Ref. [8] in Equation (7.41) on page 211. The solution is

$$BC2\,Type\,I: \quad T(r=R,t)=0$$

$$T(r,t)=\sum_{m=1}^{\infty}C_n\exp\left(-\alpha t\left(\lambda_n+\omega^2\right)\right)J_0(\lambda_n r)$$

$$C_n=\frac{2}{R^2}\frac{1}{J_1^2(\lambda_n R)}\int_{r'=0}^{R}r'F(r')J_0(\lambda_n r')dr'$$

$$\lambda_n \text{ are the roots of: } J_0(\lambda_n R)=0: n=1,2,3\ldots$$

(4.68)

Here, J_0 and J_1 are the zeroth and first order Bessel functions of the first kind and the symbol r' is a dummy variable over which the integration is performed.

4.2.2.1.2 CYLINDER WITH A HOMOGENOUS TYPE II BC2

When a zero heat flux is imposed at the outer boundary, the corresponding heat conduction solution provided in Equation (4.152) on page 153 of Ref. [9] may be adapted to state:

$$BC2\,Type\,II: \quad \left.\frac{\partial T}{\partial r}\right|_{r=R}=0$$

$$T(r,t)=\exp\left(-\alpha t\omega^2\right)\times\left(\begin{array}{c}\dfrac{2}{R^2}\displaystyle\int_{r'=0}^{R}r'F(r')dr'\\[2mm]+\displaystyle\sum_{n=1}^{\infty}C_nJ_0(\lambda_n r)\exp\left(-\alpha\lambda_n^2 t\right)\end{array}\right)$$

$$C_n=\frac{2}{R^2}\frac{1}{J_0^2(\lambda_n R)}\int_{r'=0}^{R}F(r')\cdot J_0(\lambda_n r')\cdot r'dr'$$

$$\lambda_n \text{ are the roots of: } J_1(\lambda_n R)=0: n=1,2,3\ldots$$

(4.69)

In the semi-infinite domain, the homogenous bioheat problem may be represented by the governing equations:

$$\frac{\partial^2 T}{\partial r^2}+\frac{1}{r}\frac{\partial T}{\partial r}=\frac{1}{\alpha}\frac{\partial T}{\partial t}+\omega^2 T \quad \text{in}\,[0<r<\infty],\;\; t>0$$

$$BC1: \quad T(r\to 0,t)\Rightarrow \text{finite;} \quad IC: \quad T(r,0)=F(r)$$

(4.70)

Altering the corresponding solution of the heat conduction equation provided in Equation (6.166) on page 261 of Ref. [9], the bioheat solution is represented by

$$T(r,t) = \frac{\exp(-a\omega^2 t)}{2at} \int_{r'=0}^{\infty} \exp\left(-\frac{(r^2+r'^2)}{4at}\right) \cdot I_0\left(\frac{rr'}{2at}\right) \cdot F(r') \cdot r' dr'. \qquad (4.71)$$

A special case is readily available as well in which the initial condition is a nonzero constant within a discrete radial distance.

$$\text{IC: } T(r,0) = F(r) = \begin{cases} F_O & \text{for } 0 < r < b \\ 0 & \text{for } r > b \end{cases}$$

$$T(r,t) = F_O \frac{\exp(-a\omega^2 t)}{2at} \int_{r'=0}^{b} r' \cdot \exp\left(-\frac{(r^2+r'^2)}{4at}\right) \cdot I_0\left(\frac{rr'}{2at}\right) dr' \qquad (4.72)$$

This integral must be evaluated numerically.

4.2.2.2 Bioheat in Spherical Coordinates

Next the bioheat equation in the spherical system is considered. This problem may be represented by

$$\frac{1}{\alpha}\frac{\partial T}{\partial t} = \frac{\partial^2 T}{\partial r^2} + \frac{2}{r}\frac{\partial T}{\partial r} - \omega^2 T \quad \text{in } [0 < r < R], \quad t > 0$$

$$T(r=0,t) \Rightarrow \text{finite} \qquad \text{IC's: } T(r,0) = F(r) \qquad (4.73)$$

4.2.2.2.1 SPHERE WITH A HOMOGENOUS TYPE I BC2

When the outer boundary condition is homogenous and of Type I, the solution provided in the book [8] in Equation (9.63) on page 269 may be adapted to state:

$$\text{BC2 Type I: } T(r=R,t) = 0$$

$$T(r,t) = \sum_{n=1}^{\infty} C_n \frac{\sin(\lambda_n r)}{r} \exp\left(-\alpha(\lambda_n^2 + \omega^2)t\right)$$

$$C_n = \frac{2}{R} \int_{r'=0}^{R} \sin(\lambda_n r') \cdot F(r') \cdot r' dr' \qquad (4.74)$$

$$\text{for which } \lambda_n = \frac{n\pi}{R}, \quad n = 1,2,3\ldots$$

In the special case of the uniform initial temperature $F(r) = F_O$, this solution is

$$T(r,t) = 2F_O \sum_{n=1}^{\infty} (-1)^{n+1} \frac{\sin(\lambda_n r)}{\lambda_n r} \exp\left(-\alpha(\lambda_n^2 + \omega^2)t\right). \qquad (4.75)$$

4.2.2.2.2 SPHERE WITH A HOMOGENOUS TYPE II BC2

For the case of a zero heat flux representation of thermal equilibrium at the outer boundary, the corresponding heat condition solution provided by [9] on page 199 in Equations (5.80)–(5.88) may be adapted so that:

$$\text{BC2 Type II}: \left. \frac{\partial T}{\partial r} \right|_{r=R} = 0$$

$$T(r,t) = \exp\left(-\alpha t \omega^2\right) \left[\begin{array}{c} \dfrac{3}{R^3} \displaystyle\int_{r'=0}^{R} F(r') r'^2 \mathrm{d}r \\[2mm] + \displaystyle\sum_{n=1}^{\infty} C_n \dfrac{\sin(\lambda_n r)}{r} \exp\left(-\alpha \lambda_n^2 t\right) \end{array} \right] \tag{4.76}$$

$$C_n = \frac{2\left(\lambda_n^2 + \dfrac{1}{R^2}\right)}{R\left(\lambda_n^2 + \dfrac{1}{R^2}\right) - \dfrac{1}{R}} \times \int_{r'=0}^{R} \sin(\lambda_n r') \cdot F(r') \cdot r' \mathrm{d}r'$$

λ_n are the roots of: $R\lambda_n \cot(\lambda_n R) - 1 = 0 : \quad n = 1,2,3\ldots$

4.2.2.2.3 SPHERE IN THE SEMI-INFINITE DOMAIN

The homogenous heat conduction problem in the semi-infinite spherical domain is discussed in Example 6-12 on pages 268–269 of the book [9]. The corresponding bioheat problem is represented as:

$$\frac{1}{r}\frac{\partial^2 T}{\partial r^2} = \frac{1}{\alpha}\frac{\partial T}{\partial t} + \omega^2 T \quad \text{in}\,[0 < r < \infty], \quad t > 0$$

$$T(r \to 0, t) \Rightarrow \text{finite} \quad \text{IC}: \quad T(r,0) = F(r) \tag{4.77}$$

By adapting the solution provided in Equation (6-221) of Ref. [9], the solution to Equation (4.77) is represented as:

$$T(r,t) = \frac{\exp\left(-\alpha \omega^2 t\right)}{r\sqrt{4\alpha t}} \int_{r'=0}^{\infty} F(r') \cdot \left(\exp\left(-\frac{(r-r')^2}{4\alpha t}\right) - \exp\left(\frac{(r+r')^2}{4\alpha t}\right) \right) \cdot r' \mathrm{d}r' \tag{4.78}$$

In the previous discussion, the solutions to the homogenous bioheat problem have been addressed by the SOV method. The overall outcome has been to show that an existing heat conduction solution may be modified to fit the solution of the homogenous bioheat problem. Special care should be taken to stress that this approach is not correct for the case of nonhomogenous bioheat transfer. This is addressed next.

4.3 SOLUTION APPROACHES TO NONHOMOGENOUS PROBLEMS

The SOV method requires that the problem and all boundary conditions are homogenous. This does result in an inherent limitation to the solutions provided thus far. In practical applications, it is reasonable to expect the presence of nonnegligible heat generation. Intense heat fluxes at the surface are associated with laser treatments, and high convective heat transfer at the interface between tissue and fluid results from treatments using a hot/cold water bath. Such systems would be better represented by nonhomogenous problems.

In this section, two methods of evaluating nonhomogenous solutions of the bioheat equation are discussed: superpositioning and the use of Green's functions. Superpositioning is well suited for simple nonhomogenous problems for which all of the nonhomogenous terms are steady. Its primary benefit is that the easily available solutions to the homogenous problem can be used directly. The use of Green's functions can be a bit more mathematically involved. However, in general, the method using Green's functions is much more adaptable and is regularly employed to handle transient nonhomogenous terms.

4.3.1 Method of Superpositioning

When the nonhomogenous terms are independent of time, the results of the analysis of the homogenous bioheat problem presented in Section **4.2** may still be used with slight modification. In Sections 3-4 of the book *Heat Conduction* [9], superpositioning is introduced in the evaluation of the nonhomogenous heat conduction problem. Superpositioning is equally applicable to the handling of the nonhomogenous bioheat problem.

The purpose of superpositioning is to separate the nonhomogenous bioheat problem into two subproblems: (i) a steady nonhomogenous problem that satisfies the nonhomogenous boundary conditions and energy generation terms and (ii) a homogenous problem that satisfies the homogenous bioheat problem. This allows the homogenous problem to be addressed using the SOV method. The superpositioning is accomplished by representing the temperature as the sum of homogenous components and nonhomogenous steady components.

$$T^{IH}(x,t) = \underbrace{\Psi(x)}_{\text{Nonhomogenous}} + \underbrace{T(x,t)}_{\text{Homogenous}} \tag{4.79}$$

Here, $T^{IH}(x,t)$ is the shifted temperature that is represented by the nonhomogenous bioheat problem, $\Psi(x)$ is a variable that represents the solution of the steady nonhomogenous problem, and $T(x,t)$ satisfies the homogenous (shifted) bioheat problem. The reader interested in further detail is strongly encouraged to consider the description in Ref. [9]. Note that technically superpositioning has already been used extensively in this chapter. The temperature shift introduced in Section *4.1.3* is a very simple application of superpositioning. A more involved application of the method of superposition to the bioheat problem is illustrated in the following simple example.

EXAMPLE 4.3 **NONHOMOGENOUS PROBLEM**

Consider the one dimensional transient bioheat problem that is initially at some uniform temperature. The tissue is exposed to uniform, steady energy generation. At the surface, the system has a specified temperature and at the boundary nearest the core, the system is in thermal equilibrium with the core so that there is no exchange of heat there. A uniform initial temperature distribution is

$$\frac{1}{\alpha}\frac{\partial T^{IH}}{\partial t}=\frac{\partial^2 T^{IH}}{\partial x^2}-\omega^2\left(T^{IH}-Ta\right)+\frac{g_O}{k}\quad \text{in}\,[0<x<L],\;\;t>0$$

$$\text{BC1:}\quad T^{IH}(x=0,t)=T_O\quad \text{BC2:}\quad \left.\frac{\partial T}{\partial x}\right|_{x=L}=0 \tag{4.80}$$

$$\text{IC:}\quad T^{IH}(x,t=0)=F_O$$

By using superpositioning, the temperature is represented as:

$$T^{IH}(x,t)=\Psi(x)+T(x,t) \tag{4.81}$$

Substituting Equation (4.81) into Equation (4.80), the original problem (4.80) may be represented by two subproblems. *All* of the nonhomogenous terms are satisfied by the problem of the steady variable:

$$\frac{d^2\Psi}{dx^2}-\omega^2(\Psi-Ta)+\frac{g_O}{k}=0\quad \text{in}\,[0<x<L],\;\;t>0$$

$$\text{BC1:}\quad \Psi(x=0,t)=T_O\quad \text{BC2:}\quad \left.\frac{d\Psi}{dx}\right|_{x=L}=0 \tag{4.82}$$

and whose solution is

$$\Psi(x)=Ta+\frac{g_O}{k\omega^2}+\left(\frac{g_O}{k\omega^2}+Ta-T_O\right)(\tanh(\omega L)\sinh(\omega x)-\cosh(\omega x)). \tag{4.83}$$

This allows the nonsteady component $T(x,t)$ to be represented by the homogenous problem:

$$\frac{1}{\alpha}\frac{\partial T}{\partial t}=\frac{\partial^2 T}{\partial x^2}-\omega^2 T\quad \text{in}\,[0<x<L],\;\;t>0$$

$$\text{BC1:}\quad T(x=0,t)=0\quad \text{BC2:}\quad \left.\frac{\partial T}{\partial x}\right|_{x=L}=0 \tag{4.84}$$

$$\text{IC:}\quad T(x,t=0)=F_0-\Psi(x)$$

Note that the initial condition of the homogenous term must now account for the existence of the nonhomogenous solution component. The homogenous problem of Equation (4.84) is identical to the homogenous bioheat problem introduced in Example 4.1 with one exception: the solution to the steady problem of Equation (4.83) is subtracted from the initial condition. The solution to the homogenous problem has been established in Equation (4.40) of Example 4.1 so that the solution may be represented as

$$T(x,t)=\sum_{n=1}^{\infty}\underbrace{\left(\frac{2}{L}\int_{x=0}^{L}(F_O-\Psi(x'))X_n(x')dx'\right)}_{C_n}X_n(x)\exp\left(-\alpha\left(\lambda_n{}^2+\omega^2\right)t\right) \tag{4.85}$$

which may be equivalently expressed:

$$T(x,t)=\frac{2}{L}\sum_{n=1}^{\infty}\left(\int_{x=0}^{L}F_O X_n dx\right)X_n\exp\left(-\alpha\left(\lambda_n^2+\omega^2\right)t\right)$$

$$-\frac{2}{L}\sum_{n=1}^{\infty}\left(\int_{x=0}^{L}\Psi(x)X_n dx\right)X_n\exp\left(-\alpha\left(\lambda_n^2+\omega^2\right)t\right) \tag{4.86}$$

for which the eigenfunction and eigenvalues are:

$$X_n = \sin(\lambda_n x)$$
$$\lambda_n = \frac{\pi(2n-1)}{2L}$$

(4.87)

The first integral of Equation (4.86) has already been evaluated in Equation (4.42) of Example 4.1. In order to evaluate the second integral, the relation of Equation (4.47) may be employed.

$$\int_{x=0}^{x=L} X_n \Psi dx = \frac{1}{(\omega^2 + \lambda_n^2)} \left(X_n \frac{d\Psi}{dx} - \Psi \frac{dX_n}{dx} - \frac{1}{\lambda_n^2} \frac{g_O}{k} \frac{dX_n}{dx} \right) \Bigg|_{x=0}^{x=L}$$

(4.88)

Note that the homogenous boundary conditions appearing in Equations (4.82) and (4.84) greatly simplify the result, so that the evaluation of this result may be simplified:

$$\int_{x=0}^{x=L} X_n \Psi dx = \frac{1}{(\omega^2 + \lambda_n^2)} \left(\Psi(0) + \frac{1}{\lambda_n^2} \frac{g_O}{k} \right) \frac{dX_n}{dx} \Bigg|_{x=0}$$

(4.89)

Evaluating the derivative of the eigenfunction, X_n, of Equation (4.87) and the steady component, $\Psi(x)$, Equation (4.83) at $x=0$, the above expression may be shown to be

$$\int_{x=0}^{x=L} X_n \Psi dx = \frac{\lambda_n}{(\omega^2 + \lambda_n^2)} \left(T_O + \frac{1}{\lambda_n^2} \frac{g_O}{k} \right).$$

(4.90)

Substituting the preceding and Equation (4.42) into Equation (4.86), the solution of the homogenous temperature component may be represented as

$$T(x,t) = -\frac{2}{L} \sum_{n=1}^{\infty} \left[\frac{F_O}{\lambda_n} + \frac{\lambda_n}{(\omega^2 + \lambda_n^2)} \left(T_O + \frac{1}{\lambda_n^2} \frac{g_O}{k} \right) \right] \sin(\lambda_n x) \exp(-\alpha(\lambda_n^2 + \omega^2)t).$$

(4.91)

It is important to point out that in the homogenous solution presented, the perfusion term now appears outside the transient exponential term. This is a consequence of the method of superposition which introduced the elements of the nonhomogenous boundary conditions into the homogenous solution through the steady function $\Psi(x)$. The final solution is composed by substituting the homogenous component of Equation (4.91) and the nonhomogenous component of Equation (4.83) into Equation (4.79) so that

$$T^{IH}(x,t) = Ta + \frac{g_O}{k\omega^2} + \left(\frac{g_O}{k\omega^2} + Ta - T_O \right) (\tanh(\omega L) \sinh(\omega x) - \cosh(\omega x))$$

$$- \frac{2}{L} \sum_{n=1}^{\infty} \left[\frac{F_O}{\lambda_n} + \frac{\lambda_n}{(\omega^2 + \lambda_n^2)} \left(T_O + \frac{1}{\lambda_n^2} \frac{g_O}{k} \right) \right] \sin(\lambda_n x) \cdot \exp(-\alpha(\lambda_n^2 + \omega^2)t)$$

$$\text{for } \lambda_n = \frac{\pi(2n-1)}{2L}.$$

(4.92)

4.3.2 Green's Functions

The method of superposition allows available solutions of the homogenous bioheat equation to be incorporated into problems that are nonhomogenous. Although this approach is convenient, it is limited to situations in which the nonhomogenous terms are steady. In practice, it is often of interest to predict how a system will react to transient sources of thermal energy. That is the motivation behind the use of the solution method using Green's functions.

In this section, it is shown that the published Green's functions associated with the heat conduction problem may be manipulated in order to arrive at the Green's function of the corresponding bioheat problem. In the following discussion it is shown that:

> The Green's functions associated with the transient bioheat problem may be adopted from those of the corresponding transient heat conduction problem by multiplying the Green's function of the conduction problem by the term $e^{-\alpha\omega^2(t-\tau)}$.

The application of Green's functions to problems of heat conduction is thoroughly developed in the book: *Heat Conduction Using Green's Functions* by Beck et al. [8]. The unfamiliar reader is encouraged to investigate this excellent book. In that book, the Green's functions of heat conduction problems are derived for heat conduction problems of various boundary condition types. Here, we will consider only Green's functions associated with combinations of BC Types I–III by using the general expression.

$$k_i^* \frac{\partial T}{\partial n_i} + h_i^* T = f_i(t) \tag{4.93}$$

where the temperature and its gradient are evaluated on surface "*i*." The partial derivative $\partial/\partial n_i$ represents the spatial gradient on surface "*i*" that is oriented normal to this surface and in the outward direction. This general description is a simple way to describe all three boundary condition types simultaneously. For example, to represent a Type I BC on surface 1, one would represent Equation (4.93) using the constants: $k_1^* = 0, h_1^* = 1$, and $f_1(t) = T_0(t)$. In the homogenous problem $f_i(t) = 0$ on all boundaries.

In Section 3.3.2 of the book [9] pages 47–52, a general expression of the Green's function solution equation (GFSE) is derived for governing equations which are identical to those of the bioheat problem. The result is that the solution is presented as the linear sum of the influences of the initial and nonhomogenous terms. The following GFSE is equally applicable to relate the Green's function to the solution for the bioheat problem and for the heat conduction problem.

$$
\begin{aligned}
T(x,t) = &\int_R G(\hat{r},t|\hat{r}',0) F(\hat{r}') dv' \quad \text{for the initial condition} \\
&+ \int_{\tau=0}^t \int_R \frac{\alpha}{k} G(\hat{r},t|\hat{r}',\tau) g(\hat{r}',\tau) dv' d\tau \quad \text{for energy generation} \\
&+ \int_{\tau=0}^t \sum_{i=1}^s \int_{S_i} \frac{\alpha}{k_i} f_i(\hat{r}_i,\tau) G(\hat{r},t|\hat{r}_i,\tau) ds_i' d\tau \quad \text{for BC Type II or III on surface } S_i \\
&- \int_{\tau=0}^t \sum_{j=1}^s \int_{S_j} \alpha f_j(\hat{r}_j,\tau) \frac{G}{\partial n_j'}\bigg|_{\hat{r}=\hat{r}_j'} ds_j' d\tau \quad \text{for BC Type I on surface } S_j
\end{aligned} \tag{4.94}
$$

Where $\hat{r}$ is the spatial coordinate vector, dv' is the differential volume element, and ds' is the differential surface element. The parameter S_i refers to surface "*i*."

In the following example it will be shown how the solutions to the homogenous problems that were developed in Section **4.2** can be used to derive the corresponding Green's function.

EXAMPLE 4.4 GREEN'S FUNCTION DETERMINATION

Consider the one dimensional bioheat problem with homogenous boundary conditions of Type I at $x=0$ and of Type II at $x=L$ with an unspecified initial condition:

$$\frac{1}{\alpha}\frac{\partial T}{\partial t}=\frac{\partial^2 T}{\partial x^2}-\omega^2 T \quad \text{in}\,[0<x<L], \quad t>0$$

$$\text{BC1:}\quad T(x=0,t)=0 \quad \text{BC2:}\quad \left.\frac{\partial T}{\partial x}\right|_{x=L}=0 \tag{4.95}$$

$$\text{IC:}\quad T(x,t=0)=F(x)$$

Note in the presentation of the problem described by Equation (4.95), that the temperature has already been shifted by the arterial blood temperature so that the perfusion term is homogenous. Using the SOV method, the solution to this problem was shown in Example 4.1 to be

$$T(x,t)=\sum_{n=1}^{\infty}\frac{2}{L}\left(\int_{x'=0}^{L}F(x')\sin(\lambda_n x')\mathrm{d}x'\right)\cdot \sin(\lambda_n x)\cdot \exp(-\alpha(\lambda_n^2+\omega^2)t);$$

$$\text{for } \lambda_n=\frac{(2n-1)\pi}{2L}. \tag{4.96}$$

Next consider the one dimensional Cartesian representation of the GFSE (4.94) with a BC of Type I at the surface and a BC of Type II at the core:

$$T(x,t)=\int_{x'=0}^{L}G^P(x,t|x',0)F(x')\mathrm{d}x' \quad \text{for the initial condition}$$

$$+\int_{\tau=0}^{t}\int_{x'=0}^{L}\frac{\alpha}{k}G^P(x,t|x',\tau)g(x',\tau)\mathrm{d}x'\mathrm{d}\tau \quad \text{for energy generation}$$

$$+\int_{\tau=0}^{t}\alpha f_1(\tau)\frac{G^P}{\partial x'}\bigg|_{x'=0}\mathrm{d}\tau \quad \text{for BC Type I at } x=0 \tag{4.97}$$

$$+\int_{\tau=0}^{t}\frac{\alpha}{k}f_2(\tau)G^P(x,t|L,\tau)\mathrm{d}\tau \quad \text{for BC Type II at } x=L$$

Because this problem is homogenous with no volumetric generation ($f_1=f_2=g=0$), the GFSE of Equation (4.97) reduces to:

$$T(x,t)=\int_{x'=0}^{L}G^P(x,t|x',0)F(x')\mathrm{d}x' \tag{4.98}$$

Comparing the two solutions (4.96) and (4.98), it can be shown that the Green's function for this problem evaluated at $\tau=0$ is

$$G^P(x,t|x',0)=\sum_{n=1}^{\infty}\frac{2}{L}\sin(\lambda_n x')\sin(\lambda_n x)\exp(-\alpha(\lambda_n^2+\omega^2)t). \tag{4.99}$$

The Green's function evaluated at $\tau\neq 0$ can be found by replacing $(t-0)$ in Equation (4.99) with $(t-\tau)$:

$$G^P(x,t|x',\tau)=\sum_{n=1}^{\infty}\frac{2}{L}\sin(\lambda_n x')\sin(\lambda_n x)\exp(-\alpha(\lambda_n^2+\omega^2)(t-\tau)) \tag{4.100}$$

4.3.2.1 *Transformation of the Conduction Green's Function to the Bioheat Problem*

The procedure used in Example 4.4 of evaluating Green's functions from the homogenous solution of heat conduction problems is detailed in the book [8]. In fact, that text provides a clear and organized compendium of Green's functions associated with the heat conduction problem in the Cartesian, cylindrical, and spherical coordinate systems for finite, semi-infinite, and infinite domains, and considers boundary condition Types I–V.

Because Green's functions for the heat conduction problem are readily available, it is convenient to adapt the Green's functions associated with the heat conduction problem in order to arrive at the GF of the corresponding bioheat problem.

Consider that the Green's function associated with a particular heat conduction problem may be derived from the corresponding homogenous conduction solution in a manner outlined in Example 4.4.

$$T^{HC}(\hat{r}, t) = \int_R G(\hat{r}, t|\hat{r}', 0)F(\hat{r}')dv' \tag{4.101}$$

where T^{HC} is the solution to the homogenous heat conduction problem in spatial domain R and $\hat{r}$ is a position vector. Recall Equation (4.15) which states that for transient homogenous problems, the solution to the bioheat equation is equal to that of the corresponding heat conduction problem multiplied by the term $e^{-\alpha\omega^2 t}$. Multiplying both sides of Equation (4.101) by this term results in the expression for the solution to the homogenous bioheat problem:

$$T(\hat{r}, t) = T^{HC}(\hat{r}, t)\exp(-\alpha\omega^2 t) = \int_R \underbrace{G(\hat{r}, t|\hat{r}', 0)\exp(-\alpha\omega^2 t)}_{G^P} F(\hat{r}')dv' \tag{4.102}$$

where $T(x, t)$ is the solution to the homogenous bioheat problem. So that the Green's function of the bioheat problem evaluated at $\tau = 0$ is the bracketed product in the integral of the heat conduction Green's function and the term $e^{-\alpha\omega^2(t-\tau)}$. For values of the time $\tau \neq 0$, the relationship between the Green's functions of the bioheat problem, G^P, and of the heat conduction problem G is:

$$G^P(\hat{r}, t|\hat{r}', \tau) = G(\hat{r}, t|\hat{r}', \tau)\exp(-\alpha\omega^2(t-\tau)) \tag{4.103}$$

This result is irrespective of geometry or boundary condition type. By using this relation, the Green's functions of various bioheat problems may be adapted from the corresponding heat conduction problems described in Ref. [8].

4.3.2.2 *Selected Green's Functions in the Cartesian Coordinate System*

The one dimensional nonhomogenous bioheat problem in the Cartesian coordinate system may be represented by the general problem:

$$\frac{1}{\alpha}\frac{\partial T}{\partial t} = \frac{\partial^2 T}{\partial x^2} - \omega^2 T + g(x, t)$$

BC1 $(x = 0)$

$$-k_1^*\frac{\partial T}{\partial x}\bigg| + h_1^* T = h_1^* f_1(t) \tag{4.104}$$

BC2 $(x = L)$

$$k_2^*\frac{\partial T}{\partial x}\bigg| + h_2^* T = h_2^* f_2(t)$$

IC: $T(x, t = 0) = F(x)$

Here, the boundary conditions are interpreted depending on type by the expression:

$$\begin{aligned}
\text{Type I}: \quad & h_i^* = 1, k_i^* = 0, f_i(t) = T_O(t) \\
\text{Type II}: \quad & h_i^* = 0, k_i^* = k, f_i(t) = q_O''(t) \\
\text{Type III}: & h_1^* = h, k_1^* = k, f_1(t) = hT_\infty(t)
\end{aligned} \Bigg\} \begin{aligned} x &= 0 : i = 1 \\ x &= L : i = 2 \end{aligned} \tag{4.105}$$

In the special case that the system is in thermal equilibrium at the boundary closest to the core, BC2 is homogenous so that $f_2(t) = 0$. The GFSE to the one dimensional bioheat problem may be expressed:

$$\begin{aligned}
T(x,t) = & \int_{x'=0}^{L} G^P(x,t|x',0) F(x') dx' \quad \text{for the initial condition} \\
& + \int_{\tau=0}^{t} \int_{x'=0}^{L} \frac{\alpha}{k} G^P(x,t|x',\tau) g(x',\tau) dx' d\tau \quad \text{for energy generation} \\
& + \int_{\tau=0}^{t} \alpha T_O(\tau) \frac{G^P}{\partial x'} \bigg|_{x'=0} d\tau \quad \text{if BC1 at } x = 0 \text{ is Type I} \\
& + \int_{\tau=0}^{t} \frac{\alpha}{k} q_O''(\tau) G^P(x,t|0,\tau) d\tau \quad \text{if BC1 at } x = 0 \text{ is Type II} \\
& + \int_{\tau=0}^{t} \frac{\alpha}{k} hT_\infty(\tau) G^P(x,t|0,\tau) d\tau \quad \text{if BC1 at } x = 0 \text{ is Type III}
\end{aligned} \tag{4.106}$$

Once the applicable Green's function has been identified, the GFSE can be used to evaluate the solution.

The Green's functions of the bioheat problem in the finite Cartesian coordinate system can be adopted from those of the corresponding heat conduction problem provided in Tables 4.2 and 4.3 on page 99 of the book [8] by multiplying by the term $e^{-\alpha\omega^2(t-\tau)}$ so that in $0 < x < L \quad 0 \le t$:

$$G^P(x,t|x',\tau) = \frac{C_O}{L} \cdot \exp\left(-\alpha\omega^2(t-\tau)\right) + \sum_{n=1}^{\infty} \exp\left(-\alpha\left(\lambda_n^2 + \omega^2\right)(t-\tau)\right) \frac{X_n(x) \cdot X_n(x')}{N_n} \tag{4.107}$$

The constant C_O, the eigenfunction X_n, the eigenvalues λ_n^2, and the norm N_n depend on BC type and may be taken from Table 4.1.

4.3.2.3 Semi-infinite and Infinite Cartesian Domains

The semi-infinite representation of the GFSE (4.106) differs from the discrete version in that the upper limit on the integral of the spatial coordinate, x' is evaluated at $+\infty$ instead of L. In the infinite and semi-infinite domains, the heat conduction Green's functions provided in Table 6.1 on page 143 of Ref. [8] may be modified so that they correspond to the Green's functions of the bioheat problem. These are presented according to BC1 type at $x = 0$.

BC1 Type I:

$$G^P(x,t|x',\tau) = \frac{\exp\left(-\alpha\omega^2(t-\tau)\right)}{\sqrt{4\pi\alpha(t-\tau)}} \left(\exp\left(\frac{-(x-x')^2}{4\alpha(t-\tau)}\right) - \exp\left(\frac{-(x+x')^2}{4\alpha(t-\tau)}\right) \right) \tag{4.108}$$

BC1 Type II:

$$G^P(x,t|x',\tau) = \frac{\exp(-\alpha\omega^2(t-\tau))}{\sqrt{4\pi\alpha(t-\tau)}}\left(\exp\left(\frac{-(x-x')^2}{4\alpha(t-\tau)}\right) + \exp\left(\frac{-(x+x')^2}{4\alpha(t-\tau)}\right)\right) \quad (4.109)$$

BC1 Type III:

$$\begin{aligned}G^P(x,t|x',\tau) &= \frac{\exp(-\alpha\omega^2(t-\tau))}{\sqrt{4\pi\alpha(t-\tau)}}\left(\exp\left(\frac{-(x-x')^2}{4\alpha(t-\tau)}\right) + \exp\left(\frac{-(x+x')^2}{4\alpha(t-\tau)}\right)\right)\\ &\quad - H\exp\left(H(x+x') + \alpha(H^2-\omega^2)(t-\tau)\right)erfc\left(\frac{(x+x')}{\sqrt{4\alpha(t-\tau)}} + H\sqrt{\alpha(t-\tau)}\right)\end{aligned} \quad (4.110)$$

where $H = h/k$.

In the infinite domain there are no boundary conditions. This means that in the evaluation of the GFSE (4.106), $f_1 = f_2 = 0$ and the only contributions to be considered are the terms associated with the volumetric heating with the initial condition. Furthermore, it is important to note that in the GFSE, the upper and lower bounds of integration of the spatial coordinate x' are $\pm\infty$.

The Green's function in $[-\infty < x < \infty]$, $0 \leq t$ is

$$G^P(x,t|x',\tau) = \frac{\exp(-\alpha\omega^2(t-\tau))}{\sqrt{4\pi\alpha(t-\tau)}}\exp\left(\frac{-(x-x')^2}{4\alpha(t-\tau)}\right). \quad (4.111)$$

EXAMPLE 4.5 **TRANSIENT TYPE II BC**

In this example, the ability of Green's functions to handle transient boundary conditions is illustrated. Consider the nonhomogenous bioheat problem with a uniform generation term, g_O. The system is exposed to a transient heat flux at the outer surface that can be approximated by an exponentially decaying function: $q_O'' \cdot e^{-pt}$ where p is a positive constant. Far from the surface (at $x = L$), the temperature is in equilibrium with the core. The system is initially at zero temperature.

$$\frac{1}{\alpha}\frac{\partial T}{\partial t} = \frac{\partial^2 T}{\partial x^2} - \omega^2 T + g_O \quad \text{in } [0 < x < L], \ t > 0$$

$$\text{BC1: } -k\frac{\partial T}{\partial x}\Big|_{x=0} = q_O'' \cdot \exp(-pt) \quad \text{BC2: } T(L,t) = 0 \quad (4.112)$$

$$\text{IC: } T(x,t=0) = 0$$

The transient temperature solution of this problem can be determined from the corresponding GFSE:

$$\begin{aligned}T(x,t) &= \int_{\tau=0}^{t}\int_{x'=0}^{L}\frac{\alpha}{k}G^P(x,t|x',\tau)g_O dx' d\tau \quad \text{for energy generation}\\ &\quad + \int_{\tau=0}^{t}\frac{\alpha}{k}G^P(x,t|0,\tau)q_O''\exp(-pt)d\tau \quad \text{for BC Type II at } x = 0\end{aligned} \quad (4.113)$$

The Green's function for this problem can be found in Equation (4.107) and Table 4.1:

$$G^P(x,t|x',\tau) = \frac{2}{L}\sum_{n=1}^{\infty} \cos(\lambda_n x) \cdot \cos(\lambda_n x') \cdot \exp\left(-\alpha(\lambda_n^2 + \omega^2)(t-\tau)\right)$$

(4.114)

$$\lambda_n = \frac{(2n-1)\pi}{2L}$$

In order to determine the energy generation contribution, the Green's function is substituted into the first integral and evaluated:

$$\frac{\alpha 2}{kL} g_O \sum_{n=1}^{\infty} \cos(\lambda_n x) \cdot \int_{\tau=0}^{t} \int_{x'=0}^{L} \cos(\lambda_n x') \exp\left(-\alpha(\lambda_n^2 + \omega^2)(t-\tau)\right) dx' d\tau$$

$$= \frac{\alpha 2}{kL} g_O \sum_{n=1}^{\infty} \cos(\lambda_n x) \cdot \frac{\sin(\lambda_n L)}{\alpha(\lambda_n^2 + \omega^2)\lambda_n} \cdot \left(1 - \exp\left(-\alpha(\lambda_n^2 + \omega^2)t\right)\right)$$

(4.115)

The contribution from the transient boundary condition is determined by substituting the Green's function evaluated at $x' = 0$ into the second integral and evaluating:

$$\frac{\alpha 2}{kL} q_O'' \sum_{n=1}^{\infty} \cos(\lambda_n x) \int_{\tau=0}^{t} \exp(-p\tau)\exp\left(-\alpha(\lambda_n^2 + \omega^2)(t-\tau)\right) d\tau$$

$$= \frac{\alpha 2}{kL} q_O'' \sum_{n=1}^{\infty} \cos(\lambda_n x) \frac{\exp(-pt) - \exp\left(-\alpha(\lambda_n^2 + \omega^2)(t-\tau)\right)}{\alpha(\lambda_n^2 + \omega^2) - p}$$

(4.116)

The solution may then be represented as the sum of the contributions of the generation term and of the transient boundary condition:

$$T(x,t) = g_O \frac{12}{kL} \sum_{n=1}^{\infty} \frac{\sin(\lambda_n L)}{(\lambda_n^2 + \omega^2)\lambda_n} \cdot \cos(\lambda_n x) \cdot \left(1 - \exp\left(-\alpha(\lambda_n^2 + \omega^2)t\right)\right)$$

$$+ q_O'' \frac{12}{kL} \sum_{n=1}^{\infty} \frac{1}{(\lambda_n^2 + \omega^2) - \dfrac{p}{\alpha}} \cdot \cos(\lambda_n x) \cdot \left(\exp(-pt) - \exp\left(-\alpha(\lambda_n^2 + \omega^2)t\right)\right)$$

(4.117)

4.3.2.4 Selected Green's Functions in the Cylindrical and Spherical Coordinate Systems

The most realistic applications in the cylindrical and spherical coordinate systems are representative of bioheat transfer that is entirely internal: cases in which there are no external boundaries. This is perhaps most applicable in cases of transient volumetric energy generation. For this reason the Green's functions of radial heat flow presented here are restricted to either the infinite domain, or to the discrete domain with a homogenous BC1 at $x = L$. The method of amending the existing Green's functions of the heat conduction problem by multiplying by the term $e^{-\alpha\omega^2(t-\tau)}$ may also be used in the cylindrical and spherical coordinate systems. The following provides selected Green's functions and the GFSE that have been adapted from those of the heat conduction problems presented in Ref. [8].

4.3.2.4.1 RADIAL HEAT FLOW IN THE CYLINDRICAL COORDINATE SYSTEM

A general representation of the nonhomogenous bioheat problem in the cylindrical coordinate system may be represented by

$$\frac{1}{\alpha}\frac{\partial T}{\partial t} = \frac{1}{r}\frac{\partial}{\partial r}\left(r\frac{\partial T}{\partial r}\right) - \omega^2 T + g(r,t) \quad t > 0$$

$$T(r=0,t) \Rightarrow \text{finite} \quad \text{IC:} \quad T(r,0) = F(r)$$

(4.118)

When the system is discrete, Equation (4.118) is valid in $[0 < r < R]$. In the special case that there are no internal boundaries and that the system is in thermal equilibrium at the outer boundaries, $f_2(t) = 0$, the corresponding GFSE is

$$T(x,t) = \int_{r'=0}^{R} G^P(r,t|r',0)F(r')2\pi r' dr' \quad \text{for the initial condition}$$

$$+ \int_{\tau=0}^{t}\int_{r'=0}^{R} \frac{\alpha}{k}G^P(x,t|r',\tau)g(r',\tau)2\pi r' dr' d\tau \quad \text{for energy generation}$$

(4.119)

The Green's functions for the case of a discrete radial region $0 < r < R$, may be adapted from those of the corresponding heat conduction problems presented in Ref. [8].

4.3.2.4.2 CYLINDER WITH A HOMOGENOUS TYPE I BC2

The Green's function of the bioheat problem with a Type I representation of thermal equilibrium at the outer boundary may be determined by adapting the corresponding one provided in Equations (R01.1) and (R01.2) on page 437 of Ref. [8].

$$\text{BC2 Type I:} \quad T(r=R,t) = 0$$

$$G(r,t|r',\tau) = \frac{1}{\pi R^2}\sum_{n=1}^{\infty}\frac{J_0(\lambda_n r)J_0(\lambda_n r')}{J_1^{\,2}(\lambda_n R)}\exp\left(-\alpha(t-\tau)(\lambda_n^2 + \omega^2)\right)$$

(4.120)

$$\lambda_m \text{ are the roots of:} \quad J_0(\lambda_n R) = 0 : n = 1,2,3\ldots$$

Here, J_0 and J_1 are the zeroth and first order Bessel functions of the first kind.

4.3.2.4.3 CYLINDER WITH A HOMOGENOUS TYPE II BC2

When a zero heat flux is imposed at the outer boundary, the Green's function of the corresponding heat conduction problem provided in Equation (R02.1) on page 438 of Ref. [8] may be adapted to state:

$$\text{BC2 Type II:} \quad \left.\frac{\partial T}{\partial r}\right|_{r=R} = 0$$

$$G^P(r,t|r',\tau) = \frac{1}{\pi R^2}\exp\left(-\alpha\omega^2(t-\tau)\right) \times \left(1 + \sum_{n=1}^{\infty}\frac{J_0(\lambda_n r)J_0(\lambda_n r')}{J_0^{\,2}(\lambda_n R)}\exp\left(-\alpha\lambda_n^2(t-\tau)\right)\right)$$

(4.121)

$$\lambda_n \text{ are the roots of:} \quad J_1(\lambda_n R) = 0 : n = 1,2,3\ldots$$

4.3.2.4.4 CYLINDRICAL SEMI-INFINITE PROBLEM

In the semi-infinite cylindrical domain, the bioheat problem (4.118) is valid in $[0 < r < \infty]$. The corresponding Green's function may be found by modifying Equation (7.3) on page 202 of Ref. [8]:

$$G^P(r,t|r',\tau) = \frac{\exp\left(-\alpha\omega^2(t-\tau)\right)}{4\pi\alpha(t-\tau)}\exp\left[\frac{-\left(r^2+r'^2\right)}{4\alpha(t-\tau)}\right]I_0\left[\frac{rr'}{2\alpha(t-\tau)}\right] \tag{4.122}$$

Here, I_0 is the modified Bessel Function of the first kind of zero order. A clever derivation of the corresponding heat conduction Green's function is provided in Section 7.3.2 of Ref. [8].

4.3.2.4.5 RADIAL HEAT FLOW IN THE SPHERICAL COORDINATE SYSTEM

Next the bioheat equation in the spherical system is considered. The one dimensional transient problem may be represented by

$$\frac{1}{\alpha}\frac{\partial T}{\partial t} = \frac{\partial^2 T}{\partial r^2} + \frac{2}{r}\frac{\partial T}{\partial r} - \omega^2 T + g(r,t) \quad t > 0$$

$$T(r=0,t) \Rightarrow \text{finite} \quad \text{IC:} \quad T(r,0) = F(r) \tag{4.123}$$

In the discrete spherical domain, the problem described is valid in $[0 < r < R]$. In the special case that there are no internal boundaries and that the system is in thermal equilibrium at the outer boundaries, $f_2(t) = 0$, and the GFSE is

$$T(x,t) = \int_{r'=0}^{R} G^P(r,t|r',0)F(r')4\pi(r')^2 dr' \quad \text{for the initial condition}$$

$$+ \int_{\tau=0}^{t}\int_{r'=0}^{R} \frac{\alpha}{k}G^P(x,t|r',\tau)g(r',\tau)4\pi(r')^2 dr'd\tau \quad \text{for energy generation} \tag{4.124}$$

4.3.2.4.6 SPHERE WITH A HOMOGENOUS TYPE I BC2

The spherical Green's function of the bioheat problem with an outer boundary condition of Type I may be determined from the corresponding one provided in equations (RS01.3) on page 464 of Ref. [8] to state.

$$\text{BC2 Type I:} \quad T(r=R,t) = 0$$

$$G(r,t|r',\tau) = \frac{1}{2\pi Rrr'}\sum_{n=1}^{\infty}\sin(\lambda_n r)\sin(\lambda_n r')\exp\left(-\alpha(t-\tau)\left(\lambda_n^2 + \omega^2\right)\right) \tag{4.125}$$

$$\lambda_m = \frac{n\pi}{R}$$

4.3.2.4.7 SPHERE WITH A HOMOGENOUS TYPE II BC2

When a zero heat flux is imposed at the outer boundary, the Green's function of the corresponding heat conduction problem provided in Equations (RS02.1) and (RS02.2) on page 465 of Ref. [8] may be adapted to state:

BC2 Type II: $\left.\dfrac{\partial T}{\partial r}\right|_{r=R} = 0$

$$G^P(r,t|r',\tau) = \frac{3}{4\pi R^3}\exp\left(-\alpha\omega^2(t-\tau)\right) + \frac{1}{2\pi Rrr'} \times \left(\sum_{n=1}^{\infty}\frac{\lambda_n^2 + \dfrac{1}{R^2}}{\lambda_n^2}\sin\left(\lambda_n r\right)\sin\left(\lambda_n r'\right)\exp\left(-\alpha\left(\lambda_n^2 + \omega^2\right)(t-\tau)\right)\right)$$

λ_n are the roots of: $\cot\left(\lambda_n R\right) = \dfrac{1}{\left(\lambda_n R\right)}$

(4.126)

4.3.2.4.8 SPHERICAL SEMI-INFINITE PROBLEM

For the semi-infinite spherical domain $[0 < r < \infty]$, the Green's function is adapted from Equation (9.4) on page 255 of Ref. [8]:

$$G^P(r,t|r',\tau) = \frac{\exp\left(-\alpha\omega^2(t-\tau)\right)}{8\pi rr'\sqrt{\pi\alpha(t-\tau)}} \times \left\{\exp\left[\frac{-(r-r')^2}{4\alpha(t-\tau)}\right] - \exp\left[\frac{-(r+r')^2}{4\alpha(t-\tau)}\right]\right\}$$

(4.127)

4.4 ADDITIONAL CONSIDERATIONS

The solutions and solution methods presented thus far have focused on the evaluation of a one dimensional transient bioheat problem. The primary benefit of the discussion up to this point has been to show that the development of the solution to the bioheat problem can be taken from existing works concerning the heat conduction problem published in books which act as structured compendia of solutions [6–9].

4.4.1 Bioheat Problem in Multidimensions

The existing homogenous solutions of the heat conduction problem may be multiplied by the term $e^{-\omega^2 t}$ in order to arrive at the corresponding homogenous bioheat solution. This transformation is equally applicable to the development of transient multidimensional bioheat problems. Homogenous solutions to the two and three dimensional transient solutions of the homogenous heat conduction problem in the Cartesian, cylindrical, and spherical coordinate systems are available in Refs. [6–9].

Homogenous multidimensional problems in the Cartesian coordinate system can be represented as the product of the one dimensional solutions:

$$T(x,y,z,t) = T_1(x,t)\cdot T_2(y,t)\cdot T_3(z,t)$$

(4.128)

where $T_1(x,t), T_2(y,t)$, and $T_3(z,t)$ and $T_3(z,t)$ are the transient one dimensional solutions in each of the coordinate directions, each of which address the corresponding one dimensional problem. This is called the *Product Solution* method and it is described in detail in Section 3-5 of the book [9]. The restriction of this method is that the initial condition must be able to be described as the product of "single-space variable" functions:

$$T(x,y,z,0) = F(x,y,z) = F_1(x)\cdot F_2(y)\cdot F_3(z)$$

(4.129)

In order to reconcile this restriction to some initial conditions, it may be helpful to use super-positioning of the original problem. The product solution method is not applicable in the spherical coordinate system, or in the cylindrical coordinate system with angular dependence.

Nonhomogenous multidimensional problems for which all the nonhomogenous terms are steady, can be addressed using superposition. For problems in the Cartesian coordinate system, superpositioning may be supplemented with the product solution method. Complications may arise, however, in the evaluation of the solutions' associated spatial integral.

Alternatively, the method of Green's functions can be used to address multidimensional transient nonhomogenous bioheat problems. The GFSE can be used to represent the temperature when the Green's functions are known. Fortunately, in the appendices of the book by Beck et al. [8] an indexed and organized list of the multidimensional Green's functions in the Cartesian, cylindrical, and spherical coordinate systems is provided. The corresponding Green's function of the bioheat problem may be obtained by multiplying the corresponding heat conduction Green's function by the term: $e^{-\omega^2(t-\tau)}$.

4.4.2 Short Time Convergence Issues and Integral Approximations

The solutions of the bioheat problem in discrete domains presented in this chapter are all solutions composed of infinite series and are of a type that is characterized in Ref. [8] as "large time" solutions. At larger times, fewer terms of these "large time" series are required to evaluate the solution with minimal error.

However, in some instances, it is particularly desirable to understand the behavior of a system at very early times. Evaluating the large time solution at early times can result in very slow convergence of the solution (requiring a very large number of terms). See for example the discussion in Section 3-6 of Ref. [9] and in Section 5.2 of Ref. [8]. In problems of pure conduction, the dimensionless Fourier number: $Fo = \alpha t / L^2$ is used to characterize the times above which very few terms are needed to approximate the solution.

Because the perfusion term acts to increase the magnitude of the exponential in the transient solution, it is reasonable to assume that the minimum number of terms required for high accuracy in the heat conduction solution is also valid for the minimum number of terms required for the bioheat solution.

To expedite the solution convergence at early times, alternate "small time" solutions can be used. The small time solutions require few terms for small Fo and, in general, are related to the semi-infinite solution with related boundary conditions.

In the book [8], a thorough description is provided of a method that represents the solution as the product of a small time solution and a large time solution. This is called "time partitioning." Time partitioning is applicable to the bioheat problem, although it is not discussed further here.

4.5 THE COMPOSITE BIOHEAT PROBLEM

Consider again the unshifted bioheat equation representing bioheat transfer in a region representing an isotropic tissue:

$$\frac{1}{\alpha}\frac{\partial T^*}{\partial t} = \nabla^2 T^* - \omega^2(T^* - Ta) + \frac{g}{k} \tag{4.130}$$

The application of this representation of the bioheat equation is limited to homogenous tissues for which the material properties (α and k) and the perfusion related parameters (ω and Ta) are uniform. In some instances, it is of great benefit to consider the movement of heat into a system composed of perfuse tissues that are adjacent to one another and whose thermophysical parameter values differ. A composite representation of such a system would allow for such discontinuities at the interface between the tissues. This procedure is described in the case of pure heat conduction problems (for example in chapter 10 of the book [9]).

One concern of the representation of the bioheat equation of Equation (4.130) is that it assigns a uniform rate of perfusion everywhere. In practice, this is not always the case. For example, the rapid development of tumors can result in constriction of the local blood flow. In such a case, it is anticipated that the perfusion rates within the tumor are much lower than in the adjacent healthy tissue. Sudden spatial discontinuities in the perfusion rates can also occur in healthy tissues. Consider the human skin. Heat transfer in the skin's nonperfuse outer layer, the epidermis, is governed by pure conduction, while the adjacent dermis is perfuse and may experience perfusion rates of up to five times higher than those in the deeper adjacent subcutaneous tissues. An accurate portrayal of such a system would require variations in perfusion rates. Modeling living tissue as a composite system provides the freedom to represent the perfusion coefficient, ω, with different values in different layers.

Recall that one of the criticisms of the bioheat equation is that it does not account for spatial variations in the temperature of the blood. This means that Ta has the same value at shallow locations near the surface as it does at the deep locations near the body's core. A composite representation of the tissue would allow for variations in the temperature of the perfusion related blood flow, Ta. In a composite representation, the value of Ta is no longer restricted to having a single value for all tissue layers. For example, in layers closer to the body's core, the perfusion related temperature may be set at a value closer to the body's core temperature while layers near a heat source or layers near the body's surface may have arterial blood temperatures that vary significantly from the body's core temperature. In the following discussion, the one dimensional transient solution of the composite bioheat problem in the Cartesian coordinate system is developed. Such a composite representation is depicted in Figure 4.3.

Here, a system is presented that is composed of a number, "M," of composite layers. Each of the layers can have its own unique parameter values so that in the "ith" layer, the associated

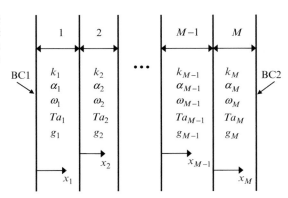

FIGURE 4.3 Composite representation of bioheat transfer in which the system is composed of M different slabs in perfect contact with one another and the system is exposed to the environment at $x_1 = 0$ and is exposed to the body's core at $x_M = L_M$.

parameter values are denoted: $k_i, \alpha_i, \omega_i,$ and Ta_i. The magnitude of each layer's volumetric heating $g_i(x_i, t)$ can vary by position and with time. Each layer has its own unique thickness denoted, L_i. To simplify the problem description, the spatial domain can be discretized so that each layer has its own spatial coordinate, $0 \leq x_i \leq L_i$. In order to fully appreciate the steps required to address this problem, we begin with the nonhomogenous unshifted representation of this composite problem. The nonhomogenous unshifted bioheat equation for each layer can be represented as:

$$\frac{1}{\alpha_i} \frac{\partial T_i^*}{\partial t} = \frac{\partial^2 T_i^*}{\partial x_i^2} - \omega^2 \left(T_i^* - Ta_i \right) + \frac{g_i(x_i, t)}{k_i} \quad \text{in } 0 \leq x_i \leq L_i, 0 < t, i = 1 \ldots M \tag{4.131}$$

It is noteworthy to stress that each layer has a corresponding governing Equation (4.131). The initial state of the system can also be represented in a discrete sense so that each layer has its own representative initial condition:

$$T_i^*(x_i, 0) = F_i^*(x_i) \quad \text{in} : 0 \leq x_i \leq L_i, i = 1 \ldots M \tag{4.132}$$

Consider briefly the interface between any two adjacent layers. In layers $i = 2 \ldots M$, the interface positions may be identified by the coordinates: $x_{i-1} = L_{i-1}$ and $x_i = 0$. Heat is conducted between adjacent layers through the interface so at each interface (for $i = 2 \ldots M$) the flux at the interface behaves as:

$$k_i \frac{\partial T_i^*}{\partial x_i}\bigg|_{x_i=0} = k_{i-1} \frac{\partial T_{i-1}^*}{\partial x_{i-1}}\bigg|_{x_{i-1}=L_{i-1}} \quad i = 2 \ldots M \tag{4.133}$$

In the realistic assumption that there is no resistance to conduction at the interface between two layers, the temperatures of the layers at their interface will be equal:

$$T_i^*(x_i = 0, t) = T_{i-1}^*(x_{i-1} = L_{i-1}, t) \quad i = 2 \ldots M \tag{4.134}$$

In the present solution, only steady nonhomogenous terms are considered. The outer boundary of layer 1 (at $x_1 = 0$) can be exposed to any one of the three BC types:

$$\text{BC1 Type I} \quad T_1^*(x_1 = 0, t) = T_O$$

$$\text{BC1 Type II} \quad -k_1 \frac{\partial T_1^*}{\partial x_1}\bigg|_{x_1=0} = q_O'' \tag{4.135}$$

$$\text{BC1 Type III} \quad -k_1 \frac{\partial T_1^*}{\partial x_1}\bigg|_{x_1=0} + hT_1^*(x_1 = 0, t) = hT_\infty$$

At the other boundary, the system is considered to be in thermal equilibrium with the body's core. This corresponds to the location in layer M $x_M = L_M$. The thermal equilibrium may be represented either by a Type I or a homogenous Type II BC:

$$\text{BC2 Type I} \quad T_M^*(x_M = L_M, t) = Ta_M$$

$$\text{BC2 Type II} \quad \frac{\partial T_M^*}{\partial x_M}\bigg|_{x_M=L_M} = 0 \tag{4.136}$$

Note that the Type I condition on boundary $x_M = L_M$ makes the realistic implication that the arterial blood temperature associated with Layer "M" is equal to the body's core temperature.

4.5.1 Homogenization of the Composite Domain

Using the SOV method will require that for each layer, the governing equation is homogenous, the interface conditions are all homogenous, and the boundary conditions of layer 1 and layer M are homogenous.

A complication that is introduced by the composite representation involves the discontinuities in the perfusion related arterial temperatures between layers. Recall that for the single layer isotropic problem, the simple temperature shift was introduced in Section *4.1.3* in order to remove the nonhomogenous perfusion related term from the governing equation. However, in the composite problem, this method is not applicable when a single arterial temperature value cannot be used in such a shift; when each layer has its own distinct value of Ta. This means that when the perfusion related temperature is not constant across the system (even if there is no generation term and the boundary conditions are homogenous) there is always a nonhomogenous component to the composite bioheat problem.

Note that the energy generation term of Equation (4.131) is free to vary in time. In the following discussion, we limit ourselves to the composite bioheat problem for which all of the nonhomogenous terms are steady. In Section *4.5.5* the Green's function solution of the composite bioheat equation is provided in order to address problems that include time dependent energy generation.

In the event that the nonhomogenous terms are all steady, a straightforward method of superposition may be implemented. In this case, it is assumed that that the solution may be considered to comprise the product of a nonhomogenous steady component and a homogenous nonsteady component. In Layer "i" the temperature is represented by this linear combination as:

$$T_i^*(x_i, t) = \underbrace{\Psi_i(x_i)}_{\text{Nonhomogenous Steady}} + \underbrace{T_i(x_i, t)}_{\text{Homogenous}} \quad i = 1 \ldots M \tag{4.137}$$

4.5.1.1 Addressing the Steady Nonhomogenous Components

The steady component addresses all of the nonhomogenous terms so that the governing equation of the steady term includes the perfusion related temperature and the nontransient generation term:

$$\frac{\partial^2 \Psi_i}{\partial x_i^2} - \omega_i^2 \Psi_i(x_i) = -\omega_i^2 Ta_i - \frac{g_i(x_i)}{k_i} \quad \text{in}: 0 \le x_i \le L_i \quad i = 1 \ldots M \tag{4.138}$$

The interface conditions of the steady nonhomogenous term are represented:

$$\left. \begin{array}{c} k_i \dfrac{d\Psi_i}{dx_i}\Big|_{x_i=0} = k_{i-1}\dfrac{d\Psi_{i-1}}{dx_{i-1}}\Big|_{x_{i-1}=L_{i-1}} \\ \Psi_i(0) = \Psi_{i-1}(L_{i-1}) \end{array} \right\} \quad \text{for } i = 2 \ldots M \tag{4.139}$$

The nonhomogenous component of layer 1 will satisfy the steady nonhomogenous terms associated with any of the three boundary nontransient condition types:

$$\begin{array}{ll} \text{BC1 Type I} & \Psi_1(0) = T_O^* \\[2mm] \text{BC1 Type II} & -k_1\dfrac{d\Psi_1}{dx_1}\Big|_{x_1=0} = q_O'' \\[2mm] \text{BC1 Type III} & k_1\dfrac{d\Psi_1}{dx_1}\Big|_{x_1=0} - h\Psi_1(0) = -hT_\infty \end{array} \tag{4.140}$$

At the inner boundary of layer M (at $x_M = L_M$), the nonhomogenous component can have a BC of Type I or of Type II:

$$\text{BC2 Type I} \quad \Psi_M(L_M) = Ta_M$$

$$\text{BC2 Type II} \quad \left.\frac{d\Psi_M}{dx_M}\right|_{x_M=L_M} = 0 \tag{4.141}$$

The evaluation of the solution of the system of linear ODEs of the problem described by Equation (4.139) is a relatively straightforward procedure. For example in the special case in which the energy generation of each layer is constant within that layer, $g_i(x_i) = \dot{q}_i$, the general solution of the problem is linear when $\omega_i = 0$ and hyperbolic when $\omega_i \neq 0$ so that:

$$\left.\begin{array}{l} \omega = 0 : \Psi_i(x_i) = -\dfrac{\dot{q}_i}{k_i}\dfrac{x_i^2}{2} + d_i x_i + e_i \\[2ex] \omega > 0 : \Psi_i(x_i) = \dfrac{\dot{q}_i}{k_i \omega_i^2} + Ta_i + d_i \cosh(\omega_i x_i) + e_i \sinh(\omega_i x_i) \end{array}\right\} \quad \text{in}: 0 \leq x_i \leq L_i \tag{4.142}$$

where the constants of integration d_i and e_i are evaluated using the boundary and interface conditions.

Special attention is given to the problem in the event that the nonhomogenous generation term is not steady. This may involve further superpositioning or the use of Green's functions and is discussed later in Section *4.5.5*.

Again, it is noted that even in the absence of any volumetric heating and when the boundary conditions are homogenous, the composite bioheat problem will *always* be nonhomogenous when the perfusion blood temperature, Ta, varies between layers. As long as there are variations between layers in Ta, superpositioning must be used and the function $\Psi_i(x_i)$ must be determined for each layer.

4.5.2 The Homogenous Composite Bioheat Problem

The composite bioheat problem representing the homogenous component of Equation (4.137) is described:

$$\frac{\partial^2 T_i}{\partial x_i^2} = \frac{1}{\alpha_i}\frac{\partial T_i}{\partial t} + \omega^2 T_i \quad \text{in}: 0 < x_i < L_i, \quad 0 < t, \quad i = 1\ldots M \tag{4.143}$$

Again, it is intentional that the spatial term has been isolated in the governing Equation (4.143) of each layer which differs from the arrangement of Equation (4.131). This was discussed in reference to the isotropic problem of Example 4.1 of Section **4.2**.

The initial condition of the homogenous problem can be determined by substituting Equation (4.137) into Equation (4.132) so that for each layer:

$$T_i(x_i, t=0) = F_i^*(x_i) - \Psi_i(x_i) \quad \text{in}: 0 < x_i < L_i, \quad i = 1\ldots M \tag{4.144}$$

This expression links the homogenous solution to any nonhomogenous terms that are expressed through the steady function, Ψ_i.

The interface conditions of the homogenous solution component can be represented:

$$\left.\begin{array}{l} k_i \left.\dfrac{\partial T_i}{\partial x_i}\right|_{x_i=0} = k_{i-1} \left.\dfrac{\partial T_{i-1}}{\partial x_{i-1}}\right|_{x_{i-1}=L_{i-1}} \\[2ex] T_i(0) = T_i(L_{i-1}) \end{array}\right\} \quad i = 2\ldots M \tag{4.145}$$

This only leaves the homogenous component's boundary conditions. The outer boundary of layer 1 can be exposed to any one of the three BC1 types:

$$\text{BC1 Type I} \quad T_1(0, t) = 0$$

$$\text{BC1 Type II} \quad \left.\frac{\partial T_1}{\partial x_1}\right|_{x_1=0} = 0$$

$$\text{BC1 Type III} \quad k_1\left.\frac{\partial T_1}{\partial x_1}\right|_{x_1=0} - hT_1(0, t) = 0$$

(4.146)

At the other boundary nearest the body's core, the homogenous Type I or Type II BC2 can be represented:

$$\text{BC2 Type I} \quad T_M(x_M = L_M, t) = 0$$

$$\text{BC2 Type II} \quad \left.\frac{\partial T_M}{\partial x_M}\right|_{x_M=L_M} = 0$$

(4.147)

4.5.3 SOV in the Composite System

Consider that the homogenous composite bioheat problem of any individual layer is very similar to the homogenous bioheat problem set in an isotropic medium: each composite layer is governed by the same equation as in the isotropic case, and in lieu of the explicit boundary conditions, adjacent layers within the composite system have interface conditions.

By applying the SOV method to the composite problem, the solution of each individual layer is assumed to be a product of a time dependent component and a spatially dependent component:

$$T_i(x_i, t) = X_i(x_i) \cdot \Gamma_i(t) \quad i = 1 \ldots M$$

(4.148)

Substituting this expression into the governing Equation (4.143) results in the equation representing the pair of equivalent ODEs:

$$\frac{1}{\alpha_i \Gamma_i} \frac{d\Gamma_i}{dt} + \omega_i^2 = \frac{1}{X_i} \frac{d^2 X_i}{dx_i^2} = -\lambda_i^2 \quad i = 1 \ldots M$$

(4.149)

This expression is valid in each of the $i = 1 \ldots M$ layers for which λ_i is the ith layer's eigenvalue. Note the similarity in this representation of the governing equation of the ith layer in Equation (4.149) to that of the single layer slab problem as presented in Equation (4.21) of Example 4.1. It is thus anticipated (and it will become evident shortly) that the solution of each individual layer is composed of a linear series of an infinite number of terms, each of which corresponds to a discrete eigenvalue.

In order to arrive at the general solutions, simply consider the ODEs as representing the spatial and transient components separately, so that taken from Equation (4.149) the transient component solution of the ith layer is governed by

$$\frac{d\Gamma_{i,n}}{dt} = -\alpha_i \Gamma_{i,n} \cdot \left(\lambda_{i,n}^2 + \omega_i^2\right) \quad i = 1 \ldots M$$

(4.150)

The subscripts, n, have been included in anticipation that the solution will be represented as the sum of an infinite number of terms, each of which is associated with a unique eigenvalue.

Consider, momentarily, that one of the characteristics is unique to the composite bioheat problem that can be seen in Equation (4.150). The parameter values α_i and ω_i^2 can be unique to each layer and are free to vary (even dramatically) between adjacent layers. This means that it is possible for the eigenvalue of the ith layer, $\lambda_{i,n}$, to be unique to that layer and to differ between adjacent layers. By establishing the relationships between the transient solution components, $\Gamma_{i,n}$, of one layer to those of another, the interface conditions can be used to establish the relationships between the eigenvalues, X_i, of adjacent layers. This can be accomplished by considering the implications of the interface conditions. Substituting product representation of Equation (4.148) into the homogenous interface conditions (4.145) will result in the expression:

$$\left. \begin{aligned} \Gamma_{i,n}(t) \cdot k_i \frac{dX_{i,n}}{dx_i}\bigg|_{x_i=0} &= \Gamma_{i-1,n}(t) \cdot k_{i-1} \frac{dX_{i-1,n}}{dx_{i-1}}\bigg|_{x_{i-1}=L_{i-1}} \\ \Gamma_{i,n}(t) \cdot X_{i,n}(0) &= \Gamma_{i-1,n}(t) \cdot X_{i-1,n}(L_{i-1}) \end{aligned} \right\} \quad i=2\ldots M \tag{4.151}$$

Because the transient time coordinate, t, is the same for all layers, the above interface conditions can only hold if the transient components of all layers have the same value at any given time:

$$\Gamma_{i,n}(t) = \Gamma_{i-1,n}(t) = \Gamma_{O,n}(t) \quad \text{in } t>0, \quad i=2\ldots M \tag{4.152}$$

This is a very useful property because it can be used to relate all the layers' eigenvalues to each other arbitrarily, or to that of a single specific reference layer (denoted in Equation (4.152) by the subscript O). The general solution of the transient solution component is

$$\Gamma_{i,n} = c \cdot \exp\left(-\alpha_i \left(\lambda_{i,n}^2 + \omega_i^2\right) t\right) \tag{4.153}$$

where c is some constant of integration that has the same value for all layers.

Because the different layers' transient solutions are all equivalent at all positions, they can be removed from the boundary and interface conditions. By substituting Equation (4.148) into Equation (4.146), the three possible types of boundary conditions of layer 1 can be represented in terms of the spatial component only:

$$\text{BC1 Type I} \quad X_{1,n}(0) = 0$$

$$\text{BC1 Type II} \quad \frac{dX_{1,n}}{dx_1}\bigg|_{x_1=0} = 0 \tag{4.154}$$

$$\text{BC1 Type III} \quad k_1 \frac{dX_{1,n}}{dx_1}\bigg|_{x_1=0} - h \cdot X_{1,n}(0) = 0$$

Noting from Equation (4.152) that, because the transient components at the interfaces are equal, the interface conditions of Equation (4.151) can now be simplified to state:

$$\left. \begin{aligned} k_i \frac{dX_i}{dx_i}\bigg|_{x_i=0} &= k_{i-1} \frac{dX_{i-1}}{dx_{i-1}}\bigg|_{x_{i-1}=L_{i-1}} \\ X_i(0) &= X_i(L_{i-1}) \end{aligned} \right\} \quad i=2\ldots M \tag{4.155}$$

Finally, the two possible BC2 types at the surface closest to the body's core ($x_M = L_M$) can be represented:

$$\begin{aligned} &\text{BC2 Type I} \quad X_M(L_M) = 0 \\ &\text{BC2 Type II} \quad \left.\frac{\mathrm{d}X_M}{\mathrm{d}x_M}\right|_{x_M=L_M} = 0 \end{aligned} \tag{4.156}$$

Looking once more to Equation (4.149), it is evident that the ODE governing the spatial solution component of the ith layer is

$$\frac{\mathrm{d}^2 X_{i,n}}{\mathrm{d}x_i^2} = -\lambda_{i,n}^2 X_{i,n} \tag{4.157}$$

The equations governing the spatial solution component (4.154)–(4.157) are used to develop the eigenfunctions that will satisfy the orthogonality condition:

$$\sum_{i=1}^{M} \frac{k_i}{\alpha_i} \int_{x_i=0}^{x_i=L_i} X_{i,n} X_{i,m} \mathrm{d}x_i = \begin{cases} 0 & n \neq m \\ N_n & n = m \end{cases} \tag{4.158}$$

where the norm is described by

$$N_n = \sum_{i=1}^{M} \frac{k_i}{\alpha_i} \int_{x_i=0}^{x_i=L_i} (X_{i,n})^2 \mathrm{d}x_i \tag{4.159}$$

The orthogonality condition has been developed for the pure conduction problem which is identical to that of the bioheat problem. A proof of the orthogonality condition is provided in appendix I of the paper [11].

A helpful relation in the evaluation of the integral in Equation (4.159) is

$$\int_{x_i=0}^{x_i=L_i} (X_{i,n})^2 \mathrm{d}x_i = \frac{1}{2} \frac{1}{\lambda_{i,n}^2} \left(\begin{array}{l} \left. X_{i,n}(0) \cdot \dfrac{\mathrm{d}X_{i,n}}{\mathrm{d}x} \right|_0 + (\lambda_{i,n} X_{i,n}(0))^2 \\[2mm] + \left(\left. \dfrac{\mathrm{d}X_{i,n}}{\mathrm{d}x} \right|_0 \right)^2 - X_{i,n}(L_i) \cdot \left. \dfrac{\mathrm{d}X_{i,n}}{\mathrm{d}x} \right|_{L_i} \end{array} \right) \tag{4.160}$$

The general solution of the temperature in any of the layers is the linear sum of all of the terms associated with each eigenvalue:

$$T_i(x_i, t) = \sum_{n=1}^{\infty} C_n X_{i,n}(x_i) \cdot \Gamma_{i,n}(t) \tag{4.161}$$

The constants of integration, C_n, are, in fact, the coefficients determined by the Fourier expansion of the initial condition (4.144), so that, for example, in the jth layer:

$$F_j^*(x_j) - \Psi_j(x_j) = \sum_{n=1}^{\infty} C_n X_{j,n}(x_j) \tag{4.162}$$

With the orthogonality condition of Equation (4.158) in mind, multiplying both sides of Equation (4.162) by $k_j/\alpha_j \cdot X_{j,m}(x_j)$ (where $X_{j,m}$ is the eigenfunction of the jth layer that is associated with the mth eigenvalue) and then integrating this, results in the equality:

$$\int_{x_j=0}^{L_j} \frac{k_j}{\alpha_j} X_{j,m}(x_j) \cdot \left(F_j^*(x_j) - \Psi_j(x_j) \right) dx_j = \int_{x_j=0}^{L_j} \frac{k_j}{\alpha_j} \sum_{n=1}^{\infty} C_n X_{j,n}(x_j) \cdot X_{j,m}(x_j) dx_j \qquad (4.163)$$

Summing these equalities over all of the $i=1\ldots M$ layers results in the expression:

$$\sum_{j=1}^{M} \int_{x_j=0}^{L_j} \frac{k_j}{\alpha_j} X_{j,m}(x_j) \cdot \left(F_j^*(x_j) - \Psi_j(x_j) \right) dx_j = \sum_{j=1}^{M} \int_{x_j=0}^{L_j} \frac{k_j}{\alpha_j} \sum_{n=1}^{\infty} C_n X_{j,n}(x_j) \cdot X_{j,m}(x_j) dx_j \qquad (4.164)$$

Note that the right side of Equation (4.164) is the orthogonality condition of Equation (4.158). This means that only a single term over the last summation on the left side of Equation (4.164) contributes to the solution: when $n=m$. Thus Equation (4.164) is better represented:

$$\sum_{j=1}^{M} \int_{x_j=0}^{L_j} \frac{k_j}{\alpha_j} X_{j,m}(x_j) \cdot \left(F_j^*(x_j) - \Psi_j(x_j) \right) dx_j = C_m \sum_{j=1}^{M} \int_{x_j=0}^{L_j} \frac{k_j}{\alpha_j} X_{j,m}(x_j) \cdot X_{j,m}(x_j) dx_j \qquad (4.165)$$

Solving this result for the constant of integration, C_m, results in:

$$C_m = \frac{1}{N_m} \sum_{j=1}^{M} \int_{x_j=0}^{L_j} \frac{k_j}{\alpha_j} X_{j,m}(x_j) \cdot \left(F_j^*(x_j) - \Psi_j(x_j) \right) dx_j \qquad (4.166)$$

where the norm associated with the mth eigenvalue is represented by Equation (4.159) which is equivalent to the right side of Equation (4.160). The relationship presented in Equation (4.47) can be very helpful in the evaluation of the integral on the right side of Equation (4.166) and will be discussed later.

The determination of the constant in Equation (4.166) allows the solution of Equation (4.161) to be represented as:

$$T_i(x_i, t) = \sum_{n=1}^{\infty} \underbrace{\left(\frac{1}{N_n} \sum_{j=1}^{M} \int_{x_j=0}^{L_j} \frac{k_j}{\alpha_j} X_{j,n}(x_j) \cdot \left(F_j^*(x_j) - \Psi_j(x_j) \right) dx_j \right)}_{C_n} \cdot X_{i,n}(x_i) \cdot \exp\left(-\alpha_i \left(\lambda_{i,n}^2 + \omega_i^2 \right) t \right)$$

$$(4.167)$$

4.5.3.1 Building the Eigenfunctions and Determining the Eigenvalues

In the previous section, some very important relationships were developed in the analysis of the composite homogenous bioheat problem. In order to complete the solution of Equation (4.167), it remains to be shown how each layer's eigenfunction, $X_{i,n}$, and eigenvalues, $\lambda_{i,n}$, are determined. In the following example, a specific application of the bioheat problem uses this information to show how the actual eigenfunctions can be determined using a method that "builds" the eigenfunctions.

EXAMPLE 4.6 HOMOGENOUS COMPOSITE BIOHEAT PROBLEM

Consider the composite homogenous bioheat problem that is composed of some number "M" of layers. A Type III BC1 is imposed at the surface and thermal equilibrium at the boundary nearest the core is represented through a Type II BC2.

$$\frac{\partial^2 T_i}{\partial x_i^2} = \frac{1}{\alpha_i}\frac{\partial T_i}{\partial t} + \omega^2 T_i \quad \text{in}: 0 < x_i < L_i, \quad 0 < t, \quad i = 1 \ldots M$$

$$\text{BC1} \quad k_1 \frac{T_1}{dx_1}\bigg|_{x_1=0} = h \cdot T_1(0,t)$$

$$\left.\begin{array}{c} \text{At the Interfaces} \quad k_i\dfrac{dT_i}{dx_i}\bigg|_{x_i=0} = k_{i-1}\dfrac{dT_{i-1}}{dx_{i-1}}\bigg|_{x_{i-1}=L_{i-1}} \\[2mm] T_i(0,t) = T_{i-1}(L_{i-1},t) \end{array}\right\} \quad i = 2 \ldots M \tag{4.168}$$

$$\text{BC2} \quad \frac{dT_M}{dx_M}\bigg|_{x_M=L_M} = 0$$

Following the results of the previous section, the solution in each layer of this system may be represented by the infinite series over the eigenvalues, so that the solution of the ith layer is

$$T_i(x_i,t) = \sum_{n=1}^{\infty} C_n X_{i,n}(x_i)\underbrace{\exp\left(-\alpha\left(\lambda_{i,n}^2 + \omega_i^2\right)t\right)}_{\Gamma_{i,n}(t)} \tag{4.169}$$

where in the ith layer, the eigenvalue, $\lambda_{i,n}$, corresponds to the nth term in the infinite series. The eigenfunction corresponding to the nth term of the series in Equation (4.169) is governed by the ODE:

$$\frac{d^2 X_{i,n}}{dx_i^2} = -X_{i,n}\lambda_{i,n}^2 \quad i = 1 \ldots M \tag{4.170}$$

And this is constrained in each layer to the corresponding boundary and interface conditions:

$$\text{BC1} \quad k_1\frac{dX_{1,n}}{dx_1}\bigg|_{x_1=0} = h \cdot X_{1,n}(0)$$

$$\left.\begin{array}{c} \text{At the Interfaces} \quad k_i\dfrac{dX_{i,n}}{dx_i}\bigg|_{x_i=0} = k_{i-1}\dfrac{dX_{i-1,n}}{dx_{i-1}}\bigg|_{x_{i-1}=L_{i-1}} \\[2mm] X_{i,n}(0) = X_{i-1,n}(L_{i-1}) \end{array}\right\} \quad i = 2 \ldots M \tag{4.171}$$

$$\text{BC2} \quad \frac{dX_{M,n}}{dx_M}\bigg|_{x_M=L_M} = 0$$

The general solution to the ith layer eigenfunction corresponding to the nth term of the series in Equation (4.169) is

$$X_{i,n}(x_i) = a_{i,n}\sin(\lambda_{i,n}x_i) + b_{i,n}\cos(\lambda_{i,n}x_i) \tag{4.172}$$

where the coefficients $a_{i,n}$ and $b_{i,n}$ are needed in order to determine the actual eigenfunction. These two coefficients (as well as the eigenvalue, $\lambda_{i,n}$) must be evaluated for each layer. Because there are M layers, this means that there are a total of $2M$ coefficients and M eigenvalues that must be determined for the nth term in the series.

Each of the BCs and interface conditions can be used to complete a specific task in the evaluation of these constants. The first layer's boundary condition (BC1) establishes the relationship between the constants within the first layer. The interface conditions establish the relationship between the constants and eigenvalues of adjacent layers. And finally the outer boundary condition (BC2) of the last layer, M, can be used directly in order to evaluate the eigenvalues. The procedure outlined in this chapter is to "build" the eigenfunctions of each layer beginning with layer 1, and then progress sequentially to layer M.

Consider first the general solution of layer 1:

$$X_{1,n}(x_1) = a_{1,n}\sin(\lambda_{1,n}x_1) + b_{1,n}\cos(\lambda_{1,n}x_1) \tag{4.173}$$

Applying the Type III BC1 to layer 1 at $x_1 = 0$ provides the relationship between the layer 1 coefficients:

$$k_1\lambda_{1,n} \cdot a_{1,n} = h \cdot b_{1,n} \tag{4.174}$$

so that the eigenfunction of layer 1 may be represented as:

$$X_{1,n}(x_1) = a_{1,n}\left(\sin(\lambda_{1,n}x_1) + \frac{k_1\lambda_{1,n}}{h}\cos(\lambda_{1,n}x_1)\right) \tag{4.175}$$

The constant of integration, C_n, in Equation (4.169) already acts as the Fourier coefficient of the initial condition. Therefore, the remaining coefficient, $a_{1,n}$, in Equation (4.175) can simply be assigned a value of unity without losing any information:

$$X_{1,n}(x_1) = \sin(\lambda_{1,n}x_1) + \frac{k_1\lambda_{1,n}}{h}\cos(\lambda_{1,n}x_1) \tag{4.176}$$

The relationships between the coefficients of the eigenfunctions of the remaining layers $i=2\ldots M$ may now be determined by substituting the general solution representing the eigenfunction of the ith layer (4.172) into the interface conditions (4.171). In doing this, it can be shown that the coefficients for layers $i=2\ldots M$ may be represented:

$$\left.\begin{array}{l} a_{i,n} = \dfrac{k_{i-1}}{k_i}\dfrac{1}{\lambda_{i,n}}\dfrac{dX_{i-1,n}}{dx_{i-1}}\bigg|_{L_{i-1}} \\[2mm] b_{i,n} = X_{i-1,n}(L_{i-1}) \end{array}\right\} \quad i=2\ldots M \tag{4.177}$$

Consider that once the eigenfunction of layer 1 is known, its first derivative can be evaluated at $x_1 = L_1$. Then, by setting $i=2$ in relations (4.177) the coefficients $a_{2,n}$ and $b_{2,n}$ can be determined. This establishes the eigenfunction of layer 2 to be represented as:

$$X_{2,n}(x_2) = \underbrace{\frac{k_1\lambda_{1,n}}{k_2\lambda_{2,n}}\left(\cos(\lambda_{1,n}L_1) - \frac{k_1\lambda_{1,n}}{h}\sin(\lambda_{1,n}L_1)\right)}_{a_2} \cdot \sin(\lambda_{2,n}x_2) + \underbrace{\frac{k_1\lambda_{1,n}}{h}\cos(\lambda_{1,n}L_1)}_{b_2} \cdot \cos(\lambda_{2,n}x_2) \tag{4.178}$$

This process is repeated for layer $i=3$ and so on until the coefficients of the last layer M have been determined.

Once the coefficients of each of the M layers' eigenfunctions have been determined, it is possible to evaluate the eigenvalues of each layer. This will be done using the BC of layer M at $x_M = L_M$. However it should be noted that, at this point, each layer has its own associated eigenvalue, so that for each term in the infinite series, the system has a total of M eigenvalues.

The transient solution (4.178) is considered between layers using the equality expressed in Equation (4.152):

$$\Gamma_{i,n} = c \cdot \exp\left(-\alpha_i \left(\lambda_{i,n}^2 + \omega_i^2\right)t\right) = c \cdot \exp\left(-\alpha_O \left(\lambda_{O,n}^2 + \omega_O^2\right)t\right) \tag{4.179}$$

From this, it is easy to show that the eigenvalue of each of the layers may be related to a single one (that of the reference layer) by the expression:

$$\lambda_{i,n} = \sqrt{\frac{\alpha_O}{\alpha_i}\left({\lambda_{O,n}}^2 + {\omega_O}^2\right) - {\omega_i}^2} \quad i = 1 \ldots M \tag{4.180}$$

where $\lambda_{O,n}$ is the reference layer eigenvalue that corresponds to the nth term in the infinite series of eigenvalues. It is recommended to choose as the reference layer, the one that has the minimum value of $\alpha_O \omega_i^2$. Any layer lacking perfusion could be assigned to be the reference layer.

Once the substitution of Equation (4.180) has been conducted in each occurrence of $\lambda_{i,n}$ in the description of the eigenfunction of layer M, the $3M$ unknowns of the system of eigenfunctions have been reduced to a single one: $\lambda_{O,n}$. Applying the final BC2 in Equation (4.171) to Layer M results in the transcendental function:

$$\tan\left(L_M \sqrt{\frac{\alpha_O}{\alpha_M}\left(\lambda_{O,n}^2 + \omega_O^2\right) - \omega_M^2}\right) = \frac{a_{M,n}}{b_{M,n}} \tag{4.181}$$

for which the eigenvalue $\lambda_{O,n}$ can represent any of an infinite number of roots. In practice, each of these roots can be evaluated numerically.

4.5.4 Solutions to Homogenous Transient Bioheat Transfer in a Composite Slab

Here, the results of the previous section are summarized to provide a general description of the solution to the composite homogenous bioheat problem described by Equations (4.143)–(4.144) When determining the solution to the composite homogenous bioheat problem, it is suggested to begin by choosing a reference layer. It is recommended to choose as the reference layer, the one that has the minimum value of $\alpha_i \omega_i^2$ but any layer lacking perfusion could be assigned to be the reference layer. In the following solution, the parameters associated with the reference layer values are denoted with the subscript O.

The SOV method for the homogenous transient bioheat equation in a composite slab of some number, M, of layers results in an expression of the temperature in the ith layer as:

$$T_i(x_i, t) = \sum_{n=1}^{\infty} C_n X_{i,n} \exp\left(-\alpha_O \left(\lambda_{O,n}^2 + \omega_O^2\right)t\right) \tag{4.182}$$

Here the parameters α_O, ω_O^2, and $\lambda_{O,n}^2$ all correspond to those of the reference layer.

In Equation (4.182) the eigenfunctions of each layer, $X_{i,n}(x_i)$, are developed sequentially beginning with the first layer. The eigenfunctions of layer 1 are contingent on BC1 and may be expressed according to BC type at $x_1 = 0$:

$$\begin{aligned}
&\text{BC1 Type I:} \quad X_{1,n} = \sin\left(\lambda_{1,n}x_1\right) \\
&\text{BC1 Type II:} \quad X_{1,n} = \cos\left(\lambda_{1,n}x_1\right) \\
&\text{BC1 Type III:} \quad X_{1,n} = \sin\left(\lambda_{1,n}x_1\right) + \frac{\lambda_{1,n}}{h}\cos\left(\lambda_{1,n}x_1\right)
\end{aligned} \tag{4.183}$$

The eigenfunctions of each of the remaining layers are represented by the expression:

$$X_{i,n}(x_i) = a_{i,n}\sin(\lambda_{i,n}x_i) + b_{i,n}\cos(\lambda_{i,n}x_i) \quad i=2\dots M \tag{4.184}$$

for which the coefficients $a_{i,n}$ and $b_{i,n}$ of each layer can be determined by building the eigenfunctions in a sequential manner. The interface conditions between layers can be used to arrive at a relationship that explicitly defines these constants based on the eigenfunction of the previous layer:

$$\left.\begin{array}{l} a_{i,n} = \dfrac{k_{i-1}}{k_i}\dfrac{1}{\lambda_{i,n}}\dfrac{dX_{i-1,n}}{dx_{i-1}}\bigg|_{L_{i-1}} \\[2mm] b_{i,n} = X_{i-1,n}(L_{i-1}) \end{array}\right\} \quad i=2\dots M \tag{4.185}$$

The eigenfunctions are built sequentially beginning with layer 2. The coefficients $a_{2,n}$ and $b_{2,n}$ are determined by setting $i=2$ in Equation (4.185) and then evaluating both the layer 1 eigenfunction and its first derivative evaluated at $x_1=L_1$. Once the layer 2 eigenfunction has been established, the process is repeated to evaluate the coefficients $a_{3,n}$ and $b_{3,n}$ of layer 3 by setting $i=3$ in Equation (4.185). This process is repeated for every subsequent layer until the coefficients $a_{M,n}$ and $b_{M,n}$ of the last layer, M, are determined.

In the development of the eigenfunctions, any appearance of the eigenvalue $\lambda_{i,n}$ should be replaced by that of the reference layer $\lambda_{O,n}$. The nth eigenvalue of each layer is related to that of the reference layer by

$$\lambda_{i,n} = \sqrt{\dfrac{\alpha_O}{\alpha_i}(\lambda_{O,n}^2+\omega_O^2)-\omega_i^2} \quad i=1\dots M. \tag{4.186}$$

Once all of the layers' eigenfunctions have been developed, the actual values of the reference layer eigenvalues, $\lambda_{O,n}$, may be determined from the BC2 at $x_M=L_M$. Here, transcendental equations are established based on BC type:

BC2 Type I, $\lambda_{O,n}$ are the roots of:

$$\tan\left(L_M\sqrt{\dfrac{\alpha_O}{\alpha_M}(\lambda_{O,n}^2+\omega_O^2)-\omega_M^2}\right) = -\dfrac{b_{M,n}}{a_{M,n}}$$

BC2 Type II, $\lambda_{O,n}$ are the roots of:
$$\cot\left(L_M\sqrt{\dfrac{\alpha_O}{\alpha_M}(\lambda_{O,n}^2+\omega_O^2)-\omega_M^2}\right) = \dfrac{b_{M,n}}{a_{M,n}} \tag{4.187}$$

And finally, the constant of integration, C_n, is developed from the application of the initial condition and for each term in the series, and is determined from the integral:

$$C_n = \dfrac{1}{N_n}\sum_{j=1}^{M}\dfrac{k_j}{\alpha_j}\int_{x_i'=0}^{L_i} X_{j,n}(x_j')\cdot\left(F_j^*(x_j')-\Psi_j(x_j')\right)dx_j' \tag{4.188}$$

The function $F_j^*(x_j')$ is the initial state of layer j at $t=0$. The function $\Psi_j(x_j')$ is the nonhomogenous steady solution component that results from the superpositioning of the nonhomogenous problem as described in Section *4.3.1*.

The norm, N_n, is the result of the sum of integrals over each layer:

$$N(\lambda_{0,n}) = \sum_{i=1}^{M} \frac{k_i}{\alpha_i} \int_{x_i=0}^{L_i} (X_{i,n})^2 dx_i \qquad (4.189)$$

The evaluation of this sum results in an expression in terms of the eigenfunctions and their first derivatives evaluated at the layers' boundaries and interfaces:

$$N(\lambda_{0,n}) = \frac{1}{2} \sum_{i=1}^{M} \frac{k_i}{\alpha_i \lambda_{i,n}^2} \left(\begin{array}{c} X_{i,n}(0) \cdot \frac{dX_{i,n}}{dx}\Big|_0 + (\lambda_{i,n} X_{i,n}(0))^2 \\ + \left(\frac{dX_{i,n}}{dx}\Big|_0\right)^2 - X_{i,n}(L_i) \cdot \frac{dX_{i,n}}{dx}\Big|_{L_i} \end{array} \right) \qquad (4.190)$$

The closure of the solution of Equation (4.188), requires the evaluation of the integral of the product of the steady components and the eigenfunction, $X_{j,n}\Psi_j$, and of the integral of the product of the initial condition and the eigenfunction $X_{j,n}F_j^*$. The steady component of each layer will satisfy the governing Equation (4.138). This allows for the result to be expressed:

$$\frac{k_j}{\alpha_j} \int_{x_j=0}^{L_j} (X_{j,n} \cdot \Psi_j) dx = \frac{k_j}{\alpha_j \left(\omega_j^2 + \lambda_{j,n}^2\right)} \left(X_{j,n} \cdot \frac{d\Psi_j}{dx_j} - \Psi_j \cdot \frac{dX_{j,n}}{dx_j} - \left(\frac{1}{\lambda_{j,n}^2} \frac{g_j}{k_j}\right) \frac{dX_{j,n}}{dx_j} \right) \Bigg|_{x_j=0}^{x_j=L_j} \qquad (4.191)$$

4.5.4.0.1 SPECIAL CASE SIMPLIFICATION

If, in the special case, the following conditions are met:

i. That the initial condition is governed by the same equation with the same parameter values as the steady component:

$$\frac{d^2 F_i^*(x_i)}{dx_i^2} - \omega_i^2 F_i^*(x_i) = -\omega_i^2 T a_i - \frac{g_i}{k_i} \quad in : 0 \le x_i \le L_i, \quad i=1\ldots M \qquad (4.192)$$

and

ii. That the interface conditions and associated parameter values of the steady function of Equation (4.139) are identical to those associated with the initial condition:

$$\left. \begin{array}{c} k_i \frac{dF_i^*}{dx_i}\Big|_{x_i=0} = k_{i-1} \frac{dF_i^*}{dx_{i-1}}\Big|_{x_{i-1}=L_{i-1}} \\ F_i^*(0) = F_i^*(L_{i-1}) \end{array} \right\} \quad i=2\ldots M \qquad (4.193)$$

Then the integrals of constant of Equation (4.188) may be simplified to state:

$$
C_n = \frac{1}{N_n} \frac{1}{\alpha_O \left(\omega_O^2 + \lambda_{O,n}^2 \right)} \times \left(\begin{array}{l} k_M X_{M,n}(L_M) \cdot \left(\left.\frac{d\Psi_M}{dx_M}\right|_{L_M} - \left.\frac{dF_M^*}{dx_M}\right|_{L_M} \right) \\[2ex] -k_M \left(\Psi_{M,n}(L_M) - F_M^*(L_M) \right) \cdot \left.\frac{dX_M}{dx_M}\right|_{L_M} \\[2ex] -k_1 X_{1,n}(0) \cdot \left(\left.\frac{d\Psi_1}{dx_1}\right|_0 - \left.\frac{dF_M^*}{dx_1}\right|_0 \right) \\[2ex] +k_1 \left(\Psi_1(0) - F_1^*(0) \right) \cdot \left.\frac{dX_1}{dx_1}\right|_0 \end{array} \right) \tag{4.194}
$$

4.5.5 Addressing Transient Energy Generation Using Green's Functions

In the implementation of superpositioning to homogenize the composite bioheat problem described in Section *4.3.1*, all of the nonhomogenous terms in each layer were steady. In this section, the composite bioheat problem is considered for which the energy generation in each layer is allowed to vary with time as well as with position. This can be accomplished through the application of both superpositioning and the method of Green's functions for the composite bioheat problem. Consider the composite bioheat problem with a homogenous perfusion term and with transient energy generation for which each layer is governed by:

$$
\frac{1}{\alpha_i} \frac{\partial T_i}{\partial t} = \frac{\partial^2 T_i}{\partial x_i^2} - \omega^2 T_i + \frac{1}{k_i} g_i(x_i, t); \quad \text{in } 0 < x_i < L_i, \ \ 0 < t, \ \ i = 1 \ldots M \tag{4.195}
$$

The temperatures at the interfaces are subject to

$$
\left. \begin{array}{l} k_i \left.\frac{\partial T_i}{\partial x_i}\right|_{x_i=0} = k_{i-1} \left.\frac{\partial T_{i-1}}{\partial x_{i-1}}\right|_{x_{i-1}=L_{i-1}} \\[2ex] T_i(0) = T_{i-1}(L_{i-1}) \end{array} \right\} \quad i = 2 \ldots M. \tag{4.196}
$$

The outer boundary of layer 1 can be exposed to any one of the following homogenous conditions:

$$
\begin{array}{l} \text{BC1 Type I} \quad T_1(0,t) = 0 \\[2ex] \text{BC1 Type II} \quad \left.\frac{\partial T_1}{\partial x_1}\right|_{x_1=0} = 0 \\[2ex] \text{BC1 Type III} \quad k_1 \left.\frac{\partial T_1}{\partial x_1}\right|_{x_1=0} - h \cdot T_1(0,t) = 0 \end{array} \tag{4.197}
$$

The boundary nearest the core may be subject to either of the homogenous conditions:

$$
\begin{array}{l} \text{BC2 Type I} \quad T_M(L_M, t) = 0 \\[2ex] \text{BC2 Type II} \quad \left.\frac{\partial T_M}{\partial x_M}\right|_{x_M=L_M} = 0 \end{array} \tag{4.198}
$$

The initial temperature profile in each layer is arbitrarily described by

$$T_i(x_i, t=0) = F_i(x_i).$$ (4.199)

Note that while the function $F_i(x_i)$ is not specified here, in practice it will be composed of (i) the function representing the actual initial condition and (ii) the steady function that is a solution to the nonhomogenous subproblem that addresses all of the steady nonhomogenous terms related to perfusion and to the boundary condition at $x_1 = 0$.

The GFSE of the composite heat conduction problem represented by Equations (4.195)–(4.199) in the absence of perfusion ($\omega = 0$) is presented in Section 10-6 of Ref. [9]. Following that description, the GFSE of the bioheat problem can be represented:

$$T_i(x_i, t) = \sum_{j=1}^{M} \int_{x'_j=0}^{L_j} G_{ij}^P\left(x_i, t | x'_j, \tau = 0\right) \cdot F\left(x'_j, \tau\right) dx'_j + \sum_{j=1}^{M} \int_{\tau=0}^{t} \int_{x'_j=0}^{L_j} G_{ij}^P\left(x_i, t | x'_j, \tau\right) \cdot \left(\frac{\alpha_j}{k_j} g_j\left(x'_j, \tau\right)\right) dx'_j, \quad i = 1 \ldots M$$

(4.200)

Recall the results of Example 4.4 in Section **4.3.2** that led to the adaptation of the single layer Green's functions of the pure heat conduction problem by the transformation of Equation (4.103). These results are equally applicable to the composite problem, so that:

$$G_{ij}^P\left(x_i, t | x'_j, \tau\right) = G_{ij}\left(x_i, t | x'_j, \tau\right) \exp\left(-\alpha_0 \omega_0^2 (t - \tau)\right)$$ (4.201)

where G_{ij} is the Green's function of the corresponding heat conduction problem. Note that the transient exponential function can be written in terms of the reference layer parameters. Using this relation, the Green's function of the composite conduction problem presented in Ref. [9] is transformed so that the Green's function of the composite bioheat problem is represented by

$$G_{ij}^P\left(x_i, t | x'_j, \tau\right) = \sum_{n=1}^{\infty} \frac{1}{N_n} X_{i,n}(x_i) \cdot \exp\left(-\alpha_0 (\lambda_{0,n}^2 + \omega_0^2)(t - \tau)\right) \cdot \left(\frac{k_j}{\alpha_j} X_{j,n}\left(x'_j\right)\right), \quad i = 1 \ldots M \quad (4.202)$$

where the eigenfunctions $X_{i,n}(x_i)$ of each layer are built in the same manner described in Equations (4.184) and (4.185), and the norm, N_n, is determined through Equation (4.190). The indices i and j are used to indicate different layers which are needed in the summations of the GFSE (4.200).

4.6 SUMMARY REMARKS

The field of bioheat transfer is dynamic and continues to be relevant. This chapter has been motivated by the high value offered by the exact solutions to the simple Penne's bioheat representation of heat transfer in perfuse living tissue. Because analytical heat conduction is well studied and documented, the approach used in this chapter is to present methods that allow the bioheat solution to be adapted from the transient solution of the corresponding heat conduction problem. In this chapter, the SOV method is used to provide some solutions to the homogenous bioheat problem. These can be used in conjunction with superpositioning in order to find the solution to the nonhomogenous problem. In order to address the presence of unsteady boundary conditions and unsteady generation terms, Green's functions have been developed for the bioheat problem. It is shown that the readily available documented Green's functions of the heat conduction problem may be amended so that they are representative of the corresponding bioheat problem.

References

[1] Pennes HH. Analysis of tissue and arterial blood temperatures in the resting human forearm. J Appl Physiol 1948;1:93–122.
[2] Nelson DA. Pennes' 1948 paper revisited. J Appl Physiol 1998;85:2–3.
[3] Wissler EH. Pennes' 1948 paper revisited. J Appl Physiol 1998;85:35–41.
[4] Xu F, Lu TJ, Seffen KA, Ng EYK. Mathematical modeling of skin bioheat transfer. Appl Mech Rev 2009;62:35.
[5] Chato JC. Heat-transfer to blood-vessels. J Biomech Eng 1980;102:110–8.
[6] Carslaw HS, Jaeger JC. Conduction of heat in solids. London: Clarendon Press; 1959.
[7] Özişik MN. Heat conduction. New York: John Wiley and Sons; 1993.
[8] Beck JV, Cole KD, Haji-Sheikh A, Litkouhi B. Heat conduction using green's functions. New York: Hemisphere Publishing Corporation; 1992.
[9] Hahn DW, Özisik MN. Heat conduction. Hoboken: Wiley; 2012.
[10] Becker S. One-dimensional transient heat conduction in composite living perfuse tissue. J Heat Transf 2013;135:071002.
[11] Becker SM, Herwig H. One dimensional transient heat conduction in segmented fin-like geometries with distinct discrete peripheral convection. Int J Therm Sci 2013;71:148–62.

Characterizing Respiratory Airflow and Aerosol Condensational Growth in Children and Adults Using an Imaging-CFD Approach

Jinxiang Xi[a], Xiuhua A. Si[b], Jong Won Kim[a]

[a]Central Michigan University, Mount Pleasant, MI, USA
[b]Calvin College, Grand Rapids, MI, USA

5.1 INTRODUCTION

Aerosol pollution particulates, which include airborne solid particles and liquid droplets, come in a range of sizes. Particulate matters smaller than 2.5 μm ($PM_{2.5}$) have been shown to pose the greatest risk to human health because they are small enough to be breathed into the lungs and enter the blood stream [1]. Among $PM_{2.5}$, fine-regime aerosols (100 nm-2.5 μm) typically deposit in the respiratory tract at a reduced rate in comparison with smaller ultrafine (<100 nm) and larger coarse (>2.5 μm) aerosols due to a minimum sum of diffusion, sedimentation, and impaction [2]. However, the deposition of hygroscopic aerosols of this size-regime can be complicated by the growth or shrinkage of such aerosols as they travel through the warm-humid respiratory tract. An aerosol particle or droplet containing water-soluble species has a reduced surface vapor pressure, and allows vapor condensation to occur even in a subsaturated environment. Many inhalable environmental and pharmaceutical aerosols are water soluble. As a result, hygroscopic effects are expected to influence the transport and deposition of such inhaled aerosols [3–8].

Fine-regime water-soluble aerosols can come from a variety of sources such as fossil fuel combustion [9], diesel exhaust (50-500 nm) [10], tobacco smoke (140-500 nm) [11,12], and radioactive decay (1-200 nm) [2]. Combustion produces large amounts of sulfur dioxide, which reacts with water vapor and other gases in the atmosphere to create sulfate aerosols [13]. Biomass burning yields smoke that is comprised mainly of organic carbon and black carbon [14]. Fine-regime airborne bioaerosols include respiratory-specific viruses such as Avian flu and SARS, which typically range from 20 to 200 nm [15]. Even indoors, cigarette smoke, cooking stoves fireplaces, and candles are sources of fine aerosols. These aerosols may deposit in the respiratory airways in discrete amounts resulting in local injury and spread of infectious diseases. Considering the extrathoracic nasal airways, which include the nasal passages, pharynx, and larynx, the deposition of submicrometer aerosols is associated with a number of detrimental health effects. The deposition of cigarette smoke particles (**CSPs**) has been quantitatively linked to the formation of respiratory tract tumors at specific sites [16]. Yang et al. [17] reported that respiratory tract cancers per unit surface area are approximately 3000 times more likely to occur in the extrathoracic airways including the larynx.

Compared to adults, children are more susceptible to respiratory risks due to their immature immune systems. Exposure to ambient toxicants in children may cause adverse effects such as asthma, sinusitis, and nasal carcinomas. Respiratory disease remains a leading contributor to childhood morbidity in the United States and other developed countries and is a leading cause of childhood deaths worldwide [18]. Moreover, children usually spend more time outdoors, breath faster, and may possibly receive a higher dose of ambient pollutants than adults. Children with a history of asthma often exhibit elevated levels of other respiratory symptoms such as cough, bronchitis, or pneumonia when living in highly polluted areas [19]. Specifically, passive tobacco smoke exposure for children may cause various health problems, such as increasing the frequency of asthma, aggravating chronic respiratory problems, and inducing recurrent ear infections [19].

Hygroscopic particle growth has been investigated experimentally for combustion particles [6], NaCl droplets [20], and pharmaceutical aerosols [21]. A number of mathematical models have also been formulated to simulate the hygroscopic growth of respiratory aerosols at relative humidity (RH) values below 100% [3,4,22,23]. These studies typically indicate a maximum size increase of approximately 400% for NaCl droplets. The hygroscopic growth of most other salts and most pharmaceutical aerosols result in size increases less than 100% at RH values of 99.5% and below. Zhang et al. [24] developed a CFD (computational fluid dynamics) model of hygroscopic growth in the upper respiratory tract. It was found that saline concentrations of 10% and higher were required for hygroscopic growth to have a noticeable impact on deposition. Finlay and Stapleton [5] applied a numerical model to show that coupling between aerosol droplets and the continuous phase was significant for droplet concentrations above 25,000 particles/cm^3.

Very few studies that evaluated the effects of supersaturated conditions (RH > 100%) on the growth rate of respiratory aerosols have been reported. Recent studies by Longest and Xi [25] and Kim et al. [26] showed that supersaturated conditions could be achieved in the respiratory tract through the inhalation of warm saturated air, which induced much higher particle growth rates by vapor condensation. Ferron et al. [27] showed that supersaturated conditions were possible in the respiratory tract under normal inhalation conditions. Significant growth of NaCl particles was observed for very localized supersaturation in the nasal cavity [27]. In a series of studies [28–30], Longest and coworkers have shown that enhanced condensational growth did improve drug delivery efficiency to the lung using an adult airway model. Kim et al. [26] evaluated the dynamic growth and deposition in a child nasal airway model and reported that hygroscopic and condensation growth of water-soluble aerosols did not necessarily lead to enhanced deposition due to the constantly changing aerosol sizes and deposition mechanisms. Nearly all previous studies concentrated on submicrometer particles, for which hygroscopic growth is believed to be most significant [25]. However, particle growth due to condensation within a supersaturated environment could be more pronounced and could affect both submicrometer and micrometer aerosols. No study has been reported that evaluated the nasal deposition of micrometer water-soluble aerosols under supersaturated conditions. An accurate knowledge of the dynamic behaviors of such aerosols and their subsequent deposition in the human upper respiratory tract is critical in order to make reliable toxicology risk assessments, establish appropriate environmental standards, and determine safe exposure limits.

In this chapter, we will describe the development of an aerosol condensation model as well as its application in several physiologically realistic respiratory airway models. The implications of particle evaporation and condensation in environmental health assessment and inhalation drug delivery will also be discussed. Specific aims include: (1) characterizing the aerodynamic (i.e., velocity and turbulence) and thermodynamic (i.e., temperature and RH) fields within the respiratory tract under different inhalation psychrometric conditions; (2) characterizing the dynamic behaviors of initially monodisperse aerosols in terms of diameter change, trajectory, and size distribution; and (3) quantifying the subsequent deposition of such hygroscopic aerosols in the nasal and mouth-lung airways on a total, regional, and localized basis.

5.2 METHODS

The respiratory airway geometries in this chapter included an adult nose, a child nose, and an adult mouth-lung. A well validated hygroscopic growth model was employed to evaluate the heat-mass transfer and particle transport in the nasal airway. Initially monodisperse aerosols ranging 0.2-2.5 μm were considered. Four inspiratory scenarios were considered which included air of cold-dry (Case 1: $T = 23\,^{\circ}C$, RH = 30%), cool-mild (Case 2: $T = 27\,^{\circ}C$, RH = 60%), warm-humid (Case 3: $T = 40\,^{\circ}C$, RH = 100%), and hot-humid (Case 4: $T = 47\,^{\circ}C$, RH = 100%), as listed in Table 5.1. The deposition of inert particles in the adult nose-throat

TABLE 5.1   Initial Temperature and Relative Humidity Conditions at the Nasal Inlets

Case	Temperature (*T*)	Relative humidity (RH)	Environment
1	23 °C (296.15 K)	30%	Cold-dry
2	27 °C (300.15 K)	60%	Cool-mild
3	40 °C (313.15 K)	100%	Warm-humid
4	47 °C (320.15 K)	100%	Hot-humid

replica cast was measured for validation purposes. More details of the nasal airway model, inhalation conditions, and flow-particle transport and growth models are provided in the following.

5.2.1 Construction of Airway Models

A physiologically accurate airway model is a critical first step for reliable analysis of inhalation deposition. Imaged-based modeling represents a remarkable improvement over conventional cadaver casting that is subject to large distortions due to the shrinkage of mucous membranes or insertion of casting materials. The adult and child airway models in this chapter were developed from MRI scans of a 53-year-old male (weight 73 kg and height 173 cm) and a 5-year-old boy (weight 21 kg and height 109 cm), respectively [31]. To construct the 3D airway model, CT scans of a healthy nonsmoking 53-year-old male were used, which were obtained with a multirow-detector helical CT scanner (GE medical systems, Discovery LS, Milwaukee, WI) with the following acquisition parameters: 0.7 mm effective slice spacing, 0.65 mm overlap, 1.2 mm pitch, and 512×512 pixel resolution. The multislice CT images were imported into MIMICS (Materialise, Ann Arbor, MI) and were segmented according to the contrast between osseous structures and intra-airway air to convert the raw image data into a set of cross-sectional contours that define the solid geometry. Based on these contours, a surface geometry was constructed in Gambit 2.4 (Ansys, Inc.). This surface geometry was then imported into ANSYS ICEM (Ansys, Inc., Canonsburg, PA) as an IGES file for meshing. To accurately represent the anatomy of the extrathoracic airway, anatomical details such as the epiglottal fold and laryngeal sinus were retained in both subjects (Figure 5.1a). These models could be used for both *in vitro* measurement and numerical analysis. A life-size adult nasal airway replica cast was manufactured using a 3-D printer (Dimension 1200es, Eden Prairie, MN). An unstructured tetrahedral mesh was created for each model using ANSYS ICEM (Ansys, Inc.) with high-resolution pentahedral elements in the near-wall region (Figure 5.1b).

In order to quantitatively evaluate the flow field and aerosol deposition, the airway models were divided into different sections (Figure 5.1a). Take the adult nasal model as an example: the divided sections include the nasal vestibule and valve (V&V) region, turbinate region (TR), nasopharynx (NP), pharynx, and larynx. On top of the nasal airway is the olfactory region (OR) where the sensory nerves are located (Figure 5.1a). The coronal passages have been further subdivided in the vertical direction, including the superior, middle, and inferior

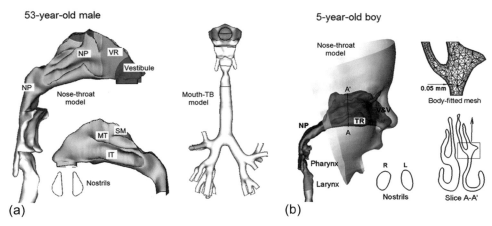

FIGURE 5.1   Image-based respiratory airway models for (a) a 53-year-old adult male and (b) a 5-year-old boy. The computational mesh of each model is composed of approximately 1.75 million unstructured tetrahedral elements and a fine near-wall pentahedral grid.

sections in Slice 1-1′ (nasal valve region) and the medial passage (MP), superior meatus (SM), middle meatus, and inferior meatus in Slice 2-2′ (middle TR).

5.2.2 Breathing and Wall Boundary Conditions

Steady inhalation was assumed for all simulations with uniform velocity profiles at both nostrils. Initial particle velocities were assumed to match the local fluid velocity. The airway surface was assumed smooth and rigid with no-slip conditions. The surface temperature and the RH were assumed to be constant and equal to mean body conditions (37 °C and 99.5%). Inhaled components considered in this chapter include air, water vapor, and water-soluble aerosols. The inhaled aerosols were assumed to have a density of 1.0 g/cm^3 and be dilute in concentration. Aerosols were assumed to deposit on the airway wall upon initial contact.

Four inspiratory psychrometric conditions were selected to evaluate the dynamic growth and deposition of water-soluble aerosols (Table 5.1). These boundary conditions were used to determine the airway temperature and RH fields by solving the conservation equations of energy and mass in the nose-throat model geometry. The first condition (Case 1) represents a cold-dry environment. In this subsaturated condition, initial evaporation of existing water from the particles or droplets is expected. Water evaporation from the airway walls will be required to recover the body RH condition and fuel hygroscopic growth of inhaled aerosols. The second condition (Case 2) represents a cool-mild environment with an RH of 60%. The third and fourth conditions (Cases 3 and 4) exemplify humid warm and hot environments, respectively. Case 3 is a saturated condition that is only 3 °C above body temperature. Case 4 provides an upper temperature boundary condition to evaluate the maximum extent of condensation growth, which is selected as 10 °C above the body temperature. With both saturated vapor conditions, cooling inside the nose will result in supersaturated RH fields and induce condensation growth of the inhaled particulates.

5.2.3 In Vitro Deposition Measurement

The deposition of inert particles in the adult nose-throat cast was measured for validation purposes. A vibrating orifice aerosol generator (Model 3050, TSI, Inc., St. Paul, MN) was used to generate the monodisperse oleic acid aerosols. An aerodynamic particle sizer (TSI, Inc., St. Paul, MN) was used to monitor and adjust the aerosol size distribution. Aerosols were neutralized by a Kr-85 source before entering the chamber. Three identical replica casts were placed at the end of the chamber so that the average deposition results could be obtained. Each replica was connected with a 25-mm cellulose filter (Tisch Environmental, Village of Cleves, OH) to collect aerosols that escape deposition. To mimic the natural mucus lining airway surfaces, a grease coating was applied on the surface of the replica by filling up and draining out with silicon oil (Dow Corning 550 Fluid, Dow Corning, Inc., Midland, MI). After the test, the replicas were rinsed with a solution of 50% isopropyl and 50% distilled water to wash out oleic acid particles that deposited in the replica. The relative concentrations of fluorescent tracers in the solutions were measured with a fluorometer (Model 450, Sequoia-Turner Corp., Mountain View, CA). Detailed procedures on aerosol preparation and sampling can be found in Zhou et al. [32].

5.2.4 Continuous and Discrete Particle Transport Equations

The bulk flow considered in this chapter was assumed to be isothermal and incompressible. Multiregime flow dynamics can coexist in the respiratory tract due to its unique physiology. To resolve the possible laminar-transitional-turbulent flow conditions, the low Reynolds number (LRN) k-ω model [33] was selected based on its ability to accurately predict pressure drop, velocity profiles, and shear stress for transitional and turbulent flows. Moreover, the LRN k-ω model was shown to provide an accurate solution for laminar flow as the turbulent viscosity approaches zero [33]. The governing equations for the conservation of mass and momentum are

$$\frac{\partial u_i}{\partial x_i} = 0 \tag{5.1}$$

$$\frac{\partial u_i}{\partial t} + \overline{u}_j \frac{\partial u_i}{\partial x_j} = -\frac{1}{\rho}\frac{\partial p}{\partial x_i} + \frac{\partial}{\partial x_j}\left[(\nu + \nu_T)\left(\frac{\partial u_i}{\partial x_j} + \frac{\partial u_j}{\partial x_i}\right)\right] \tag{5.2}$$

where x_i is the coordinates in three directions, u_i is the velocity in three coordinate directions (i.e., i or $j=1$, 2, and 3), p is the time-averaged pressure, ρ is the fluid density, and ν is the kinematic viscosity. The turbulent viscosity ν_T is defined as $\nu_T = \alpha^* k/\omega$. For the LRN k-ω approximation, which models turbulence through the viscous sublayer, the α^* parameter in the previous expression for turbulent viscosity is evaluated as [33]:

$$\alpha^* = \frac{0.024 + k/6\nu\omega}{1.0 + k/6\nu\omega} \tag{5.3}$$

For laminar flow, ν_T is zero and only Equations (5.1) and (5.2) are solved. Transport equations governing the turbulent kinetic energy (k) and the specific dissipation rate (ω) are

$$\frac{\partial k}{\partial t} + \bar{u}_j \frac{\partial k}{\partial x_j} = \tau_{ij} \frac{\partial \bar{u}_i}{\partial x_j} - \varepsilon_k + \frac{\partial}{\partial x_j}\left[(\nu + 0.5\nu_T)\left(\frac{\partial k}{\partial x_j}\right)\right] \tag{5.4}$$

$$\frac{\partial \omega}{\partial t} + \bar{u}_j \frac{\partial \omega}{\partial x_j} = \frac{13}{25}\frac{\omega}{k}\tau_{ij} \frac{\partial \bar{u}_i}{\partial x_j} - \varepsilon_\omega + \frac{\partial}{\partial x_j}\left[(\nu + 0.5\nu_T)\left(\frac{\partial \omega}{\partial x_j}\right)\right] \tag{5.5}$$

In the previous equations, τ_{ij} is the shear stress tensor and ε_k and ε_ω represent the dissipation of k and ω, respectively [33].

The field distribution of temperature (T) and vapor (Y_v) in the nasal airway are governed by the following heat and mass transfer equations [34]:

$$\rho C_p \frac{\partial T}{\partial t} + \rho C_p \frac{\partial u_j T}{\partial x_j} = \frac{\partial}{\partial x_j}\left[\left(\kappa_g + \frac{\rho C_p \nu_T}{Pr_T}\right)\left(\frac{\partial T}{\partial x_j}\right) + \sum_s h_s\left(\rho \tilde{D}_v + \frac{\rho \nu_T}{Sc_T}\right)\frac{\partial Y_s}{\partial x_j}\right] \tag{5.6}$$

$$\frac{\partial Y_v}{\partial t} + \frac{\partial u_j Y_v}{\partial x_j} = \frac{\partial}{\partial x_j}\left[\left(\tilde{D}_v + \frac{\nu_T}{Sc_T}\right)\left(\frac{\partial Y_v}{\partial x_j}\right)\right] + S_v \tag{5.7}$$

In the above expressions, C_p is the constant specific heat, κ_g is the air thermal conductivity, Y_v is the mass fraction of water vapor, $\tilde{D}_v$ is the binary diffusion coefficient of water vapor in air, Pr_T is the turbulent Prandtl number ($Pr_T = 0.9$) and Sc_T is the turbulent Schmidt number ($Sc_T = 0.9$). The enthalpy of the air and water vapor is represented as h_s. In the mass transport relation, the transport of thermal energy due to diffusion was excluded based on Lewis numbers close to one for both air and water vapor. For the two species considered, the mass fraction of air was evaluated as $Y_a = 1.0 - Y_v$.

The RH of the mixture entering the upper airway depends on the local temperature and mixture density. The local RH will influence water vapor evaporation and condensation on the surface of droplets and on the walls of the geometry. RH of the ideal gas mixture can be expressed as:

$$RH = \frac{P_v}{P_{v,sat}} = \frac{Y_v \rho_m R_v T}{Y_{v,sat}\rho R_v T} = \frac{Y_v}{Y_{v,sat}} \tag{5.8}$$

where R_v is the gas constant of water vapor and ρ_m is the mixture density. The saturation vapor pressure $P_{v,sat}$ is computed from the Antoine equation [35]. The mass flux of water vapor at the wall depends on the local RH conditions. For near-wall RH values greater than 1, condensation onto the wall surface occurs. If the near-wall RH is less than one, then evaporation from the surface is assumed.

The hygroscopic growth of initially submicrometer aerosols in the upper airway is expected to result in a polydisperse distribution of aerosols. The transport and deposition of these hygroscopic particulates or droplets were simulated using a well-tested discrete Lagrangian tracking model enhanced with near-wall treatment. In our previous studies, the Lagrangian tracking model enhanced with user-defined routines was shown to provide close agreement with experimental deposition data in upper respiratory airways for both submicrometer [36] and micrometer particles [37]. The discrete Lagrangian transport equations can be expressed as:

$$\frac{dv_i}{dt} = \alpha \frac{\partial u_i}{\partial t} + \frac{f}{\tau_p C_c}(u_i - v_i) + g_i(1 - \alpha) + f_{i,\text{Brownian}} \tag{5.9}$$

$$\frac{dx_i}{dt} = v_i(t) \tag{5.10}$$

where v_i is the particle velocity and α is the air-particle density ratio. The particle residence time τ_p is defined as $\rho_p d_p^2 / 18\,\mu$, with μ being the air viscosity and d_p the particle diameter [38]. The drag factor f, which represents the ratio of the drag coefficient C_D to Stokes drag, is based on the expression of Morsi and Alexander [39]. The Brownian force is of the form [40]

$$f_{i,\text{Brownian}} = \frac{\varsigma_i}{m_d}\sqrt{\frac{1}{\widetilde{D}_p}\frac{2k_B^2 T^2}{\Delta t}} \tag{5.11}$$

where ς_i is a zero mean variant from a Gaussian probability density function [40], k_B is the Boltzmann constant, Δt is the time-step for particle integration, and m_d is the mass of the droplet. Assuming dilute concentrations of spherical particles, the Stokes-Einstein equation was used to determine the diffusion coefficients for various size particles as:

$$\widetilde{D}_p = \frac{k_B T C_c}{3\pi \mu d_p} \tag{5.12}$$

where $k_B = 1.38 \times 10^{-16}$ cm^{2}g/s is the Boltzmann constant in cgs units. The Cunningham correction factor C_c was computed using the expression of Allen and Raabe [41]

$$C_c = 1 + \frac{\lambda}{d_p}\left(2.34 + 1.05\,\exp\left(-0.39\frac{d_p}{\lambda}\right)\right) \tag{5.13}$$

where λ is the mean free path of the dry air, assumed to be 65 nm. The previous expression has been reported to be valid for all particle sizes [38]. The influence of nonuniform fluctuations in the near-wall region was taken into account by implementing an anisotropic turbulence model proposed by Matida et al. [42]. A near-wall interpolation algorithm was implemented to precisely determine the fluid velocity at the particle location. This near-wall treatment has been shown to provide an effective approach to accurately predict submicrometer depositions in the respiratory tract [43].

5.2.5 Droplet Evaporation and Condensation Model

A distinction can be made between condensation and hygroscopic growth, as illustrated in Figure 5.2. Droplet growth due to condensation, or heterogeneous nucleation, is the general phenomena arising from RH values greater than 100% [38], where there is more vapor in the mixture than the air can possibly hold (Figure 5.2). A special case of condensation growth arises from droplets that contain a soluble component. The inclusion of the water-soluble species reduces the water vapor pressure on the droplet surface, which allows condensation to take

FIGURE 5.2 Psychrometric chart for the four test cases: (a) Case 1: $T=23\,°C$, RH=30%; (b) Case 2: $T=27\,°C$, RH= 60%; (c) Case 3: $T=40\,°C$, RH=100%; (d) Case 4: $T=47\,°C$, RH=100%. *Condensation* occurs when the vapor pressure is larger than the saturation pressure, while *evaporation* occurs when the vapor pressure is less than the saturation pressure. Water-soluble contents reduce the droplet surface vapor pressure and cause *hygroscopic* growth even below the droplet saturation pressure. Humidity ratio denotes the air water content (g water/kg dry air).

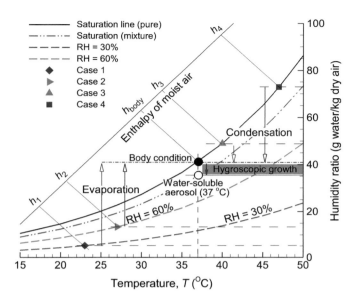

place at RH values even below saturation conditions. Condensation onto a droplet at RH values below 100% due to a reduction in surface vapor pressure is referred to as hygroscopic growth in this chapter (gray area in Figure 5.2). Most available CFD packages, including Fluent (Ansys, Inc., Canonsburg, PA) do not simulate the condensation of a vapor species into a discrete phase. In this chapter, we implemented a specific user-defined function (UDF) that had been developed to simulate condensation of water vapor onto soluble particles and droplets [25]. The hygroscopic rapid-mixing model employed here is based on previous approximations for salts [3,4,38], CSPs [6,22], and multicomponent evaporating liquid droplets in the respiratory tract [44]. In this model, the inhaled aerosols are assumed to consist of liquid water (90% mole fraction), water-soluble components, and nonsoluble components, among which only water is volatile. Following Raoult's Law [45], the droplet vapor pressure will be reduced by 10% (red dash-dot-dot line in Figure 5.2) and creates the potential for hygroscopic growth at RH values less than 100% (gray area in Figure 5.2). The associated reduction in surface vapor pressure will also increase the rate of condensation growth at RH values greater than 100%.

Conservation of energy and mass for an immersed droplet, indicated by the subscript d, under rapid mixing model (RMM) conditions can be expressed [44]:

$$\frac{dT}{dt}m_d C_{pd} = -\int_{surf} q_{conv}\, dA - \int_{surf} n_v L_v\, dA = -\bar{q}_{conv}\cdot A - \bar{n}_v L_v \cdot A \qquad (5.14)$$

$$\frac{d(m_d)}{dt} = -\int_{surf} n_v dA = -\bar{n}_v \cdot A \qquad (5.15)$$

In the previous equations, m_d is the droplet mass, C_{pd} is the composite liquid specific heat, q_{conv} is the convective heat flux, n_v is the mass flux of the evaporating water vapor at the droplet surface, and L_v is the latent specific heat of the water vapor component.

The concentration of water vapor on the droplet surface can be approximated using Raoult's Law for dilute concentrations of soluble components:

$$Y_{v,\,surf} = \frac{X_w K P_{v,sat}(T_d)}{\rho R_v T_d} \tag{5.16}$$

In the previous expression, $P_{v,sat}(T_d)$ is the temperature dependent saturation pressure of water vapor and the X_w is the mole fraction of liquid water.

5.2.6 Numerical Method and Convergence Sensitivity Analysis

To solve the governing momentum, energy, and mass conservation equations in each of the cases considered, the CFD package ANSYS Fluent was employed. User-supplied FORTRAN and C programs were implemented for the calculations of inlet flow and particle profiles, particle transport and deposition locations, grid convergence, and deposition enhancement factors. A specific set of UDFs was also applied that considered the anisotropic turbulence effect [43], the near-wall velocity interpolation [46], and the hygroscopic growth model [25]. All transport equations were discretized to be at least second order accurate in space. A segregated implicit solver was employed to evaluate the resulting linear system of equations. This solver uses the Gauss-Seidel method in conjunction with an algebraic multigrid approach [47] for improving the calculation performance on tetrahedral meshes. Convergence of the flow field solution was assumed when the global mass residual was reduced from its original value by five orders of magnitude and when the residual-reduction rates for both mass and momentum were sufficiently small.

The computational mesh was generated with ANSYS ICEM 12.0 (Ansys, Inc.). For each model, an unstructured tetrahedral mesh was generated with high-resolution prismatic cells in the near-wall region (Figure 5.1b) to accommodate the high complexity of the model geometry. A grid sensitivity analysis was conducted in the adult nose-throat model by testing the effects of different mesh densities with approximately 0.54million (M) control volumes, 1 M control volumes, 1.75 M control volumes, and 2.81 M control volumes while keeping the near-wall cell height constant at 0.05 mm [48,49]. Increasing grid resolution from 1.75 to 2.81 M control volumes resulted in total deposition changes of less than 1%. As a result, the final grid for reporting flow field and deposition conditions consisted of approximately 1.75 M cells. For discrete Lagrangian tracking, the number of seeded particles required to produce count-independent depositions was tested. Particle count sensitivity analysis was performed by incrementally releasing groups of 10,000 particles. The number of groups was increased until the deposition rate changed by less than 1%. Due to low deposition rates, fine-regime aerosols require a larger amount of particles to produce count-independent results relative to micrometer aerosols. The final number of particles tracked was 60,000.

5.3 RESULTS

5.3.1 Child-Adult Discrepancies

The airway differences between the 5-year-old boy and the 53-year-old male are apparent, both in morphology and dimension (Figure 5.3). Considering the airway morphology, the child has smaller nostrils, shorter turbinates, a slenderer NP, and a thinner pharynx-larynx.

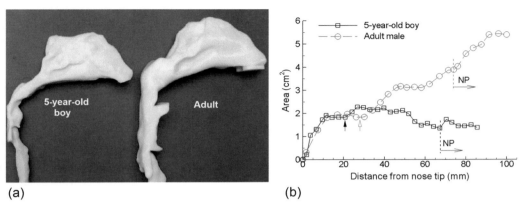

FIGURE 5.3 Comparison of the child and adult nasal models: (a) geometry; (b) cross-sectional area as a function of the distance from the nose tip.

Moreover, the nostril shape appears oval in the child compared to a wedge shape in the adult (Figure 5.3a). The nasal airway volume and surface area of the 5-year-old child is approximately 40.3% and 65.7% that of the adult, respectively. It is expected that as a human grows from birth to adulthood, the airway geometry keeps changing, which in turn alters the nasal aerodynamics as well as the deposition patterns of inhaled aerosols.

A quantitative comparison of airway dimensions between these two subjects is shown in Figure 5.3b in terms of coronal cross-sectional area as a function of distance from the nose tip. Two major child-adult differences are noted. First, the child has a much narrower and smaller NP lumen. From Figure 5.3b, the NP cross-sectional area of the child ($A = 170$ mm^2) is only one-third that of the adult ($A = 535$ mm^2). Second, the distance to the nasal valve (the minimum cross-sectional area) is similar between the child (20.8 mm, black arrow) and adult (27.2 mm, red arrow). This observation may imply that the nasal valve matures at early ages (around 5 years old). A child's airway is not merely a smaller version of an adult one: its growth could vary at different parts, with certain parts maturing earlier than others. As a result, the age-specific airway models are necessary for reliable predictions of airflow and aerosol depositions in children and adults.

5.3.2 Hygroscopic Growth Model Testing

To evaluate the accuracy of the hygroscopic RMM, predicted results have been validated with the experimental data of Li and Hopke [6] and Ishizu et al. [50]. Figure 5.4a shows the growth rate (d/d_0) of CSPs across a range of initial particle sizes (d_0) in comparison with the experiential data of Li and Hopke [6] at an RH of 99.75%. The properties used for mainstream (MS) and sidestream (SS) CSPs are given in Table 5.2. For both MS and SS particles, predictions of the hygroscopic RMM appear to match well with the trend of experimental data and therefore provide a reasonable estimate of the size increase of hygroscopic aerosols. As the particle increases in size, the competing influences of the surface tension (Kelvin effect), hygroscopic properties, and dilution reach an equilibrium resulting in the observed steady state particle diameters.

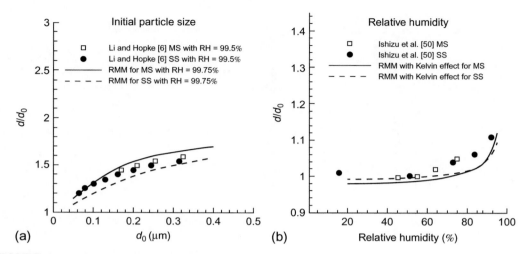

FIGURE 5.4   Comparison of predicted equilibrium growth rate as a function of (a) initial particle size (d_0) and (b) relative humidity (RH) with existing experimental data.

TABLE 5.2   Initial Properties of Mainstream (MS) and Sidestream (SS) Tobacco Smoke Particles

Smoke condition	Mole. Wt. of soluble component (kg/kmol)	Initial water mass fraction (%)	Initial mass ratio of soluble to nonsoluble components (%)	Density, ρ_d (g/cm^3)
MS	340 (Ref. [50])	8-25 (Ref. [50])	60 (Ref. [6])	1.0
SS	450 (Ref. [50])	6-25 (Ref. [50])	54 (Ref. [6])	1.0

To further evaluate the effect of RH on the hygroscopic particle growth, predicted results are compared with the experimental data of Ishizu et al. [50] in Figure 5.4b for RH ranging from 18% to 100%. Based on the experimental data for MS and SS particles, relatively little growth is observed for RH values below 60%. In contrast, equilibrium particle growth ratios increase exponentially as the RH approaches 100%. Results of the numerical model appear to match the experimental data to a high degree across the range of RH values considered (Figure 5.4b). Furthermore, relatively little difference is observed in the growth ratios of MS and SS particles for both the experimental results and model predictions.

5.3.3 Adult Nasal Airway Model

5.3.3.1 Airflow, Temperature, and RH Field

Detailed knowledge of aerodynamics is crucial in predicting the behavior and fates of inhaled agents. Airflows in the child and adult nasal airways are visualized in Figure 5.5 in terms of streamlines, turbulent viscosity ratio, and snapshots of particle transport under normal breathing conditions. In both models, curvature streamlines of the core flow are

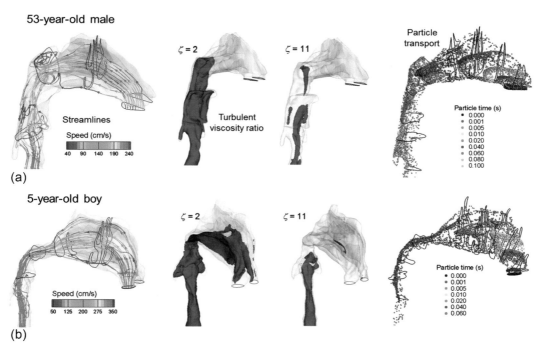

FIGURE 5.5 Visualization of inhaled airflow in nasal airways of (a) an adult and (b) a 5-year-old child in terms of streamlines, turbulent viscosity ratio, and messless fluid particles.

apparent from the nostril through the NP, forming nearly 180° bends. A recirculation zone is observed in the NP as a result of sudden area expansion. In light of particle transport, faster transport and deeper penetration of aerosols are observed in the MPs while slow-moving particles are found near the airway walls. Particles reach the OR at about 40 ms after inhalation (red solid arrow in Figure 5.5). The seemingly random particle distributions after 40 ms indicate enhanced turbulent mixing downstream of the NP. Particles are transported at a higher rate in the pharyngo-laryngeal region, which reaches the mid-pharynx at 70 ms and the glottal aperture at 100 ms. The turbulent characteristics within the nasal airway are also displayed in Figure 5.5. Airflow in both child and adult nasal airways are laminar under normal breathing conditions, with turbulence occurring mainly in the nasal V&V and pharyngo-laryngeal regions. This is reasonable considering that the nasal valve and glottal aperture are the two flow-limiting segments and have the maximum local Reynolds numbers [51]. However, Reynolds numbers in other regions are significantly lower because of larger effective flow areas and decreasing average velocities. The main passages of the nasal cavity are still dominated by laminar flows.

Steady state temperature and RH fields in the nasal airway are shown in Figure 5.6 for both subsaturated (Case 2: $T=27\,°C$ and $RH=60\%$) and saturated (Case 3: $T=40\,°C$ and $RH=100\%$) inhalation conditions. In both cases, the inhaled airflow is observed to quickly adapt to body conditions in the nasal cavity where the watery vascular-rich mucosa lines

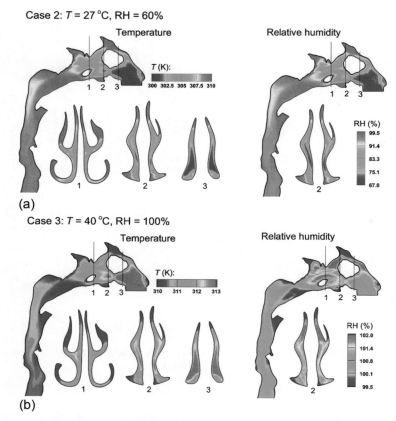

FIGURE 5.6 Steady state profiles of temperature, wall heat flux, and relative humidity (RH) for (a) Case 2 ($T = 27\,^{\circ}C$, RH = 60%) and (b) Case 3 ($T = 40\,^{\circ}C$, RH = 100%). Cooling and condensation occurs in Case 3, leading to localized supersaturated regions in the nasal cavity. In contrast, warming and moistening of the inhaled air will occur when inhaling cold airs, with evaporation of droplets in subsaturated regions.

the nasal epithelium. This air-conditioning function of the nose is critical to the health of human lungs because respiratory epitheliums are highly sensitive to cold, hot, or dry air, which may invoke adverse respiratory symptoms such as inflammation or injury. When cold, dry air is inhaled, it is warmed and humidified from the heat/mass exchange between the capillary-rich turbinate mucosa and airflow (Figure 5.6a). The temperature and RH are observed to be around 30 °C (303 K) and 78% at the middle passage (Slice 2-2′) and above 35 °C (308 K) and 91% at the end of the TR (Figure 5.6a). When the inhaled air reaches the nasopahrynx, it has approached the body condition (37 °C and 99.5% RH). This progressive increase of temperature and RH is also illustrated in the three coronal slices as shown in the right panels of Figure 5.6a.

Considering Case 3 ($T = 40\,^{\circ}C$ and RH = 100%), the initially warm saturated air is cooled down due to the lower body temperature (Figure 5.6b). Heat exchange between airflow and mucosa in this case is much weaker relative to Case 2 because of a smaller temperature difference (3 °C). The airstream temperature is reduced to approximately 38.5 °C (311.5 K) near the middle passage and is further reduced to the body condition near the posterior TR (Figure 5.6b). Cooling of the inhaled air lowers the water vapor pressure, resulting in supersaturated conditions with RH greater than 100% (Figure 5.6b). This level of supersaturation is observed to persist through a major portion of the nasal cavity. Localized elevated humidity is also noticed deep into the narrow curved meatus (Slice 2 in Figure 5.6b). A significant amount of vapor will condense onto the airway wall or inhaled aerosol particles, the latter of which can invoke significant particle growth.

5.3.3.2 *Baseline Case: Aerosol Transport and Deposition of Inert Particles*

Nasal depositions of inert aerosols (with constant diameter) are shown in Figure 5.7 in comparison with available *in vitro* deposition data for both submicrometer [52,53] and micrometer particles [54,55]. Good agreement has been achieved between the numerical predictions and *in vitro* measurements in terms of both magnitude and trend for particles ranging from 1 nm to 25 μm. Specifically, the predicted results for micrometer particles agree with the *in vitro* measurements to a high degree (Figure 5.7b). It is noted that the CFD airway model and the *in vitro* cast are based on the same set of images originally reported in Guilmette et al. [56]. Therefore, a direct comparison of deposition results between the simulations and measurements is possible. For submicrometer particles, the deposition rate decreases with increasing particle size and flow rates because of reduced molecular diffusivity and particle residence times, respectively. In contrast, for micrometer particles where inertial impaction is the most effective mechanism, larger particle size and higher inhalation flow rate both yield increased deposition because of increased particle inertia. For fine aerosols ranging from 100 nm to 1 μm, the deposition rate is quite low and is of similar magnitude for both flow rates considered. Minimum deposition rate is predicted for particles around 400 nm. More

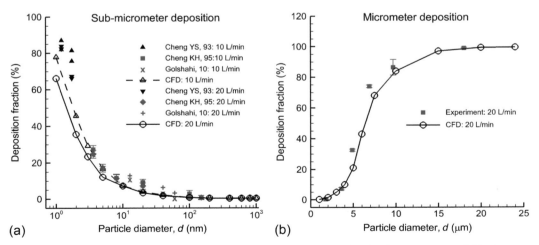

(a)   (b)

FIGURE 5.7   Good agreement between predictions and measurements for (a) submicrometer and (b) micrometer particles. The deposition rate is minimum around 400 nm and increases dramatically within the range of 2-7 μm. The nonhygroscopic results will serve as the baseline case (Case 0) in comparison to hygroscopic depositions under varying inhaled psychrometric conditions.

importantly, the deposition rate exhibits high sensitivity to particle size in the range of 2-7 μm. These results will be used later as the baseline case (Case 0) to evaluate the influences of hygroscopic growth on nasal depositions.

5.3.3.3 *Hygroscopic Behavior of Individual Particles in Equilibrium Humidity*

Before evaluating the localized heterogeneous thermohumidity conditions as discussed earlier, particle growth in equilibrium humidity will first be elucidated in terms of three underlying mechanisms: evaporation, hygroscopic growth, and condensation growth. Figures 5.8a and 5.8b show the time histories of aerosol size change under equilibrium sub-saturated and saturated conditions, respectively. The initial size of the droplets modeled in these simulations is 400 nm. Considering evaporation at RH values less than 100%, shrinkage occurs after a very short duration, i.e., within the first 0.05 ms (Figure 5.8a). Relatively little evaporation is observed at a RH of 90%. For RH values of 80% and below, the droplets evaporate very quickly to an equilibrium diameter.

In comparison to evaporation, hygroscopic or condensation growth for RH values of 99.75% and above appears to be a much slower process. An equilibrium droplet size is reached at about 7 ms after its release in a surrounding of RH = 99.75% and at 10 ms in a saturated environment (RH = 100%). For a circumstance with RH of 101%, an equilibrium droplet size is not attained and condensation growth will continue indefinitely. Comparison of results between RH = 99.75% and 100% indicates that a relative minor change in RH (0.25%) can have a noticeable impact on the particle growth. Furthermore, the particle growth rate increases nonlinearly as the RH increases above 99.75%. It is observed that the droplet at an RH of 101% increases in size by a factor of 4 within a 20 ms period (Figure 5.8b). As discussed in Figure 5.5 (particle transport), the average particle residence time through the nasal airway is typically longer than 20 ms. As a result, significant particle growth is expected for even relatively minor supersaturation conditions.

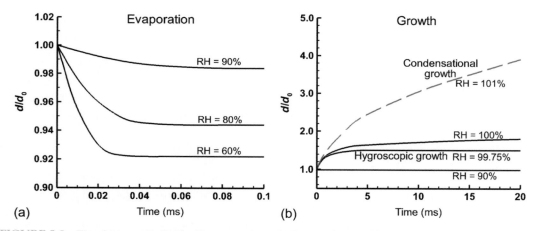

FIGURE 5.8 Time history of individual hygroscopic particulates within equilibrium humidity for (a) evaporation and (b) condensation or hygroscopic growth. Equilibrium diameters will be reached in less than 0.1 ms for evaporation but will take much longer for hygroscopic growth and condensation.

5.3.3.4 *Particle Growth and Deposition in Nonequilibrium Nasal Environments*

The growth of the particles (initially 200 nm) is illustrated in Figure 5.9 at a constant flow rate of 20 L/min under four psychrometric conditions. Particles are observed to quickly evaporate to around 165 nm upon entering the nostrils for both subsaturated cases (Figure 5.9a and b). Hygroscopic growth starts approximately at the NP where the mean flow RH recovers to the body condition due to evaporations from airway walls. The average particle size exiting the upper trachea is around 208 nm for Case 1 and 224 nm for Case 2. It is interesting to note that some particles passing the SM experience a more dramatic growth (red colored) than others, presumably due to localized condensation growth. The inhaled cold air lowers the temperature near the airway walls, which could generate a thin film of supersaturated air in those regions. Overall, RH values below 100% produce relatively minor hygroscopic growth ratios.

Significant condensation growth occurs under the supersaturated conditions, which results in a particle size increase to 2.5 μm in Case 3 (Figure 5.9c) and 3.2 μm in Case 4 (Figure 5.9d). Due to a higher vapor saturation level, more particles in Case 4 are observed to deposit than in Case 3, as indicated by the arrows in Figure 5.9d. The larger growth observed in the SM results from the longer particle residence times through this convoluted

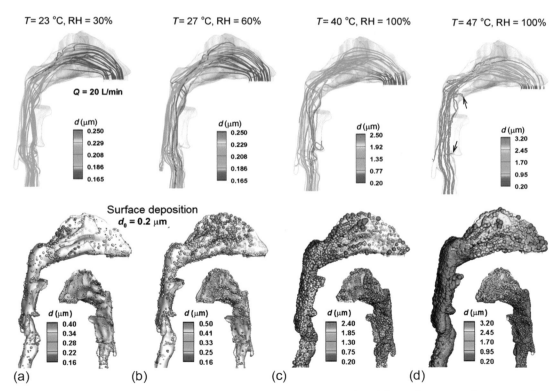

FIGURE 5.9 Particle condensation growth and surface deposition in the adult nasal airway under four psychrometric inhalation conditions for initially 200 nm particles: (a) Case 1, (b) Case 2, (c) Case 3, (d) Case 4.

region. Insignificant growth is observed in the pharyngeal region because the supersaturated condition is mainly limited to the nasal passages.

Due to the dynamic variations of inhaled particle size and associated deposition mechanisms, the surface deposition patterns among the four inhalation conditions are significantly different, as displayed in Figure 5.9 (lower panel). The initial aerosol size herein is 200 nm and the inhalation flow rate is 20 L/min. The final diameters are denoted with both color and size. Generally, the deposition color map migrates from blue to red in the mean flow direction, indicating a steady particle growth in the nasal airway. The influence of psychrometric conditions on depositions can be vividly viewed by the disparity in both deposition locations and color mapping. For Cases 1 and 2 (Figure 5.9a and b), one major difference exists in the TR, with substantially higher particle growth and deposition in Case 2. For Cases 3 and 4, particles of larger diameters (2.4-3.2 μm) are noted to accumulate densely in the region from the NP to larynx, presumably from the increased particle inertia. However, the deposition patterns in the nose are quite different between the two (Figure 5.9c vs. d). In contrast to Case 3 ($T = 40\,°C$) where scarce depositions are spotted in the anterior and middle nasal passages (Figure 5.9c), more enhanced depositions occur in these regions in Case 4 (Figure 5.9d), indicating significant condensational growth in and before these regions.

It is observed that the saturation level and initial particle size are the two major factors that determine the particle growth rate (d/d_0), which increases at higher saturation levels and decreases for larger initial particle sizes. As shown in Figure 5.10, an empirical correlation for condensation growth that includes the influences of these two factors is suggested in the following:

$$\frac{d}{d_0} = 1 + \frac{a\left(p_v - p_{v,0}\right)^b}{d_0^c} \tag{5.17}$$

where $p_{v,0}$ is the droplet surface vapor pressure at 37 °C, and ($p_v - p_{v,0}$) signifies the saturation level of the inhalation. The coefficients a, b, and c are to be determined, with b and c

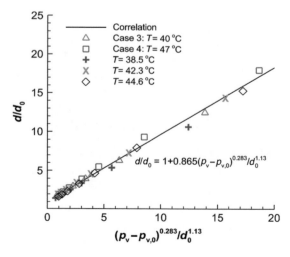

FIGURE 5.10 Empirical correlation for particle condensation growth rate in the adult nasal cavity. Units: p_v (mbar) and d_0 (μm).

representing the appropriate dependence of d/d_0 on $(p_v - p_{v,0})$ and d_0, respectively. For this purpose, three additional inhalation conditions have been considered with saturated air at varying temperatures: 38.5 °C ($p_v = 68.13$ mbar), 42.3 °C ($p_v = 83.38$ mbar), and 44.6 °C ($p_v = 93.98$ mbar). The best-fit correlation was found using a nonlinear least-square fitting algorithm [57] as shown in Figure 5.10:

$$\frac{d}{d_0} = 1 + \frac{0.865 \left(p_v - p_{v,0}\right)^{0.283}}{d_0^{1.13}} \tag{5.18}$$

with $R^2 = 0.93$. Even though this correlation has been developed based on inhalations of saturated air only (RH = 100%), it is applicable to inhalation scenarios as long as $(p_v - p_{v,0})$ is positive (condensation growth).

Cumulative deposition fractions (DFs) in the adult nose along the axial direction are shown in Figure 5.11a and b for 0.2 and 2.5 μm particles, respectively. To highlight the effect of particle growth, cumulative DFs at initial and final particle sizes are also plotted. For particles with an initial size of 0.2 μm, deposition rates are similar with and without particle growth in the nasal V&V regions. Condensation effect on deposition begins to become apparent in the anterior TR and continues to increase until the NP. For initially larger size particles (2.5 μm), condensation effect manifests itself much earlier and in much larger magnitude, as evidenced in the abrupt change of deposition in the nasal valve region ($x = 2$ cm, Figure 5.11b). For both particle sizes considered, the resulting cumulative deposition profile for the condensational growth and constant final-sized particles are similar, with approximately 0.4% difference for 0.2 μm particles and 3% for 2.5 μm particles.

Nasal depositions across a range of initial particle sizes (0.2-2.5 μm) are further shown in Figure 5.11b. Depositions of constant-diameter aerosols (Case 0) are also plotted in Figure 5.11 to highlight the influences from hygroscopic growth. Compared to Case 0, large differences arise from the particle condensation growth. Without growth, the DF in Case 0 exhibits a minimum value around 0.4 μm, and is generally small across the particle sizes considered (i.e., $\leq 2.6\%$, Figure 5.11b). When the condensation growth is present, significantly elevated depositions are obtained due to increased inertial impaction of condensational particles. Specifically, the condensation-induced DF for initially 2.5 μm aerosols is 12.8% in Case 3 and 21% in Case 4, which are five and eight times greater than without particle growth, respectively. In contrast, the effect of hygroscopic growth on subsaturated DF appears to be insignificant, with slightly reduced DF in Case 1 and equivalent DF in Case 2 for the particle-size range considered. One exception is the 0.2 μm particles whose evaporative shrinkage in Case 1 amplifies the molecular diffusion and gives a higher DF, whereas its hygroscopic growth toward 0.4 μm in Case 2 depresses the diffusive effect and leads to a lower deposition (Figure 5.11c).

5.3.4 Five-Year-Old Child Nose-Throat Model

The effect of evaporation and condensation on nasally inhaled monodisperse 200 nm particles in the nose model of the 5-year-old child is illustrated in Figure 5.12. The aerosol considered was assumed to have the same initial hygroscopic conditions as the MS smoke particles shown in Table 5.2. Similar to the adult nose model, particles inhaled with cool

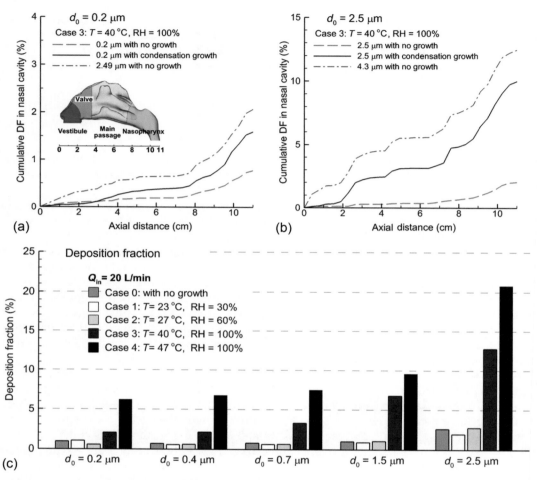

FIGURE 5.11   Cumulative deposition fractions along the axial direction in the adult nasal airway model for aerosol with an initial size of (a) 0.2 μm and (b) 2.5 μm. Deposition fractions across a range of initial particle sizes are shown in (c).

subsaturated air quickly evaporate to 150 nm near the nostrils. For the supersaturated conditions in Cases 3 and 4, significant condensation growth is observed (Figure 5.12a and b). Considering Case 3, condensation growth results in a particle size increase to approximately 2.5 μm (Figure 5.12a). Particles under these conditions will continue to grow and can reach approximately 4 μm before entering the lung. Even more dramatic particle growth is observed in Case 4 (Figure 5.12b). With this upper boundary of inhaled temperature, growth up to 3.5 μm is observed throughout the nasal airway including the superior TR, where noticeable growth is absent in Case 3. Even more interestingly, particles that have made their way to the superior turbinate are observed to deposit in or slightly downstream of that region, as indicated by the two arrows in Figure 5.12b. It appears that these particles grow so large

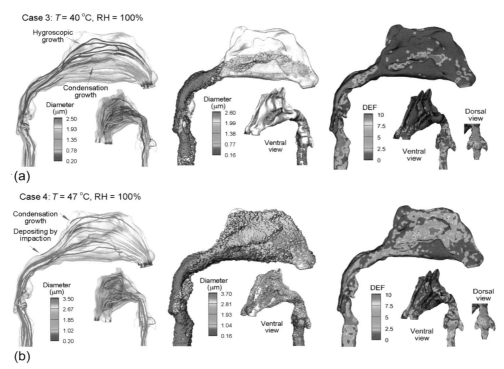

FIGURE 5.12 Particle condensation growth in the 5-year-old child nasal airway for the Case 3 (RH=100%, $T=40\,°C$) and Case 4 (RH=100%, $T=47\,°C$) inhalation conditions: (a) particle growth colored according to the transient size, (b) surface deposition, and (c) deposition enchantment factor (DEF) for initially 200 nm particles.

that they cannot keep with the airstream any further and deposit onto the dorsal wall by inertial impaction. As discussed before, significant size increase is expected for relatively minor super saturation conditions given that particles have a sufficiently long time to grow. It is concluded that the large growth near the OR in Case 4 and the lack of growth in Case 3 result from a different vapor saturation level as well as a different growth mechanism between the two conditions. When the particles are slowly transported through the superior turbinate, they experience a *hygroscopic* growth in Case 3 (RH ≤ 100%) which reaches the equilibrium diameters (∼0.3 μm). Whereas in Case 4, particles undergo a *condensational* growth (RH ≥ 101%) and will grow indefinitely as time permits until they become too large to be entrained within the airstream. Clearly, the initial higher gradient of temperature ($\Delta T=10\,°C$ in Case 4 vs. $\Delta T=3\,°C$ in Case 3) is responsible for the elevated saturation level near the OR in Case 4 which triggers the *condensational* growth. This finding has great potential for improved intranasal drug delivery that target medications at the OR for neurological disorder interventions, and for risk assessment of environment exposures where high temperature and/or high humidity prevail.

Condensational growth leads to strikingly different particle deposition patterns between Cases 3 and 4. In contrast to Case 3 ($T=40\,°C$) where the turbinate deposition is largely limited to the inferior nasal passage (Figure 5.12, middle panel), increased deposition occurs in the middle and superior passages, indicating significant condensational growth in and before these passages. Furthermore, significant hot spots are observed in the dorsal walls of the pharynx and larynx for both cases. For Case 4 ($T=47\,°C$), a more scattered map of hot spots is noted in the whole TR (Figure 5.12, left panel). However, such hot spot distribution is missing in Case 3, as discussed before. Compared to Cases 1 and 2, condensation growth in Cases 3 and 4 produces more localized deposition on the laryngeal dorsal wall as a result of increased impaction deposition. Furthermore, condensation growth in Case 4 is observed to remarkably increase the extent of hot spots, especially in the superior turbinate, compared to the other three inhalation conditions.

Figure 5.13a shows the comparison of DFs among the four inhalation conditions. Deposition with constant-diameter particles (Case 0) is also plotted in Figure 5.13a to highlight the effect from hygroscopic growth. The three inhalation flow rates considered represent sedentary (5 L/min), light activity (15 L/min), and heavy activity (30 L/min) breathing conditions, respectively. From Figure 5.13a, the effect of flow rate appears to be not significant. However, dramatically varying effects are observed from the hygroscopic growth. As expected, hygroscopic growth under Case 4 conditions results in a significant increase in total deposition rate, which is about two to three times that of the other four cases. Surprisingly, despite dramatic variations in particle growth among Cases 0-3, the deposition rates are similar. It is, therefore, noted that a larger possible condensation growth does not necessarily imply a larger potential deposition.

Response of the olfactory deposition to the different inhalation conditions is shown in Figure 5.13b. A remarkably enhanced deposition is observed for the hot saturated inhalation condition (Case 4: $T=47\,°C$, RH $=100\%$), which is more than one order of magnitude larger (about 10 times) than the other inhalation scenarios. Again, Case 3 gives the lowest

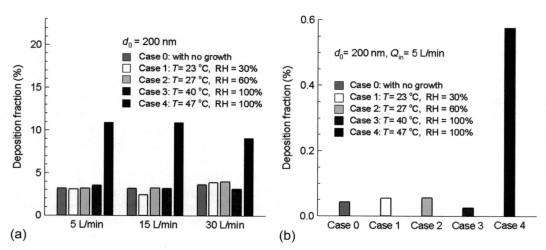

(a)   (b)

FIGURE 5.13   Deposition fractions in the 5-year-old child nasal airway under different inhalation psychrometric conditions for initially 200 nm particles: (a) total deposition and (b) olfactory deposition. Significant enhanced olfactory deposition occurs at the inhalation of hot-humid air compared with other inhalation conditions.

deposition. Due to the small initial temperature gradient (i.e., $\Delta T = 3\ ^\circ C$) in this condition, the humidity near the OR is only slightly oversaturated; particle growth hereof is expected to be not significant, and may increase the initial 200 nm particle size towards the range of 400-500 nm where minimum depositions typically occur. By contrast, evaporation in Cases 1 and 2 shrinks the initial 200 nm particles, which magnifies the molecular diffusion and produces an olfactory deposition larger than both Cases 3 and 0 (no growth).

5.3.5 Adult Mouth-Lung Model

The steady state RH conditions in the mouth-lung model are shown in Figure 5.14a for warm-humid inhalation conditions (Case 3). Similar to nasal airways, the cooling of the initially warm saturated air increases the RH to supersaturated conditions greater than 101%. This level of supersaturation is observed to persist through a major portion of the trachea. However, absorption of the supersaturated water vapor onto airway walls acts to reduce the RH values in the deeper tracheobronchial (TB) region. Reduced supersaturated conditions below 101% are observed near the first carina and persist through the main bronchi. Subsaturated conditions are not observed until the more distal bifurcations. Cross-sectional slices provide further details regarding the RH conditions throughout the mouth-throat (MT) and TB regions. In the MT, a significant portion of the cross-sectional flow field is observed to have RH values greater than 101% (Figure 5.14a). Supersaturated conditions are then observed to persist through a majority of the flow field into the main and lobar bronchi of the left and right lungs.

The dynamic growth of hygroscopic aerosols in the mouth-lung model is shown in Figure 5.14b for an initial particle size of 200 nm. Condensation growth in the MT region results in a particle size increase to approximately 2.5 μm (Figure 5.14b). Particles under these conditions continue to grow and reach approximately 4 μm near the geometry outlet. Due to enhanced particle deposition, fewer particles exit the respiratory model. Even more dramatic

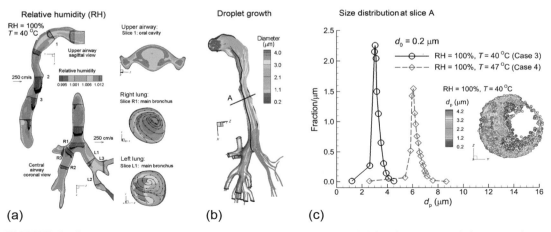

(a)   (b)   (c)

FIGURE 5.14 Particle condensation growth in the mouth-throat model for the Case 3 inhalation condition (RH = 100%, $T = 40\ ^\circ C$): (a) relative humidity field, (b) particle growth colored according to the transient size, and (c) size distribution at the slice A presented as mass fraction per micrometer for initially 200 nm particles.

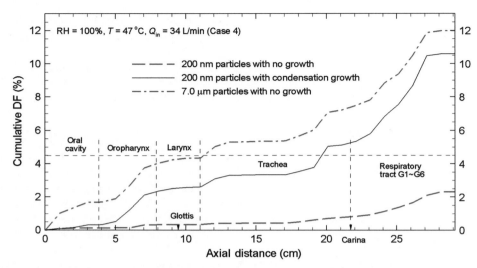

FIGURE 5.15   Cumulative particle deposition fraction (DF) along the airflow direction in the mouth-lung geometry for the Case 4 inhalation condition (RH = 100%, T = 47 °C).

particle growth is observed under Case 4 conditions (Figure 5.14c). With this upper boundary of inhaled temperature, growth up to larger than 7 µm is observed after the larynx (Slice A in Figure 5.14b).

Cumulative DFs along the axial direction in the mouth-throat model are shown in Figure 5.15 for initially 200 nm particles and hot-saturated (Case 4) inhalation conditions. Results are also presented for a constant particle diameter that is representative of the growth observed in Case 4, which is 7 µm. As expected, hygroscopic growth under Case 4 conditions results in a significant increase in cumulative deposition throughout the geometry (Figure 5.15). Condensation growth is observed to initially enhance the deposition of 200 nm particles in the oropharynx region. Thereafter, deposition of the condensation aerosols appears to be consistent with, but approximately 2% below, the deposition of a constant diameter 7 µm aerosol. The resulting total deposition for the 200 nm condensation aerosol and constant 7 µm particles is approximately 10-12%. However, the trend in deposition is sharply increasing at the outlet of the upper TB geometry. As a result, it appears reasonable that a total rate of deposition in the full 16 generations of the TB geometry may be greater than 50% due to condensation growth effects.

5.4 DISCUSSION

The primary implication of the results in this chapter is that condensation growth may play a more significant role than previously assumed in the deposition of fine-regime and small micrometer aerosols. The results show that the condensation-enhanced deposition for initially 2.5 µm aerosols is 12.8% for Case 3 (T = 40 °C, RH = 100%) and 21% for Case 4 (T = 47 °C, RH = 100%), in comparison to 2.6% without particle growth (Case 0). Considering the large deposition increase as well as the prevalence of ambient pollutants of this size range, these findings are worthy of special attention and could have profound implications both in health risk assessment of environmental hazards and intranasal delivery of therapeutic aerosols. When a warm-humid environment is concerned, condensation effect should be included for accurate assessment of pollutant accumulation or medication dosage. Examples of such environments include cigarette smoking, manufacturing workplaces, humidified hospital units, tropical geographical locations, and summer seasons. The highly elevated depositions on a hot-humid day may indicate higher health risks to humans than on a regular day. Likewise, people who routinely work in high temperature and/or high humidity conditions, such as industries of paper, textile, mining, food, etc., will have a higher possibility of developing respiratory distress symptoms. Another phenomenon that is likely attributed to particle condensational growth is the seemingly higher deposition of CSPs where the typically submicrometer CSP have been observed to deposit in the upper airways like a much larger 6-7 μm aerosol [58–61]. It has also been suggested that the small size increase due to hygroscopic growth under body conditions (RH=99.5%) cannot be responsible for the significant enhancements in CSP deposition [22,58,59,62]. Results in this chapter suggest that when inhaling warm/hot saturated airs, the RH may exceed 100% in both nasal and mouth-lung models. As a result, significant particle growth in the supersaturated airways was found to be possible. Interestingly, for the relatively mild inhalation temperature of 3 °C above average body conditions (Case 3), initially 200 nm particles were observed to increase in size to above 3 μm in the mouth-lung model (Figure 5.14c). Even more striking is the fact that the upper boundary inhalation temperature conditions of Case 4 resulted in a 7-8 μm particle entering the trachea. Clearly, these results do not prove that the enhanced deposition of CSPs is purely a result of condensation growth. However, this chapter does highlight condensation growth as a potentially significant mechanism in the deposition of smoke particles under initially supersaturated conditions. It appears possible that condensation growth could be responsible for smoke particles depositing like 3-8 μm aerosols in the upper TB region, as observed by Black and Pritchard [63] and Phalen et al. [59]. Evaporation in the deeper lung at a subsaturated RH of 99.5% could then result in submicrometer aerosols, which is consistent with the deposition findings of Black and Pritchard [63]. Nevertheless, there is currently no experimental evidence that directly proves significant condensation growth of CSPs in the upper respiratory airways, as reported in the current study.

The hypothesis that significant condensation growth occurs for hygroscopic particles in the upper airways and subsequent evaporation occurs in the lower airways does not contradict the *in vivo* observations of Hicks et al. [64]. The study of Hicks et al. [64] reported that inhaled dense MS CSPs only increased in size by a factor of 1.7 based on measurements made at exhalation. However, Hicks et al. [64] were not able to assess size change during inhalation as the CSPs passed through the upper airways. It is reasonable to assume that significant size change did occur during the initial inhalation of CSPs in the experiment of Hicks et al. [64]. Partial evaporation in the deeper lung and during exhalation, as described above, could then explain the relatively small growth ratios that were measured in the experiment.

Deposition of hygroscopic aerosols is an intricate dynamic process with constantly changing particle diameters and behaviors. The results in this chapter show that the degree of saturation and the initial particle size are the two factors that have predominant impacts on the aerosol growth. However, a larger condensation growth rate does not necessarily imply a correspondingly larger deposition due to the complex interplay between the relevant growth, transport, and deposition mechanisms. Hygroscopic particle transport is also a dynamic process. The trajectory of such a particle is largely affected by its instantaneous properties, such as diameter (i.e., diffusion or impaction), particle velocity (i.e., particle residence time, or how long it can grow), and position (i.e., local temperature, humidity, or how fast it can grow). Considering these interacting factors, it is not surprising to observe the seemingly erratic DF profiles as a function of RH and d_0 in Figure 5.11.

An interesting difference has been noticed between the nasally and orally inhaled aerosols in their respective growth rates within the upper airway. Aerosols (initially at 200 nm) that were inhaled through the mouth were observed to grow in size to above 8 μm before leaving the larynx, as shown in Section **5.3.5**. This is much larger than the final size of 3.6 μm nasally inhaled aerosols in the nasal airways of both the adult (Section 5.3.3) and child (Section **5.3.4**). This difference may result from the nose's better capacity for warming and humidifying the inhaled air to the body condition. This capacity is weaker in the mouth, which explains the feeling of a dry throat after a period of oral breathing [65]. Considering Case 4, a highly supersaturated condition arises from the air cooling within the anterior nasal passage. However, due to quick vapor absorptions to the turbinate mucosa, the over-saturated RH fields start to decrease from the middle turbinate, which fall below 101% within the NP and approach the mucus surface condition of ~99.5% RH before entering the trachea. Similar results were reported by Rouadi et al. [66] who measured relatively constant humidity levels in the NP for different flow rates. As a result, significant condensation growth may not be expected in the pharyngeal region. This was also the case under orally inhaled conditions [25]. Compared to the nasal route, the oral airway cannot condition the inhaled hot-humid air in Case 4 to body temperature as quickly, therefore allowing supersaturated conditions in the whole upper airway. Accordingly, particles have a much longer residence time for condensation growth and thus give rise to larger growth rates before they enter the lungs, which are about 3 μm for Case 3, and 7.5 μm for Case 4 [25]. By contrast, the nasally inhaled particles grow only to 2.5 μm for Case 3 and 3.6 μm for Case 4 when they leave the nasal TR.

Considering the age effect on the hygroscopic growth, the results in a 5-year-old boy were compared to those in an adult male [29]. With an inhalation condition similar to Case 3 in this chapter (warm saturated air) and initial 0.9 μm particles, Longest et al. [29] reported a size of approximately 2 μm at the exit of the adult airway geometry, which is slightly smaller than the size of 2.5 μm. Multiple factors can contribute to this difference, such as initial particle size, inhalation flow rate, geometry difference, and the associated flow field variation. In particular, the inhalation flow rate adopted for the child herein is 5 L/min (i.e., quiet breathing) whereas those for the adult are 20-40 L/min (i.e., light to heavy activity breathing). The higher flow rates result in shorter particle residence times to allow condensation growth to occur within the adult nasal airway [29]. In light of the age-related geometry effect, further studies are necessary to examine the discrepancies in airflow, temperature, humidity, and particle growth between children and adults.

Uniform distributions of inlet velocity and particles were assumed in this chapter. However, the actual distributions could be quite different due to flow development and nonuniform particle transport before the nostrils [67,68]. In addition, transient breathing could alter particle deposition efficiencies in comparison to steady breathing and has been shown to exert different effects depending on the size of inhaled particles [69,70]. Considering particles larger than 1 μm, Zhang et al. [71,72] showed that cyclic flow increased overall deposition due to increased inertia during peak inspiratory flow. For monodisperse ultrafine particles (<0.1 μm), Dendo et al. [73] showed that transient flow decreased deposition, potentially as a result of reduced residence time for particle diffusion. Considering the dynamic growth of the initial fine-regime particles, which range from 0.2 to 5 μm with a concurrent deposition mechanism of inertia impaction and diffusion, transient flow is expected to slightly decrease the deposition under the two subsaturated conditions and increase the deposition under the two saturated conditions.

Other limitations of this chapter include the assumptions of simplified inlet conditions, dilute aerosols, and an idealized airway model. Other studies have highlighted the significance of transient breathing [74], inlet velocity profiles [75], aerosol colligative effects [76,77], and nasal wall motion [78]. Evaporation and condensation can noticeably impact the temperature and saturation ratio of the airflow, rendering the hygroscopic growth a two-way coupled process [5,79]. Moreover, the nasal model in this chapter is based on images of a single subject acquired at the end of exhalation and therefore does not account for the variability between different subjects [80,81] or between different instants of a breathing cycle in one subject [82,83]. Therefore, future studies are needed that should be orientated toward: (1) improving numerical realism and (2) including a broader population group. Our knowledge of nasal deposition is currently lacking in subpopulations such as pediatrics, geriatrics, and patients with respiratory diseases. Due to physiological development, aging, or disease stages, the airway anatomy can be remarkably different from that of a healthy adult. Concentrating on these specific subpopulations will help to clarify intergroup and interindividual variability and will allow for the design of more efficient pharmaceutical formulations and drug delivery protocols. In particular, future *in vivo* and *in vitro* studies on hygroscopic growth of inhaled aerosols are needed to cross-validate numerical predictions before results in this chapter can be applied to clinical applications.

5.5 CONCLUSION

This chapter reviewed some of the latest advances in modeling and simulations of the thermohumidity phenomena in human upper airways with a coupled imaging-CFD approach. Specifically, hygroscopic effects on particle growth, transport, and deposition were evaluated in the nasal airways of an adult and a 5-year-old child, as well as in an adult mouth-lung model for initially monodisperse aerosols. Various psychrometric inhalation conditions that represent cold-dry, cool mild, warm-humid, and hot-humid environments were considered under quiet, light activity, and heavy activity breathing conditions. Highlights of this chapter include:

1. For both the child and adult, a supersaturated humid environment is possible in the TR and can induce a significant increase in the size of inhaled aerosols by condensation. The average growth rate (d/d_0) in the nose is 12.5 and 17.5 for warm and hot saturated conditions (Cases 3 and 4), respectively.
2. Depositions in the upper airway can be noticeably enhanced by condensation growth depending on the inhalation temperature and humidity. For an inhalation of 47 °C

saturated air, condensation growth of initial 200 nm aerosols increases the nasal deposition by a factor of 2.5.

3. The thermohumidity distributions, particle growth, and particle deposition are significantly different between the adult and the 5-year-old child. Specifically, a dramatic deposition increase (a factor of 11) was observed in the child OR under hot-humid conditions, which was not observed in the adult.

4. For inhalation conditions normally encountered in our daily life (cold/cool unsaturated or warm saturated), evaporation and hygroscopic growth are found to have a small impact on the total deposition rate of inhaled aerosols.

5. While the growth of orally inhaled aerosols can occur in both the mouth and trachea, the growth of nasally inhaled aerosols occurs predominantly in the anterior nose. As a result, for a given initial particle size and inhalation condition, a smaller size growth of nasally inhaled aerosols is expected compared to when inhaled through the mouth.

6. Results in this chapter have great importance in health risk assessment of exposure to fine-regime airborne hazards under severely high temperature and humid environments that may result in much higher possibilities of injury to the human respiratory system than previously assumed.

References

[1] Bell ML, Dominici F, Ebisu K, Zeger SL, Samet JM. Spatial and temporal variation in PM2.5 chemical composition in the United States for health effects studies. Environ Health Perspect 2007;115:989–95.
[2] ICRP. Human respiratory tract model for radiological protection. New York: Elsevier Science Ltd; 1994.
[3] Ferron GA. The size of soluble aerosol particles as a function of the humidity of the air: application to the human respiratory tract. J Aerosol Sci 1977;3:251–67.
[4] Ferron GA, Kreyling WG, Haider B. Inhalation of salt aerosol particles—II. Growth and deposition in the human respiratory tract. J Aerosol Sci 1988;19:611–31.
[5] Finlay WH, Stapleton KW. The effect on regional lung deposition of coupled heat and mass-transfer between hygroscopic droplets and their surrounding phase. J Aerosol Sci 1995;26:655–70.
[6] Li W, Hopke PK. Initial size distributions and hygroscopicity of indoor combustion aerosol particles. Aerosol Sci Tech 1993;19:305–16.
[7] Martonen TB, Bell KA, Phalen RF, Wilson AF, Ho A. Growth rate measurements and deposition modeling of hygroscopic aerosols in human tracheobronchial models. In: Walton WH, editor. Inhaled particles V. Oxford: Pergamon Press; 1982. p. 93–107.
[8] Zhang Z, Kleinstreuer C, Kim CS. Water vapor transport and its effects on the deposition of hygroscopic droplets in a human upper airway model. Aerosol Sci Tech 2006;40:52–67.
[9] Chow JC, Watson JG. Review of PM2.5 and PM10 apportionment for fossil fuel combustion and other sources by the chemical mass balance receptor model. Energy Fuel 2002;16:222–60.
[10] Kittelson DB. Engines and nanoparticles: a review. J Aerosol Sci 1998;29:575–88.
[11] Bernstein GM. A review of the influence of particle size, puff volume, and inhalation pattern on the deposition of cigarette smoke particles in the respiratory tract. Inhal Toxicol 2004;16:675–89.
[12] Keith CH. Particle size studies on tobacco smoke. Beitr Tabakforsch 1982;11:123–31.
[13] Jacobson MZ. Short-term effects of controlling fossil-fuel soot, biofuel soot and gases, and methane on climate, Arctic ice, and air pollution health. J Geophys Res-Atmos 2010;115:D14209 (1–24).
[14] Slater JF, Currie LA, Dibb JE, Benner BA. Distinguishing the relative contribution of fossil fuel and biomass combustion aerosols deposited at Summit, Greenland through isotopic and molecular characterization of insoluble carbon. Atmos Environ 2002;36:4463–77.
[15] Mandell GL, Bennett JE, Bolin RD. Principles and practices of infectious diseases. New York: Churchill Livingstone; 2004.

[16] Martonen TB. Surrogate experimental models for studying particle deposition in the human respiratory tract: an overview. In: Lee SD, editor. Aerosols. Chelsea, MI: Lewis Publishers; 1986.

[17] Yang CP, Callagher RP, Weiss NS, Band PR, Thomas DB, Russel DA. Differences in incidence rates of cancers of the respiratory tract by anatomic subside and histological type: an etiologic implication. J Natl Cancer Inst 1989;81:1828–31.

[18] U.S. Department of Health and Human Services. Children and secondhand smoke exposure-excerpts from the health consequences of involuntary exposure to tobacco smoke: a report of the Surgeon General: U.S. Surgeon General Report. Atlanta, GA: U.S. Department of Health and Human Services; 2007.

[19] California EPA. Proposed identification of environmental tobacco smoke as a toxic air contaminant. Part B: health effects. Sacramento, CA: California Environmental Protection Agency, Office of Environmental Health Hazard Assessment; 2005.

[20] Cinkotai FF. The behavior of sodium chloride particles in moist air. J Aerosol Sci 1971;2:325–9.

[21] Peng C, Chow AHL, Chan CK. Study of the hygroscopic properties of selected pharmaceutical aerosols using single particle levitation. Pharm Res 2000;17:1104–9.

[22] Robinson R, Yu CP. Theoretical analysis of hygroscopic growth rate of mainstream and sidestream cigarette smoke particles in the human respiratory tract. Aerosol Sci Tech 1998;28:21–32.

[23] Ferron GA, Oberdorster G, Hennenberg R. Estimation of the deposition of aerosolised drugs in the human respiratory tract due to hygroscopic growth. J Aerosol Med 1989;2:271.

[24] Zhang Z, Kleinstreuer C, Kim CS. Isotonic and hypertonic saline droplet deposition in a human upper airway model. J Aerosol Med 2006;19:184–98.

[25] Longest PW, Xi JX. Condensational growth may contribute to the enhanced deposition of cigarette smoke particles in the upper respiratory tract. Aerosol Sci Tech 2008;42:579–602.

[26] Kim JW, Xi J, Si XA. Dynamic growth and deposition of hygroscopic aerosols in the nasal airway of a 5-year-old child. Int J Numer Meth Biomed Eng 2013;29:17–39.

[27] Ferron GA, Haider B, Kreyling WG. Conditions for measuring supersaturation in the human lung using aerosols. J Aerosol Sci 1984;15:211–5.

[28] Longest PW, Hindle M. Numerical model to characterize the size increase of combination drug and hygroscopic excipient nanoparticle aerosols. Aerosol Sci Tech 2011;45:884–99.

[29] Longest PW, Tian G, Hindle M. Improving the lung delivery of nasally administered aerosols during noninvasive ventilation—an application of enhanced condensational growth (ECG). J Aerosol Med Pulm Drug Deliv 2011;24:103–18.

[30] Tian G, Longest PW, Su GG, Hindle M. Characterization of respiratory drug delivery with enhanced condensational growth using an individual path model of the entire tracheobronchial airways. Ann Biomed Eng 2011;39:1136–53.

[31] Xi J, Longest PW. Characterization of submicrometer aerosol deposition in extrathoracic airways during nasal exhalation. Aerosol Sci Tech 2009;43:808–27.

[32] Zhou Y, Xi J, Simpson J, Irshad H, Cheng YS. Aerosol deposition in a nasopharyngolaryngeal replica of a 5-year-old child. Aerosol Sci Tech 2013;47:275–82.

[33] Wilcox DC. Turbulence modeling for CFD. 2nd ed. California: DCW Industries, Inc.; 1998

[34] Bird RB, Steward WE, Lightfoot EN. Transport phenomena. New York: John Wiley & Sons; 1960.

[35] Green DW. Perry's chemical engineers' handbook. New York: McGraw-Hill; 1997.

[36] Xi J, Longest PW, Martonen TB. Effects of the laryngeal jet on nano- and microparticle transport and deposition in an approximate model of the upper tracheobronchial airways. J Appl Physiol 2008;104:1761–77.

[37] Xi J, Longest PW. Transport and deposition of micro-aerosols in realistic and simplified models of the oral airway. Ann Biomed Eng 2007;35:560–81.

[38] Hinds WC. Aerosol technology: properties, behavior, and measurement of airborne particles. New York: John Wiley and Sons; 1999.

[39] Morsi SA, Alexander AJ. An investigation of particle trajectories in two-phase flow systems. J Fluid Mech 1972;55:193–208.

[40] Li A, Ahmadi G. Dispersion and deposition of spherical particles from point sources in a turbulent channel flow. Aerosol Sci Tech 1992;16:209–26.

[41] Allen MD, Raabe OG. Slip correction measurements of spherical solid aerosol particles in an improved Millikan apparatus. Aerosol Sci Tech 1985;4:269–86.

[42] Matida EA, Finlay WH, Grgic LB. Improved numerical simulation of aerosol deposition in an idealized mouth-throat. J Aerosol Sci 2004;35:1–19.

[43] Longest PW, Xi J. Effectiveness of direct Lagrangian tracking models for simulating nanoparticle deposition in the upper airways. Aerosol Sci Tech 2007;41:380–97.

[44] Longest PW, Kleinstreuer C. Computational models for simulating multicomponent aerosol evaporation in the upper respiratory airways. Aerosol Sci Tech 2005;39:124–38.

[45] Atkins P, de Paula J. Physical chemistry. Oxford: Oxford University Press; 2006.

[46] Xi J, Longest PW. Evaluation of a drift flux model for simulating submicrometer aerosol dynamics in human upper tracheobronchial airways. Ann Biomed Eng 2008;36:1714–34.

[47] Patankar S. Numerical heat transfer and fluid flow. Boca Raton, FL: Taylor & Francis; 1980.

[48] Xi J, Longest PW. Numerical predictions of submicrometer aerosol deposition in the nasal cavity using a novel drift flux approach. Int J Heat Mass Tran 2008;51:5562–77.

[49] Xi J, Longest PW. Effects of oral airway geometry characteristics on the diffusional deposition of inhaled nanoparticles. J Biomech Eng-T ASME 2008;130:16.

[50] Ishizu Y, Ohta K, Okada T. The effect of moisture on the growth of cigarette smoke particles. Beitr Tabakforsch 1980;10:161–8.

[51] Wexler DB, Davidson TM. The nasal valve: a review of the anatomy, imaging, and physiology. Am J Rhinol 2004;18:143–50.

[52] Cheng K-H, Cheng Y-S, Yeh H-C, Swift DL. Deposition of ultrafine aerosols in the head airways during natural breathing and during simulated breath holding using replicate human upper airway casts. Aerosol Sci Tech 1995;23:465–74.

[53] Cheng YS, Su YF, Yeh HC, Swift DL. Deposition of thoron progeny in human head airways. Aerosol Sci Tech 1993;18:359–75.

[54] Cheng YS. Aerosol deposition in the extrathoracic region. Aerosol Sci Tech 2003;37:659–71.

[55] Garcia GJM, Tewksbury EW, Wong BA, Kimbell JS. Interindividual variability in nasal filtration as a function of nasal cavity geometry. J Aerosol Med Pulm Drug Deliv 2009;22:139–55.

[56] Guilmette RA, Wicks JD, Wolff RK. Morphometry of human nasal airways in vivo using magnetic resonance imaging. J Aerosol Med 1989;2:365–77.

[57] Gill PE, Murray W. Algorithms for the solution of the nonlinear least-squares problem. SIAM J Numer Anal 1978;15:977–92.

[58] Martonen TB. Deposition patterns of cigarette smoke in human airways. Am Ind Hyg Assoc J 1992;53:6–18.

[59] Phalen RF, Oldham MJ, Mannix RC. Cigarette smoke deposition in the tracheobronchial tree: evidence for colligative effects. Aerosol Sci Tech 1994;20:215–26.

[60] Pritchard JN, Black A. An estimation of the tar particulate material depositing in the respiratory tracts of healthy male middle- and low-tar cigarette smokers. In: Liu C, Pue D, Fissan H, editors. Aerosols. New York: Elsevier Science; 1984. p. 989–92.

[61] Schlesinger RB, Gurman JL, Lippmann M. Particle deposition within bronchial airways: comparisons using constant and cyclic inspiratory flows. Ann Occup Hyg 1982;26:47–64.

[62] Ingebrethsen BJ. The physical properties of mainstream cigarette smoke and their relationship to deposition in the respiratory tract. In: Crapo JD, Smolko ED, Miller FJ, Graham JA, Hayes AW, editors. Extrapolation of dosimetric relationships for inhaled particles and gases. San Diego: Academic Press; 1989.

[63] Black A, Pritchard JN. A comparison of the regional deposition and short-term clearance of tar particulate material from cigarette smoke, with that of 2.5 micrometer polystyrene microspheres. J Aerosol Sci 1984;15:224–7.

[64] Hicks JF, Prichard JN, Black A, Megaw WJ. Experimental evaluation of aerosol growth in the human respiratory tract. In: Schikarski W, Fissan HJ, Friedlander SK, editors. Aerosols: formation and reactivity. Oxford: Pergamon Press; 1986. p. 244–7.

[65] Doorly DJ, Taylor DJ, Gambaruto AM, Schroter RC, Tolley N. Nasal architecture: form and flow. Philos Trans A Math Phys Eng Sci 2008;366:3225–46.

[66] Rouadi P, Baroody FM, Abbott D, Naureckas E, Solway J, Naclerio RM. A technique to measure the ability of the human nose to warm and humidify air. J Appl Physiol 1999;87:400–6.

[67] Kleinstreuer C, Zhang Z. Laminar-to-turbulent fluid-particle flows in a human airway model. Int J Multiphase Flow 2003;29:271–89.

[68] Kleinstreuer C, Zhang Z, Li Z. Modeling airflow and particle transport/deposition in pulmonary airways. Respir Physiol Neurobiol 2008;163:128–38.

[69] Gurman JL, Lippmann M, Schlesinger RB. Particle deposition in replicate casts of the human upper trancheo-bronchial tree under constant and cyclic inspiratory flow. I. Experimental. Aerosol Sci Tech 1984;3:245–52.

[70] Kim CS, Garcia L. Particle deposition in cyclic bifurcating tube flow. Aerosol Sci Tech 1991;14:302–15.

[71] Zhang Z, Kleinstreuer C, Kim CS. Cyclic micron-size particle inhalation and deposition in a triple bifurcation lung airway model. J Aerosol Sci 2002;33:257–81.

[72] Zhang Z, Kleinstreuer C. Transient airflow structures and particle transport in a sequentially branching lung airway model. Phys Fluids 2002;14:862–80.

[73] Dendo RI, Phalen RF, Mannix RC, Oldham MJ. Effects of breathing parameters on sidestream cigarette smoke deposition in a hollow tracheobronchial model. Am Ind Hyg Assoc J 1998;59:381–7.

[74] Shi H, Kleinstreuer C, Zhang Z. Laminar airflow and nanoparticle or vapor deposition in a human nasal cavity model. J Biomech Eng 2006;128:697–706.

[75] Keyhani K, Scherer PW, Mozell MM. Numerical simulation of airflow in the human nasal cavity. J Biomech Eng 1995;117:429–41.

[76] Robinson RJ, Yu CP. Deposition of cigarette smoke particles in the human respiratory tract. Aerosol Sci Tech 2001;34:202–15.

[77] Subramaniam RP, Richardson RB, Morgan KT, Kimbell JS, Guilmette RA. Computational fluid dynamics simulations of inspiratory airflow in the human nose and nasopharynx. Inhal Toxicol 1998;10:91–120.

[78] Fodil R, Brugel-Ribere L, Croce C, Sbirlea-Apiou G, Larger C, Papon JF, et al. Inspiratory flow in the nose: a model coupling flow and vasoerectile tissue distensibility. J Appl Physiol 2005;98:288–95.

[79] Finlay WH. Estimating the type of hygroscopic behavior exhibited by aqueous droplets. J Aerosol Med 1998;11:221–9.

[80] Hilberg O, Jensen FT, Pedersen OF. Nasal airway geometry: comparison between acoustic reflections and magnetic resonance scanning. J Appl Physiol 1993;75:2811–9.

[81] Pickering DN, Beardsmore CS. Nasal flow limitation in children. Pediatr Pulmonol 1999;27:32–6.

[82] Ohki M, Ogoshi T, Yuasa T, Kawano K, Kawano M. Extended observation of the nasal cycle using a portable rhinoflowmeter. J Otolaryngol 2005;34:346–9.

[83] Mirza N, Kroger H, Doty RL. Influence of age on the 'nasal cycle'. Laryngoscope 1997;107:62–6.

Transport in the Microbiome

R.J. Clarke

University of Auckland, Auckland, New Zealand

6.1 INTRODUCTION

In this chapter we shall discuss the current theoretical understanding of the transport of microorganisms, such as those found in unimaginable numbers within the human body. Many of these microorganisms are able to swim through their fluid environment, although the means by which they propel themselves differs substantially from much larger organisms, such as birds or fish. We shall, therefore, begin by discussing the biomechanics of distinct types of microorganisms, and outline the hydrodynamic principles at work in each case.

Because microorganisms tend to be present as part of large populations, rather than as individual swimmers, we shall go on to discuss models for suspension of swimmers, which have been shown to produce highly-complex displays of collective motion. This is currently an area of intense research activity, drawing in ideas not only from Fluid and Solid Mechanics, but also Nonequilibrium Statistical Mechanics, Soft Matter Physics, and Granular Flow Theory. It is hoped that this chapter will provide the reader with an appreciation of both the relevant mechanical considerations, as well as a flavor of the on-going theoretical challenges that are being addressed at the frontier of the subject.

6.2 THE HUMAN MICROBIOME

Our bodies are said to contain ten times more microorganism cells than human cells, and combined, these possess one hundred times as many genes than our own genome [1]. According to the findings of the Human Genome Project, there are likely over two hundred genes that actually originate from the bacteria inhabiting our bodies [2]. This microbiome, or collection of microorganisms within our bodies, performs an array of essential functions, such as allowing us to digest carbohydrates (e.g., *B. thetaiotaomicron*), maintain the health of our skin (e.g., *M. luteus*), and even influence healthy brain development through their ability to regulate hormones such as dopamine and serotonin [3]. There are also the microorganisms produced by the human body itself, the *Spermatozoa*, that allow us to pass on our genetic material to the next generation. At the same time, however, occasional members of the microbiome constitute some of the most lethal pathogens ever encountered. These include *Vibrio cholerae* (cholera), *Yersinia pestis* (bubonic plague), *Shigella* (dysentry), *Trypanosoma brucei* (sleeping sickness), and *Plasmodium* (malaria).

Over much of human history, however, we were completely unaware of the existence of such microscopic forms of life. For a long time, all known life was multicellular in nature (Metazoic), and classed as either Animal, or Plant. However, by the time this notion was formalized by Carl Linnaeus in the mid-eighteenth century, the field of microscopy was beginning to bring into question its validity. The work of Antonie van Leeuwenhoek a little earlier had uncovered the existence of this hitherto unknown microscopic world of single-celled organisms. These microorganisms would increasingly be shown to not easily fit neatly into either the plant kingdom, nor the animal kingdom. For instance, one of the newly-observed microorganisms reported by van Leeuwenhoek, *Euglena*, can both photosynthesize inorganic carbon like a plant and use organic carbon for growth, like animals. By the mid-nineteenth century it had become apparent that the two kingdom perspective did not stand on particularly solid ground. A third kingdom, the Protists, was therefore proposed by Ernst Haeckel. This contained microorganisms with animal-like characteristics, the Protozoa ("First Animals"), and those with plant-like behaviors, the Protophyta ("First Plants").

However, advances in microscopy once again proved the catalyst for change. Electron Microscopy in the 1930s revealed that some Protists contain a distinct nucleus (karyon) within which their genetic information is stored (as do the cells of animals and plants), while other Protists do not. This ultimately lead to another high level reorganization of organisms into the eukaryotes, whose cells contain nuclei, and the prokaryotes whose genetic material is dispersed within the cell (and which are almost exclusively unicellular). Although the prokaryotes are dominated by the Bacteria and Archaea (once both classed under the kingdom of Monera, until their distinct evolutionary histories were uncovered), the eukaryotes are more diverse, consisting of both unicellular organisms, the Protists, as well as multicellular organisms such as the plants, fungi, and animals (see Figure 6.1).

Multicellular organisms, of course, evolved from unicellular ones, which were the only forms of life for the three billions years before the Cambrian Explosion (believed to have occured approximately 540 million years ago). The evidence now points to the choanoflagellates, or "collared flagellates," as being our closest unicellular relatives. (Their collars, visible in Figure 6.2 Left, are, in fact, a ring of microvilla used to filter nutrients from the surrounding

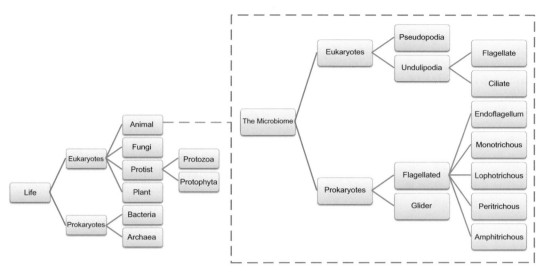

FIGURE 6.1   Taxonomy of living organisms based on six kingdom classifications, with the microbiome organized in terms of methods of motility.

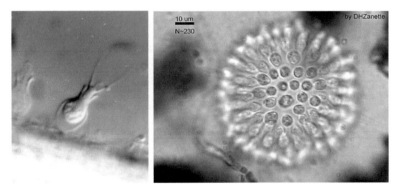

FIGURE 6.2   Choanoflagellates, or "collared flagellates," believed to be the closest unicellular organisms to humans. (Left) A single *Salpingoeca Amphoridium* tethered to a surface and generating feeding currents by beating an elongated appendage, known as a flagellum (reproduced from [4] with permission); (Right) Colony of the choano-flagellate *Sphaeroeca*. [*Wikicommons. Photo by Dhzanette, September 6, 2002.*]

liquid environment.) The beginnings of multicellularity are indeed evident in the behavior of choanaflagellates, which have been shown to form colonies to improve their survival prospects (see Figure 6.2 Right).

Indeed, our evolutionary beginnings are evident in the very structure of our cells. There is good evidence to suggest that the mitochondria, which provide our cells with energy, as well as those of other eukaryotes, originated from proteobacteria being absorbed into the cells of early unicellular ancestors. This is not the only example of this kind of endosymbiosis. For instance, it is believed that the Eukaryotic Algae, which now produce around three-quarters of the oxygen in the atmosphere, and which are a base organism in the aquatic food chain, obtained their photosynthesizing capabilities (chloroplasts) from one of the earliest forms of life on Earth, the Cyanobacteria. These same Cyanobacteria are believed to have been responsible for transforming the toxic atmosphere of the early Earth into an oxygenated one that can support the current diversity of life. Later, in an act of secondary endosymbiosis, some of these algae are believed to have been absorbed by other Protozoa (including the *Euglena* discovered by van Leewenhoek), thereby allowing them to also photosynthesize.

Whatever their form and function, the microorganisms that inhabit our bodies face many of the same survival pressures encountered by their hosts: the need to feed, avoid predators, and seek out favorable environments. For these reasons, many are motile which, given the high fluid content of the human body (over ten litres of extracellular fluid), often amounts to the ability to swim.

6.3 SWIMMING MICROORGANISMS

The challenges facing any microorganism wishing to swim through its fluid environment are well documented. For swimmers that are microscopic in size, it is virtually impossible to generate flows that possess any appreciable inertia. To provide a comparison, a 10 μm-sized microorganism, swimming at one body length per second through water is analogous to a human trying to swim through honey (which is up to ten thousand times as viscous as water). In the absence of any appreciable inertia, fluid forces such as pressures, viscous stresses (i.e., frictional forces), as well as any body forces present in the system, must therefore be in balance. Under such circumstances, the swimming hydrodynamics are governed by

$$-\nabla p + \nabla\cdot\boldsymbol{\tau} + \boldsymbol{F} = \mathbf{0}, \quad \nabla\cdot\boldsymbol{u} = 0, \tag{6.1}$$

where p is flow pressure, $\boldsymbol{F}$ represents any body forces on the fluid, and $\boldsymbol{\tau}$ is the deviatoric stress tensor. For a Newtonian fluid, such as water, the stress is linearly proportional to the flow's rate of strain $\dot{\gamma}$, i.e.

$$\boldsymbol{\tau} = 2\mu\dot{\boldsymbol{\gamma}} \tag{6.2}$$

where μ is dynamic viscosity, and

$$\dot{\boldsymbol{\gamma}} = \frac{1}{2}\left(\nabla\mathbf{u} + (\nabla\mathbf{u})^T\right) \tag{6.3}$$

where u is the flow velocity. This flow regime is known as incompressible Stokes (or Creeping) Flow. (It is worth noting that numerous fluids in the body are actually non-Newtonian, and we shall return to this issue a little later.)

Importantly, the lack of inertia at these scales means that the microorganism itself can experience no net force or torque. Any force or torque that it exerts upon the surrounding fluid must be countered by an equal and opposite hydrodynamic force or torque from the fluid itself. This simple realization of Newton's Third Law of Motion is the basic means by which swimming microorganisms propel themselves.

However, there is a complication, because zero-inertia (Newtonian) flow is also reversible. This means that reciprocal motions (whereby the return stroke is a retrace of the forward stroke) generate no net displacement. This rules out many of the fluid dynamical strategies used by larger organisms, such as flipping (fish), or flapping (birds). A famous example, popularized by Purcell [5], concerns the swimming action of a scallop, which swims by opening and closing its shell to jet out water. That approach simply does not work for microorganisms (at least in Newtonian liquids, like water), because it is a reversible action. Were it to exist, a microscopic scallop would move forward when closing its shell, but then return to its original starting position upon re-opening the shell. This is due to the fact that the opening action here is the exact time reversal of the closing action.

Consequently, swimming microorganisms have tailored specific methods for propelling themselves through their highly-viscous fluid environments. As might be expected, given their distinct evolutionary developments, the eukaryotes and the prokaryotes have both evolved different swimming apparati to overcome this fluid-dynamical dilemma. However, in what is perhaps a striking example of convergent evolution, the end designs both bear a remarkable similarity: an appendage known as a flagellum (the Latin word for "whip").

6.3.1 Flagellar Biomechanics and Hydrodynamics

For both eukaryotes and prokaryotes, the effectiveness of their flagellum (-a) as swimming instruments in this regime rests upon a single basic principle. A thin body moving in a Stokes Flow regime experiences approximately double the hydrodynamic resistance when moving perpendicular to its axis, as when moving parallel to its axis. This is readily seen from the resistance coefficients (the force per unit length on the cylinder in each direction resulting from prescribed unit velocity in that same direction) for an isolated finite-length slender cylinder of length l and radius a [6]

$$K_T = \frac{4\pi\mu}{(\log(4/\epsilon^2) - 1)}, \quad K_N = \frac{8\pi\mu}{(\log(4/\epsilon^2) + 1)}, \quad (6.4)$$

where $\epsilon = a/l \ll 1$ is a measure of the cylinder's slenderness. Here K_T relates to motions tangential to the cylinder axis, while K_N relates to motions perpendicular to the axis (because the cylinder is considered to be in unbounded flow, there is little coupling between parallel and perpendicular motions and forces). As can be seen, approximately double the force per unit length is needed to generate unit velocity in a direction normal to the cylinder axis, as needed for motion parallel to the cylinder axis (because the logarithmic term dominates the denominator for small ϵ).

There are, however, two major structural differences between the flagella evolved by prokaryotes and those evolved by eukaryotes. First, eukaryotic flagella are much larger, having diameters in the hundreds of nanometers, and hence observable using optical microscopy. Prokaryotic flagella, on the other hand, are much thinner (with diameters that are typically tens of nanometers) and generally require the levels of magnification offered by Electron Microscopy in order to be observed. Nonetheless, both are sufficiently small so as to be governed by the same hydrodynamics (i.e., Stokes Flow).

The second major difference between prokaryotic and eukaryotic flagella relates to their biomechanical structure, and this does lead to some important hydrodynamic (and elastohydrodynamic) consequences.

6.3.1.1 Prokaryotic Flagella

The flagella used by prokaryotes are inert filaments made of the protein *flagellin* that are driven by a biological motor embedded in the cell membrane (a "basal" motor, see Figure 6.3 left). This rotates the attached flagellum, possibly at hundreds of hertz, fuelled by an ion gradient (Bacteria) or ATP (Archaea). This driving mechanism therefore enables prokaryotic flagella to rotate independently of the cell body.

In flow models, it is straightforward to prescribe the motion of prokaryotic flagella as a function of its shape and beating frequency. One of the simplest (one-dimensional) demonstrations of how such a flagellum can generate thrust is provided by an infinite two-dimensional elastic sheet model (see Figure 6.4 Top Left). Here the elastic sheet undergoes prescribed undulations in the $x - y$ plane, in a rough approximation to a planar-beating flagellum

$$y = A \sin (kx + \omega t) \tag{6.5a}$$

(where the oscillation amplitude A, wavelength $2\pi/k$ and frequency ω are all appropriately-sized). A relatively simple flow calculation reveals that, in a laboratory frame of reference, the sheet would translate with speed [9]

$$V = \frac{1}{2}A^2 k \omega \tag{6.5b}$$

in the x-direction.

In practice, of course, microorganism flagella more closely resemble filaments than infinite elastic sheets, and Resistive Force Theory [10] is capable of providing a more geometrically faithful representation. This approach entails approximating the slender, curved flagellum by

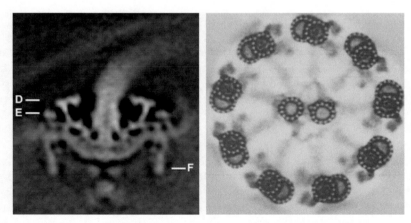

FIGURE 6.3   (Left) Cryoelectron Tomography image of the biological motor embedded in the cell body of *T. pallidum*, that rotates its prokaryotic flagellum. Labels D, E and F indicate the ring structures responsible for generating rotation (*reproduced from [7] with permission*). (Right) Cross-section through a eukaryotic flagellum, clearly showing the 9+2 microtubular structure of the axoneme (*reproduced from [8] with permission*).

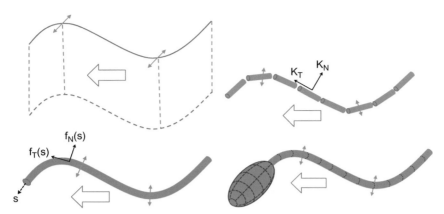

FIGURE 6.4 Hydrodynamic models for microorganism swimming of increasing complexity from (Top Left) Infinite Sheet, (Top Right) Resistive Force Theory, (Bottom Left) Slender Body Theory to (Bottom Right) Hybrid Boundary Element-Slender Body Theory. Solid arrows illustrate the local flagellum velocity and hollow arrows illustrate resultant thrust.

a series of straight cylindrical rods (see Figure 6.4 Top Right) of radius a and length $2l$ ($a \ll l \ll L$, where L is the total length of the flagellum), and by assuming that each segment behaves hydrodynamically like an isolated circular cylinder. Due to the linearity of the Stokes Flow equations, the resistance coefficients, K_T and K_N given earlier for the cylinder (Equation 6.4), then yield the resultant forces acting on a flagellum segment (cylinder) due to any prescribed velocities. In particular, by considering the local velocity on a segment of a flagellum undergoing prescribed beating motion (at its midpoint, say), the resistive force coefficients immediately provide the resulting force on that segment. By then summing these forces over all segments, the overall hydrodynamic thrust produced by that beating action can be estimated. Since Stokes Flow is inertialess, Newton's Third Law states that an equal-and-opposite force must be exerted by the fluid onto the flagellum, and it is this which generates forward motion through the fluid. By equating the propulsive force from the flagellum with the drag that would be induced on the swimming microorganism, it is possible to obtain an estimate for swimming speeds. This has been done for both planar beating [10] as well as helical beating [11] (where no net torque must also be observed).

However, the type of local force-velocity relationship exploited in Resistive Force Theory is only a reasonable approximation when the element length is sufficiently small with respect to the wavelength of the wave, and the beat amplitudes are sufficiently small [12]. If this is not the case, then it may be necessary to turn to a more sophisticated hydrodynamic treatment. Slender Body Theory is a more general flow approximation whereby forces applied at any point along the flagellum (or any other elongated body) can affect the velocities induced along the entire length of the flagellum [13]. Expressed mathematically

$$u_i(\mathbf{x}_0) = \int_{\mathcal{L}} G_{ij}(\mathbf{x}, \mathbf{x}_0) f_j(\mathbf{x}) ds(\mathbf{x}) \tag{6.6}$$

($i, j = 1 \ldots 3$ represent x, y, and z directions, respectively) where $\mathcal{L}$ represents the flagellum's centerline, G_{ij} represent one or more singularly forced (e.g., Delta function forced) solutions to the Stokes Flow equations (e.g., Stokeslets, Rotlets, Potential Dipoles [14]), f is the force per unit length, and u the resulting flow velocity at a point x_0 in the flow (which will coincide with the motion of the flagellum, when x_0 lies on the flagella surface). The Slender Body Theory formulation can be thought of as approximating the flow by distributing Point Forces (and Point Torques for cases where the flagellum is being rotated) along the centerline of the flagellum. Different beating patterns can be prescribed, and the above integral equations solved (in general, numerically) to determine the propulsive thrust that such a waveform can generate. Rudimentary cell bodies, such as spheres, can also easily be accounted for in the Slender Body Theory approximation through an appropriate choice of image systems that automatically enforce zero flow on the cell body. The same is true for any nearby solid surface [15].

If a more physiologically realistic cell body is required, however, then it often becomes necessary to use an exact integral representation for Stokes Flow. The Boundary Integral Representation can be thought of as a generalization of the Slender Body Theory, capable of representing Stokes flows generated by arbitrarily-shaped bodies, by distributing Fundamental Solutions S_{ij} and T_{jik} over the entire surface of that body, A (rather than just along some one-dimensional centerline) [14]

$$u_i(x_0) = -\int_A S_{ij}(x, x_0) f_j(x) dS(x) + \int_A T_{jik}(x, x_0) u_j(x) n_k(x) dS(x)$$

where n is a unit normal to the surface, S_{ij} represents the flow due to point forces (i.e., Stokeslets), with T_{ijk} being the associated stress tensor. Because the Boundary Integral formation requires a two-dimensional surface distribution of singularities, as compared with the Slender Body Theory approximation which only requires a one-dimensional axial distribution, the increased accuracy comes at a computational price. One common tactic to reduce computational burden has been to use a hybrid Boundary Integral-Slender Body Theory approach, where the hydrodynamics of the cell body are captured using a Boundary Integral representation, whereas those generated by the flagellum can still effectively be approximated using Slender Body Theory, as illustrated in Figure 6.4 Bottom Right. Using this approach, it has been possible to show that microorganisms have seemingly evolved body shapes that are designed to be hydrodynamically efficient [16] (rather than the evolutionary priority of high maneuverability, say).

The previous Slender Body and Boundary Element formulations are readily able to model the hydrodynamics of microorganisms with multiple flagella (albeit at a computational cost). This is relevant because the number of flagella employed for swimming can vary from microorganism to microorganism. *Spermatazoa*, for example, employ just a single posterior flagellum (see Figure 6.5 Top Left) and is therefore described as being *monotrichous*. The bacterium *Streptococcus* (strains of which are implicated in illnesses such as pneumonia and meningitis), however, is endowed with multiple flagella located at different locations over its body surface (see Figure 6.5 Top Right), and is termed *peritrichous*. When multiple flagella emanate from the same location on the cell surface, as is the case with the sulphur producing bacterium *Chromatium*, the microorganism is termed *lophotrichous*. Less commonly, there are also the *amphitrichous* microorganisms like *Campylobacter jejuni*, the bacterium responsible in a high percentage of food poisoning incidents, which have a flagellum at either end of their cell body (see Figure 6.5 Bottom). They only actuate one of these flagella at a time, which dictates the direction of travel.

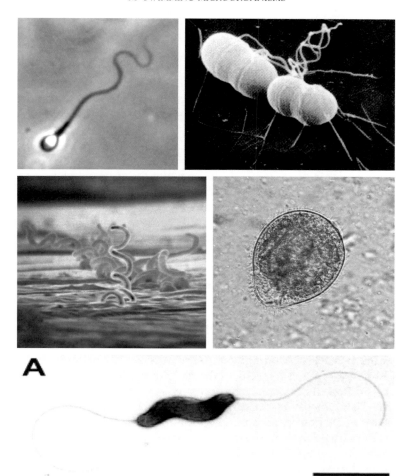

FIGURE 6.5 Four (potential) members of the microbiome: (Top Left) Human spermatazoon, that swims with a single eukaryotic flagellum (*reproduced from [17] with permission*), (Top Right) Bacterium *Streptococcus*, associated with pneumonia and meningitis, endowed with multiple prokaryotic flagella over its surface (*reproduced from [18] with permission*), (Middle Left) Spirochete *Treponema pallidum*, associated with the disease syphilis, and which swims by twisting its entire body (*Wikicommons: Centers for Disease Control and Prevention's Public Health Image Library: Image Number 1977*), (Middle Right) Ciliated Protozoan *Balantidium coli* associated with an intestinal infection in humans [Wikicommons Photo by Euthman, November 25, 2006], (Bottom) Amphitrichous bacterium *Campylobacter jejuni*, associated with many cases of food poisoning [19].

Prokaryotes with multiple rotating flagella can also undergo Run-and-Tumble swimming. This describes the situation where a microorganism swims in approximately a straight line, by rotating all flagella in the same direction, before suddenly rotating one of its flagellum the opposite way. This causes the flagella to become tangled, or *bundled*, resulting in an abrupt halt to the microorganism's progress, and a random change in its orientation. The microorganism then reverts to rotating all flagella in the same direction again, to swim in this new direction. Run-and-Tumble allows the microorganism to survey a region for nutrient gradients (because their cell sizes are typically too small to span any detectable chemical gradients, unlike the generally larger eukaryotic microorganisms), and follow positive nutrient gradients by lowering the frequency of tumbles. As a result, Run-and-Tumble motion takes on the appearance of a biased Random Walk, and shall be discussed more in the section on Continuum Models. (Note that some eukaryotes, such as the algae *Chlamydomonas reinhardtii*, can undergo similar Run-and-Tumble motions by beating their pair of forward facing flagella either synchronously, or asynchronously.)

6.3.1.2 Spirochetes

It is also worth mentioning the distinctive spirochetes, bacteria which use their prokaryotic flagella in an altogether different way. Rather than protruding outwards from the cell body, the spirochete flagella are wrapped around the cell body, and trapped between two membranes. This configuration is known as an endoflagellum. These flagella are attached to motors at both ends of the cell body, so that when they are rotated the whole cell body twists in a helical motion, thereby propelling itself through its fluid environment with the whole body effectively acting as the swimming instrument [20]. Figure 6.5 Middle Left) shows an image of the spirochete *Treponema pallidum*, which is connected with the venereal disease syphilis.

6.3.1.3 Eukaryotic Flagella

Eukaryotes, in contrast to prokaryotes, possess no such basal motor to drive their flagella. Instead, their flagella actively bend along their entire length. In fact, even before the biological mechanisms behind this active bending were fully understood, the theoretical investigations of Machin [21] had shown active bending of the flagellum to be necessary in order to account for the waveforms of sea-urchin spermatazoa as reported by Gray [22].

Microbiologists have since confirmed this hypothesis, showing that there is indeed active actuation along the entire flagellum. This actuation is facilitated by a structure known as the 9+2 Axoneme, so-named because it consists of a central microtubule doublet surrounded by a ring of nine outer microtubule doublets (see Figure 6.3 Right). Eukaryotic flagella bend through the active sliding (rather than contracting) of these outer microtubules relative to each other, via ATP-powered protein motors called dynein that are located along the length of the microtubules. These dynein motors can engage and disengage with the microtubules in response to loading, which is now understood to play an important part in conferring a particular waveform upon a beating flagellum. The outer microtubules are elastically connected to each other via protein links (nexin), thus enabling the sliding between microtubules to result in overall bending of the flagellum

Microorganisms endowed with this structure are sometimes referred to as Undulipodia, and those that use this kind of flagella to swim are known as flagellates (a term which, perhaps confusingly, does not include microorganisms that swim with prokaryotic flagella). The eukaryotic flagella can be beat in a variety of different ways, with some undergoing a planar wave (see Figure 6.6 Top Left) and others a helical one, in a similar fashion to prokaryotic flagella (see Figure 6.6 Top Right, Bottom Left). However, because the eukaryotic flagellum, unlike the prokaryotic flagellum, does not rotate independently of the body, helical beating can lead to rolling of the entire cell during swimming (to generate a torque that negates that produced by the flagellum). This has been observed for some spermatozoa that swim with

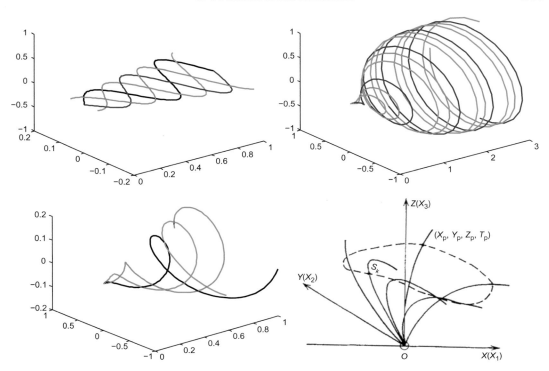

FIGURE 6.6 (Top Left) Planar beat $y(x) = b \sin(kx - t)$ with $k = 3\pi$ at times $t = 0, \pi/2, \pi, 3\pi/2$ as shown by black, green, blue, and red (respectively) $b = 0.1087x + 0.0543$, (Top Right) Helix beat $y(x) = (1 - \exp(-x^2))\cos(kx - t)$, $z(x) = (1 - \exp(-x^2))\sin(kx - t)$ [23], (Bottom Left) Conical helical beat $y(x) = x \cos(kx - t)$, $z(x) = -0.2x \sin(kx - t)$ [24], (Bottom Right) Cilia beat (taken from [25], with permission).

such an action. Some microorganisms, like *Euglena*, utilize both types of wave motion, with a planar-beating flagellum behind the cell, and a second helical-beating flagellum around the circumference of the cell body. It is also worth mentioning the algae *Chlamydomonas reinhardtii*, which has a pair of forward facing (apical) eukaryotic flagella that it can beat either synchronously or asynchronously to undergo a type of Run-and-Tumble swimming motion discussed earlier.

The active bending of Eukaryotic flagella means that the shape of their beating flagella (waveforms) emerge as a product of driven fluid-structure interactions. It is therefore of interest to see whether models can produce the observed waveforms as an emergent property, rather than through direct prescription. The models discussed earlier for describing flagellar hydrodynamics (Resistive Force Theory, Slender Body Theory, and the Boundary Element Method) are still applicable. However, they must now be coupled to an elasticity model that mimics the bending of the flagellum as well as a constitutive model for the active forcing due to the dynein motors.

A simple model for predicting the emergent beating patterns of a flagellum involves an Euler-Bernoulli beam description of the elastic bending, subject to hydrodynamic damping prescribed by Resistive Force Theory coefficients [26]. Assuming small deflections (a similar, albeit more involved, expression can be derived for finite deflections), in frequency space (i.e., upon taking Fourier Transforms in time), the flagellum slope $\hat{\psi}(s, \omega)$, is governed by

$$E\frac{\partial^4 \hat{\psi}}{\partial s^4} + i\omega K_N \hat{\psi} - \frac{\partial^2 \hat{f}}{\partial s^2} = 0, \tag{6.7}$$

where s represents an arclength parameter along the length of the flagellum, E is the elastic modulus of the flagellum, K_N is the normal resistance coefficient, and f is the active actuation along the flagellum. This forcing f must reflect the ability of the dynein motors to slide the microtubules in the axoneme past each other. If $\Delta(s)$ is used to represent the displacement between microtubules, a simple geometrical argument provides a connection with the slope of the flagellum

$$\hat{\Delta}(s) = \hat{\Delta}_0 + a(\hat{\psi}(s) - \hat{\psi}(0)) \tag{6.8}$$

where a is the microtubule radius, and $\hat{\Delta}_0$ corresponds to sliding displacement at the base of the flagellum, with a proposed relationship between $\hat{f}$ and $\hat{\Delta}_0$ given as [27]

$$(i\omega c_0 + k_0)\hat{\Delta}_0 = -\int_0^L \hat{f} ds \tag{6.9}$$

where c_0 and k_0 are coefficients of basal friction and stiffness. Finally, in order to close the model, a relationship between $\hat{f}$ and $\hat{\Delta}$ is needed, with a proposed linear relationship given by

$$\hat{f}(s) = \chi \hat{\Delta}(s), \tag{6.10}$$

where χ is known as the dynamic stiffness. This may have imaginary parts that correspond to a phase lag between activation force and sliding, as well as possible s dependence (i.e., a force-sliding relationship that varies along the length of the flagellum). Substituting into Equation (6.7) results in the following governing equation for the flagellum slope:

$$E\frac{\partial^4 \hat{\psi}}{\partial s^4} + i\omega K_N \hat{\psi} - a\chi \frac{\partial^2 \hat{\psi}}{\partial s^2} = 0, \tag{6.11}$$

with the dynamic stiffness, χ, used as a fitting parameter. Although relatively simple, the above model was shown by Riedel and colleagues [27] to well-approximate the flagellar waveforms generated by bull spermatazoa reported by Gray [28] (see Figure 6.7).

6.3.1.4 Ciliates

Microorganisms that have a dense surface covering of short eukaryotic flagella, known as cilia (typically $5 - 10\mu$ m in length), are accordingly known as ciliates (see for example, *Balantidium coli* shown in Figure 6.5 Middle Right, which is a protozoan associated with intestinal infection in humans). Their cilia also beat in a distinctive manner, compared with other eukaryotic flagella. This consists first of an Effective Stroke, where the cilia moves largely perpendicularly to a surface, in order to generate significant hydrodynamic forces (and hence generate thrust). This is followed by a Recovery Stroke, where the cilia returns to its original position via a flagella motion that is more tangential to the surface, thereby minimizing the fluid forces and reducing the amount of reverse motion of the cell body (see Figure 6.6

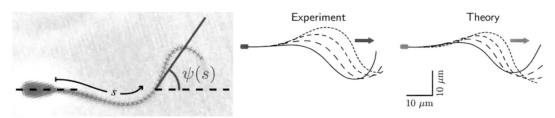

FIGURE 6.7   Images taken from the work of Riedel et al. [27], showing (Left) Experimental observations of a swimming bull spermatazoa (Middle) Centerline tracking of the flagellum, alongside (Right) predictions from flagellar activiation models of the kind given by Equation (6.11), at four uniformly spaced time intervals during the first quarter of the flagellum beat period.

Bottom Right). Each cilium beats in a cooperative way with its neighbors, synchronizing its beat to produce a wave that travels across the layer of cilia. The waves can travel in the same direction as that in which the individual cilia beat during their effective strong (symplectic) waves, but also in the opposite direction (antiplectic) and perpendicular (diaplectic). (Note that cilia are also used by other non-motile eukaryotic cells and are believed to be involved in an array of different functions including cell signaling and mechanosensing. However, these non-motile cilia typically lack the central microtubules, consisting instead of a 9+0 axoneme.)

Explicitly accounting for the hydrodynamics of each cilia is computationally demanding. For ciliates that use a symplectic beat, the Envelope Model has proved a popular effective approximation. Here, the surface coating of the beating cilia averaged over multiple beats is represented as an effective instantaneous tangential flow induced at the surface of the microorganism [29]

$$u_r = \sum_{n=0}^{\infty} A_n P_n(\cos\theta), \quad u_\theta = \sum_{n=0}^{\infty} B_n V_n(\cos\theta) \tag{6.12}$$

($V_n = 2\sin\theta(dP/d\theta)/n(n+1)$) where θ is a spherical polar coordinate, u_r and u_θ are radial and azimuthal velocities, and P_n are the Legendre Polynomials. An appropriate mathematical form for the coefficients A_n and B_n have been found by fitting to experimentally-observed cilia beating patterns [30]. Equipped with these surface velocities, the resulting hydrodynamics can be then determined using techniques such as the Boundary Element Method. The model microorganisms using this swimming action are sometimes referred to as *Squirmers* [31, 32], and these have been shown to be a reasonable theoretical approximation to certain ciliated microorganisms [33] (although it has recently been suggested that flow effects on a timescale shorter than the period of beating could be important in some situations [34, 35]).

For ciliates that use an antiplectic beat, an Envelope Model is not appropriate, and a Sublayer Model can instead be used (which, it should be noted, can also be used for symplectic waveforms). Here, an array of cilia is modelled, the hydrodynamics of each approximated using Slender Body Theory, and net flow quantities are obtained by integrating across the array. For an infinite sheet of cilia, this method predicts a propulsive velocity given by [25]

$$U_i = \frac{1}{ab\mu} \left\langle \int_0^L \omega(s,t) f_i(s) y(s) ds \right\rangle \qquad (6.13)$$

where angular brackets denote that time and spatial averages have been taken, L is the length of the cilia, s is an arclength parameter along the cilia centerline, $y(s)$ is the vertical height of the cilium at point s along its centerline, a and b represent horizontal and vertical spacing between the individual cilia, f is the force per unit length that appears in the Slender Body Theory flow representation (Equation 6.6). The form of weighting function, $\omega(s,t)$, is determined by whether the cilia wave is symplectic or antiplectic.

The Lighthill-Gueron-Liron (LGL) Theorem approaches the hydrodynamics of cilia beating in a similar fashion and is used not only in the context of microorganism propulsion, but also in areas such as cilia-driven transport (e.g., of mucus) (see [36] for more details).

6.3.1.4.1 *PUSHERS* AND *PULLERS*

Irrespective of whether an individual swimmer is propelled by a eukaryotic flagella or a prokaryotic one, there is one distinction which we shall see is highly relevant to the collective behavior of microorganism suspensions. This concerns whether the microorganism is propelled from behind by its flagellum (-a) (*pushers*), as are spermatozoa and many types of bacteria (see Figure 6.8, Left, for an illustration), or whether it is pulled through the fluid by its flagellum (-a) (*pullers*), as for instance is the algae *Chlamydomonas reinhardtti* (see Figure 6.8 Right for an illustration). This collective behavior comes about due to hydrodynamic interactions of a swimming microorganism with both its surrounding environment and other swimming individuals.

6.3.1.5 *Non-Newtonian Effects*

The previous models assume that the microorganism is swimming in a Newtonian fluid. However, in a physiological setting this is often not the case. For instance, mucus has been shown to exhibit strongly non-Newtonian behavior. Gastric mucus, which lines the stomach wall and protects it from digestive acids, as well as mucus linings in the duodenum, female reproductive tract, and lung airways (where it is associated with particle clearance), have

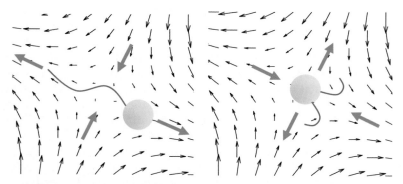

FIGURE 6.8   Illustrations of (Left) *pusher,* such as the bacterium *Vibrio cholerae* which is propelled from behind by a single prokaryotic flagellum, and (Right) a *puller,* such as the algae *Chlamydomonas reinhardtii,* which is pulled by a pair of forward-facing (apical) eukaryotic flagella. Both here are aligned to the extensional component of the local flow. The pusher reinforces the local flow, while the puller opposes it.

been reported to have a yield stress and exhibit shear-thinning behavior [37]. Other examples of non-Newtonian physiological fluids include amniotic fluids, as well as chyme within the intestines [38].

Constitutive laws for non-Newtonian fluids are more involved, introducing nonlinearities into the governing equations, and this necessarily complicates the analysis. For example, an Oldroyd-B fluid, which models a fluid containing polymer chains, has the following stress–rate-of-strain relationship:

$$\boldsymbol{\tau} + \lambda_1 \overset{\triangledown}{\boldsymbol{\tau}} = 2(\mu + \mu')\left(\dot{\boldsymbol{\gamma}} + \lambda_2 \overset{\triangledown}{\dot{\boldsymbol{\gamma}}}\right) \tag{6.14}$$

where μ' is an effective viscosity for the polymers, λ_1 and λ_2 are relaxation coefficients, and $\triangledown$ denotes an Upper Convected Derivative (whereby a tensor is both advected and deformed by the flow). Reanalyzing the undulating infinite sheet in such a fluid (Equation 6.5), the swimming speed obtained in this non-Newtonian fluid becomes [39]

$$V' = \frac{1}{2}A^2 k\omega \frac{1 + \mu\omega^2\lambda_1^2/\mu'}{1 + \omega^2\lambda_1^2} \tag{6.15}$$

Because $\mu < \mu'$, $V' < \frac{1}{2}A^2 k\omega$ (the propulsive velocity in Newtonian fluid for the same swimming action, Equation 6.5b). Hence, mobility is predicted to be reduced in this non-Newtonian fluid.

It is also worth noting that flow reversibility does not necessarily apply in a non-Newtonian fluid. Hence, the scallop's method of swimming, by opening and closing its shell, could lead to net motion in a microscopic non-Newtonian fluid environment.

6.3.2 Flow-Induced Motion

The hydrodynamics of a swimming microorganism do more than simply provide it with a propulsive force. They also generate flowfields which couple the microorganism to its surroundings, including nearby solid surfaces, and its neighbors. These can significantly change the swimming behavior of each individual, as well as the collective behavior of a suspension of many swimming microorganisms (with the latter increasingly becoming an intense area of research interest).

6.3.2.1 Boundary Effects

One of the most evident examples of the influence of hydrodynamic coupling occurs when a microorganism swims close to a solid surface. In fact, flow coupling between the microscope slide and the swimming microorganisms under observation are believed to be responsible for some of the discrepancies between the early predictions of Resistive Force Theory and experimental measurements. For a flagellum beating with a given planar waveform, the presence of a solid surface leads (perhaps counterintuitively) to an increase in the propulsive force, although more power is required by the swimmer to beat its flagellum [40]. Moreover, close to solid surfaces bacteria have been observed to exhibit circular swimming motions (see Figure 6.9 Top Left, for example). For prokaryotic microorganisms, such as bacteria, this relates to the fact that the cell body rotates in the opposite direction to the flagellum, in order

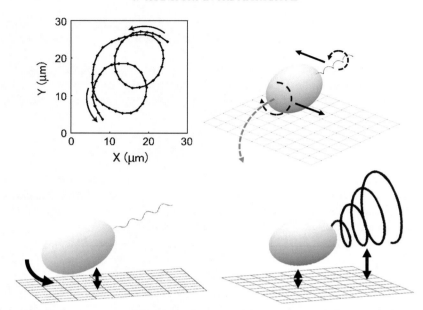

FIGURE 6.9   (Top Left) Top-down tracking of *V. alginolyticus* (a bacterium with a single flagalla that can rotate in either direction to swim either forward or backward) swimming backward at about 10 μ m above a glass slide. Taken from Goto et al. [41] (Reproduced with permission). (Top Right) Illustration of the hydrodynamic turning mechanism for prokaryotes. Dashed arrows indicate the opposite directions of rotation of the cell body and flagellum, with solids arrows indicating resulting forces on both. The dashed gray line indicates the circular trajectory under the net yawing force. (Bottom Left) Illustration of the trapping mechanism for prokaryotes. Torque on the head orientates the swimming direction toward the wall, with lubricating forces between the cell and the wall preventing contact (the hydrodynamics of the small flagellum do not play a significant part). (Bottom Right) Illustration of the hydrodynamic trapping mechanisms for a eukaryote beating with a conical helical waveform. Here, both wall–cell and wall–flagellum interactions are important, with the long edge of the flagellum becoming aligned with the wall (as reported in [24]).

to ensure that there is no net torque on the microorganism. Close to a solid surface, however, rotational motions couple with translational hydrodynamic forces, and vice versa. Because the cell body and flagellum rotate at different angular velocities and are of different sizes, the resulting translational forces on the cell body and flagellum are different. This yields a net torque which induces a turning motion about an axis perpedicular to the swimming direction (i.e., a yawing motion, see Figure 6.9 Top Right for an illustration), which consequently leads to a circular swimming motion.

As well as swimming in circular paths, it has also been observed that swimming prokaryotes tend to spend relatively long periods of time swimming in close proximity to solid surfaces, effectively having been hydrodynamically trapped. This effect is believed to be important in a number of scenarios, not least of which being the formation of biofilms. Biofilms are bacterial colonies aggregated as a thin layer attached to a surface. These biofilms are highly resistant to removal and antibiotics, with dental plaque on teeth constituting one of the better-known examples. They can also occur on the lining of the gastrointestinal tract, where they are associated with some inflammatory bowel diseases, as well as in a range of other physiological situations. Because prokaryotes have relatively small flagella, wall trapping is believed to be dominated by the hydrodynamic interactions between the surface and cell body, and again stem from the coupling between translations and rotations [41]. When the microorganism swims close to a wall, a torque induced on the cell body causes the swimming direction to orientate toward the wall. Actual contact with the wall is prevented, however, by strong hydrodynamic lubrication forces that increase as separation distance decreases. The result is that the microorganism swims at a distance where these two competing effects are in balance (see Figure 6.9 Bottom Left).

In situations where the bacteria occupy just a thin liqud film next to a solid surface (e.g., a biofilm), one might expect especially high hydrodynamic forces to act as a barrier against mobility. In such a case, some bacteria (such as *Salmonella* and *Escherichia*) adopt *swarming* behavior [42], whereby they release a lubricant that can also have surfactant properties. (Interestingly, surfactant production only appears to occur when there are sufficiently many bacteria present to produce effective quantities, thereby suggesting some form of cell–cell chemical signaling between the bacteria, i.e. quorum sensing.) Furthermore, the bacteria can undergo morphological changes, developing additional flagella to increase their propulsive power. In addition, *rafting* of bacteria has been observed, for example with *Proteus mirabilis*, a bacterium that forms part of the human gut flora, but which also has strains associated with certain disease states. Here, the bacteria within the film swim as groups, possibly bundling their flagella together across individuals. As a result of these strategies, bacteria have been observed to be capable of transporting themselves at $10 \mu m$ per second in such a setting; moreover, the release of surfactant is hypothesized to be responsible for the observed overall spreading of the film itself.

Wall trapping of eukaryotes also occurs, with perhaps most attention given to the behavior of spermatozoa near surfaces, due to the likely implications for the mobility of the sperm cells that have entered the female reproducive tract. The hydrodynamics here, though, are more complex, due to the larger size of the eukaryotic flagella, which now plays a part in the hydrodynamic interactions between the microorganism and the wall. In fact, recent BEM-SBT simulations of both planar- and helicoid-beating spermatozoa predict that trapping will only occur over certain ranges of beating frequencies. These same simulations also demonstrate that conical helical flagella can become hydrodynamically aligned with the surface [24] (see Figure 6.9 Bottom Right for an illustration), as earlier noted in the experimental observations of Woolley [43].

For hydrodynamic trapping next to physiological surfaces, however, compliant deformations of surfaces may be important. This has been modelled through the use of the Immersed Boundary Method (IBM) which can account for fluid-structure interactions. Similarly to the Regular Boundary Element Method described earlier, the presence of surfaces in the flow are represented through distributions of point forces. However, the method requires flow quantities throughout the domain to be computed simultaneously, and not just those on the domain boundary as in the standard Boundary Element Method. The flow equations can be the full (forced) incompressible Navier–Stokes equations (and so the method can also handle flow inertia, if required)

$$\rho \frac{D\boldsymbol{u}}{\partial t} = -\nabla p + \mu \nabla^2 \boldsymbol{u} + \boldsymbol{F}, \quad \nabla \cdot \boldsymbol{u} = 0 \tag{6.16}$$

where $\boldsymbol{u}$, p, and ρ represent flow velocity, pressure, and fluid density, respectively, and

$$F(x) = \int_\Omega f(s)\delta(x-s)\mathrm{d}s \qquad (6.17)$$

(where $\delta(x)$ is the Delta function), represents the forcing due to the presence of a distribution of forces f over the boundary surface(s), Ω, as parameterized by s. The position of compliant surfaces can be updated through the kinematic condition

$$\frac{\partial\Omega}{\partial t} = u(\Omega). \qquad (6.18)$$

Finally, the strength of the force distribution, which dictates the influence of the boundary upon the flow, is determined through an appropriate choice of constitutive law. For the case of a flagellated microorganism close to an elastic wall with surface Ω_w, Fauci and MacDonald [44] utilized a Hookean model to relate force on the wall $f(\Omega_w)$ to its displacement away from its non-deformed configuration (denoted as $\overline{\Omega}_w$), and an energy functional $E(\Omega_0)$ for the force $f(\Omega_0)$ due to the presence of the microorganism (with surface Ω_0), in other words,

$$f(\Omega_w) = k\big(\Omega_w - \overline{\Omega}_w\big), \quad f(\Omega_0) = -\nabla E \qquad (6.19)$$

where k is some spring constant. When this method is used to simulate one or more (monotrichous) flagellated microorganisms swimming through an elastic channel, a number of emergent behaviors are predicted [44]: i) the flagella waveform is altered by the presence of the channel wall, ii) the channel wall itself can be sucked inwards as a result of its hydrodynamic interactions with the microorganism(s), and (iii) multiple microorganisms appear to adjust their swimming so that their flagella become phase locked, (i.e., beat in synchrony).

6.3.2.2 Cell–Cell Hydrodynamics

The phase locking of flagella constitute just one example of how hydrodynamics can couple together the motion of swimming individuals. This hydrodynamic coupling can occur even in the absence of compliant effects, and has been shown to lead to some extremely rich dynamics, making it an area of intense current research interest.

Modeling this behavior, however, is extremely challenging. The low-inertia character of the flow regime means that the hydrodynamics are long-range, thereby coupling together the dynamics of swimmers that are many body lengths apart. Furthermore, the flows themselves are highly complex (see Figure 6.10), meaning that even pairwise interactions can be complex, as shown by explicit Boundary Element computations of model spherical squirmer representations of two ciliated microorganisms [46]. These simulations demonstrate an array of possible pairwise interactions. Depending upon the relative orientation, squirming action, and distance between the squirmers, these interactions might include scattering-type behavior, as well as *waltzing*-type interactions, where swimmers circle each other in a similar fashion to that observed between individual *Volvox*. (*Volvox* are colonies of algae cells which propel themselves somewhat like a ciliated microorganism, although using the synchronized flagella of many inward facing individuals, rather than beating cilia). Even at relatively dilute suspensions, in situations where two-dimensionality in the flow is expected to be promoted, such as in a thin film, it has been suggested that hydrodynamic interactions between potentially distant cells can remain non-negligible [47].

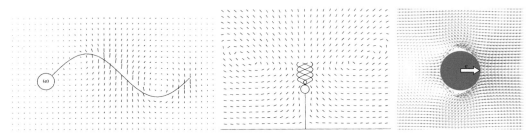

FIGURE 6.10   Flowfields generated by model microorganisms with a spherical cell body, simulated using Slender Body Theory, when: (Left) Swimming using planar beating flagellum in infinite fluid, as computed by Higdon [45] (*reproduced with permission*); (Middle) Generating feeding currents while tethered to a wall, using helical flagellum beating, as computed by Higdon [23] (*reproduced with permission*); (Right) Flowfield generated by a model spherical ciliate (*squirmer*), as determined by the Boundary Element computations of Ishikawa et al. [46] (*reproduced with permission*).

As the number of swimming microorganisms increases, the influence of hydrodynamic interactions compound, altering the transport properties of the collection as a whole. Experiments using tracer particles have shown that relatively high density suspensions of swimming microorganisms can spread superdiffusively in the short-to-medium term [48]. Simulating such large numbers of swimming microorganisms is, of course, extremely challenging computationally. Some smaller-scale simulations involving suspensions of several dozen model ciliates (squirmers) have shown indications of anomalous diffusion on the shorter timescales [49], transitioning to diffusive behavior in the long-time limit. In this long-time diffusive limit, translational diffusivity (of neutrally buoyant microorganisms, so excluding some bottom heavy cells, such as those of some algae) was predicted to decrease as cell concentrations increase. By contrast, rotational diffusivity was seen to increase to levels significantly higher than the values measured for the alga *Chlamydomonas nivali* in more dilute regimes, where hydrodynamic interactions might be expected to be less influential [50].

Given that rotational diffusion is a relaxation process, one might expect distributions of microorganism swimming directions to become more isotropic as cell concentrations increase. However, the complete opposite is sometimes observed, as is well-exemplified by the experiments of Dombrowksi and colleagues [51]. These experiments involved a high-density population of the peritrichous bacterium *B. subtilis* suspended in a droplet (a microorganism usually associated with soil, but now also believed to form part of the human gut flora). Far from observing a suspension with an isotropic distribution of swimming directions, aggregations of microorganisms were observed swimming in broadly-aligned directions. Moreoover, these aggregations, each much larger in size than an individual swimmer, were seen to undergo highly-complex spatiotemporal dynamics (see Figure 6.11 Left). So much so, in fact, that the term *slow turbulence* has been coined to describe this behavior (although the generated flows themselves, of course, are inertialess). These dynamics only occur by virtue of the fact that the microorganisms are swimming, and if they are rendered inert, then the patterns disappear.

In a similar manner to the generation of classical flow turbulence, it is thought that *slow turbulence* may also arise through the onset of instabilities that stem from the reduced stabilizing influence of viscosity [54]. It has long been known that a suspension of passive particles,

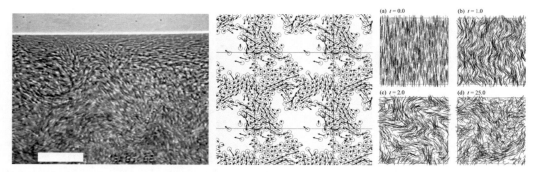

FIGURE 6.11   Collective dynamics observed in dense suspensions of microorganisms: (Left) Experiments of Dombrowski et al. involving a suspension of *B. Subtilis* in a drop (*Reprinted figure with permission from [51]. Copyright 2004 by the American Physical Society*). (Middle) Simulations of Ishikawa and Pedley that investigate the stability of an isotropic suspension (i.e., no initial orientational order) of squirmers (e.g., model ciliates) that are *pullers* (*Reprinted figure with permission from [52]. Copyright 2008 by the American Physical Society*). (Right) Simulations of Saintillan and Shelley that analyze the stability of suspensions of *pushers* with initial orientational order, using rod-like swimmers, (e.g., model bacteria) (*Reprinted figure with permission from [53]. Copyright 2007 by the American Physical Society*).

for example a colloidal suspension, often has a larger effective viscosity than the same volume of clear fluid. This is due to the fact that the solid particles do not deform in the same manner as the fluid volume that they have displaced. When the bodies within the suspension actively swim, however, this effect upon the effective viscosity can either be enhanced, or reduced, depending upon whether the swimmer is a *pusher* (i.e., propelled from behind, as is the case for most bacteria) or a *puller* (pulled through the fluid, as is the case for most algae). This is due to the fact that (non-spherical) swimming microorganisms will generally align with the extensional axis of the surrounding local flow, and either reinforce it, or oppose it, depending upon whether they are *pushers* or *pullers*, as shown in Figure 6.8. This idea is seemingly borne out by the explicit simulations of Underhill et al. [55], which used a dumbbell-shaped model microorganism (two spheres joined by a connecting rod, with a force acting upon one of the spheres to mimic the thrust from a flagellum), as a simple finite-sized generator of the force-dipole (stresslet) that arises from equal-and-opposite forces on a force-free swimmer. These simulations show that significantly enhanced transport occurs within a suspension of model swimmers that are dumbbell-shaped *pushers*, but not so for a suspension of *pullers*. (The same dichotomy between *pusher* and *puller* behavior is also predicted by some continuum models, as we shall discuss shortly.) However, if the *puller* is spherical, as is the case for the squirmer simulations conducted by Ishikawa et al. [52], then coherent structures have been shown to emerge from an isotropic initial state, as shown in Figure 6.11 Middle (which may tally with the relatively weak ability of neutrally-buoyant squirmers to affect effective viscosity of the suspension [56]).

In the converse situation, a state that has some form of orientational order has also been shown to be unstable for both *pullers* and *pushers*. Saintillan and Shelley [53] considered a suspension of rod-like model microorganisms, the hydrodynamics of which they approximated using Slender Body Theory, with their simulations demonstrating that an initially aligned (*nematic*) configuration quickly evolves into a non-aligned state that exhibits some of the structures observed in *slow turbulence*, rather than being isotopic (see Figure 6.11 Right). It was also noted that this destabilization occurs irrespective of whether all microorganisms initially swim in the same direction (a *polar* suspension), or not (an *apolar* suspension).

Even with simplified representations of the swimming microorganisms, such as squirmer models, swimming rods, or dumbbells (with Table 6.1 giving a by-no-means exhaustive list of such models), explicit computation of microorganism populations that contain the numbers of individuals expected in practice remains out of reach by some distance. To put the scale of the challenge into perspective, many of the computational studies discussed above simulated populations of tens to hundreds of individuals. Microorganism suspensions in which

TABLE 6.1 Summary of Some Studies which Explicitly Model the Hydrodynamics of Swimming Microorganisms

Year	Study	M-O	Wave	# Cells	Model	Body	Focus
1951	Taylor [9]	F	P-P-R	1	∞ (Plane)	N	Flagella Propulsion
1952	Lighthill [31]	C	C-P-R	1	∞ (Cyl)	–	Cilia Propulsion
	Taylor [57]	F	H-P-R	1	∞ (Cyl)	–	Flagella Propulsion
1953	Hancock [13]	F	(P,H)-P-R	1	SBT	N	Flagella Propulsion
1955	Gray & Hancock [10]	F	P-P-R	1	RFT	P	Flagella Propulsion
1958	Machin [21]	F	P-E-E	1	RFT	N	Flagella Activation
1963	Holwill & Burge [58]	F	(H,P)-P-R	1	RFT	P & E	Flagella Role
1965	Reynolds [59]	F	P-P-E	1	∞	N	Elastic & Wall Effects
1968	Tuck [32]	C	C-P-R	1	SQ	–	Cilia Propulsion
1971	Blake [29]	C	C-P-R	1	SQ	–	Cilia Propulsion
	Blake [60]	C	C-P-R	1	SQ	–	Cilia Propulsion
	Chwang & Wu [11]	F	H-P-R	1	RFT	I	Flagella Propulsion
	Wang & Jang [61]	S	H-P-R	1	∞ (Cyl)	–	Spirochete Propulsion
	Brokaw [62]		P-E-E	1	RFT	N	Flagella Activation
1972	Coakley & Holwill [63]	F	HC-P-R	1	RFT	I	Flagella Propulsion
	Silvester & Holwill [64]	F	P-P-R	1	RFT	N	Flagella Propulsion
1973	Blake [65]	C	C-P-N	1	SL	–	Cilia Propulsion
1974	Katz & Blake [66]	F	P-P-R	1	∞ (Plane)	–	Wall Effects

Continued

TABLE 6.1   Summary of Some Studies which Explicitly Model the Hydrodynamics of Swimming Microorganisms—Cont'd

Year	Study	M-O	Wave	# Cells	Model	Body	Focus
	Chwang et al. [67]	S	H-P-R	1		−	Spirochete Propulsion
	Pironneau & Katz [68]	F	P-P-R	1	RFT	I	Flagella Propulsion
	Brennen [69]	C	C-P-N	1	SQ	−	Cilia Propulsion
1976	Winet & Keller [70]	F	HC-P-R	1	RFT	E (Helix)	Flagella Propulsion
	Higdon [45]	F	P-P-R	1	SBT	I	Propulsion
	Higdon [71]	F	P-P-R	1	SBT	I	Feeding Currents
	Higdon [23]	F	H-P-R	1	SBT	I	Flagella Propulsion
1980	Dresdner & Katz [72]	F	P-P-R	1	SBT	E	Flagella Propulsion
1987	Phan-Thien [16]	F	H-P-R	1	BEM-SBT	E	Propulsion
1993	Ramia et al. [73]	F	H		BEM-SBT	Y	Wall Bounded
1995	Fauci & MacDonald [44]	F	P-P-E	≤ 2	IBM	E	(Elastic) Wall Effects
2000	Camalet and Julicher [26]	F	P-E-E	1	RFT	N	Flagella Activation
2006	Lauga et al. [40]	F	H-P-R	1	RFT	I	Wall Effects
2006	Ishikawa et al. [46]	C	C-P-N	2	BEM	−	Pairwise Hydrodynamics
2007	Ishikawa et al. [49]	C	C-P-N	≤ 64	BEM	−	Enhanced Diffusion
	Ishikawa et al. [56]	C	C-P-N	≤ 64	BEM	−	Modified Rheology
	Saintillan and Shelley [53]	F	O-P-R	2500	SBT	N	Collective Behavior
2008	Ishikawa et al. (2008a) [52]	C	C-P-N	≤ 196	BEM	−	Collective Behavior
2010	Smith et al. [24]	F	HC-P-E	1	BEM-SBT	E	Wall Trapping
2011	Lauga [39]	F	P-P-R	1	∞ (Plane)	N	Non-Newtonian Effects
2012	Underhill et al. [55]	F	P-P-R	≤ 6400	DB	P	Collective Behavior

The following key is applicable to each column: Third: Microorgansims (M-O): Flagellated (F), Ciliated (C), or Spirochetes (S); Fourth: Beating Action (Wave): Letter before hyphen indicates waveform: Planar Wave (P), Helical Wave (H), Conical Helix (CH), Cilia Wave (C), Spirochete (S), or Other (O). Letter after the first hyphen indicates whether the waveform is Prescribed (P) or Emergent (E). Letter after the second hyphen indicates whether flagellar is modelled as Rigid (R), Elastic (E) or Not-Applicable (N), (i.e. for Ciliates); Fifth: Number of microorganisms being simulated. Sixth: Type of flow model, either Infinite Sheet (∞ Plane), Infinite Cylinder ∞ (Cyl) Resistive Force Theory (RFT), Slender Body Theory (SBT), Boundary Element Method (BEM), Hybrid Boundary Element-Slender Body Method (BEM-SBT), Immersed Boundary Method (IBM), Envelope Cilia Model (Squirmer, SQ), Sub-Layer Cilia Model (SL), or Dumb-bell model (DB). Seventh: Cell Body (not applicable for ciliates or spirochetes). Passive, (i.e. as a drag force only) (P), Explicit inclusion (E), Inclusion via an image system (I), None (N). Eighth: Main focus of the model.

collective behavior has been observed have densities of around 10^9 cells per cubic centimeter. For these reasons, effective continuum descriptions of microorganism suspensions have long-been sought, and remain the focus of intense current research efforts.

6.4 CONTINUUM DESCRIPTIONS

Rather than considering each individual microorganism, continuum models deal with volume-averaged quantities, such as cell densities. The greatest obstacle to an effective continuum-level description, however, remains the cell–cell hydrodynamic interactions. In situations where there are sufficiently many organisms that a continuum description is legitimate, but where the suspension remains sufficiently dilute such that hydrodynamic interactions between cells are expected to be of secondary importance, it might reasonably be supposed that transport is dominated by advection and diffusion processes. This scenario is known as the semi-dilute regime.

6.4.1 Semi-Dilute Suspensions

A simple cell conservation argument leads to the following Advection–Diffusion (Smoluchowski equation) description for the density of microorganisms, $n(x,t)$ [74],

$$\frac{\partial n}{\partial t} + \nabla \cdot \mathbf{J} = 0, \quad \mathbf{J} = n(\mathbf{u} + \langle \mathbf{U} \rangle) - \mathbf{D}_T \cdot \nabla n \tag{6.20}$$

where $\mathbf{u}$ represents any background flow which will advect the microorganism and $\mathbf{D}_T$ is the tensor for translational diffusivity, which accounts for inherent randomness present in biological systems. Moreover, the microorganism will be transported by its own swimming velocity, and because it is rational to expect elements of randomness in the swimming motion from individual to individual, an average value is taken $\langle \mathbf{U} \rangle$. One way in which this can be factored into the model is to assume that the swimming direction, $\mathbf{p}(x)$, is sampled from some distribution $f(\mathbf{p})$ which satisfies the steady Fokker-Planck equation (dots denote derivatives with respect to time)

$$\nabla \cdot (\dot{\mathbf{p}} f) = D_R \nabla^2 f, \tag{6.21}$$

where D_R is a coefficient of rotational diffusivity (which might come about due to slight variations in the flagellar motor, or flagellar bundle, between individuals). The average swimming velocity $\langle \mathbf{U} \rangle$ is then given by

$$\langle \mathbf{U} \rangle = U_0 \int \mathbf{p} f(\mathbf{p}) d\mathbf{p}, \tag{6.22}$$

assuming the microorganisms swim with the same speed, U_0, in each direction (and similarly for other quantities averaged over orientation space, $\langle \cdot \rangle$). Importantly, the inertialess nature of the flow regime means that $\mathbf{p}$ is constrained so as to produce no net hydrodynamic force, or torque. Classical viscous flow theory tells us that for a solid body within a background flow with vorticity (local rotation) $\boldsymbol{\omega} = \nabla \times \boldsymbol{u}$, and rate of strain tensor $\dot{\boldsymbol{\gamma}}$ (see Equation 6.3), the torque free rotation satisfies

$$\dot{\boldsymbol{p}} = \frac{1}{2}\boldsymbol{\omega} \times \boldsymbol{p} + \Gamma \boldsymbol{p} \cdot \dot{\boldsymbol{\gamma}} \cdot (\boldsymbol{I} - \boldsymbol{p} \otimes \boldsymbol{p}) \tag{6.23}$$

($\boldsymbol{I}$ being the appropriately-sized identity matrix) where Γ is a shape parameter (for a prolate spheroid, $\Gamma = (\alpha_0^2 - 1)/(\alpha_0^2 + 1)$ where α_0 represents the ratio of major and minor spheroid axes). Note also that the tensor $\boldsymbol{p} \otimes \boldsymbol{p}$ refers to the outer product of the orientation vector $\boldsymbol{p} = (p_1, p_2, p_3)$, and has components $(p_i p_j)$.

In characterizing observations of collective dynamics, such as *slow turbulence*, orientations and alignments of microorganisms are an important characterizing feature. In order to extract this level of detail from the models, it can be useful to consider cell distributions not only as a function of spatial location $\boldsymbol{x}$, but also orientation $\boldsymbol{p}$, in other words, $\Psi(\boldsymbol{x}, \boldsymbol{p}, t)$ (this can be thought of as a probability density for finding a cell at location $\boldsymbol{x}$, with orientation $\boldsymbol{p}$). Transport of microorgansism cells with a particular orientation, through advection and diffusion processes, is then governed by [75]

$$\frac{\partial \Psi}{\partial t} + \nabla \cdot \boldsymbol{J}_T + \nabla_p \cdot \boldsymbol{J}_R = 0, \tag{6.24a}$$

with

$$\boldsymbol{J}_T = \Psi(\boldsymbol{u} + \langle \boldsymbol{U} \rangle) - \boldsymbol{D}_\mathbf{T} \cdot \nabla \Psi, \quad \boldsymbol{J}_R = \dot{\boldsymbol{p}} \Psi - D_R \nabla_p \Psi \tag{6.24b}$$

where ∇_p represents the gradient operator in orientation space, and translational and rotational diffusivities might account for a number of different effects, including hydrodynamic dispersion, variability in the flagellar motor, as well as Run-and-Tumble swimming strategies.

Although the torque-free requirement exerts itself through the constraints (Equation 6.23) placed upon $\dot{\boldsymbol{p}}$, the force-free nature of a swimming microorganisms makes its presence felt through its influence upon the rheology of the fluid itself. This can be reflected through the addition of an active stress field into the flow equations, Σ' (sometimes referred to as the *bacterial stress*). Polymer Theory suggests the following form, which aggregates the combined effect of the force-dipoles generated by the swimming microorganisms [76]

$$\Sigma' = -\alpha n \mu \int \Psi(\boldsymbol{x}, \boldsymbol{p}) \left(\boldsymbol{p} \otimes \boldsymbol{p} - \frac{1}{3}\boldsymbol{I} \right) \mathrm{d}\boldsymbol{p} \tag{6.25}$$

where μ is the dynamic viscosity of the fluid. Here $-\alpha(\boldsymbol{p} \otimes \boldsymbol{p} - \boldsymbol{I}/3)$ is the force-dipole (stresslet) generated by an individual swimmer. The constant α depends upon the details of the swimming method, whereby $\alpha > 0$ corresponds to *pushers*, and $\alpha < 0$ corresponds to *pullers*. The active stress here can then be interpreted as the averaged stresslet contribution across all swimmers. However, it does not take into account cell–cell interactions and, as such, is only expected to strictly be legitimate in the semi-dilute regime. Here the active stress is expected to scale with the cell concentration, c, with cell–cell interactions making a higher-order contribution of size c^2. (For squirmers, Ishikawa and Pedley [56] have computed this higher-order, cell–cell contribution to the stress tensor through simulation.)

The flow subjected to this additional stress consequently must satisfy

$$\rho\frac{Du}{Dt} = -\nabla p + \nabla^2 u + \nabla\cdot\Sigma' + F, \quad \nabla\cdot u = 0, \tag{6.26}$$

where F represents any body forces, such as buoyancy (important in the consideration of bottom heavy swimmers, such as some algae, which are observed to produce bioconvective patterns similar to thermal cells seen in the classical Rayleigh-Benard instability; the interested reader is directed to the review article by Hill and Pedley [77] for more details).

These continuum models contain the competing mechanisms of rotational diffusion which, as previously discussed, can have a stabilizing influence upon a suspension, as well as the ability of the swimming microorganisms to change the effective viscosity of the suspension. The latter has the potential to be either stabilizing or destabilizing, depending upon whether the swimmer is a *pusher* or a *puller*. The question of stability of a particular suspension configuration can then be investigated through a linear stability analysis of these continuum models. By linearly perturbing a suspension in which rod-like swimmers are initially aligned, Simha and Ramaswarmy [54] were able to demonstrate that such orientational ordering is unstable for both *pushers* and *pullers*, as later confirmed in the (previously discussed) Slender Body Theory simulations of Saintillan and Shelley [53]. Using the continuum models above, Saintillan and Shelley [76] also examined the reasons why rotational diffusivity does not necessarily lead to suspensions that are orientationally isotropic, finding that such a state is unstable for *pushers* (but not *pullers*). In doing so, they demonstrated a plausible mechanism through which the homogenizing influence of diffusion can be countered, to produce the type of collective dynamics seen in some experiments.

However, the above analysis somewhat underestimates the influence of rotational decoherence mechanisms, by representing them as purely diffusive processes that decay at large spatial distances. This leads to the prediction that instability of the suspensions occur even at very low concentrations. This is in apparent contradiction to the experiments of Wu and Libchaber [48], and Dombrowski et al. [51]. For prokaryotes that execute Run-and-Tumble swimming, however, a more rational treatment that explicitly includes changes to the orientation due to tumbling can be incorporated into the continuum models, through the addition of appropriate extra terms [75, 78]

$$\frac{\partial\Psi}{\partial t} + \nabla\cdot J_T + \nabla_p\cdot J_R + \Lambda(p)\Psi - \int\Psi(p)\Lambda(p')T(p,p')dp' = 0 \tag{6.27}$$

where J_T and J_R are as before (Equation 6.24b), and $T(p,p')$ is the probability of a swimmer tumbling from orientation p to orientation p' (in cases where tumbling can be considered isotropic, then $T(p,p') = 1/4\pi$), and $\Lambda(p)$ is the rate of tumbling. Given that a purpose of Run-and-Tumble can be to follow gradients of some quantity, C (such as nutrients, light, or oxygen), it can sometimes be appropriate to set

$$\Lambda(p) = \Lambda_0 + \beta\nabla C\cdot p, \tag{6.28}$$

where Λ_0 is some unbiased tumbling rate, and β some measure of tumbling bias towards favorable gradients in C. In the absence of any such preference in swimming direction (taxis), in other words, $\Lambda(p) = \Lambda_0$, Subramanian and Koch (2009) used the explicit inclusion of tumbling in the above model to find a threshold concentration for the onset of long-wave (compared with cell size) instabilities to an isotropic suspension. The critical concentration, n_c, was shown to be

$$n_c = (5\Lambda_0 + 30D_R)/\alpha \tag{6.29}$$

with α as in Equation (6.25). Hence it can be seen that an instability occurs only for *pushers*, where $\alpha > 0$. This, combined with the fact that the instability seems to depend solely upon the characteristics of individual swimmers, specifically their tumbling rate (Λ_0), rotational diffusivity (D_R), and swimming parameters (α), points to the reduction of local effective viscosity as the instability mechanism (as illustrated earlier in Figure 6.8). The maximum growth rate of $0.2n\alpha$ is achieved in the asymptotically-large wavelength limit of the imposed perturbation, meaning that nonlinear effects are likely important in determining the emergent structures from the instability. There is a second, nonlocal mode of long-wavelength instability connected to the influence of the swimming microorganisms upon the bulk stress of the suspension through Σ', although this instability mechanism is seen to be weaker. Perturbations with wavelengths, λ, between

$$U_0/(0.17n\alpha) > \lambda > U_0/(0.57n\alpha) \tag{6.30}$$

(where U_0 is the swimming speed) take the form of unstable traveling waves. Any perturbations with wavelengths $\lambda < U_0/(0.57n\alpha)$ simply decay, and those with wavelengths greater than $U/0.17\alpha$ have growing stationary modes.

6.4.2 Cell–Cell Collisions

The above semi-dilute models show that collective dynamics can emerge through the combined hydrodynamics of effectively isolated individual swimming microorganisms, without the need for cell–cell interactions. However, in many of the experiments where such behavior is observed, cell concentrations are typically very high (with an example concentration of 10^9 cells per cubic centimeter previously quoted), and it becomes difficult to argue against the inclusion of cell–cell interactions in such situations.

Incorporating hydrodynamic interactions remains hugely challenging. This is due to their highly complex form, as illustrated by the explicit simulations of Ishikawa et al., [46] and others. Models which account for cell–cell interactions through collisions, rather than hydrodynamic interctions *per se*, have been developed out of the literature on Non-Equilibrium Statistical Physics. For instance, the Cell Conservation model (Equation 6.24) can be modified to the form:

$$\frac{\partial \Psi}{\partial t} + \nabla \cdot \mathbf{J}_T + \nabla_p \cdot \mathbf{J}_R - \Pi(\mathbf{x}, \mathbf{p}, t) = 0 \tag{6.31}$$

where $\mathbf{J}_T$ and $\mathbf{J}_R$ are, in principle, as before (Equation 6.24b), while Π is a collision operator. There have been a number of different choices for the form of this term, including Volume Excluded Interactions [79]

$$\Pi = \nabla - (D_T \cdot (\Psi \nabla V)) - D_R \nabla_p \cdot (\Psi \nabla_p V), \tag{6.32}$$

where V is some potential function. When only pairwise interactions are considered, the potential V takes the form

$$V(\mathbf{x}, \mathbf{p}, t) = \int\int V_{ex}(\hat{\mathbf{x}}, \mathbf{p}, \mathbf{p}_1) \Psi(\mathbf{x}_1, \mathbf{p}_1, t) \, d\mathbf{x}_1 d\mathbf{p}_1 \tag{6.33}$$

($\hat{x} = x - x_1$) in other words, an integral over all possible (center of mass) positions, x_1, and orientations, p_1 of the swimmers. Here, $V_{ex}(\hat{x}, p, p_1) = 1$ if both swimmers at x and x_1 intersect when their respective orientations are p and p_1, and takes the value zero otherwise.

An alternative model considers instead fully inelastic collisions between pairs of microorganisms, and derives from granular flow theory. This follows the Kinetic Theory approach, whereby two microorganisms with pre-scattering positions and orientations (x_1, p_1), (x_2, p_2) collide, and leave the collision with average position $\bar{x} = (x_1 + x_2)/2$ as a pair with average orientation $\bar{p} = (p_1 + p_2)/2$. Expressed mathematically [80]

$$\Pi = \iiiint W(x_1, x_2) \Psi(x_1, p_1) \Psi(x_2, p_2) \times \ldots$$
$$\times \left(\delta(x_2 - x, p_2 - p) - \delta(x - \bar{x}, p - \bar{p}) \right) dp_1 dp_2 dx_1 dx_2 \tag{6.34}$$

where W is some localization function, such as a Gaussian, which decays as a function of separation distance between the two swimmers, and δ are Delta functions.

6.4.3 Coarse Graining

Because the probability density, $\Psi(x, p, t)$, is not a directly observable quantity, it can often be useful to work with coarse-grained quantities, which are moments of $\Psi(x, p, t)$. The zeroth moment simply returns cell concentration

$$n(x, t) = \int \Psi(x, p, t) dp, \tag{6.35}$$

whereas the first moment yields the polarization vector

$$P(x, t) = \int p \Psi(x, p, t) dp, \tag{6.36}$$

which gauges whether there is any net flux of cells in a suspension, due to a preferred swimming direction (i.e., due to a nutrient gradient). The second moment, the nematic tensor Q, is defined by

$$Q(x, t) = \int \left(p \otimes p - \frac{1}{2} I \right) \Psi(x, p, t) dp, \tag{6.37}$$

and measures the extent to which rod-like swimmers are aligned, hence the degree of orientational order within the suspension.

For the Run-and-Tumble model given in Equation (6.27), in the presence of a vertical chemical gradient, $\nabla C = -\hat{z}$ (where $\hat{z}$ represents the unit vector in the vertical direction), the coarse-grained equations in the absence of a background flow ($u = 0$, $E = 0$), and no translational or rotational diffusion (i.e., $D_T = D_R = 0$), take the form [78]

$$\frac{\partial n}{\partial t} = -\nabla \cdot P \tag{6.38}$$

$$\frac{\partial P}{\partial t} = U_0 \left(\Lambda_0 \beta \hat{k} - U_0 \nabla \right) \cdot \left(Q + \frac{1}{2} In \right) + \omega \times P - \Lambda_0 P \tag{6.39}$$

with the higher-order moment, the nematic tensor, evidently appearing, but without an evolution equation in its own right. This is a common issue in coarse graining, and in order to close the system of equations, some heuristic approximation must be invoked in order to connect the highest-order moment to the lower-order ones, (i.e., in the case above, a relationship between Q in terms of n and P). This might arise through consideration of the system in certain limits, such as weak gradients, or long wavelength effects.

As another example, for the case of inelastic collisions (Equation 6.34), the coarse grained equations take the form [80]

$$\frac{\partial n}{\partial t} + \nabla \cdot (n\boldsymbol{u}) = D_T \nabla^2 n - U_0 \pi \nabla \cdot \boldsymbol{P}, \tag{6.40}$$

$$\frac{D\boldsymbol{P}}{Dt} + \boldsymbol{U} \cdot \nabla \boldsymbol{P} + \frac{1}{2}\omega \times \boldsymbol{P} = (\in n - 1)\boldsymbol{P} - A_0(\boldsymbol{P} \cdot \boldsymbol{P})\boldsymbol{P} + D_1 \nabla^2 \boldsymbol{P} + D_2 \nabla(\nabla \cdot \boldsymbol{P}) - \frac{U_0}{4\pi}\nabla n \tag{6.41}$$

where $\in$ and A_0 are known constants, and D_T, D_1, and D_2 are appropriate diffusion coefficients. These can be coupled to the flow equations for $\boldsymbol{u}$, which may take into account modified rheology due to bacterial stress, as discussed earlier.

6.5 DISCUSSION

In this chapter, we have given an overview of some of the prevalent models of the swimming microorganisms that inhabit our bodies in such vast numbers. We have focused here on the principal mode of microorganism motility, namely swimming, although it is worth mentioning that other means of transport are employed, especially for migration over surfaces. *Twitching* is a form of motion whereby the microorganism uses micropili to drag themselves over a surface [81]. Microorganisms can also coordinate cell binding and unbinding to transport themselves over a surface, in a process sometimes referred to as *gliding* [82].

The field of microorganism hydrodynamics has progressed a great deal in the sixty years or so since Taylor, Lighthill, Gray, and Hancock set out to better understand the means by which such small individual organisms can propel themselves through their highly-viscous fluid environments. More recent hydrodynamic treatments involving multiple swimming microorganisms have revealed how the combined swimming forces of many individuals can change the rheology of the fluid, leading to instabilities that produce coordinated large-scale motions. Observation of such *slow turbulence* over the past decade has stimulated a whole new level of interest in the subject. The observed phase transitions from ordered to disordered states has caught the attention of those with a background in Non-Equilibrium Statistical Mechanics and Granular Flows, and the novel rheological properties of microorganism suspensions has similarly interested the Soft Matter Physicists. The Hydrodynamicists also continue to be heavily active in the field, with simulation playing an ever greater role.

The computational challenge of simulating realistic population sizes, however, remains daunting, and this has further stimulated the development of appropriate continuum models. In the case of semi-dilute suspensions, where cell–cell hydrodynamics are expected to be of secondary importance (subject to some caveats [47]), Semi-Dilute Cell Conservation Models seem appropriate. For higher density suspensions, Kinetic Models which take into account orientational ordering (i.e., alignment) between microorganisms due to collisions have been proposed. However, continuum models which comprehensively include the complex cell–cell hydrodynamics between swimming microorganisms remain elusive (although there are some who argue that such near-field hydrodynamics are actually unnecessary, due to the decohering effects of Brownian motion in the far-field, and the expectation that collisions will dominate over hydrodynamics in the near-field [83]).

As well as these issues, the ability to fully explore the ramifications of microorganism dynamics in a physiological setting such as the human digestive system, or reproductive system, awaits certain other modeling obstacles to be overcome. These include appropriate boundary conditions on surfaces, as well as incorporation of biochemistry. For example, both could be important for theoretically predicting biofilm formation on surfaces, where there is evidence for the importance of chemical signaling (quorum sensing) between bacteria. Also, hyperactivation of spermatozoa (whereby the beating frequency of their flagella increases substantially under certain environmental conditions) is a further area where biochemistry is likely to play an important role. Another highly relevant area, which has only begun to be explored, is the behavior of microorganisms in non-Newtonian fluids which, as was demonstrated earlier for the simple elastic sheet model, can differ substantially from that within Newtonian flow.

For these reasons, the field of microorganism swimming continues to grow, attracting researchers from a range of different disciplines, and is consequently progressing at an ever increasing pace. It is a subject that has benefitted substantially from ingenious modeling strategies in the past, and is one that awaits equally creative solutions to some of the significant modeling challenges that lie ahead.

References

[1] Sears CL. A dynamic partnership: Celebrating our gut flora. Anaerobe 2005;11:247–51.
[2] Relman DA, Falkow S. The meaning and impact of the human genome sequence for microbiology. Trends Microbiol 2001;9:206–8.
[3] Heijtz RD, Wang S, Anuar F, Qian Y, BjÃrkholm B, Samuelsson A, et al. Normal gut microbiota modulates brain development and behavior. PNAS 2011;108(7):3047–52.
[4] Orme BA, Blake JR, Otto SR. Modelling the motion of particles around choanoflagellates. J Fluid Mech 2003;475:333–55.
[5] Purcell EM. Life at Low Reynolds Number. Am J Phys 1976;45:3–11.
[6] Gray J. Undulatory propulsion. Q J Microsc Sci 1953;94:551–78.
[7] Liu J, Howell JK, Bradley SD, Zheng Y, Zhou ZH, Norris SJ. Cellular Architecture of Treponema pallidum: Novel Flagellum, Periplasmic Cone, and Cell Envelope as Revealed by Cryo Electron Tomography. J Mol Biol 2010;403:546–61.
[8] Afzelius BA, Dallai R, Lanzavecchia S, Bellon PL. Flagellar structure in normal human spermatozoa and in spermatozoa that lack dynein arms. Tissue Cell 1995;27:241–7.
[9] Taylor GI. Analysis of the swimming of microscopic organisms. Proc R Soc Lond A 1951;209:447–61.
[10] Gray J, Hancock GJ. The propulsion of sea-urchin spermatozoa. J Exp Biol 1955;32:802–14.
[11] Chwang JC, Wu TY. A note on the helical movement of microorganisms. Proc R Soc Lond B 1971;178:322.
[12] Lighthill MJ. Mathematical Biofluiddynamics. SIAM Rev 1975;19:576–7.
[13] Hancock GJ. The self-propulsion of microscopic organisms through liquids. Proc R Soc Lond A 1953;217:96–121.
[14] Pozrikidis C. Boundary integral and singularity methods for linearized viscous flow. USA: Cambridge University Press; 1992.

[15] Clarke RJ, Jensen OE, Billingham J, Williams PM. Three-dimensional flow due to a microcantilever oscillating near a wall: an unsteady slender-body analysis. Proc R Soc Lond A 2006;462:913–33.

[16] Phan-Thien N, Tran-Cong T, Ramia M. A boundary-element analysis of flagellar propulsion. J Fluid Mech 1987;185:533–49.

[17] Smith DJ, Gaffney EA, Gadelha H, Kapur N, Kirkman-Brown JC. Bend propagation in the flagella of migrating human sperm, and its modulation by viscosity. Cell Motil Cyto 2009;66:220–36.

[18] Manson MD, Tedesco PM, Berg HC. Energetics of flagellar rotation in bacteria. J Mol Biol 1980;138:541–61.

[19] Balaban M, Hendrixson DR. Polar Flagellar Biosynthesis and a Regulator of Flagellar Number Influence Spatial Parameters of Cell Division in Campylobacter jejuni. PLoS Pathog 2011;7(12):e1002420.

[20] Canale-Parola E. Motility and chemotaxis of spirochetes. Annu Rev Microbiol 1978;32:69–99.

[21] Machin KE. Wave propagation along flagella. J Exp Biol 1958;35:796–806.

[22] Gray J. The movement of sea urchin spermatozoa. J Exp Biol 1955;32:775801.

[23] Higdon JJL. The hydrodynamics of flagellar propulsion: helical waves. J Fluid Mech 1979;94:331–51.

[24] Smith DJ, Gaffney EA, Blake JR, Kirkman-Brown JC. Human sperm accumulation near surfaces: a simulation study. J Fluid Mech 2010;621:289–320.

[25] Blake JR. A model for the micro-structure in ciliated organisms. J Fluid Mech 1972;55:1–23.

[26] Camalet S, Julicher F. Generic aspects of axonemal beating. New J Phys 2000;24:1–23.

[27] Riedel-Kruse IH, Hilfinger A, Howard J, Julicher F. How molecular motors shape the flagellar beat. HFSP J 2007;1:192–208.

[28] Gray J. The movement of the spermatozoa of the bull. J Exp Biol 1958;35:96108.

[29] Blake JR. A spherical envelope approach to ciliary propulsion. J Fluid Mech 1971;46:199–208.

[30] Sleigh MA. Patterns of ciliary beating. Symp Soc Exp Biol 1968;22:131–150.

[31] Lighthill MJ. On the squirming motion of nearly spherical deformable bodies through liquids at very small Reynolds numbers. Commun Pure Appl Math 1952;5:109–18.

[32] Tuck EO. A note on the swimming problem. J Fluid Mech 1968;31:305–8.

[33] Ishikawa T, Hota M. Interaction of two swimming Paramecia. J Exp Biol 2006;22:4452–63.

[34] Pooley CM, Alexander GP, Yeomans JM. Hydrodynamic Interaction between Two Swimmers at Low Reynolds Number. Phys Rev Lett 2007;99:228103.

[35] Giacche D, Ishikawa T. Hydrodynamic interaction of two unsteady model microorganisms. J Theor Biol 2010;267:252–63.

[36] Liron N. The LGL (Lighthill-Gueron-Liron) Theorem - historical perspective and critique. Math Meth Appl Sci 2001;24:1533–40.

[37] Roselli RJ, Diller KR. Biotransport: principles and applications. New York: Springer; 2011.

[38] Lew HS, Fung YC, Lowenstein CB. Peristaltic carrying and mixing of chyme in small intestine. J Biomechanics 1971;4:297–315.

[39] Lauga E. Propulsion in a viscoelastic fluid. Phys Fluids 2011;19:083104.

[40] Lauga E. Life at high Deborah Number. EPL 2006;86:64001.

[41] Goto T, Nakata K, Baba K, Nishimura M, Magariyama Yl. A fluid-dynamical interpretation of the asymmetric motion of singly flagellated bacteria swimming close to a boundary. Biophys J 2005;89:3771–9.

[42] Kearns DB. A field guide to bacterial swarming motility. Nat Rev Microbiol 2010;8:634–44.

[43] Woolley DM. Motility of spermatozoa at surfaces. Reproduction 2003;126:259–70.

[44] Fauci LJ, MacDonald A. Sperm motility in the presence of boundaries. Bull Math Biol 1995;57:679–99.

[45] Higdon JJL. A hydrodynamic analysis of flagellar propulsion. J Fluid Mech 1979;90:685–711.

[46] Ishikawa T, Simmonds MP, Pedley TJ. Hydrodynamic interaction of two swimming model micro-organisms. J Fluid Mech 2006;568:119–60.

[47] Clarke RJ, Finn MD, MacDonald M. Hydrodynamic persistence within very dilute two-dimensional suspensions of squirmers. Proc R Soc Lond A 2014;470:20130508.

[48] Wu XL, Libchaber A. Particle diffusion in a quasi two-dimensional bacterial bath. Phys Rev Lett 2000; 84:3017–20.

[49] Ishikawa T, Pedley TJ. Diffusion of swimming model micro-organisms in a semi-dilute suspension. J Fluid Mech 2007;588:437–62.

[50] Vladimirov VA, Denissenko PV, Pedley TJ, Wu M, Moskalev IS. Algal motility measured by a laser-based tracking method. Mar Freshwater Res 2000;51:589–600.

[51] Dombrowski C, Cisneros L, Chatkaew S, Goldstein RE, Kessler JO. Self-Concentration and Large-Scale Coherence in Bacterial Dynamics. Phys Rev Lett 2004;93:098103.

[52] Ishikawa T, Pedley TJ. Coherent structures in monolayers of swimming particles. Phys Rev Lett 2008;100:088103.

[53] Saintillan D, Shelley MJ. Orientational Order and Instabilities in Suspensions of Self-Locomoting Rods. Phys Rev Lett 2007;99:058102.

[54] Simha RA, Ramaswarmy S. Hydrodynamic fluctuations and instabilities in ordered suspensions of self-propelled particles. Phys Rev Lett 2002;89:058101.

[55] Underhill PT, Hernandez-Ortiz JP, Graham MD. Diffusion and spatial correlations in suspensions of swimming particles. Phys Rev Lett 2008;100:248101.

[56] Ishikawa T, Pedley TJ. The rheology of a semi-dilute suspension of swimming model micro-organisms. J Fluid Mech 2007;588:399–435.

[57] Taylor GI. The action of waving cylindrical tails in propelling microscopic organisms. Proc R Soc Lond A 1952;211:225–39.

[58] Holwill MEJ, Burge RE. A hydrodynamic study of the motility of flagellated bacteria. Arch Biochem Biophys 1963;101:249–60.

[59] Reynolds AJ. The swimming of minute organisms. J Fluid Mech 1965;23:241–60.

[60] Blake JR. Infinite models for ciliary propulsion. J Fluid Mech 1971;49:209–22.

[61] Wang C, Jahn TL. A theory for the locomotion of spirochetes. J Theor Biol 1972;36:53–60.

[62] Brokaw C. Bend propagation by a sliding filament model for flagella. J Exp Biol 1971;55:289–304.

[63] Coakley CJ, Holwill ME. Propulsion of micro-organisms by three-dimensional flagellar waves. J Theor Biol 1972;35:525–42.

[64] Silvester NR, Holwill MEJ. An analysis of hypothetical flagellar waveform. J Theor Biol 1972;35:505–33.

[65] Blake JR. A finite model for ciliated microorganisms. J Biomech 1973;6:133–40.

[66] Katz DF, Blake JR. On the propulsion of micro-organisms near solid boundaries. J Fluid Mech 1974;64:33–49.

[67] Chwang AT, Winet H, Wu TY. A theoretical mechanism for spirochetal location. J Mechanochem Cell Motil 1974;3:69–76.

[68] Pironneau O, Katz DF. Optimal swimming of flagellated micro-organisms. J Fluid Mech 1974;66:391–415.

[69] Brennen C. An oscillating boundary-layer theory for ciliary propulsion. J Fluid Mech 1974;65:799–824.

[70] Winet H, Keller SR. Spirillum swimming. Theory and observation of propulsion by the flagellear bundle. J Exp Biol 1976;65:577–602.

[71] Higdon JJL. The generation of feeding currents by flagellar motions. J Fluid Mech 1979;94:305–30.

[72] Dresdner RD, Katz DF, Berger SA. The propulsion by large amplitude waves of uniflagellar micro-organisms of finite length. J Fluid Mech 1980;97:591621.

[73] Ramia M, Tullock DL, Phan-Thien N. The role of hydrodynamic interaction in the locomotion of microorganisms. Biophys J 1993;65:755–78.

[74] Pedley TJ, Kessler JO. A new continuum model for suspensions of gyrotactic micro-organisms. J Fluid Mech 1990;212:155–82.

[75] Subramanian G, Koch DL. Critical bacteria concentration for the onset of collective swimming. J Fluid Mech 2009;632:359–400.

[76] Saintillan D, Shelley MJ. Instabilities and pattern formation in active particle suspensions: kinetic theory and continuum simulations. Phys Rev Lett 2008;100:178103.

[77] Hill NA, Pedley TJ. Bioconvection. Fluid Dynamics Research 2005;37:1–20.

[78] Bearon RN, Pedley TJ. Modelling Run-and-Tumble chemotaxis in a shear flow. Bull Math Biol 2000;62:775–91.

[79] Baskaran A, Marchetti MC. Hydrodynamics of self-propelled hard rods. Phys Rev E 2008;77:011920.

[80] Aranson IS, Sokolov A, Kessler JO, Goldstein RE. Model for dynamical coherence in thin films of self-propelled microorganisms. Phys Rev E 2007;75:040901.

[81] Mattick JS. Type IV pili and twitching motility. Annu Rev Biochem 2002;56:289–314.

[82] Mignot T. The elusive engine in Myxococcus xanthus gliding motility. Cell Mol Life Sci 2007;64:2733–45.

[83] Drescher K, Dunkel J, Cisneros LH, Sujoy G, Goldstein RE. Fluid dynamics and noise in bacterial cell-cell and cell-surface scattering. PNAS 2011;108:10940–5.

A Critical Review of Experimental and Modeling Research on the Leftward Flow Leading to Left-Right Symmetry Breaking in the Embryonic Node

I.A. Kuznetsov[a], A.V. Kuznetsov[b]

[a]Johns Hopkins University, Baltimore, MD, USA
[b]North Carolina State University, Raleigh, NC, USA

Nomenclature

C^* concentration of the morphogen (molecules/m^3)
C dimensionless concentration of the morphogen, $\frac{C^* D^*}{j_0^* H^*}$
D^* morphogen diffusivity (m^2/s)
$f(x^*)$ function defined in Equation (7.4) that describes the deviation of vorticity at the edge of the ciliated layer from a constant value of ω_0^*
H^* cavity depth (m)
j_0^* constant flux of the morphogen at the $y^* = 0$ boundary resulting from the morphogen secretion at the floor of the node (molecules/(m^{2}s))
K geometric factor characterizing the cavity aspect ratio, $\frac{L^*}{H^*}$
L^* cavity half-width (m)
p^* pressure (Pa)
Pe Péclet number, $\frac{\omega_0^* H^{*2}}{D^*}$
S dimensionless parameter characterizing the morphogen half-life, $\frac{\omega_0^* T_{1/2}^*}{\ln(2)}$
$T_{1/2}^*$ half-life of the morphogen (s)
u^* streamwise velocity (m/s)
u dimensionless streamwise velocity, $\frac{u^*}{\omega_0^* H^*}$
$\mathbf{u}^*$ flow velocity vector, (u^*, v^*) (m/s)
v^* wall-normal velocity (m/s)
v dimensionless wall-normal velocity, $\frac{v^*}{\omega_0^* H^*}$
x^* streamwise coordinate (m)
x dimensionless streamwise coordinate, $\frac{x^*}{L^*}$
y^* wall-normal coordinate (m)
y dimensionless wall-normal coordinate, $\frac{y^*}{H^*}$

Greek Symbols

μ^* dynamic viscosity of the extraembryonic fluid (Pa s)
ν^* kinematic viscosity (m^2/s)
Ψ^* stream function (m^2/s)
Ψ dimensionless stream function, $\frac{\Psi^*}{\omega_0^* H^{*2}}$

ω^* z^*-component of vorticity (s^{-1})
ω_0^* reference value of the z^*-component of vorticity at the edge of the ciliated layer (s^{-1})

Superscripts

* dimensional quantity

Abbreviations

AP anterior-posterior
DV dorsal-ventral
LR left-right
NVP nodal vesicular parcel

7.1 INTRODUCTION

Cilia are small, hair-like, microtubule-based organelles that serve a number of physiological functions, including sensory and propulsary functions [1,2]. Cilia are divided into motile and primary. Motile cilia induce the propulsion force by beating. Surfaces covered with motile cilia can induce fluid motion [1,3]. Examples of the latter include movement of mucus in the trachea [4] and transport of gametes and embryos in fallopian tubes [5]. The axoneme of a motile cilium typically contains a ring of nine microtubule (MT) doublets positioned along the outer ring of the axoneme plus two MTs in the center (9 + 2 axoneme) [6]. The axoneme of a primary cilium (also called a monocilium) lacks the central pair of MTs. Until the work of Nonaka et al. [7], it was believed that all primary cilia are immotile, and probably function as sensors. However, Nonaka and colleagues [7] demonstrated that some primary cilia covering the floor of a mouse ventral node rotate [8]. The cilia that rotate are tilted toward the posterior side of the node [9], and their rotational motion causes the leftward flow. Nonaka and colleagues [7,10] demonstrated a surprising developmental function of this cilia-induced fluid flow. Apparently, in order to break the left-right (LR) symmetry (at least in mice), nature has designed a microfluidic device, a ventral node. We will now discuss the details of experimental and numerical research leading to this groundbreaking discovery.

7.2 EXPERIMENTAL RESEARCH ON THE LEFTWARD NODAL FLOW AND LR SYMMETRY BREAKING

The development of LR asymmetry is preceded by the development of anterior-posterior (AP) and dorsal-ventral (DV) asymmetries [11]. However, until the groundbreaking work by Nonaka and colleagues at Hirokawa's lab [7], the process of translation of the AP and DV asymmetries into the LR asymmetry was not elucidated. Nonaka et al. [7] reported that in mice the rotation of primary cilia in a small triangular cavity, which is called the ventral node (hereafter "node"), induces leftward flow above the ciliated surface. The velocity of nodal flow can range from 5 to 50 μm/s [12]. The nodal flow lasts a few hours; it develops before any other evidence of LR asymmetry can be found [13].

A mouse node is ~50 μm across and ~10 μm deep. The floor of the nodal pit is covered by 200-300 monocilia whose rotation is responsible for the leftward flow in the node. The cilia are tilted toward the posterior side of the node; the tilting means the AP asymmetry is already in place before the LR symmetry is broken [9]. The tilt is explained as follows. Cells at the floor of a mouse node have dome-like curvature. Because the roots of the cilia (the basal bodies) are shifted to the posterior side of the cell membranes [14], the cilia are tilted toward the posterior of the node. Recent research identified genes that direct the cilia's posterior tilting [15].

The cilia are 2-3 μm in length and rotate clockwise (as viewed from above) at ~10 Hz [16,17] (Figures 7.1 and 7.2). The nodal cavity forms a closed space filled with extraembryonic fluid; the top of the cavity is covered by the Reichert's membrane.

A connection between the establishment of LR asymmetry and cilia motility is elucidated by studying disorders associated with a lack of cilia motility, such as Kartagener's syndrome [18,19]. In Kartagener's syndrome, cilia immobility occurs simultaneously with partial or complete *situs inversus* [7,20,21]. If the LR symmetry is broken correctly, the heart and stomach are on the left side and liver and gall bladder are on the right side of the body (*situs solitus*), but in *situs inversus*, the position of the internal organs is reversed [22]. A connection between *situs inversus* and defects in cilia motility suggests that cilia-induced flow is related to the development of the LR asymmetry. Indeed, it has been established that the loss of nodal cilia

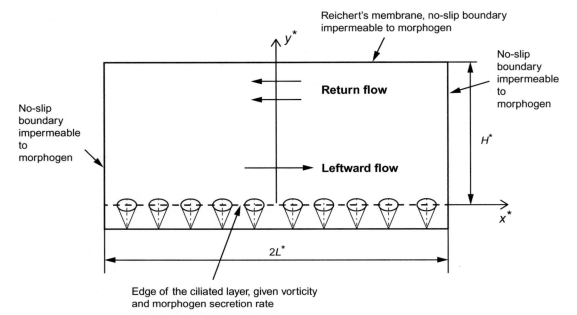

FIGURE 7.1   Schematic diagram of a mouse ventral node showing the ciliated layer at the nodal floor and the flow circulation induced by this ciliated layer. The diagram also shows boundary conditions for the flow velocity and morphogen concentration assumed at the node boundaries for the modeling approach which is based on assuming a given vorticity at the edge of the ciliated layer.

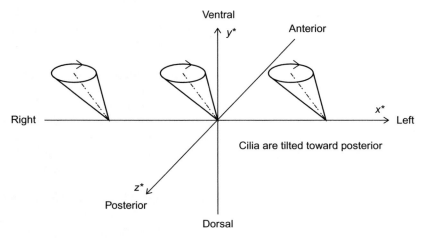

FIGURE 7.2   The orientation of cilia's rotational axes in the node.

motility causes randomization of LR patterning [23]. Okada et al. [13] established a link between the abnormal nodal flow and *situs inversus*.

In order to prove that the leftward nodal flow causes the breaking of the LR symmetry and does not simply accompany the event, Nonaka et al. [10] carefully removed the membranes covering the nodes and subjected mouse embryos to an artificial rightward flow. They demonstrated that this indeed induced *situs inversus*. In addition, subjecting embryos with immotile cilia (such embryos normally exhibit randomized LR patterning) to a leftward flow led to normal LR patterning in such embryos. Watanabe et al. [24] have shown that normal LR patterning in mouse embryos with the mutated inversin protein, which is believed to be required for the correct movement of nodal cilia, can be restored by subjecting them to an artificial fast leftward flow.

As noted in McGrath and Brueckner [25], what makes the LR symmetry breaking mechanism based on the nodal flow really interesting is the fact that it is based on mechanical motion rather than on pure biochemistry.

The mechanism that relies on cilia-generated flow to break the LR symmetry is not unique to mice. Ciliated organs similar to the ventral node in mice have also been found in zebrafish, rabbits, and frogs [14,26–30]. For example, in zebrafish, an organ corresponding to the ventral node in mice is called the Kupffer's vesicle [31]. Interestingly, the nodal flow velocity and the node size are not conserved properties across species. For example, in rabbits, the flow velocity is much smaller and the node is much larger than in mice [14]. No mechanism of symmetry breaking based on cilia-generated flow has been detected in chicks [11]. This brings up the question of why the nodal flow is involved in the LR symmetry breaking mechanism in some, but seemingly not all, vertebrates.

7.3 MODELING RESEARCH ON THE NODAL FLOW

As pointed out by Purcell [32], for low Reynolds number biological flows, due to the lack of inertia, the flow is entirely determined by forces acting on the fluid at the moment. Propulsion forces generated by flagella are analyzed in Purcell [33]. Cartwright et al. [17] pioneered modeling the leftward flow and morphogen transport in the node. In their original model, cilia were simulated as elementary point rotators (rotlets). Effects of bending of individual cilia are discussed in Brokaw [34]. The application of the slender body theory to modeling nodal flow induced by discrete cilia was further developed in Smith et al. [8,35,36]. An extension to the regularized Stokeslet method to cilia-driven flows was discussed in Smith [37]. Research on the simulation of cilia-driven nodal flow, including relevant biology, physical assumptions, and modeling approaches, was reviewed in Smith et al. [38].

Unlike most cilia-driven flows [3], the nodal flow is generated not by cilia beating, but rather by cilia rotation. There is experimental evidence suggesting that cilium rotation does not occur at a constant velocity; a cilium moves faster during the power (leftward) portion of the stroke and slower during the return (rightward) portion of the stroke [22,39]. However, Smith et al. [36,40] argued that the physics of leftward flow generation is not related to the difference in cilium velocity during the leftward and rightward portions of the stroke. Indeed, for a low Reynolds number situation, the flow inertia is negligible. The slower velocity during the rightward portion of the stroke would not matter because the rightward motion will occur for a longer period of time. The physical reason for the flow to occur in the leftward direction is the interaction between the flow produced by cilia rotation and the floor of the node. Because the axis of the rotating cilium is tilted, the cilium moves much closer to the floor during the rightward portion of the stroke, and this dampens the rightward flow. The leftward flow is dampened to a lesser extend because during the leftward portion of the stroke the cilium is farther away from the nodal floor. The average flow near the ciliated surface thus occurs in the leftward direction. The tilt of the cilium's axis toward the posterior of the node has been established experimentally, for example, in Nonaka et al. [23].

7.4 LEFTWARD FLOW OR FLOW RECIRCULATION?

In this section we discuss macroscopic flow in the nodal pit, which is generated by several hundred rotating cilia at the nodal floor. This is different from the flow generated by a single cilium that is discussed at the end of the previous section.

There seems to be a contradiction between experimental and numerical published reports in the sense that experiments indicate just the leftward flow while numerical results suggest the presence of a circulation composed of the leftward flow (closer to the nodal floor) and a return, rightward flow (closer to Reichert's membrane covering the node). The explanation seems to be in the difference between experimental and modeling geometries. In experiments, the Reichert's membrane had to be removed to allow flow visualization and measurement. This increased the fluid volume and resulted in a much smaller velocity of the return flow [41], and for this reason, the return flow could not be detected. On the contrary, numerical simulations faithfully reproduced *in vivo* conditions, in which the node was closed by the Reichert's membrane. As discussed in Cartwright et al. [41,42], for a node covered by the Reichert's membrane, the macroscopic return flow is not associated with rightward return currents generated by the rotating cilia, but rather is a consequence of mass conservation and the fact that the extraembryonic fluid is incompressible. The return flow just completes the circulation loop. Because the thickness of the boundary layer adjacent to the floor of the node (the surface with rotating cilia) is smaller than the thickness of the boundary layer adjacent to Reichert's membrane, even in a closed node the velocity of the leftward flow is larger than the velocity of the return flow [43].

7.5 SENSING OF THE FLOW: MECHANOSENSING OR CHEMOSENSING?

In order to break the LR symmetry, it is not sufficient to generate the leftward flow, and it is also necessary to sense it. There are two hypotheses explaining how this can happen: mechanosensing and chemosensing. The mechanosensing hypothesis [44] postulates that it is the mechanical stress generated by the flow that is sensed. Mechanosensing can be accomplished by immotile cilia, which, along with rotating cilia, are also present in the nodal floor [45–48]. If this hypothesis is correct, cilia are entirely responsible for breaking the LR symmetry—the leftward flow is produced by the motile population of cilia and is then sensed by the nonmotile population.

Chen et al. [48] tested the mechanosensing hypothesis by using a computer model. They concluded that the bending of the immotile cilia caused by the leftward flow would be large enough to induce the chemical signaling cascade in the cells. If the mechanosensing hypothesis is correct, the mechanism of flow sensing is probably associated with calcium signaling. There is data indicating that bending of the immotile primary cilium increases the calcium influx into the cell [49–51].

One of the difficulties of the mechanosensing hypothesis is associated with the following argument. According to Cartwright et al. [17,41], for creeping flow conditions, the magnitude of the shear stress is the same across the node. The main difficulty of the mechanosensing hypothesis is, therefore, that the cilium must be able to detect the direction of the flow, not just its magnitude, and it is not clear how a single monocilium can do this. A bundle of stereocilia in the inner ear can determine the direction [52], but there are no cilia bundles in the floor of the node.

Although experimental data obtained by Yoshiba et al. [45] seem to suggest that the leftward flow is indeed sensed by cilia, it is not known at this point whether cilia sense chemicals that are transported by the flow or they sense the mechanical force induced by the flow.

The chemosensing hypothesis postulates that the leftward flow transports a yet unknown signaling protein (a morphogen), and the side with a higher concentration of the morphogen becomes the left side of the node. According to Hirokawa et al. [16], the most likely candidates for the morphogen are proteins with a mass of 20-40 kDa. For smaller proteins, diffusion will dominate and that will produce a uniform morphogen distribution in the cavity. For larger proteins, advection will dominate, and such proteins will recirculate in the cavity. However, for proteins with a mass of 20-40 kDa, providing that they have a finite half-life, the leftward flow with experimentally measured velocities will produce a stationary concentration gradient, which, in theory, can be sensed. Possible candidates for signaling molecules are FGF, Nodal, and SSH. Cartwright et al. [17] suggested that immotile cilia in the nodal floor are chemoreceptors for a morphogen rather than mechanoreceptors. Experimental results reported in Kamura et al. [53] suggest that the motile cilia in the Kupffer's vesicle of medaka fish, in addition to generating the flow, are also capable of acting as chemosensors.

One particular difficulty of the chemosensing hypothesis is expressed in the argument presented in Pazour and Witman [44]. As mentioned earlier, it has been shown by Nonaka et al. [10] that subjecting mouse embryos to an artificial rightward flow can induce *situs inversus*. This procedure requires pumping a large amount of fluid over the embryos. The question, then, is how a morphogen concentration gradient sufficient to distinguish left from right can be established in such conditions.

However, there is another possibility. It may be that the morphogen molecules are transported inside small extracellular vesicles which are called nodal vesicular parcels (NVPs) [26,54,55]. NVPs are small vesicles with diameters 0.3-5 μm [54]. They have virtually no inertia, and in terms of advection, their transport can be viewed as transport of passive tracers. However, due to their size, their diffusivity is negligibly small. Even in the absence of diffusion, mixing will occur because of a process that is called chaotic advection [17]. The theory of chaotic advection was developed by Hassan Aref [56,57]. For discussion of the relevance of chaotic advection to the nodal flow see, for example, Cartwright et al. [41] and Smith et al. [38]. Cartwright and colleagues [42] simulated transport of NVPs and demonstrated that due to chaotic advection, trajectories of NVPs that were initially very close can become very distinct.

The recirculation of NVPs is avoided by the fact that NVPs rupture when they reach the left side of the node. As NVPs rupture, they release their content [54]. Theoretical results of Cartwright et al. [42] suggested that the NVP rupture mechanism is biochemical rather than mechanical. This is because for a low Reynolds number flow, which exists in the node, the collision force between the NVP and a cilium or the NVP and a wall is estimated to be very small. The idea of Cartwright et al. [42] that the NVP rupture mechanism is biochemical is supported by the observed swelling of NVPs before their rupture, which was reported in Hirokawa et al. [58].

Results recently reported by Shinohara et al. [59] suggest that whatever the flow sensing mechanism is, it must be extremely sensitive to the leftward flow. Indeed, just two rotating motile cilia in the nodal cavity are sufficient to break the LR symmetry, and only a few non-motile cilia are required to sense the flow. It is also interesting that two rotating cilia tend to synchronize their rotations due to hydrodynamic interactions. The interactions result in phase locking between the rotating cilia [60].

Finally, after breaking the LR symmetry in the node, signals must be transmitted from the node to induce asymmetric morphogenesis. This process is discussed, for example, in Yamamoto et al. [61]. Growth factors Nodal and Lefty are involved in signal transduction [62,63]. Other important issues are the establishment of the midline [64] and how transport of asymmetric signals to the wrong side of the developing organism is prevented [65,66]. Another significant question is how asymmetric signaling translates into asymmetric morphology [11].

7.6 MODELING THE EFFECT OF A CILIATED SURFACE BY IMPOSING A GIVEN VORTICITY AT THE EDGE OF THE CILIATED LAYER

For the leftward flow above the ciliated layer, Kuznetsov et al. [67,43] developed a large-scale model that accounted for the effect of cilia rotation in an effective, averaged sense. They considered the leftward flow above the cilia tips (Figure 7.1) and asked the following question: what boundary condition needs to be imposed at the interface between the ciliated layer and a clear fluid in order to effectively model the leftward flow in the clear fluid (above the interface)? Kuznetsov et al. [67,43] proposed that the leftward flow (more precisely, the circulation in the nodal pit) can be effectively modeled by imposing a given vorticity (more precisely, the z^*-component of the vorticity vector) at the edge of the ciliated layer. The other two components of the vorticity vector at the edge of the ciliated layer were set to zero. The z^*-axis is directed from the anterior to posterior of the node (Figure 7.2), and prescribing the z^*-component of vorticity at the edge of the ciliated layer generates a circulation in the x^*-y^* plane. This approach greatly simplifies the numerical solution because it does not require modeling the effects of individual cilia.

Because surfaces covered with cilia are common in living organisms, developing effective methods for modeling fluid flows induced by such surfaces is important. The approach developed by Kuznetsov and colleagues [67,43] that suggests modeling the nodal flow by prescribing a vorticity value (or distribution) at the ciliated surface may not be restricted to modeling the flow in the embryonic node but may also be useful in modeling various bioflows. Furthermore, Evans et al. [68] and Shields et al. [69] described flow induced by biomimetic cilia arrays that rotate and thus generate flow similar to the nodal flow. Flows generated by biomimetic cilia can also be modeled by prescribing a given vorticity at the ciliated surface; such flows may be relevant to various microfluidic systems.

In this section, we review the main ideas developed in Kuznetsov et al. [43,67]. These papers consider a two-dimensional flow situation in a rectangular cavity, as shown in Figure 7.1. The x^*-axis is parallel to the ciliated surface, its direction coincides with the direction of the flow. The y^*-axis is normal to the ciliated surface. The ciliated layer is at the bottom of the cavity. The Reichert's membrane covering the top of the cavity and the side walls of the cavity were modeled as no-slip boundaries. The extraembryonic fluid filling the cavity was modeled as a Newtonian fluid. Because the characteristic Reynolds number is $\sim 10^{-3}$ [17], viscosity dominates over inertia. Therefore, the flow was modeled as a creeping flow by using the Stokes equations:

$$-\nabla^* p^* + \mu^* \nabla^{*2} \mathbf{u}^* = 0, \tag{7.1}$$

$$\nabla^* \cdot \mathbf{u}^* = 0, \tag{7.2}$$

where $\mathbf{u}^*$ is the flow velocity vector, (u^*, v^*) (m/s); u^* and v^* are the x^* and y^* velocity components, respectively, (m/s); p^* is the pressure (Pa); and μ^* is the dynamic viscosity of the extraembryonic fluid (Pa s).

Equations (7.1) and (7.2) were solved subject to the following boundary conditions:

$$\mathbf{u}^* = 0 \quad \text{at } x^* = \pm L^*, \tag{7.3}$$

$$v^* = 0, \omega^* = \frac{\partial v^*}{\partial x^*} - \frac{\partial u^*}{\partial y^*} = \omega_0^* f(x^*) \quad \text{at } y^* = 0, \tag{7.4}$$

$$\mathbf{u}^* = 0 \quad \text{at } y^* = H^*, \tag{7.5}$$

where ω^* is the z^*-component of vorticity and ω_0^* is the reference value of the z^*-component of vorticity at the edge of the ciliated layer (s^{-1}). The function $f(x^*)$ in Equation (7.4) accounts for the possibility that the vorticity at the ciliated surface is not constant but rather exhibits a periodic variation. If the vorticity at the ciliated surface is constant, then $f(x^*) = 1$.

The imposition of the z^*-component of vorticity at the ciliated surface is justified by the following argument. The flow induced by a rotating cilium tilted toward the posterior of the node (Figure 7.2) produces net flow in the x^*-direction (Figure 7.1). This happens because when the cilium passes near the nodal floor, the amount of fluid entrained by the cilium is reduced due to zero fluid velocity at the floor [58,69]. A single large cilium (a propeller) would also have caused rotational movement of the fluid and would have produced some flow in the z^*-direction. However, because there are 200-300 cilia on the floor of the node, most of the flow in the z^*-direction caused by one cilium is cancelled by a neighboring cilium, and the macroscopic flow is approximately 2D and occurs mostly in the x^*-y^* plane. The assumption about the negligible flow in the z^*-direction is supported by experimental observations, reviewed in Hirokawa et al. [16], which indicate that the dominating time-averaged flow is the leftward flow in the x^*-direction (see Figure 7.1).

Introducing the stream function Ψ^* (m^2/s), Equations (7.1) and (7.2) were transformed as follows:

$$\frac{\partial^4 \Psi^*}{\partial x^{*4}} + 2\frac{\partial^4 \Psi^*}{\partial x^{*2}\partial y^{*2}} + \frac{\partial^4 \Psi^*}{\partial y^{*4}} = 0, \tag{7.6}$$

where

$$u^* = \frac{\partial \Psi^*}{\partial y^*}, v^* = -\frac{\partial \Psi^*}{\partial x^*}. \tag{7.7}$$

Equation (7.6) was solved subject to the following boundary conditions:

$$\Psi^* = \frac{\partial \Psi^*}{\partial x^*} = 0 \ \text{at} \ x^* = \pm L^*, \tag{7.8}$$

$$\Psi^* = 0, \frac{\partial^2 \Psi^*}{\partial y^{*2}} = -\omega^* = -\omega_0^* f(x^*) \ \text{at} \ y^* = 0, \tag{7.9}$$

$$\Psi^* = \frac{\partial \Psi^*}{\partial y^*} = 0 \ \text{at} \ y^* = H^*, \tag{7.10}$$

where L^* is the half-width of the cavity and H^* is the depth of the cavity. The boundary condition at $y^* = 0$ is a consequence of the following equation:

$$\omega^* = -\Delta^* \Psi^*, \tag{7.11}$$

which links the stream function and vorticity.

The dimensionless forms of Equations (7.6)–(7.10) are

$$\frac{\partial^4 \Psi}{\partial x^4} + 2K^2 \frac{\partial^4 \Psi}{\partial x^2 \partial y^2} + K^4 \frac{\partial^4 \Psi}{\partial y^4} = 0, \tag{7.12}$$

$$\Psi = \frac{\partial \Psi}{\partial x} = 0 \ \text{at} \ x = \pm 1, \tag{7.13}$$

$$\Psi = 0, \frac{\partial^2 \Psi}{\partial y^2} = -f(x) \ \text{at} \ y = 0, \tag{7.14}$$

$$\Psi = \frac{\partial \Psi}{\partial y} = 0 \ \text{at} \ y = 1, \tag{7.15}$$

In Equations (7.12)–(7.15)

$$\Psi = \frac{\Psi^*}{\omega_0^* H^{*2}}, K = \frac{L^*}{H^*}, x = \frac{x^*}{L^*}, y = \frac{y^*}{H^*}. \tag{7.16}$$

In Equation (7.16), K is the geometric factor, the ratio of the cavity half-width to its depth. K characterizes the cavity aspect ratio.

The dimensionless velocity components, $u = \frac{u^*}{\omega_0^* H^*}$ and $v = \frac{v^*}{\omega_0^* H^*}$, are related to the dimensionless stream function by the following equations:

$$u = \frac{\partial \Psi}{\partial y}, v = -\frac{1}{K} \frac{\partial \Psi}{\partial x}. \tag{7.17}$$

In the chemosensing hypothesis, the left side of the node is determined by a higher concentration of morphogen. According to Cartwright et al. [17], the contribution of diffusion to morphogen transport is expected to be important. The following convection-diffusion equation is used to describe transport of the morphogen:

$$u^* \frac{\partial C^*}{\partial x^*} + v^* \frac{\partial C^*}{\partial y^*} = D^* \left(\frac{\partial^2 C^*}{\partial x^{*2}} + \frac{\partial^2 C^*}{\partial y^{*2}} \right) - \frac{C^* \ln(2)}{T^*_{1/2}}, \tag{7.18}$$

where D^* is the morphogen diffusivity (m^2/s) and $T^*_{1/2}$ is the half-life of the morphogen (s). The Péclet number, which shows the relative importance of advection and diffusion, is estimated to be in the range 6-600 for morphogen transport in the node [17]. This indicates that advection is expected to dominate, but diffusion is also important because morphogen transport is effectively chaotic advection, in which diffusion provides additional mixing [17].

A constant and uniform flux of morphogen at the bottom boundary due to morphogen secretion was assumed (Figure 7.1). This boundary condition can be justified as follows. According to Shields et al. [69], flow below the cilia tips (in the ciliated layer) is dominated by local vortices which are induced by rotation of individual cilia. These vortices enhance mixing in the ciliated layer. This supports the assumption that the morphogen, which is presumably secreted at the floor of the node and then mixed in the ciliated layer, reaches the top of the ciliated layer at a constant rate.

The other three walls were assumed to be impermeable to morphogen (Figure 7.1). These assumptions lead to the following boundary conditions for the morphogen concentration:

$$\frac{\partial C^*}{\partial x^*} = 0 \quad \text{at } x^* = -L^*, x^* = L^*, \tag{7.19}$$

$$-D^* \frac{\partial C^*}{\partial y^*} = j_0^* \quad \text{at } y^* = 0, \tag{7.20}$$

$$\frac{\partial C^*}{\partial y^*} = 0 \quad \text{at } y^* = H^*, \tag{7.21}$$

where j_0^* is the constant flux of the morphogen at the $y^* = 0$ boundary, which results from morphogen secretion at the floor of the node (molecules/(m^{2}s)).

A dimensionless form of Equation (7.18) is

$$\frac{\partial \Psi}{\partial y} \frac{\partial C}{\partial x} - \frac{\partial \Psi}{\partial x} \frac{\partial C}{\partial y} = (\text{Pe}K)^{-1} \left(\frac{\partial^2 C}{\partial x^2} + K^2 \frac{\partial^2 C}{\partial y^2} \right) - \frac{KC}{S}, \tag{7.22}$$

where

$$C = \frac{C^* D^*}{j_0^* H^*}, \text{Pe} = \frac{\omega_0^* H^{*2}}{D^*}, S = \frac{\omega_0^* T^*_{1/2}}{\ln(2)}. \tag{7.23}$$

In Equation (7.22), C is the dimensionless morphogen concentration; Pe is the Péclet number, which characterizes the ratio of the rate of morphogen transport by advection to the rate of morphogen transport by diffusion; and S is a dimensionless parameter characterizing the morphogen half-life.

The dimensionless forms of boundary conditions (19)-(21) are

$$\frac{\partial C}{\partial x} = 0 \quad \text{at } x = -1, x = 1, \tag{7.24}$$

$$\frac{\partial C}{\partial y} = -1 \quad \text{at } y = 0, \tag{7.25}$$

$$\frac{\partial C}{\partial y} = 0 \quad \text{at } y = 1. \tag{7.26}$$

Kuznetsov et al. [43,67] solved Equations (7.12) and (7.22) subject to boundary conditions (7.13)–(7.15) and (7.24)–(7.26) by using a proper generalized decomposition (PGD) method [70,71]. The utilization of the PGD technique made it possible to obtain semianalytical solutions for both the velocity (which characterizes the circulation in the cavity) and the morphogen concentration. The preceding numerical problem can also be solved by any other numerical method.

The developed method, which is based on prescribing a given vorticity at the edge of the ciliated layer and simulating the macroscopic flow, can also be used to describe morphogen transport if the morphogen is transported inside NVPs. In this case, the diffusivity of NVPs should be set to a small value, probably zero, and a model for NVP rupture should be added. The method can also be used to simulate the mechanosensing hypothesis. In the latter case, one needs to assume that resolving the macroscopic flow is sufficient to accurately describe average forces on the cilia.

7.7 SUMMARY OF RELEVANT PARAMETERS DESCRIBING THE NODAL FLOW AND ESTIMATES OF THEIR VALUES

One of the benefits of the model that is based on specifying vorticity at the edge of the ciliated layer [43,67] is the small number of parameters that are involved in the model (in this sense, this model can be viewed as a minimal model of the nodal flow). The values and ranges of parameters involved in the model are summarized in Table 7.1.

TABLE 7.1   Estimated Values of Parameters Involved in the Model Based on Specifying Vorticity at the Edge of the Ciliated Layer

Parameter	Symbol	Value or range	References
Morphogen diffusivity	D^*	$10^1\text{-}10^3\,\mu\text{m}^2/\text{s}$	[14,16,17]
Depth of the cavity	H^*	$10\text{-}20\,\mu\text{m}$	[16,17,38,42]
Half-width of the cavity	L^*	$25\text{-}50\,\mu\text{m}$	[16,17,38,42]
Morphogen half-life	$T^*_{1/2}$	$8\text{-}250\,\text{min}$	[72,73]
Dynamic viscosity of the extraembryonic fluid	μ^*	$10^{-3}\,\text{Pa}\,\text{s}$	[17,39,42]
Reference value of vorticity at the edge of the ciliated layer	ω^*_0	$60\,\text{s}^{-1}$	[43,67]

Because the identity of the morphogen responsible for the LR symmetry breaking is not established at this point, the range for $T^*_{1/2}$ in Table 7.1 was estimated based on half-lives of representative morphogens regulating patterning and growth in various tissues, by using data reported in Kicheva et al. [72,73]. The value of ω_0^* was estimated in Kuznetsov et al. [43,67] from the results of direct numerical simulation of flow in a cavity induced by 81 rotating cilia.

Data reported in Table 7.1 were used to estimate values of dimensionless parameters involved in the dimensionless formulation of the model, which is given by Equations (7.12)– (7.15), (7.17), (7.22), and (7.24)–(7.26). There are three dimensionless parameters that are involved in these equations. These are the geometric factor $K = \frac{L^*}{H^*}$, the Péclet number $\text{Pe} = \frac{\omega_0^* H^{*2}}{D^*}$, and the dimensionless morphogen half-life $S = \frac{\omega_0^* T_{1/2}}{\ln(2)}$. The estimated ranges for these parameters are $1.25 \leq K \leq 5$, $6 \leq \text{Pe} \leq 2400$, and $4.5 \times 10^4 \leq S \leq 1.3 \times 10^6$.

7.8 NUMERICAL RESULTS OBTAINED ASSUMING A CONSTANT VORTICITY AT THE EDGE OF THE CILIATED LAYER

Kuznetsov et al. [43,67] presented extensive comparisons between the direct numerical simulation, performed by ANSYS CFX 13.0 software package (ANSYS, Canonsburg, PA), and approximate solutions obtained by prescribing a constant vorticity at the edge of the ciliated layer. The direct numerical simulation was based on modeling the flow induced by rotation of 81 cilia (9×9 arrangement) in a closed rectangular cavity with dimensions $50 \times 50 \times 10$ μm. The frequency of cilia rotation was 10 Hz and the length of the cilia was 3 μm. The path of the rotating cilium formed a cone. The half angle of the cone opening was 45°, and the tilt of the cone axis from the vertical toward posterior was 25° (see Figure 7.2). The length of cilia and their tilt toward the posterior side of the node were based on Cartwright et al. [42].

For the CFX solution, no-slip conditions were used on all surfaces, including the surfaces of rotating cilia. The effect of cilia rotation in Kuznetsov et al. [43] was modeled as follows. The position and velocity vectors of grid cells composing a cilium, for a given time, were calculated by using the fact that cilia rotate with a constant angular velocity. The velocity components were then transferred to CFX via a user defined function. Steady-state results were obtained by averaging the CFX solution over each rotation cycle of cilia.

Kuznetsov et al. [43,67] also investigated the effect of imposing different vorticity distributions at the edge of the ciliated surface. The direct numerical simulation predicts that vorticity at the edge of the ciliated surface oscillates, with a period equal to the distance between the two neighboring cilia. However, when Kuznetsov and colleagues tested the method with (i) the oscillatory vorticity distribution taken from the direct numerical simulation and (ii) a constant vorticity calculated as the average of the above oscillatory distribution, they found very little difference in the results. They concluded that assuming a constant vorticity at the edge of the ciliated layer is sufficiently accurate for simulating flow and morphogen transport in the node.

The problem of predicting flow in the nodal cavity is thus reduced to the problem of predicting a creeping flow in a cavity with three no-slip walls and a fourth wall where a constant vorticity is imposed (see Figure 7.1). Due to the reduction of the parameter space in the dimensionless formulation, the magnitude of vorticity at the edge of the ciliated layer, ω_0^*, is not present in the dimensionless equations describing a steady circulation in the cavity. The predicted circulation was found to be in good qualitative agreement with that reported in Cartwright et al. [42].

For the ease of physical interpretation, the numerical results are shown in the dimensional form. Figure 7.3 shows representative results obtained by the method developed in Kuznetsov et al. [43,67]. This figure is computed for $H^* = 10\,\mu m$, $L^* = 25\,\mu m$, and $\omega_0^* = 60\,s^{-1}$. It displays distributions of the streamwise velocity component, u^*, versus the distance from the edge of the ciliated layer, y^*. The results are shown in three different locations in terms of x^*. The location $y^* = 0$ corresponds to the edge of the ciliated layer; at this location, a constant vorticity is imposed. In this location, the streamwise velocity is the largest. In the region close to the ciliated surface (at approximately $y^*/H^* < 0.35$), the flow is leftward, while closer to the top of the nodal pit, the flow is rightward. As expected, the region with rightward velocity is wider and velocity in this region is smaller. The leftward velocity at the edge of the ciliated layer is the largest at $x^*/L^* = 0$ because this location is most distant from the side walls of the node.

A representative distribution of the dimensionless morphogen concentration in the nodal cavity is shown in Figure 7.4. It is computed for $D^* = 10\,\mu m^2/s$, $H^* = 10\,\mu m$, $L^* = 25\,\mu m$, $T_{1/2}^* = 40\,min$, and $\omega_0^* = 60\,s^{-1}$ (this corresponds to the following dimensionless parameters values: $K = 2.5$, $Pe = 600$, and $S = 208,000$). For numerical work, the PGD method developed in Kuznetsov et al. [43,67] was utilized. The distribution of the morphogen concentration shows a complicated interplay between advection and diffusion. The morphogen concentration on the left side of the node is clearly larger. This is because advection dominates, due to a large value of the Péclet number, and morphogen is transported toward the left side of the node by the leftward flow.

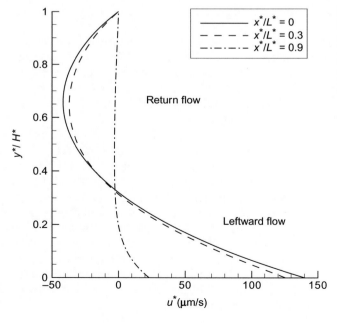

FIGURE 7.3 Representaive results for the flow in the nodal cavity obtained by assuming a constant vorticity at the edge of the ciliated layer. The x^*-velocity component, u^*, versus the distance from the ciliated layer, y^*/H^*, in three different locations. $H^* = 10\,\mu m$, $L^* = 25\,\mu m$, and $\omega_0^* = 60\,s^{-1}$.

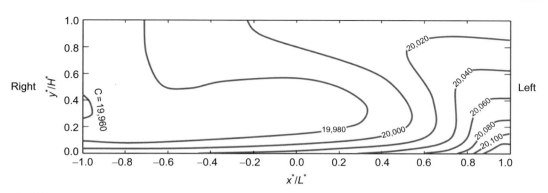

FIGURE 7.4 Representaive results for the dimensionless morphogen concentration in the nodal cavity obtained by assuming a constant vorticity at the edge of the ciliated layer. $D^* = 10\,\mu m^2/s$, $H^* = 10\,\mu m$, $L^* = 25\,\mu m$, $T^*_{1/2} = 40$ min, $\omega^*_0 = 60\,s^{-1}$ (this corresponds to $K = 2.5$, Pe $= 600$, $S = 208{,}000$).

7.9 CONCLUSIONS

In this chapter, we reviewed recent research on the leftward flow in a mouse ventral node. This flow lasts for several hours and presents the first evidence of the LR asymmetry before any molecular evidence of the LR asymmetry is developed [25,74]. It is believed that the leftward flow is the mechanism of translation of the AP and DV asymmetries, which develop first, into the LR asymmetry. The mouse ventral node is thus an elegant microfluidic device developed during the process of evolution in order to translate one type of asymmetry into another type of asymmetry.

The flow is generated by the rotation of primary cilia with axes that are tilted toward the posterior of the node. Two flow-sensing mechanisms are feasible: mechanosensing and chemosensing. The mechanosensing hypothesis assumes that the shear stress produced by the flow is sensed. The chemosensing hypothesis assumes that the flow produces a concentration gradient of a yet undetermined morphogen, and the side of the node with a higher concentration of morphogen becomes the left side. A variant of the chemosensing hypothesis assumes that the morphogen is transported within small vesicles, called NVPs, which rupture when they reach the left side of the node. The difference between the two variants of the chemosensing hypothesis is the role of diffusion. In the traditional chemosensing hypothesis, both diffusion and advection affect morphogen transport, while in the hypothesis involving NVP transport, diffusion is negligible due to the large size of NVPs (0.3-5 μm). It is believed that the role of sensors is played by the immotile population of primary cilia, which, along with rotating primary cilia, are present in the nodal floor. Recent experimental results indicate that even a very weak leftward flow produced by rotation of just two primary cilia can be sensed and is sufficient to break the LR symmetry.

Previous computational work on modeling the nodal flow simulated the effects of rotation of many cilia located at the floor of the node; each cilium was modeled individually. This resulted in a numerical problem of great complexity. Recently, the present authors, in collaboration with Drs. A.A. Avramenko, D.G. Blinov, I.V. Shevchuk, and A.I. Tyrinov, suggested a large-scale approach to modeling the leftward flow and morphogen transport in the nodal cavity by imposing a given vorticity distribution at the edge of the ciliated layer. This approach greatly simplifies the numerical formulation and makes it possible to concentrate on the effects of parameters describing the geometry and size of the node and the flow velocity. This is important because the geometry of the organ that plays the role of the ventral node in mice is different in different vertebrates.

Acknowledgments

We are grateful to Drs. A.A. Avramenko, D.G. Blinov, I.V. Shevchuk, and A.I. Tyrinov for many helpful discussions.

References

[1] Blake J. Flow in tubules due to ciliary activity. Bull Math Biol 1973;35:513–23.
[2] Yu X, Lau D, Ng CP, Roy S. Cilia-driven fluid flow as an epigenetic cue for otolith biomineralization on sensory hair cells of the inner ear. Development 2011;138:487–94.
[3] Smith DJ, Gaffney EA, Blake JR. Mathematical modelling of cilia-driven transport of biological fluids. Proc R Soc A Math Phys Eng Sci 2009;465:2417–39.
[4] Matsui H, Randell S, Peretti S, Davis C, Boucher R. Coordinated clearance of periciliary liquid and mucus from airway surfaces. J Clin Invest 1998;102:1125–31.
[5] Lyons RA, Saridogan E, Djahanbakhch O. The reproductive significance of human fallopian tube cilia. Hum Reprod Update 2006;12:363–72.
[6] Fawcett D. Mammalian spermatozoon. Dev Biol 1975;44:394–436.
[7] Nonaka S, Tanaka Y, Okada Y, Takeda S, Harada A, Kanai Y, et al. Randomization of left-right asymmetry due to loss of nodal cilia generating leftward flow of extraembryonic fluid in mice lacking KIF3B motor protein. Cell 1998;95:829–37.
[8] Montenegro-Johnson TD, Smith AA, Smith DJ, Loghin D, Blake JR. Modelling the fluid mechanics of cilia and flagella in reproduction and development. Eur Phys J E 2012;35:111.
[9] Hashimoto M, Shinohara K, Wang J, Ikeuchi S, Yoshiba S, Meno C, et al. Planar polarization of node cells determines the rotational axis of node cilia. Nat Cell Biol 2010;12:170–6.
[10] Nonaka S, Shiratori H, Saijoh Y, Hamada H. Determination of left-right patterning of the mouse embryo by artificial nodal flow. Nature 2002;418:96–9.
[11] Hamada H. Breakthroughs and future challenges in left-right patterning. Dev Growth Differ 2008;50:S71–8.
[12] Basu B, Bruedner M. Cilia: multifunctional organelles at the center of vertebrate left-right asymmetry. Ciliary Funct Mamm Dev 2008;85:151–74.
[13] Okada Y, Nonaka S, Tanaka Y, Saijoh Y, Hamada H, Hirokawa N. Abnormal nodal flow precedes situs inversus in iv and inv mice. Mol Cell 1999;4:459–68.
[14] Okada Y, Takeda S, Tanaka Y, Belmonte J, Hirokawa N. Mechanism of nodal flow: a conserved symmetry breaking event in left-right axis determination. Cell 2005;121:633–44.
[15] Borovina A, Superina S, Voskas D, Ciruna B. Vangl2 directs the posterior tilting and asymmetric localization of motile primary cilia. Nat Cell Biol 2010;12:407–12.
[16] Hirokawa N, Okada Y, Tanaka Y. Fluid dynamic mechanism responsible for breaking the left-right symmetry of the human body: the nodal flow. Annu Rev Fluid Mech 2009;41:53–72.
[17] Cartwright J, Piro O, Tuval I. Fluid-dynamical basis of the embryonic development of left-right asymmetry in vertebrates. Proc Natl Acad Sci USA 2004;101:7234–9.
[18] Afzelius B. Human syndrome caused by immotile cilia. Science 1976;193:317–9.
[19] Bartoloni L, Blouin J, Pan Y, Gehrig C, Maiti A, Scamuffa N, et al. Mutations in the DNAH11 (axonemal heavy chain dynein type 11) gene cause one form of situs inversus totalis and most likely primary ciliary dyskinesia. Proc Natl Acad Sci USA 2002;99:10282–6.
[20] Supp D, Witte D, Potter S, Brueckner M. Mutation of an axonemal dynein affects left right asymmetry in inversus viscerum mice. Nature 1997;389:963–6.
[21] Beddington R, Robertson E. Axis development and early asymmetry in mammals. Cell 1999;96:195–209.

[22] Raya A, Belmonte J. Left-right asymmetry in the vertebrate embryo: from early information to higher-level integration. Nat Rev Genet 2006;7:283–93.

[23] Nonaka S, Yoshiba S, Watanabe D, Ikeuchi S, Goto T, Marshall WF, et al. De novo formation of left-right asymmetry by posterior tilt of nodal cilia. Plos Biol 2005;3:1467–72.

[24] Watanabe D, Saijoh Y, Nonaka S, Sasaki G, Ikawa Y, Yokoyama T, et al. The left-right determinant inversin is a component of node monocilia and other 9+0 cilia. Development 2003;130:1725–34.

[25] McGrath J, Brueckner M. Cilia are at the heart of vertebrate left-right asymmetry. Curr Opin Genet Dev 2003;13:385–92.

[26] Bisgrove BW, Makova S, Yost HJ, Brueckner M. RFX2 is essential in the ciliated organ of asymmetry and an RFX2 transgene identifies a population of ciliated cells sufficient for fluid flow. Dev Biol 2012;363:166–78.

[27] Essner J, Amack J, Nyholm M, Harris E, Yost J. Kupffer's vesicle is a ciliated organ of asymmetry in the zebrafish embryo that initiates left-right development of the brain, heart and gut. Development 2005;132:1247–60.

[28] Kramer-Zucker A, Olale F, Haycraft C, Yoder B, Schier A, Drummond I. Cilia-driven fluid flow in the zebrafish pronephros, brain and kupffer's vesicle is required for normal organogenesis. Development 2005;132:1907–21.

[29] Schweickert A, Weber T, Beyer T, Vick P, Bogusch S, Feistel K, et al. Cilia-driven leftward flow determines laterality in *Xenopus*. Curr Biol 2007;17:60–6.

[30] Schweickert A, Walentek P, Thumberger T, Danilchik M. Linking early determinants and cilia-driven leftward flow in left-right axis specification of *Xenopus laevis*: a theoretical approach. Differentiation 2012;83:S67–77.

[31] Oishi I, Kawakami Y, Raya A, Callol-Massot C, Belmonte JCI. Regulation of primary cilia formation and left-right patterning in zebrafish by a noncanonical wnt signaling mediator, duboraya. Nat Genet 2006;38:1316–22.

[32] Purcell EM. Life at low reynolds number. Am J Phys 1977;45:3–11.

[33] Purcell E. The efficiency of propulsion by a rotating flagellum. Proc Natl Acad Sci USA 1997;94:11307–11.

[34] Brokaw CJ. Computer simulation of flagellar movement IX. oscillation and symmetry breaking in a model for short flagella and nodal cilia. Cell Motil Cytoskeleton 2005;60:35–47.

[35] Smith DJ, Gaffney EA, Blake JR. Discrete cilia modelling with singularity distributions: application to the embryonic node and the airway surface liquid. Bull Math Biol 2007;69:1477–510.

[36] Smith DJ, Blake JR, Gaffney EA. Fluid mechanics of nodal flow due to embryonic primary cilia. J R Soc Interface 2008;5:567–73.

[37] Smith DJ. A boundary element regularized stokeslet method applied to cilia- and flagella-driven flow. Proc R Soc A Math Phys Eng Sci 2009;465:3605–26.

[38] Smith DJ, Smith AA, Blake JR. Mathematical embryology: the fluid mechanics of nodal cilia. J Eng Math 2011;70:255–79.

[39] Buceta J, Ibanes M, Rasskin-Gutman D, Okada Y, Hirokawa N, Izpisua-Belmonte J. Nodal cilia dynamics and the specification of the left/right axis in early vertebrate embryo development. Biophys J 2005;89:2199–209.

[40] Smith AA, Johnson TD, Smith DJ, Blake JR. Symmetry breaking cilia-driven flow in the zebrafish embryo. J Fluid Mech 2012;705:26–45.

[41] Cartwright JHE, Piro N, Piro O, Tuval I. Fluid dynamics of nodal flow and left-right patterning in development. Dev Dyn 2008;237:3477–90.

[42] Cartwright JHE, Piro N, Piro O, Tuval I. Embryonic nodal flow and the dynamics of nodal vesicular parcels. J R Soc Interface 2007;4:49–55.

[43] Kuznetsov AV, Blinov DG, Avramenko AA, Shevchuk IV, Tyrinov AI, Kuznetsov IA. Approximate modeling of the leftward flow and morphogen transport in the embryonic node by specifying vorticity at the ciliated surface. J Fluid Mech 2014;738:492–521.

[44] Pazour G, Witman G. The vertebrate primary cilium is a sensory organelle. Curr Opin Cell Biol 2003;15:105–10.

[45] Yoshiba S, Shiratori H, Kuo IY, Kawasumi A, Shinohara K, Nonaka S, et al. Cilia at the node of mouse embryos sense fluid flow for left-right determination via Pkd2. Science 2012;338:226–31.

[46] McGrath J, Somlo S, Makova S, Tian X, Brueckner M. Two populations of node monocilia initiate left-right asymmetry in the mouse. Cell 2003;114:61–73.

[47] Tabin C, Vogan K. A two-cilia model for vertebrate left-right axis specification. Genes Dev 2003;17:1–6.

[48] Chen D, Norris D, Ventikos Y. Ciliary behaviour and mechano-transduction in the embryonic node: Computational testing of hypotheses. Med Eng Phys 2011;33:857–67.

[49] Praetorius H, Spring K. Bending the MDCK cell primary cilium increases intracellular calcium. J Membr Biol 2001;184:71–9.

[50] Praetorius H, Frokiaer J, Nielsen S, Spring K. Bending the primary cilium opens Ca^{2+}-sensitive intermediate-conductance K^+ channels in MDCK cells. J Membr Biol 2003;191:193–200.

[51] Shiba D, Takamatsu T, Yokoyama T. Primary cilia of inv/inv mouse renal epithelial cells sense physiological fluid flow: Bending of primary cilia and Ca^{2+} influx. Cell Struct Funct 2005;30:93–100.

[52] Zetes D, Steele C. Fluid-structure interaction of the stereocilia bundle in relation to mechanotransduction. J Acoust Soc Am 1997;101:3593–601.

[53] Kamura K, Kobayashi D, Uehara Y, Koshida S, Iijima N, Kudo A, et al. Pkd1l1 complexes with Pkd2 on motile cilia and functions to establish the left-right axis. Development 2011;138:1121–9.

[54] Tanaka Y, Okada Y, Hirokawa N. FGF-induced vesicular release of sonic hedgehog and retinoic acid in leftward nodal flow is critical for left-right determination. Nature 2005;435:172–7.

[55] Hirokawa N, Tanaka Y, Okada Y. Left-right determination: Involvement of molecular motor KIF3, cilia, and nodal flow. Cold Spring Harb Perspect Biol 2009;1:a000802.

[56] Aref H. Stirring by chaotic advection. J Fluid Mech 1984;143:1–21.

[57] Meleshko V, Aref H. A blinking rotlet model for chaotic advection. Phys Fluids 1996;8:3215–7.

[58] Hirokawa N, Tanaka Y, Okada Y, Takeda S. Nodal flow and the generation of left-right asymmetry. Cell 2006;125:33–45.

[59] Shinohara K, Kawasumi A, Takamatsu A, Yoshiba S, Botilde Y, Motoyama N, et al. Two rotating cilia in the node cavity are sufficient to break left-right symmetry in the mouse embryo. Nat Commun 2012;3:622.

[60] Takamatsu A, Shinohara K, Ishikawa T, Hamada H. Hydrodynamic phase locking in mouse node cilia. Phys Rev Lett 2013;110:248107.

[61] Yamamoto M, Mine N, Mochida K, Sakai Y, Saijoh Y, Meno C, et al. Nodal signaling induces the midline barrier by activating nodal expression in the lateral plate. Development 2003;130:1795–804.

[62] Kawasumi A, Nakamura T, Iwai N, Yashiro K, Saijoh Y, Belo JA, et al. Left-right asymmetry in the level of active nodal protein produced in the node is translated into left-right asymmetry in the lateral plate of mouse embryos. Dev Biol 2011;353:321–30.

[63] Nakamura T, Hamada H. Left-right patterning: conserved and divergent mechanisms. Development 2012;139:3257–62.

[64] Aw S, Levin M. What's left in asymmetry? Dev Dyn 2008;237:3453–63.

[65] Bisgrove B, Essner J, Yost H. Regulation of midline development by antagonism of lefty and nodal signaling. Development 1999;126:3253–62.

[66] Kato Y. The multiple roles of notch signaling during left-right patterning. Cell Mol Life Sci 2011;68:2555–67.

[67] Kuznetsov A.V., Blinov D.G., Avramenko A.A., Shevchuk I.V., Tyrinov A.I., Kuznetsov I.A., Modeling leftward flow in the embryonic node. In: Proceedings of the ASME 2013 International Mechanical Engineering Congress & Exposition, ASME, 2013, IMECE2013-62503.

[68] Evans BA, Shields AR, Carroll RL, Washburn S, Falvo MR, Superfine R. Magnetically actuated nanorod arrays as biomimetic cilia. Nano Lett 2007;7:1428–34.

[69] Shields AR, Fiser BL, Evans BA, Falvo MR, Washburn S, Superfine R. Biomimetic cilia arrays generate simultaneous pumping and mixing regimes. Proc Natl Acad Sci USA 2010;107:15670–5.

[70] Nouy A. A priori model reduction through proper generalized decomposition for solving time-dependent partial differential equations. Comput Methods Appl Mech Eng 2010;199:1603–26.

[71] Chinesta F, Ammar A, Cueto E. Recent advances and new challenges in the use of the proper generalized decomposition for solving multidimensional models. Arch Comput Methods Eng 2010;17:327–50.

[72] Kicheva A, Bollenbach T, Wartlick O, Jülicher F, Gonzalez-Gaitan M. Investigating the principles of morphogen gradient formation: From tissues to cells. Curr Opin Genet Dev 2012;22:527–32.

[73] Kicheva A, Pantazis P, Bollenbach T, Kalaidzidis Y, Bittig T, Juelicher F, et al. Kinetics of morphogen gradient formation. Science 2007;315:521–5.

[74] Shiratori H, Hamada H. The left-right axis in the mouse: From origin to morphology. Development 2006; 133:2095–104.

8

Fluid-Biofilm Interactions in Porous Media

George E. Kapellos, Terpsichori S. Alexiou, Stavros Pavlou

University of Patras, Patras, Greece

8.1 MICROBIAL BIOFILMS IN POROUS MEDIA

A biofilm is a multiphase, complex, cellular, biological medium, which is usually found attached to a material interface and consists of a microbial consortium dispersed in a matrix of extracellular polymers (Figure 8.1). The extracellular polymeric substances (EPS) are mainly polysaccharides (e.g., alginates, xanthan, cellulose) and proteins, which are either actively secreted by the microbial cells or accumulate in the extracellular space after cell lysis [3,4]. The primary role of the EPS matrix is to provide cohesiveness between neighboring cells and to maintain the structural integrity of the entire biofilm. Additionally, the matrix might also serve a number of other important functions, including protection from predation, anti-microbials, and desiccation [5]. Multispecies biofilms grown in physical habitats might also contain a significant amount of inorganic and other abiotic colloidal particles which are passively trapped or chemically adsorbed within the EPS matrix [6]. In natural ecosystems, most microbial cells are able to attach, grow, and eventually form biofilms at the interface between

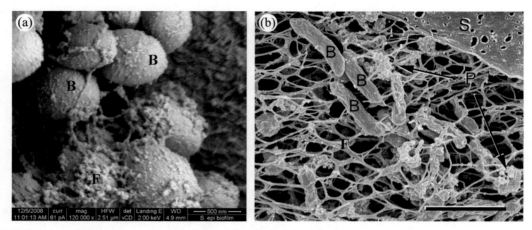

FIGURE 8.1   Indicative scanning electron microscopy (SEM) images of microbial biofilms: (a) *Staphylococcus epidermidis* and (b) *Pseudomonas fluorescens*; scale bar = 2 μm. The symbols on the images denote: B, bacterium; F, extracellular polymers; P, particulate matter; S, substrate. *(a) Reprinted from Williams and Bloebaum [1] with permission from Cambridge University Press and (b) reprinted from Baum et al. [2] under the terms of the Creative Commons Attribution (http://creativecommons.org/licences/by/2.0).*

an aqueous phase and another fluid, solid, or even porous material, under favorable local environmental conditions (Figure 8.2). The ubiquity of biofilms can be attributed to a number of unique survival advantages, such as the EPS support and cell-cell interactions, which are associated with the sessile mode of growth (i.e., attached cells) as compared to the planktonic mode of growth (i.e., floating cells) [7,8].

A porous medium whose solid matrix is composed of pores of characteristic sizes that are larger than several micrometers provides an environment to host biofilm-forming bacteria because of its high specific surface (i.e., the ratio of the matrix wetted surface area to the matrix volume). The process of biofilm formation in porous media is of key importance in several natural phenomena and technological applications, including the biodegradation of organic contaminants by indigenous bacteria in soil and aquifers [9,10], as well as in the marine environment [11]; the construction of permeable reactive barriers for the confinement and attenuation of contaminant plumes [12]; the treatment of water and wastewater in bioreactors [13,14]; the infection of vasculature and porous tissues by pathogenic bacteria (e.g., cystic fibrosis of the lungs by *Pseudomonas aeruginosa* [15,16]), as well as others.

The analysis of biofilm growth in porous media is challenging because the structure of the system exhibits a hierarchy of characteristic length scales that span several orders of magnitude (for example, from several nanometers in the EPS up to a few hundreds of meters at the aquifer scale) (Figure 8.3) and, further, there exists an intricate interplay of hydrodynamic, physicochemical, and biological processes occurring at different characteristic time scales (Figure 8.4). In addition, each structural level might be heterogeneous with respect to geometrical and topological characteristics (e.g., pore and grain size, shape and connectivity), physical properties (e.g., fluid density and viscosity), chemical composition (e.g., mineralogy of the solid matrix), as well as biological composition and activity (e.g., number and physiological state of cells).

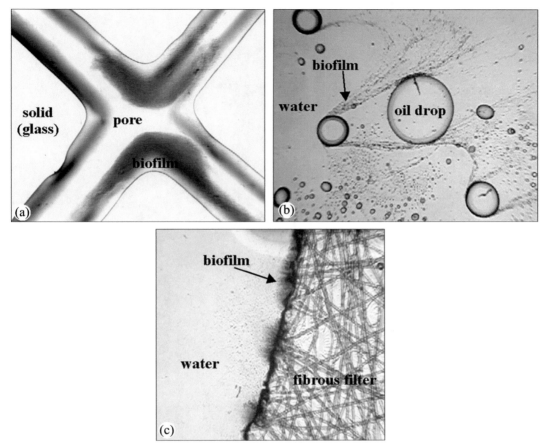

FIGURE 8.2   Microbial biofilms grown on interfaces: (a) solid-water interface; (b) oil-water interface (*n*-heptane oil); (c) interface between an aqueous solution and a fibrous porous material.

The objective of this chapter is twofold; first, to present a comprehensive review on biofilm formation in porous media with emphasis at the pore-scale processes and by including both experimental and modeling aspects (Sections 8.1–8.3) and second, to present a hybrid hierarchical simulator for the prediction of the pattern of evolution and the rate of growth of heterogeneous biofilms within the pore space of porous materials (Section 8.4). The simulator combines continuum-based descriptions of fluid flow and solute transport with particle-based descriptions of biofilm growth and detachment. As a case study, the interactions between fluid flow and growing biofilms are examined in the context of biological clogging of various porous structures (single pore, granular 2D core, consolidated 3D core) and under different flow regimes (constant flow rate versus constant pressure drop) (Section 8.5).

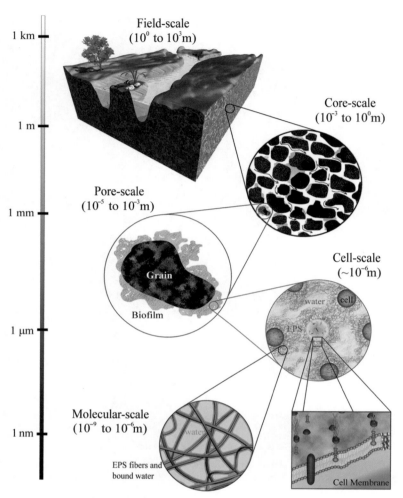

FIGURE 8.3   Biofilms in porous media: hierarchy of characteristic lengths.

8.1.1 Historical Stepping-Stones

According to the scientific literature, the first documented observations of sessile bacteria, those attached to a surface, were made in the early 1930s. The bacteriologist Arthur T. Henrici [17] was interested in isolating and examining under the microscope some of the algae that were growing on the walls of his aquarium. To this end, he immersed glass slides into the aquarium. To his own surprise, he discovered on the glass slides, aside from the algae, a layer of attached bacteria covered by a sheath of slime. Among other early studies, the work of the marine microbiologist Claude E. ZoBell [18] on the bacterial fouling of submerged surfaces in marine environments stands out. ZoBell observed that the majority of marine bacteria are found attached to abiotic or other living surfaces, and he showed that the number of attached

FIGURE 8.4 Biofilms in porous media: hierarchy of characteristic time scales for fluid flow and solute transport at the pore-scale. Here, L_p is the characteristic pore size, U_f is the characteristic velocity of the fluid in the pores, ρ_f is the density of the fluid, μ_f is the viscosity of the fluid, k_b is the hydraulic permeability of the porous biofilm, D_A is the diffusion coefficient of the Ath solute, C_A is the characteristic concentration of the Ath solute, and R_A is the characteristic reaction rate of the Ath solute in the biofilm. The four dimensionless numbers are the Péclet number Pe_A for the Ath solute, the Thiele number Th_A for the Ath solute, the Reynolds number Re, and the Darcy number Da.

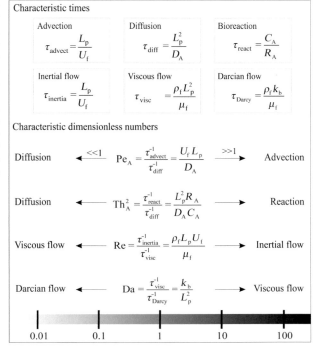

bacteria increases with increasing specific surface area (the so-called volume or bottle effect). He proposed that the affinity of bacterial cells for a surface is enhanced, and presumably stimulated, by prior adsorption of nutritive substances on that surface. ZoBell also observed that the sessile cells were encapsulated in a slimy substance, but the technology required to determine the nature of that substance was not available at that time.

The mystery of the slime that encapsulates sessile bacterial cells was solved three decades later with the use of electron microscopy [19–23]. In particular, Jones et al. [23] obtained excellent TEM microphotographs and showed that the bacterial cells are held together, and also to other surfaces, by a fibrous matrix which is structurally similar to polysaccharides. Since that observation, the research on the formation of microbial slimes gradually expanded toward both biological and physicochemical aspects of the phenomenon. The first mechanistic model for the attachment of a microbial cell on a surface was set forth by Marshall et al. [24] as a two-step process. In a first step, the microbial cell approaches and reversibly attaches to the surface in accordance with the balance of physicochemical forces (electric double layer, London-van der Waals). In a second step, the microbial cell becomes irreversibly attached to the surface via extracellular filamentous macromolecules.

The term *biofilm* first appeared in the 1970s [19,25] and prevailed over other introduced terms (e.g., microbial slime, colony, aggregate, plaque) in spite of some criticism related to whether the attached cells literally form *films* or *colonies*. A first systematic effort at the rigorous mathematical description and quantification of the physical processes involved in biofilm formation was made by the chemical engineer William G. Characklis [6,26–28]. Mass transfer

1930s	Observation of sessile bacteria on surfaces	Henrici (1933)
	The "volume" or "bottle" effect is observed	ZoBell (1943)
1970s	Visualization of the extracellular matrix with TEM	Jones (1969)
	Mechanistic explanation of cell attachment to surfaces	Marshall et al. (1971)
	The term "biofilm" is introduced	Costerton et al. (1978)
	Prevalence of biofilms in natural streams	Geesey et al. (1978)
	Mathematical modeling and quantification	Characklis (1973a,b; 1982);
		McCarty et al. (1981); Wanner & Gujer (1986)
1990s	3D nondestructive visualization with CLSM	Lawrence et al. (1991)
	Observation of water channels	Costerton et al. (1994);
		Stoodley et al. (1994); Massol-Deya et al. (1995)
	Structural, functional, chemical heterogeneity	Zhang & Bishop (1994)
		de Beer et al. (1994); Stewart et al. (1995)
	Cell-to-cell communication	Davies et al. (1998)
	Computational simulation of biofilm growth	Wimpenny & Colasanti (1997)
		Kreft et al. (1998); Picioreanu et al. (1998)
2000	Biofilm phenotypic adaptation	O'Toole & Kolter (1998a,b); Sauer et al. (2002)

FIGURE 8.5   Timeline of selected significant contributions in biofilm research.

limitations were recognized early as essential factors for causing vertical stratification of activity and density in biofilms and these limitations were incorporated into the first biofilm models [29–31].

A landmark in biofilm research is the nondestructive visualization of fully hydrated biofilms with the use of confocal laser scanning microscopy [32], which led to a paradigm shift in the conceptualization of biofilms from layered films of uniform thickness to heterogeneous three-dimensional structures. The combination of confocal microscopy with fluorescent molecular probes [33] and microelectrodes revealed a high degree of heterogeneity with respect to the spatial arrangement and functional activity of microbial cells and EPS [34–39], as well as the existence of intrabiofilm water channels [40–42] in many biofilms. Selected contributions that led to significant advances in biofilm research are summarized in Figure 8.5.

8.1.2 Of Polymers and Cells: Processes Involved in Biofilm Formation

The survival advantage associated with the biofilm mode of growth against the planktonic mode of growth is well-captured in the following remark by Costerton and Lappin-Scott [43]:

> Once bacterial cells had evolved, the planktonic (floating) mode of growth would deliver them from one habitat to another until they perished in the first non-permissive locus. The sessile mode of growth as attached bacteria would allow these primitive organisms to colonize a permissive habitat and persist therein.

The accumulation of microbial cells and the formation of biofilms on material interfaces is the outcome of the complex interaction between several physical, chemical, and biological processes (Figure 8.6) [6,44,45]. A generic description of the main processes that are involved in biofilm development is given in the following paragraphs.

Suspended bacteria usually travel within the pore space under the action of hydrodynamic, gravitational, Brownian, London-van der Waals, and electrostatic forces. Numerous bacteria possess the ability of autonomous, self-propelled motion with the use of filamentous

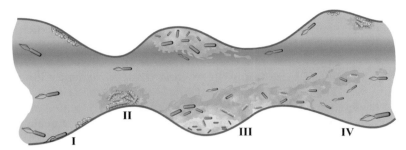

FIGURE 8.6 Main processes involved in the formation of microbial biofilms on a solid-water interface: (I) attachment and adaptation, (II) cell proliferation and synthesis of extracellular polymers, (III) biofilm growth and maturation (mass transfer limitations, quorum sensing, etc.), (IV) detachment of cells and biofilm fragments.

extracellular structures, which extend from the cell body and are referred to as flagella [46]. Among the motile bacteria, many are also able to detect specific chemical substances and set their swimming direction in accordance with the concentration gradient of those substances [47,48]. This phenomenon is known as *chemotaxis*. If the bacterium moves toward the region of low/high concentration of a chemical substance (i.e., down/up the concentration gradient), then we refer to negative/positive chemotaxis, respectively. Detailed review of bacterial transport in porous media is given in Ref. [49].

As they travel, the bacteria approach and interact with the surface of the solid matrix. This interaction depends strongly on physicochemical properties (hydrophobicity, electrokinetic potential) of the bacterial surface [50], the presence of extracellular polymeric filaments extending from the cell membrane [51], and on the physicochemical properties and the architecture of the abiotic surface [52]. Under favorable surface interactions, some of the bacteria temporarily attach to the solid surface. Thereafter, a bacterium might: (a) move away from the surface either by using its flagella or because of fluid lift force, (b) roll over the surface because of fluid drag force, (c) actively glide on the surface using filamentous extracellular structures, such as type IV pili, or other cellular machinery [53,54], or (d) become permanently attached.

In the last case, the attached bacteria adapt to the local environmental conditions and, usually, alter their genotype (expression of genes) and phenotype (size and shape) [54–56]. Moreover, if sufficient concentrations of nutrients, low concentrations of toxins, and low flow shear stress conditions persist, the bacteria grow, proliferate, and assimilate nutrients to synthesize new cellular material and other products. Some of the synthesized macromolecules (mainly polysaccharides and proteins) are excreted to the environment and form the EPS matrix which has been proposed to serve several functions, including cell adhesion, structural integrity, protection (from predators, antimicrobials, and mechanical stresses), and nutrient reserve [5]. Conceptual models have provided theoretical support and quantification for certain functions of the EPS matrix. For example, an extracellular layer of polymers serves as a diffusive barrier by reducing both the concentration and the mobility of solute molecules that approach or leave the cell membrane [57]. Furthermore, for a biofilm that is exposed to fluid flow, a compressible EPS matrix might also serve as a bumper and reduce the magnitude of the overall shear stress (viscous and elastic) acting on the surface of the cells, as compared to the fluid shear stress that would be exerted in its absence [58].

Over time, further growth of the initial colonies results in the formation of continuous or patchy biofilms on the surface. The colony growth is restrained if the nutrients are depleted, or if large flow-related shear stresses develop. The shear stresses can cause biofilm deformation and frequent detachment of biofilm fragments (erosion), or even detachment of the entire biofilm (sloughing) [59,60]. As the biofilm grows and matures, some exciting cell-cell interactions come into play. For instance, several bacterial species secrete chemical signaling molecules (such as acylated homoserine lactones or small peptides), which accumulate in the surrounding environment as the population density increases. If a critical concentration of the chemical signal is reached, the expression or repression of certain sets of genes is triggered. This type of cell-to-cell communication has been termed *quorum sensing* [61,62]. It has been proposed that quorum sensing is also involved in many processes that affect biofilm formation (attachment, detachment, physiological state, production of EPS, etc.) [63].

8.1.3 Morphology of Biofilms in Porous Media

Until recently, the analysis of processes that involved the formation of biofilms was based on an *a priori* hypothesis for the geometry of the biofilm either as a continuous film of uniform thickness or as a discrete colony of prescribed shape [64–67]. However, the nondestructive visualization of irregular biofilm patterns over planar surfaces and simple flow channels [7,40] put under question the validity of this important hypothesis. Of course, the consequences of this hypothesis are escalated for biofilm growth in porous media systems because of the high specific area and the complicated geometry and topology of the fluid-solid or other fluid-fluid interfaces that are colonized by microbial cells. An additional challenge to the research of biofilms is that most natural and synthetic porous media are opaque. To this end, planar pore networks etched in glass [68–70] and flow chambers packed with glass beads [71–73], which permit direct visual observation of the pore- and core-scale phenomena, have served as model porous media in several experimental studies of biofilm growth. In a few studies, methods based on confocal microscopy [74,75] and magnetic resonance [76] have been properly adapted for the investigation of biofilm growth in porous media.

A rich spectrum of biofilm patterns has been observed in porous media depending on the bacterial species, the characteristics of the porous structure, and the prevailing hydrodynamic and nutritional conditions. For example, Figures 8.7 and 8.8 show various forms of *Pseudomonas fluorescens* biofilms that have been observed in microfluidic pore networks [77]. Paulsen et al. [69] replicated the porous structure of sandstone on glass plates by means of photolithography and observed three distinct stages during biofilm growth: (a) initially, smooth biofilms formed on the pore walls, (b) gradually, the biofilms became nonuniform, and (c) finally, biofilm strands spanned through the pores (web-like structure). Both continuous, smooth biofilms and patchy, irregular aggregates were observed by Dupin and McCarty [68] to grow simultaneously within the pores of a square lattice of interconnected pores. A pore network of the chamber-and-throat type etched on glass was used by Vayenas et al. [70] to investigate the biodegradation of a mixture of organic pollutants during biofilm growth. One of the interesting findings of that study was that the average thickness of biofilms showed a positive correlation with the size of the pores, with larger pores hosting thicker biofilms.

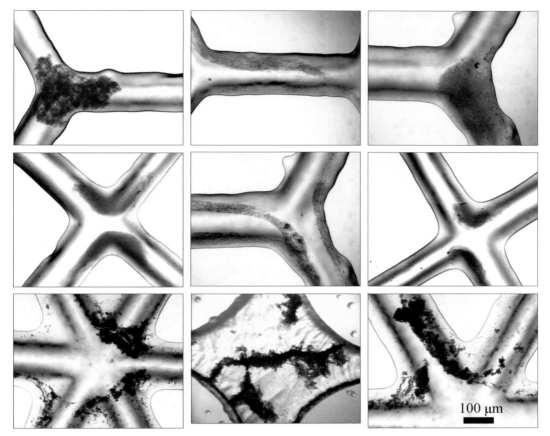

FIGURE 8.7 Morphology of *Pseudomonas fluorescens* biofilms grown in various microfluidic pore networks under constant flow rate. *(Reprinted from Kapellos [77]).*

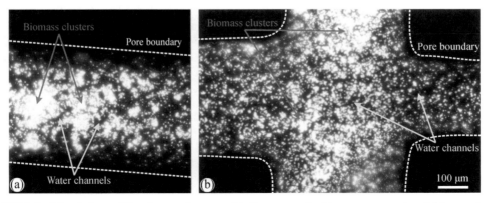

FIGURE 8.8 Morphology of *Pseudomonas fluorescens* biofilms grown in: (a) a pore segment, and (b) a pore junction of a microfluidic pore network under constant pressure drop. *(Reprinted from Kapellos [77]).*

Sharp et al. [72] used a bed of glass beads to investigate the effects of biofilm formation on the flow pattern and the transport of a nonreacting solute. Interestingly, they observed that preferential flowpaths developed during the first stages of biofilm formation and, although they did appear to fluctuate in size and shape, the general flow pattern persisted throughout the experiment. In another study, Stoodley et al. [73] observed biofilms flowing and creating streamers along the direction of local fluid flow within a bed of glass beads. Under a turbulent flow regime, the biofilms were found to concentrate around contact points between neighboring beads, where the magnitude of fluid shear stress is locally minimized.

8.2 A MOTIVATING PROBLEM: BIOFILMS AND THE FATE OF CONTAMINANTS IN SOIL

The ability of certain biofilm-forming microbial species to degrade detrimental chemical species or synthesize useful ones has been driving the interest of scientists and engineers in the study and development of techniques based on microbial biofilms. For example, this ability of microbes is observed during the natural attenuation of pollutants in the subsurface and is exploited during the treatment of contaminated and waste water in porous medium-based bioreactors (e.g., packed and fluidized beds, hollow-fiber membranes). In this section, the fate of organic contaminants in soil is briefly introduced as an example to demonstrate qualitatively the multitude of roles that biofilms might play in a process.

Accidental or deliberate anthropogenic pollution of soil and aquifers often begins with the discharge of hazardous wastes and materials to surface water streams, inappropriate land-fills, and leaking underground storage tanks. Other important sources of soil pollution are fine atmospheric particles (for instance, soot from industrial chimneys), which deposit on top soil, dissolve in rain water, and seep deeper into the ground. Toxic organic compounds that originate from petroleum oil, liquid fuels, organic solvents, lubricants, biocides, insulating or hydraulic fluids, etc., usually leak into the ground in the form of an oily phase, which is also referred to as nonaqueous phase liquid or simply NAPL. Both aqueous and nonaqueous liquids move downward under the action of gravity and migrate faster and in greater depth through networks of interconnected fractures, if present, in the subsurface. Part of the oily phase is trapped in the form of individual ganglia under capillary action within the unsaturated zone. Depending on its density, the rest of the oily phase either accumulates above the water table (light NAPLs, e.g., aliphatic and aromatic hydrocarbons) or penetrates through the aquifer and continues moving underneath it (dense NAPLs, e.g., chlorinated solvents like trichloroethylene). Dissolution, emulsification, and biodegradation of organic compounds occur at the water-NAPL interfaces. Microbial cells and biofilms have been found to affect both the concentration and the mobility of pollutants in the subsurface [9,10,12,78,79].

8.2.1 Impact of Biofilms on Fluid Composition

Growth of microorganisms and synthesis of EPS can be achieved with the availability of energy and carbon sources, the presence of several inorganic elements (N, P, Fe, Ca, etc.), and the prevalence of favorable environmental conditions (temperature, pressure, pH, salinity, and the like). Of particular interest is the case in which indigenous microorganisms assimilate harmful organic compounds in order to fulfill their requirements for carbon and/or energy, thus reducing the concentration of those compounds in their physical habitat. This process is known as natural biodegradation and constitutes a self-cleaning mechanism of the physical environment. Under conditions in which natural biodegradation is effective, the need for human intervention is minimized. For example, it has been reported that natural biodegradation is sufficiently effective in reducing the concentration of several petroleum hydrocarbons and BTEX (benzene, toluene, ethylene, xylenes) in groundwater [79]. Besides biodegradation, biosorption is another effective mechanism of reducing both the concentration and the mobility of harmful chemical substances in soil and groundwater. In particular, the EPS of microbial biofilms has been found to be very efficient in this task [80].

However, natural biodegradation might not be feasible at all or it might take place at a very slow rate under circumstances related to shortage of specific nutrients or the absence of suitable microorganisms. A common situation is the rapid depletion of dissolved oxygen in the saturated zone. In such cases, human intervention for enhancing the biodegradation process is deemed necessary [9]. For instance, the aerobic biodegradation of petroleum hydrocarbons can be enhanced by the enrichment of groundwater with oxygen, which is effected through the supply of either gaseous oxygen (air sparging), or oxygenated water [10]. An interesting example is the enhanced dechlorination of trichloroethene (TCE, 16th in ATSDR list of 2005) in the subsurface of the Dover air force base (Delaware, USA) [81]. In a first step, lactate was supplied from injection wells in order to promote the production of hydrogen from the anaerobic fermentation of lactic acid by indigenous bacteria. Subsequently, hydrogen serves as an electron donor for the reductive dechlorination of TCE to cDCE (*cis*-1,2-dichloroethene), which is a less harmful substance (270th in ATSDR list of 2005). In a second step, cultures of nonindigenous bacteria that had been isolated from another natural habitat and had the ability to fully dechlorinate TCE were injected into the subsurface. After sufficient time (adaptation phase), full dechlorination of TCE to the nonharmful ethylene was reported.

8.2.2 Impact of Biofilms on Fluid Mobility

Microbial cells and biofilms can also strongly affect the mobility of aqueous and nonaqueous liquid contaminants in soil. First of all, the formation of cell aggregates and biofilms alters the geometric and topological characteristics of the pore structure and this in turn modifies the local flow field. For example, the extensive clogging of large pores and fractures results in flow diversion to smaller pores and the creation of preferential flow paths [12]. Furthermore, the biosynthesis of surface active substances (lipopeptides, rhamnolipids, fatty acids, etc.), which decrease the surface tension of oil-water interfaces, promotes the emulsification and solubilization of the oily phase [82]. In this way, the bioaccessibility and potential biodegradation of organic contaminants is augmented. There are several other activities of specific microorganisms that might affect the motion and saturation of fluids in the subsurface, including the biologically mediated formation of gas bubbles, precipitation of salts, and corrosion of the solid matrix [83,84]. For example, some soil bacteria can excrete low-molecular fatty acids which dissolve carbonate precipitates from the solid matrix and enlarge the pores, especially in calcareous rock formations.

8.3 MODELS OF BIOFILM GROWTH AND PATTERN FORMATION IN QUIESCENT FLUIDS

Over the last two decades, the conceptual model of biofilms evolved from a homogeneous, bioactive, layered film to a dynamically evolving, highly heterogeneous, three-dimensional structure. This conceptual leap was attained through experimental observations of several fascinating features including, among others, the formation of irregular morphological patterns like mushrooms and streamers, the existence of intrabiofilm water channels, the heterogeneity with respect to the spatial distribution and functional activity of cells and EPS, and the demonstration of societal consciousness by microbial cells. The concurrent progress in computer hardware and numerical methods enabled the development and implementation of advanced mathematical models and, mainly, computational simulators for the description of biofilm growth and pattern formation.

In this section, the major approaches for modeling and simulating biofilm formation are concisely presented and discussed. These approaches are based upon the assumption that growth forces are more important than any other physical forces and, therefore, the surrounding fluid is essentially considered to be quiescent. Methods of accounting for various interactions between growing biofilms and flowing fluids are presented later in Section 8.4. Depending on the adopted hypothesis for the material nature of the biofilm, mathematical models of biofilm formation can be distinguished in two fundamentally different categories: (a) continuum-based and (b) discrete-based models (Figure 8.9).

8.3.1 Continuum-Based Models

Theoretical models, which represent a cellular biological medium as a continuum body, are formulated in terms of spatial average quantities (e.g., cell volume fraction) that are assumed to be continuous functions of space and time. Features of the system at finer spatial scales can be taken into account only implicitly and, in principle, this approach is reasonable for systems with large populations of cells. The process of growth is mathematically described using standard forms of mass and momentum balances for either monophasic or multiphasic (mixture-type) materials. For instance, if the biofilm is considered as a mixture of n components (e.g., various cell species, different types of EPS, and free-fluid) then the volume balance for the biofilm is:

$$\sum_{i=1}^{n} V_i(t) = V_B(t) \tag{8.1}$$

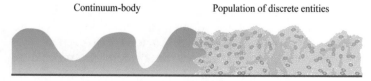

Continuum-body   Population of discrete entities

FIGURE 8.9   Schematic illustration of the continuum-versus-discrete description of biofilms.

where V_i is the volume of the ith component within the biofilm and V_B is the total volume of biofilm. The volume balance must be satisfied at any instant in time t. The volume fraction of the ith component in the biofilm is defined as:

$$\varphi_i = \frac{V_i}{V_B} \qquad (8.2)$$

From the volume balance given in Equation (8.1), the following condition is obtained for the volume fractions:

$$\sum_{i=1}^{n} \varphi_i(t) = 1 \qquad (8.3)$$

The mass and momentum balances for the ith component can be expressed as follows [85,86]:

$$\frac{\partial \varphi_i \rho_i}{\partial t} + \nabla \cdot (\varphi_i \rho_i \mathbf{v}_i) = q_i \qquad (8.4)$$

$$0 = \nabla \cdot (\varphi_i \boldsymbol{\sigma}_i) + \mathbf{F}_{m \rightarrow i} \qquad (8.5)$$

respectively. Here, ρ_i is the density, $\mathbf{v}_i$ is the velocity, q_i is the rate of mass production or consumption per unit volume, $\boldsymbol{\sigma}_i$ is the Cauchy stress tensor, and $\mathbf{F}_{m \rightarrow i}$ is the interaction force between the ith component and all the other components of the mixture. During biofilm growth, inertial forces are considered to be much less important than viscous-elastic and interaction forces and, thus, are not included in the momentum balance Equation (8.5). The definition of a growth-associated stress tensor in Equation (8.5) is an issue of primary difficulty, and several different approaches have been proposed to either confront or circumvent this issue in the biofilm literature.

In a first approach, the volumetric expansion or shrinkage of the biofilm is treated as a one-dimensional process with the direction of growth prescribed as normal to the substrate. This assumption supersedes the momentum balance. Furthermore, upon considering that the density of each component is constant, the velocity of the fluid-biofilm interface, v_{fbi}, can be obtained by integrating the mass balance equation for the mixture [31]:

$$v_{fbi}(t) = \int_0^{L(t)} \left(\sum_{i=1}^{n} \frac{q_i(z)}{\rho_i} \right) dz \qquad (8.6)$$

where L is the thickness of the biofilm at time instant t. In a second approach, biofilm growth is modeled as a multidimensional process, but the momentum balance is circumvented again by imposing a constitutive relation for the velocity of each component or of the entire biofilm. For example, Eberl et al. [87] considered biofilm growth as analogous to solute diffusion and introduced an expression of the following form:

$$\mathbf{v} = -D_B(c_b) \nabla c_b \qquad (8.7)$$

where $c_b = \sum_{i=1}^{n} \varphi_i \rho_i$ is the apparent density of the biofilm, and D_B is a density-dependent diffusion coefficient. In another study, Dockery and Klapper [88] considered biofilm growth as analogous to viscous flow through a porous medium and related the velocity of the biofilm to a biomass pseudopressure, P_B, as follows:

$$\mathbf{v} = -\lambda \nabla P_B \qquad (8.8)$$

where λ is a so-called mobility coefficient that might also be viewed as the reciprocal of the viscosity of the biofilm. In both of these approaches, it is assumed that all the components move with the same local velocity. In a third approach, Cogan and Keener [89] considered that the biofilm behaves as a highly viscous two-component fluid ($n=2$), and its expansion or shrinkage can be regarded as analogous to osmotic flow. They introduced expressions of the following form for the stress tensors and the interaction force:

$$\boldsymbol{\sigma}_w = -P_w \mathbf{I} + \mu_w \left[\nabla \mathbf{v}_w + (\nabla \mathbf{v}_w)^T \right] \qquad (8.9)$$

$$\boldsymbol{\sigma}_f = -P_f(\varphi_f)\mathbf{I} + \mu_f \left[\nabla \mathbf{v}_f + (\nabla \mathbf{v}_f)^T \right] \qquad (8.10)$$

$$\mathbf{F}_{f \rightarrow w} = P_{wf} \nabla \varphi_w - \mathbf{R} \cdot (\mathbf{v}_w - \mathbf{v}_f) \qquad (8.11)$$

Here, the subscript w denotes the free-fluid within the biofilm and the subscript f denotes the matrix of EPS along with embedded cells. Furthermore, μ_i is the viscosity of the ith component, P_w is the pressure of the free-fluid, P_f is the osmotic pressure of the matrix (function of the volume fraction), P_{wf} is an interphase pressure, and $\mathbf{R}$ is a tensor associated with the resistance in the relative motion between the two components. Further details are available in Refs. [85,89].

8.3.2 Discrete-Based Models

In computer simulations, which represent the cellular biological medium as a population of discrete entities, the macroscopically observed form and function emerge from the interactions between entities and their environment. An entity might represent a single microbial cell or a coarse-grained particle made up of several cells and, perhaps, also EPS. Typically, each entity is assigned three vectors: position, velocity, and a state vector that contains information about other properties, such as volume, age, and metabolic activity. The behavior of each entity is defined by a set of locally applied rules that are based on information about the entity and its neighborhood. The rules might be deterministic or stochastic and are defined on the basis of:

- Fundamental physical laws (e.g., mass balance for the cell)
- Experimental observations from the biological system under consideration (e.g., if the volume of a cell reaches a critical value, then the cell is divided into two daughter cells)
- Work hypotheses based on experience from other multiparticle physical systems (e.g., cells push their neighbors along a path of least resistance).

There are two general methods of discrete-based modeling: (1) the method of cellular automata, and (2) the method of individual-based modeling (also referred to as agent- or particle-based). Cellular automata are defined as rule-based dynamic systems, which are discrete in both space and time. The computational domain is discretized with a simple or complex grid and every grid-cell might be empty or contain one or more entities (cells and/or EPS particles). The exact position of an entity in the grid-cell is not defined. In individual-based models, each entity moves in space (continuous or discrete) and interacts with other entities and the environment in pursuit of specific objectives (e.g., maximization of nutrient availability, minimization of shear stress).

The first computer-aided simulator of biofilm growth based on cellular automata was developed by Wimpenny and Colasanti [90]. The two-dimensional space was discretized with a square lattice and each lattice-cell could be either empty or occupied by a microbial cell. The nutrients were modeled as small particles performing random walks on the cellular automaton lattice. Microbial cells were considered to take up nutritive particles from their local environment and were allowed to divide only if an adjacent lattice-cell was empty. The strong effect of the nutrient availability on biofilm structure was shown in Ref. [90]. High nutrient concentration resulted in compact biofilms, while low nutrient concentration resulted in porous biofilms with dendritic structure. A few years later, Chang et al. [91] developed a cellular-automaton simulator that can be viewed as a significant extension of that of Wimpenny and Colasanti [90]. The three-dimensional space was discretized with a cubic lattice and divisions of microbial cells resulted in cell-by-cell displacements along a path of least resistance. The production of a growth inhibitor by the microbial cells (antagonistic behavior) results also in biofilms with heterogeneous porous structure [91].

The first individual-based (off-lattice) simulator was developed by Kreft et al. [92] and each microbial cell was represented by a sphere in a planar computational domain. Binary fission was modeled by the replacement of a spherical mother cell with two spherical daughter cells of equal total volume. Typically, this event created an overlap between neighboring spheres. Overlapping volumes resulted in repulsive forces and local rearrangement of the spheres so as to diminish the overlap volume. At the same time, Picioreanu et al. [93] developed the first hybrid simulator by combining coarse-grained cellular-automata for biofilm growth with a continuum-based model for the diffusion and consumption of nutrients. Each lattice-cell could be either empty or contain biomass (microbial cells and EPS). Volumetric growth was modeled with cell-by-cell displacements of biomass from an overflowing lattice-cell toward the nearest empty cell along a random path. Both studies in Refs. [92,93] confirmed that nutrient limitation results in irregular biofilm patterns. Furthermore, comparative tests between those two simulators showed that there is satisfactory agreement for the prediction of the total biomass with some (expected) differences in the detailed biofilm structures [94].

Hermanowicz [95] argued that cell-by-cell displacements are random during cell proliferation and introduced the concept of least mechanical resistance in order to define the path along which biomass is displaced from an overflowing lattice-cell to a near empty lattice-cell. He also employed a heuristic approach in order to account for shear-induced detachment of biomass and showed that increased shear results in more compact biofilms. The distinction between active and inert biomass in coarse-grained cellular automata was introduced by Laspidou and Rittmann [96] who showed that long-term consolidation of inert biomass results in stratification within biofilms.

8.4 COMPUTATIONAL SIMULATION OF FLUID-BIOFILM INTERACTIONS IN POROUS MEDIA

The development of physically based, mechanistic models for fluid-biofilm interactions represents a formidable challenge for a number of reasons. First and foremost, the emergent mechanical behavior of a biofilm might range from fluid-like to solid-like, depending on the composition and the detailed structural arrangement of cells, EPS, and free/bound water at finer spatial scales. As a consequence, there exists no unique, universally valid, simple constitutive model for the description of biofilm mechanics at a macroscopic scale of observation (i.e., B-scale referring to Figure 8.3). Among various models that have been proposed and implemented in the relevant literature, the biofilm has been considered as highly viscous Newtonian fluid, viscoelastic fluid, elastic solid, or biphasic poroelastic material with regard to its mechanical response under the action of external (non-growth-associated) physical stresses [86]. Furthermore, the experimentally reported flow-induced modulation of various cellular functions suggests a strong coupling between hydrodynamic and biological processes during biofilm formation. This coupling can be relaxed to some extent on the basis of the large disparity in characteristic times associated with viscous flow and biofilm growth. Finally, the mechanisms of flow-induced fracture and detachment in biofilms pose significant additional degrees of complexity to the process and of difficulty to the modelers.

Significant advances on modeling fluid-biofilm interactions have been achieved in the context of computer-aided simulators that were developed to investigate the mechanisms of biological clogging in porous media, at the core scale (a few to hundreds of pores). The first notable works used pore-network simulation [65,67,97,98], while some more recent efforts rely on direct numerical simulation in realistic porous structures [99–102]. In the context of the pore-network simulation, the porous structure is represented by a network of interconnected pores. The process of interest is first analyzed for a representative pore with simplified geometric features, and then suitable considerations are made to ensure that the overall mass, energy, and momentum are conserved at nodal points of the network where two or more pores join together. Usually, the representative pore is a cylindrical tube that in some cases (but not always) includes constriction. For instance, Dupin et al. [65,97] developed a remarkable pore-network simulator to investigate the impact of biofilm morphology (shape, size) on flow through porous media. The biofilm was treated as a hyperelastic solid with regard to the growth-induced expansion and deformation. In particular, the proliferation of microbial cells and the production of EPS resulted in the development of internal stresses which deformed the biofilm until a new mechanical equilibrium state was reached. It was shown that the shape of the biofilm strongly affected the permeability-porosity correlation during bioclogging, with patchy aggregates causing a larger reduction in the permeability than continuous films of uniform thickness.

In the following discussion, we present the improved version of our hybrid hierarchical simulator for the prediction of the pattern of evolution and the rate of growth of heterogeneous biofilms within the pore space of porous materials [77,99]. This was the first hybrid simulator for biofilm formation in porous structures that accounted *explicitly* for fluid-biofilm interactions via appropriate mechanistic models of flow-induced detachment-reattachment, flow-modulated growth, and flow through porous biofilms.

From a programmer's perspective, the simulator has three important features: it is hierarchical, hybrid, and modular. First, the simulator is characterized as hierarchical because it provides predictions about the behavior of the system at the core-scale (O[cm]) by simulating directly the main processes at the pore- and cell-scales (O[μm]) and by taking into account implicitly the properties of the system at the EPS-scale (O[nm]). Second, the simulator is characterized as hybrid because it combines continuum-based descriptions of fluid flow and solute transport with particle-based descriptions of biofilm growth and detachment. Finally, the simulator is characterized as *modular* because it consists of a set of interacting, yet independent, modules each of which is associated with a specific process or structure. As scientific research on biofilms progresses, each module can be updated or replaced accordingly. On the basis of these features, we refer to the simulator with the acronym HiBioSim-PM (Hierarchical Biofilm Simulator in Porous Materials). The simulation starts with the generation of the porous structure and the inoculation with bacterial cells (the first "colonists"). Thereafter, the following main processes are modeled:

- Cell proliferation and EPS spreading
- Fluid flow and flow-related biofilm stresses
- Detachment, migration, reattachment of cells and EPS
- Solute transport.

A concise description of the main modules of the simulator is given in the following paragraphs, and more details can be found in Refs. [77,99].

8.4.1 Generation and Initial Colonization of the Porous Structure

Any type of computer-generated virtual structure or digitized representations of real porous media (for instance obtained with computed microtomography) can be used in the simulator. The structure is divided into voxels with prespecified resolution and each voxel is represented by a binary variable that takes the value zero for solid and one for bulk fluid. The sites of initial settlement by microbes are determined as follows. First, the flow field within the porous space is determined (Section 8.4.4). Then, a single microbial cell is inserted randomly at the inflow boundary of the virtual porous medium and moves along the streamlines (Section 8.4.5) until it is captured at a grain surface or exits the system. This procedure is repeated until the number of initially attached microbial cells (first colonists) equals a prescribed value.

8.4.2 Cell Proliferation and EPS Spreading

With regard to the biological processes, the biofilm is treated as a population of interacting individual microbial cells, which consume nutrients, grow, proliferate, and synthesize EPS. The growth and proliferation of microbial cells within the biofilm is modeled using a 3D cubic lattice of unit biomass compartments (UBCs). A UBC might contain a single microbial cell (single occupancy UBC) or more than one microbial cells of the same species (multiple occupancy UBC). A state vector is assigned to each UBC. The state vector contains information about the species (for mixed biofilms), the size, and the physiological status (active, dormant, or apoptotic) of the microbial cell that occupies the UCB, as well as the volume fraction and the intrinsic porosity of the EPS matrix. The active microbial cells assimilate nutrients and synthesize new cellular mass at a rate proportional to their mass. Part of the cellular mass is used for maintenance purposes (endogenous metabolism). Simultaneously, they synthesize and secrete EPS within their UBC at a rate proportional to their growth rate. Part of the EPS matrix lyses (maybe due to enzymatic or hydrolytic action) at a rate proportional to its mass. Based on these assumptions, the mass balances for bacterial cells and EPS within a UBC are

$$\frac{dX_\kappa}{dt} = \mu_{\kappa,g} X_\kappa - \mu_{\kappa,m} X_\kappa \tag{8.12}$$

$$\frac{dX_\pi}{dt} = Y_{\pi/\kappa} \mu_{\kappa,g} X_\kappa - k_{lys} X_\pi \tag{8.13}$$

where X_κ is the mass of cells and X_π is the mass of EPS contained in the UBC volume, and $\mu_{\kappa,g}$, $\mu_{\kappa,m}$, $Y_{\pi/\kappa}$, k_{lys} are kinetic parameters which might be defined as functions of the local environmental conditions (nutrient concentrations, mechanical stresses, temperature, pH, etc.). For the simulation results presented in this chapter, the following expression has been used for the local specific growth rate of cells:

$$\mu_{\kappa,g} = F_n(C_n) F_\sigma(I_\beta) \tag{8.14}$$

$$F_n(C_n) = \mu_{max} \frac{C_n}{K_C + C_n} \tag{8.15}$$

$$F_\sigma(I_\beta) = \begin{cases} 1 - I_\beta/I_{\beta,crit} & \text{if } I_\beta \leq I_{\beta,crit} \\ 0 & \text{otherwise} \end{cases} \tag{8.16}$$

Here, C_n is the local concentration of a growth-limiting nutrient, $I_\beta = \mathrm{tr}\boldsymbol{\sigma}_s$ is the first invariant of the local stress tensor for the solid components of the biofilm, and μ_{max}, K_C, $I_{\beta,crit}$ are kinetic parameters.

The fate of microbial cells is based on the following conceptual model of cell life-cycle. If the specific growth rate is greater than a critical value, $\mu_{\kappa,g}^{crit}$, then an active bacterial cell continuously increases its mass until a prescribed upper threshold value, $m_{\kappa,crit}^{+}$, is reached. Then the active cell divides into two equal daughter cells (the number of cells within the UBC doubles and the mass of each cell halves). If the specific growth rate becomes lower than the critical value, the cell enters the dormancy state during which the metabolic activity is halted and only consumption of cellular mass for maintenance purposes takes place. Dormant cells may be reactivated if the specific growth rate is restored to a value greater than the critical. During dormancy, cellular mass decreases continuously until a prescribed lower threshold value, $m_{\kappa,crit}^{-}$, is reached. Then the cells enter the apoptosis state (programmed cell death), which is irreversible and lysis of the cells occurs with probability

$$p_{ap}(\tau_{ap}) = 1 - \exp(-k_{ap}\tau_{ap}) \tag{8.17}$$

where τ_{ap} is the time since the cells became apoptotic and k_{ap} is the apoptosis rate constant.

The microbial cells and EPS particles move within the biofilm because of the internal stresses that develop during the division of cells and excretion of EPS. In this work, the cellular motion that is caused by cell divisions is modeled on the 3D cubic lattice of UBCs, by implementing a least-action principle. In particular, if the mass within a UBC exceeds a prescribed maximum value, then biomass is displaced from the overflowing-UBC toward the nearest empty-UBC along the path that minimizes the energy of displacement. First, a random walk procedure is used to generate a large number of paths that connect the overflowing-UBC and the nearest empty-UBC, without passing over solid obstacles. The shortest path is chosen. If there is more than one path with the same minimum length, then the path that corresponds to the minimum EPS content is chosen. Then, starting with a randomly selected fraction (40-60%) of biomass from the overflowing-UBC, a sequence of successive displacements of biomass is performed between neighboring UBCs along the chosen path.

8.4.3 Fluid-Flow and Flow-Induced Biofilm Stresses

With regard to momentum and mass transport processes, the biofilm is treated as a biphasic continuum with poroelastic material behavior. A single-domain approach is used to describe the flow of an incompressible Newtonian fluid (say a dilute aqueous solution) within the pore space of the porous medium, which is occupied partly by free-fluid and partly by biofilms. Within the fluid regions, the Navier-Stokes equations and the continuity equation result from the formulation of the momentum and mass balance, respectively. Within the permeable biofilm, Brinkman's extension of Darcy's law is considered as an appropriate equation to describe the flow along with the conservation of total fluid mass. Further, linear elastic behavior is considered for the solid components of the biofilm [103]. The final equations are:

$$\nabla \cdot \mathbf{v}_f = 0 \qquad (8.18)$$

$$\rho_f \frac{\partial \mathbf{v}_f}{\partial t} + \alpha_c \rho_f \nabla \cdot \left(\mathbf{v}_f \mathbf{v}_f / \varepsilon_\beta\right) = -\varepsilon_\beta \nabla P_f + \mu_f \nabla^2 \mathbf{v}_f - (1 - \alpha_c) \varepsilon_\beta \frac{\mu_f}{k_\beta} \mathbf{v}_f \qquad (8.19)$$

$$0 = \nabla \cdot \boldsymbol{\sigma}_s + \mathbf{F}_{f \to s} \qquad (8.20)$$

$$\boldsymbol{\sigma}_s = -\left(1 - \varepsilon_\beta\right) P_f + \lambda_s (\nabla \cdot \mathbf{u}_s) + \mu_s \left[\nabla \mathbf{u}_s + (\nabla \mathbf{u}_s)^{\mathrm{T}}\right] \qquad (8.21)$$

$$\mathbf{F}_{f \to s} = \varepsilon_\beta \frac{\mu_f}{k_\beta} \mathbf{v}_f \qquad (8.22)$$

Here, $\mathbf{v}_f$ is the local superficial velocity of the fluid, P_f is the intrinsic pressure of the fluid, $\mathbf{u}_s$ is the local displacement of the solids in the biofilm, λ_s, μ_s are the Lamé parameters for the solid, μ_f is the fluid viscosity, ρ_f is the fluid density, ε_β is the local volume fraction of fluid, k_β is the local hydraulic permeability (defined only within the regions of porous biofilms), and α_c is a computational parameter that equals unity within regions of fluid and zero within biofilms. The local hydraulic permeability of the biofilm is calculated as a function of the volume fractions of cells, EPS, and water; the average diameter of cells; the average diameter of EPS fibers; and the internal porosity of the EPS [77].

Equations (8.18) and (8.19) are valid everywhere in the pore space (fluid and biofilm regions), while Equation (8.20) applies only in the biofilm regions. These equations are solved numerically by combining finite differences and finite element methods as follows. First, Equations (8.18) and (8.19) are solved numerically using a staggered grid for the spatial discretization, central finite differences for the viscous and pressure terms, and a higher-order upwinding scheme for the inertial terms. Then, Equation (8.20) is solved using the Galerkin finite element method on a structured mesh of hexahedral elements with C0-quadratic interpolation functions for the displacement.

8.4.4 Detachment, Migration, and Reattachment of cells and EPS

For each UBC that is adjacent to free-fluid, the average shear stress acting on the surfaces exposed to fluid is calculated. If the exerted shear stress exceeds a designated critical value, then the UBC is considered to lose the cohesiveness with adjacent UBCs or solid surfaces. Afterward, the UBC begins to move along the fluid streamlines as if a fluid element (as a first approximation, the effects of gravity and drag forces are neglected based on the facts that biofilm is highly porous and its density is very close to that of the aqueous solution). The trajectory of the UBC within the pore space is calculated from the numerical integration of:

$$\frac{d\mathbf{r}_p}{dt} = \mathbf{v}_f \tag{8.23}$$

where $\mathbf{r}_p$ is the position of the mass center of the UBC at time t. The UBC stops moving if it passes over the outflow boundary of the porous medium or if it becomes reattached to a grain or biofilm surface, which is exposed to shear stress lower than the critical value. If at least one UBC has been detached, the flow field is updated.

8.4.5 Solute Transport

The fate of a dissolved substance, denoted by "A," (nutrient, chemical signal, etc.) in the pore space is determined from the advection-diffusion-reaction equation:

$$\frac{\partial}{\partial t}\left(K_{A,\beta/f}C_A\right) + \nabla\cdot(\mathbf{v}_f C_A) = \nabla\cdot[\mathbf{D}_{A,\,\text{eff}}\cdot\nabla C_A] + R_A \tag{8.24}$$

where C_A is the concentration, $K_{A,\beta/f}$ is the partition coefficient between the biofilm and the aqueous solution, $\mathbf{D}_{A,\text{eff}} = D_{A,\text{eff}}\mathbf{I}$ is the effective diffusivity tensor, and R_A is the local reaction rate of the dissolved substance A. The diffusion coefficient in the biofilm is calculated using the following, recently developed, correlation [57]:

$$\frac{D_{A,\,\text{eff}}}{D_{Av}} = \frac{(1-\delta)\lambda_\kappa \Phi_{11} + \left(\xi_\mu + 1\right)\lambda_\pi \Phi_{12}}{(1-\delta)\lambda_\kappa \Phi_{21} + \left(\xi_\mu + 1\right)\lambda_\pi \Phi_{22}} \tag{8.25}$$

with:

$$\Phi_{11} = 2\varphi_v\varphi_\pi + 3(1-\varphi_v)[\varphi_\pi + 3(1-\delta)\varphi_\kappa]\lambda_\pi + \varphi_v(\varphi_\pi + 3\varphi_\kappa)\lambda_\pi \tag{8.26a}$$

$$\Phi_{12} = 2\varphi_v(2\varphi_\pi + 3\varphi_\kappa) + 2\varphi_\pi(3 - 2\varphi_v)\lambda_\pi \tag{8.26b}$$

$$\Phi_{21} = \varphi_\pi(3 - \varphi_\upsilon) + \varphi_\upsilon(\varphi_\pi + 3\varphi_\kappa)\lambda_\pi \tag{8.26c}$$

$$\Phi_{22} = (3 - \varphi_\upsilon)(2\varphi_\pi + 3\varphi_\kappa) + 2\varphi_\upsilon\varphi_\pi\lambda_\pi \tag{8.26d}$$

$$\lambda_\kappa = K_{A,\pi/\upsilon}K_{A,\kappa/\pi}D_{A\kappa}/D_{A\upsilon}, \lambda_\pi = D_{A\pi}/D_{A\upsilon}, \xi_\mu = \frac{K_{A,\kappa/\pi}D_{A\kappa}}{r_\kappa J_{A\mu}} \tag{8.26e}$$

Here, $D_{A,eff}$ is the effective diffusion coefficient in the biofilm, $D_{A\upsilon}$ is the diffusion coefficient in large interstitial pores filled with free-fluid, $D_{A\pi}$ is the effective diffusion coefficient in the EPS matrix, $D_{A\kappa}$ is the diffusion coefficient in the cellular phase, φ_α is the volume fraction of the αth phase ($\alpha = \kappa, \pi, \upsilon$), $K_{A,\alpha/\omega}$ is the equilibrium partition coefficient between the phases α and ω, $J_{A\mu}$ is the membrane permeability, $\delta = \delta_\mu/r_\kappa$ is the dimensionless membrane thickness, and r_κ is the cell size.

Equation (8.24) is solved using a fractional step method, in which the solution procedure is split up into independent steps corresponding to the advection, diffusion, and reaction processes and each step is solved independently. An explicit in time, higher-order upwinding scheme is used for the advective terms, implicit central differencing is used for the diffusive terms, and the explicit fourth-order Runge-Kutta method is used for the reaction terms.

8.5 MECHANISMS OF BIOLOGICAL CLOGGING IN POROUS MEDIA

Improved understanding and quantification of fluid-biofilm interactions are of essential importance in unraveling the mechanisms of clogging during biofilm growth in porous materials. In this section, after a very concise exposition of the current knowledge on biological clogging, the impact of two factors, namely the applied flow regime and the detachment of biofilms, is investigated with our computer simulator HiBioSim-PM. The term *biological clogging* was conceived in order to describe the reduction in the saturation (porosity) and permeability of water in a porous medium that is caused by *any* activity of microorganisms (e.g., production of gas components, precipitation of insoluble salts, synthesis of EPS) [104–106]. For instance, the formation of biofilms might significantly alter the geometric and topological characteristics of a porous structure. As a consequence, the values of various transport coefficients, including the hydraulic permeability, the coefficients of diffusion, dispersion, mass transfer, adsorption, etc., will also be affected to a certain extent. In particular, it has long been reported that the hydraulic permeability of a porous material can be reduced up to three or four orders of magnitude because of biofilm formation. The observed rate and degree of the permeability reduction have been found to show significant dependence on multiple variables associated with characteristics of the microorganisms, the porous structure, the hydrodynamic and nutritive conditions, and the chemical composition of the aqueous and solid phases, among others [83]. To this end, the analysis of the underlying mechanisms of biological clogging remains active and interesting.

8.5.1 Experimental Observations and Conceptual Models

The variation of permeability with respect to time, which is caused by the proliferation and accumulation of microorganisms in a porous material, can be qualitatively described by the curve shown in Figure 8.10. For an initial period of time, the permeability barely varies, if at all

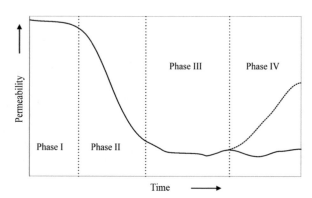

FIGURE 8.10   Anticipated impact of biofilm formation on the permeability of a porous medium.

(phase I). During this phase, the microorganisms that are found attached on the solid-fluid interface are either isolated or form small colonies. The duration of this phase is strongly dependent on the initial concentration of microorganisms and the availability of nutrients. Subsequently, a rapid decrease of the permeability usually occurs (phase II) and is followed by a gradual stabilization with occasional fluctuations around an average value (phase III). The rapid decrease of the permeability observed in phase II can be attributed to: (1) the production of EPS and the formation of biofilm or aggregates that occlude mainly the entrance of the porous material [106–108], (2) the formation of air bubbles that is caused by the oversaturation of the aqueous solution with respect to gaseous products of the biological activity of the microorganisms, such as nitrogen [109] and methane [110], (3) the precipitation of insoluble salts, such as sulfuric iron that is formed by microorganisms utilizing sulfuric anion as electron acceptor [83], and finally, (4) some other factor or any combination of the previous factors.

Various hypotheses have been suggested with regard to phase III, but none seems to be valid under all circumstances. This phenomenon (i.e., the stabilization of the permeability) was first observed by Frankenberger et al. [111] and later confirmed by several other researchers [71,104,106,112]. Taylor and Jaffé [104] put forth two alternative hypotheses for the interpretation of the observed behavior during phase III. According to a first hypothesis, which is referred to as the "closed pore model," the pores become completely clogged by porous, permeable biofilms. In this case, the minimum value of the permeability of the porous material corresponds to the intrinsic permeability of the biofilm. According to a second hypothesis, which is referred to as the "open pore model," the shear stress exerted by the fluid on the biofilm causes continuous detachment of biofilm fragments, and thus, the pores never become completely clogged. In this case, the minimum value of the permeability of the porous material is determined by the balance between growth and detachment. Taylor and Jaffé [104] leaned toward the open pore model as a more satisfactory explanation of their experimental results. Finally, it is possible that partial or complete restoration of the permeability might occur (phase IV), if: (1) the growth-limiting nutrients are depleted, (2) growth-inhibitors or biomass-oxidizers are supplied, or (3) the pressure-drop or flow-rate is drastically increased so as to detach the biofilm. It is worth mentioning that if the formed biofilms contain a large amount of EPS, then the depletion of nutrients or the administration of antimicrobial agents might cease cell growth without restoring the permeability. In that case, agents that promote enzymatic of chemical lysis and dissolution of the EPS matrix would be required.

8.5.2 Single-Pore Simulations Under Two Different Flow Regimes

The applied flow regime has been an overlooked factor in the analysis of biological clogging in porous media. In many experimental setups, the flow is driven by a pump that maintains constant the average volumetric flow-rate through the system [104,106,112]. Typically, the instantaneous flow-rate exhibits fluctuations with amplitude and intensity that depend on the exact features of the pump and tubing being used. The device-related flow pulsation is considered as an undesirable side-effect and, in a few cases, has been attenuated with the use of special flow dampeners. There also exist several studies where the flow was driven by appropriate setups that maintain constant the pressure-drop across the system [71,107,108]. This flow regime is considered to be closer to the natural flow through soil. However, there is an unwanted effect associated with such systems as well: the contamination of the inflow tubing and nutrient vessel by bacteria. The use of suitable filters can limit bacterial spreading, but the pressure drop across the filter must also be taken into account in the design and analysis of the experiments. After some time, the surface of the filter would eventually suffer from bacterial fouling and the filter along with any contaminated tubing should be replaced, if possible without disrupting the processes occurring within the porous medium. It is because of this troublesome technical issue that the use of experimental setups with constant pressure-drop is not widespread for the study of bioclogging.

Here, we present indicative results from computer simulations with regard to the impact of the applied flow regime on biofilm formation in a single straight pore. The values of the physicochemical and biological parameters are the same with those used in Ref. [99]. The entire surface of the pore is initially inoculated with microbial cells in order to exclude any effects associated with random inoculation. In addition, the dimensions of the channel along with the nutrient concentrations at the inflow boundary are selected so as to minimize mass transfer limitations. Therefore, in the following two scenarios, the key factors that determine the pattern of the biofilms within the pore are: unlimited growth and flow-induced detachment. Figure 8.11 shows snapshots of the spatiotemporal evolution of biofilms within a straight pore under *constant flow-rate* and with a uniform velocity profile at the entrance of the pore. In this scenario, the biofilm grows as a uniform layer on the pore surfaces. The thickness of the biofilm gradually increases up to a specific value, which signifies a balance between growth and detachment. The mechanism underlying this scenario is explained as follows. Biofilm growth results in reduction of the effective cross-section that is available to the flowing bulk fluid within the pore. Under a constant flow-rate, both the local velocity of the bulk fluid and the shear stress at the fluid-biofilm interface increase with decreasing cross-section. If the applied shear stress exceeds the intrabiofilm cohesive strength, then microbial cells and EPS fragments become detached and roll away from the biofilm surface. The local balance between the growth rate and the detachment rate determines the final size and shape of the biofilm and whether the pore will remain open or become clogged.

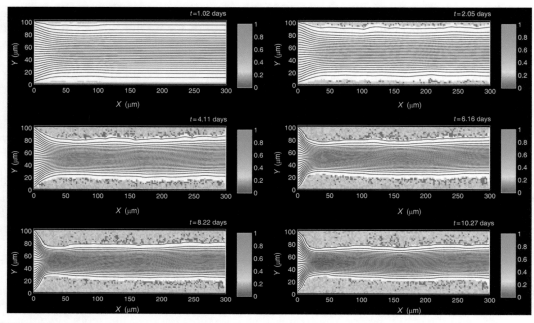

FIGURE 8.11   Snapshots from a computer-aided simulation of biofilm formation in a straight pore under *constant flow-rate* (the color corresponds to the volume fraction of biomass).

Figure 8.12 shows snapshots of the spatiotemporal evolution of biofilms within a straight pore under *constant pressure-drop*. In this scenario, the biofilm forms nonuniform layers which grow, merge, and plug the pore. The biofilm pattern exhibits two interesting features in this case. First, the thickness of the biofilm becomes nonuniform along the pore length, which implies that a gradient in nutrient concentration is built up. Second, as the two biofilm layers approach each other, the symmetry at the longitudinal axis of the pore breaks down. This behavior can be attributed to random detachment events of loose EPS fragments (areas in red color) from the top surface of each biofilm layer. The clogging mechanism associated with this scenario can be explained as follows. Again, the effective cross-section of the pore is reduced as the biofilms expand. However, under constant pressure-drop, both the local velocity of the bulk fluid and the shear stress at the fluid-biofilm interface decrease with decreasing cross-section. This leads to two important consequences. First, as the velocity of the fluid decreases, mass transfer limitations come into play especially along the longitudinal direction. Second, as the shear stress at the fluid-biofilm interface decreases, shear-induced detachment also diminishes. Only pressure-induced detachment could occur and, in this scenario, was set to be weak and able to detach only very loose EPS fragments. Consequently, the growth rate prevails over the detachment rate and the pore becomes clogged. Upon these conditions, the intrinsic permeability of the biofilm is a critical parameter and determines the flow and mass transfer through the plug. In accordance with experimental observations, the plug is located near the entrance of the pore. The results from these two simulations exemplify and generalize the conceptual models of "open pore" and "closed pore," as originally proposed by Taylor and Jaffé [104] but, most importantly, support the fact that both of these scenarios are admissible under certain conditions.

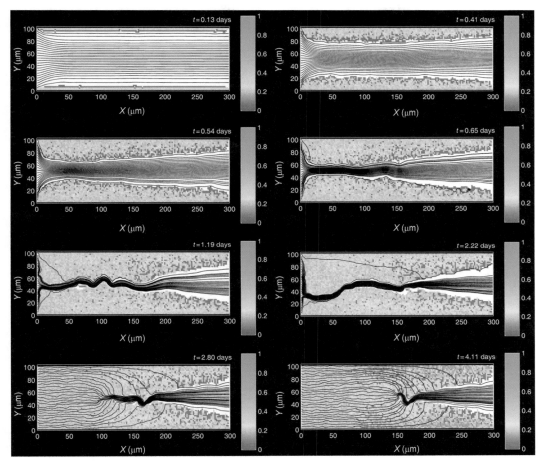

FIGURE 8.12   Snapshots from a computer-aided simulation of biofilm formation in a straight pore under *constant pressure-drop* (the color corresponds to the volume fraction of biomass).

8.5.3 Detachment and Downstream Migration of Biofilms

The detachment of individual cells and EPS fragments, or even of large biomass clusters, has long been recognized as an extremely important process in the formation of biofilms. Usually, its effect is taken into account implicitly with the use of phenomenological approaches that are based on some kind of "average" detachment rate. The first, physically based, mechanistic model of flow-induced detachment that also accounts for the fate of the detached particles was introduced as a module of the simulator HiBioSim-PM [99], which is outlined in Section 8.4.

Here, we present results from computer simulations with regard to the impact of detachment and downstream reattachment on the migration and pattern formation of biofilms in porous media. The values of the physicochemical, biological, and operational parameters are the same as those used in Ref. [99]. The sites of initial colonization by

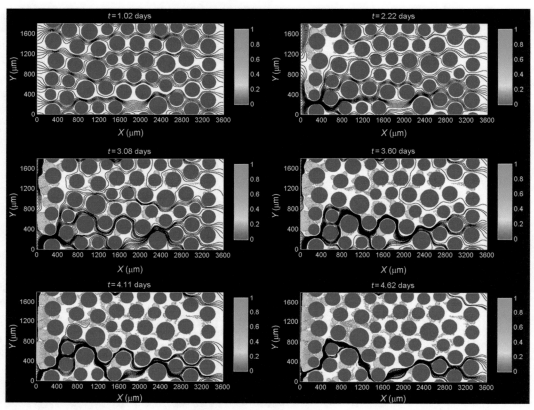

FIGURE 8.13   Computer-aided simulation of the downstream migration of biofilms in a 2D granular core (the color corresponds to the volume fraction of biomass).

microbial cells are selected randomly, but close to the inflow boundary (within 600 μm). Figure 8.13 shows snapshots of the spatiotemporal evolution of biofilms within a two-dimensional granular porous structure. Initially, small colonies are formed at the sites of settlement by microbial cells. Thereafter, these first colonies expand further and, also, new colonies appear downstream. The new colonies are formed by the reattachment of cells that detached from biofilms near the inflow boundary. As time goes by, the biofilms spread over the grain surfaces and, frequently, merge to occlude several pores. Under conditions of constant flow-rate through the porous structure, the reduction of the clear pore sections results in increased fluid velocities which, in turn, cause increased shear stresses at the fluid-biofilm interface. High shear stresses cause continuous erosion of the biofilm surface and, thus, maintain a single primary stream of fluid unplugged. Qualitatively similar behavior is also observed for a three-dimensional consolidated core in Figure 8.14.

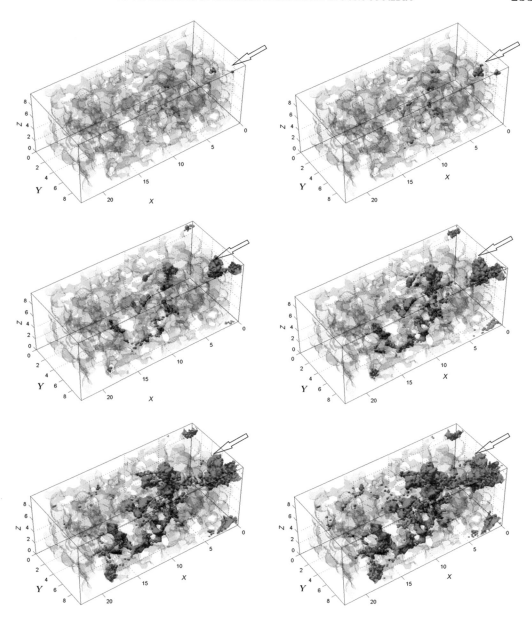

FIGURE 8.14 Computer-aided simulation of the downstream migration of biofilms (green color) in a 3D consolidated core. The arrow indicates the direction of flow.

8.6 SUMMARY

In this chapter, the formation of microbial biofilms in porous media is reviewed in depth and with emphasis on the processes occurring at the pore-scale. A computer-aided simulator is also presented for the prediction of the pattern of evolution and the rate of growth of heterogeneous biofilms within the pore space of porous materials. The simulator combines continuum-based descriptions of fluid flow and solute transport with particle-based descriptions of biofilm growth and detachment. Fluid-biofilm interactions are taken into account explicitly via appropriate mechanistic models of flow-induced detachment-reattachment, flow-modulated growth, and flow through porous biofilms. The flow regime, which is imposed in the form of either a constant flow-rate or a constant pressure-drop, is shown to be a key determinant of the balance between the processes of biofilm growth and detachment and affects strongly the degree of clogging of the porous medium. Furthermore, the flow-induced detachment and reattachment processes are shown to be of paramount importance for the downstream migration of biofilms in porous media. Future research toward improved understanding and quantification of biofilm mechanics and fluid-biofilm interactions is required across a wide spectrum of length-scales, time-scales, and force-scales considering both growth- and nongrowth associated physical stresses. To this end, two-pronged approaches combining careful experiments with appropriate theoretical analyses should be pursued.

References

[1] Williams DL, Bloebaum RD. Observing the biofilm matrix of *Staphylococcus epidermidis* ATCC 35984 grown using the CDC biofilm reactor. Microsc Microanal 2010;16:143–52.
[2] Baum MM, Kainović A, O'Keeffe T, Pandita R, McDonald K, Wu S, et al. Characterization of structures in biofilms formed by *Pseudomonas fluorescens* isolated from soil. BMC Microbiol 2009;9:103.
[3] Sutherland IW. The biofilm matrix—an immobilized but dynamic microbial environment. Trends Microbiol 2001;9(5):222–7.
[4] Wingender J, Jaeger K-E, Flemming H-C. Interaction between extracellular polysaccharides and enzymes. In: Wingender J, Neu TR, Flemming H-C, editors. Microbial extracellular polymeric substances: characterization, structure and function. Berlin: Heidelberg; 1999. p. 231–51.
[5] Wolfaardt GM, Lawrence JR, Korber DR. Function of EPS. In: Wingender J, Neu TR, Flemming H-C, editors. Microbial extracellular polymeric substances: characterization, structure and function. Berlin: Heidelberg; 1999. p. 171–200.
[6] Characklis WG. Biofilm processes. In: Characklis WG, Marshall KC, editors. Biofilms. New York: Willey; 1990. p. 195–231.
[7] Costerton JW, Lewandowski Z, Caldwell DE, Korber DR, Lappin-Scott HM. Microbial biofilms. Annu Rev Microbiol 1995;49:711–45.
[8] Davey ME, O'Toole GA. Microbial biofilms: from ecology to molecular genetics. Microbiol Mol Biol Rev 2000;64(4):847–67.
[9] Englert A, Hubbard SS, Williams KH, Li L, Steefel CI. Feedbacks between hydrological heterogeneity and bioremediation induced biogeochemical transformations. Environ Sci Technol 2009;43:5197–204.
[10] Jørgensen K. *In situ* bioremediation. Adv Appl Microbiol 2007;61:285–305.
[11] McGenity TJ, Folwell BD, McKew BA, Sanni GO. Marine crude-oil biodegradation: a central role for interspecies interactions. Aquat Biosyst 2012;8:10.
[12] Kalin RM. Engineered passive bioreactive barriers: risk-managing the legacy of industrial soil and groundwater pollution. Curr Opin Microbiol 2004;7:227–38.

[13] Casey E, Glennon B, Hamer G. Review of membrane aerated biofilm reactors. Resour Conserv Recycl 1999;27:203–15.
[14] Lazarova V, Manem J. Innovative biofilm treatment technologies for water and wastewater treatment. In: Bryers JD, editor. Biofilms II: process analysis and applications. New York: Willey-Liss; 2000. p. 159–206.
[15] Hall-Stoodley L, Costerton JW, Stoodley P. Bacterial biofilms: from the natural environment to infectious diseases. Nat Rev Microbiol 2004;2:95–108.
[16] Singh PK, Schaefer AL, Parsek MR, Moninger TO, Welsh MJ, Greenberg EP. Quorum-sensing signals indicate that cystic fibrosis lungs are infected with bacterial biofilms. Nature 2000;407:762–4.
[17] Henrici AT. Studies of freshwater bacteria. I. A direct microscopic technique. J Bacteriol 1933;25:277–86.
[18] Zobell CE. The effect of solid surfaces upon bacterial activity. J Bacteriol 1943;46:39–56.
[19] Costerton JW, Geesey GG, Cheng K-J. How bacteria stick. Sci Am 1978;238:86–95.
[20] Costerton JW, Irvin RT, Cheng K-J. The bacterial glycocalyx in nature and disease. Annu Rev Microbiol 1981;35:299–324.
[21] Eighmy TT, Maratea D, Bishop PL. Electron microscopic examination of wastewater biofilm formation and structural components. Appl Environ Microbiol 1983;45(6):1921–31.
[22] Geesey GG, Mutch R, Costerton JW, Green RB. Sessile bacteria: an important component of the microbial population in small mountain streams. Limnol Oceanogr 1978;23(6):1214–23.
[23] Jones HC, Roth IL, Sanders III WM. Electron microscopic study of a slime layer. J Bacteriol 1969;99:316–25.
[24] Marshall KC, Stout R, Mitchell R. Mechanism of the initial events in the sorption of marine bacteria to surfaces. J Gen Microbiol 1971;68:337–48.
[25] Williamson K, McCarty PL. Verification studies of the biofilm model for bacterial substrate utilization. J Water Pollut Control Federation 1976;48(2):281–96.
[26] Characklis WG. Attached microbial growths—I. Attachment and growth. Water Res 1973;7:1113–27.
[27] Characklis WG. Attached microbial growths—II. Frictional resistance due to microbial slimes. Water Res 1973;7:1249–58.
[28] Characklis WG. Fouling biofilm development: a process analysis. Biotechnol Bioeng 1981;XXIII:1923–60.
[29] LaMotta EJ. Internal diffusion and reaction in biological films. Environ Sci Technol 1976;10(8):765–9.
[30] McCarty PL, Reinhard M, Rittmann BE. Trace organics in groundwater. Environ Sci Technol 1981;15(1): 40–51.
[31] Wanner O, Gujer W. A multispecies biofilm model. Biotechnol Bioeng 1986;28(3):314–28.
[32] Lawrence JR, Korber DR, Hoyle BD, Costerton JW, Caldwell DE. Optical sectioning of microbial biofilms. J Bacteriol 1991;173(20):6558–67.
[33] Amann RI, Krumholz L, Stahl DA. Fluorescent-oligonucleotide probing of whole cells for determinative, phylogenetic, and environmental studies in microbiology. J Bacteriol 1990;172(2):762–70.
[34] de Beer D, Stoodley P, Roe F, Lewandowski Z. Effects of biofilm structures on oxygen distribution and mass transport. Biotechnol Bioeng 1994;43:1131–8.
[35] Lawrence JR, Neu TR, Swerhone GDW. Application of multiple parameter imaging for the quantification of algal, bacterial, and exopolymer components of microbial biofilms. J Microbiol Methods 1998;32:253–61.
[36] Lawrence JR, Swerhone GDW, Leppard GG, Araki T, Zhang X, West MM, et al. Scanning Transmission X-ray, laser scanning, and transmission electron microscopy mapping of the exopolymeric matrix of microbial biofilms. Appl Environ Microbiol 2003;69(9):5543–54.
[37] Stewart PS, Murga R, Srinivasan R, de Beer D. Biofilm structural heterogeneity visualized by three microscopic methods. Water Res 1995;29(8):2006–9.
[38] Wimpenny J, Manz W, Szewzyk U. Heterogeneity in biofilms. FEMS Microbiol Rev 2000;24:661–71.
[39] Zhang TC, Bishop PL. Density, porosity and pore structure of biofilms. Water Res 1994;28(11):2267–77.
[40] Costerton JW, Lewandowski Z, DeBeer D, Caldwell D, Korber D, James G. Biofilms, the customized microniche. J Bacteriol 1994;176(8):2137–42.
[41] Massol-Deyá AA, Whallon J, Hickey RF, Tiedje JM. Channel structures in aerobic biofilms of fixed-film reactors treating contaminated groundwater. Appl Environ Microbiol 1995;61(2):769–77.
[42] Stoodley P, de Beer D, Lewandowski Z. Liquid flow in biofilm systems. Appl Environ Microbiol 1994;60 (8):2711–6.
[43] Costerton JW, Lappin-Scott HM. Introduction to microbial biofilms. In: Lappin-Scott HM, Costerton JW, editors. Microbial biofilms. Cambridge: Cambridge University Press; 1995. p. 1–14.

[44] Bryers JD. Biologically active surfaces: processes governing the formation and persistence of biofilms. Biotechnol Prog 1987;3(2):57–68.

[45] Bryers JD. Biofilm formation and persistence. In: Bryers JD, editor. Biofilms II: process analysis and applications. New York: Willey-Liss; 2000. p. 45–88.

[46] Berg HC, Anderson RA. Bacteria swim by rotating their flagellar filaments. Nature 1973;245:380–4.

[47] Adler J. Chemotaxis in bacteria. Science 1966;153:708–16.

[48] Berg HC, Brown DA. Chemotaxis in Escherichia coli analyzed by three-dimensional tracking. Nature 1972;239:500–4.

[49] Ginn TR, Wood BD, Nelson KE, Scheibe TD, Murphy EM, Clement TP. Processes in microbial transport in the natural subsurface. Adv Water Resour 2002;25:1017–42.

[50] van Loosdrecht MCM, Lyklema J, Norde W, Schraa G, Zehnder AJB. Electrophoretic mobility and hydrophobicity as a measure to predict the initial steps of bacterial adhesion. Appl Environ Microbiol 1987;53(8):1898–901.

[51] O'Toole GA, Kaplan HB, Kolter R. Biofilm formation as microbial development. Annu Rev Microbiol 2000;54:49–79.

[52] Bazaka K, Jacob MV, Crawford RJ, Ivanova EP. Plasma-assisted surface modification of organic biopolymers to prevent bacterial attachment. Acta Biomater 2011;7:2015–28.

[53] McBride MJ. Bacterial gliding motility: multiple mechanisms for cell movement over surfaces. Annu Rev Mcrobiol 2001;55:49–75.

[54] O'Toole GA, Kolter R. Flagellar and twitching motility are necessary for *Pseudomonas aeruginosa* biofilm development. Mol Microbiol 1998;30(2):295–304.

[55] O'Toole GA, Kolter R. Initiation of biofilm formation in Pseudomonas fluorescens WCS365 proceeds via multiple, convergent signaling pathways: a genetic analysis. Mol Microbiol 1998;28(3):449–61.

[56] Sauer K, Camper AK, Ehrlich GD, Costerton JW, Davies DG. Pseudomonas aeruginosa displays multiple phenotypes during development as a biofilm. J Bacteriol 2002;184(4):1140–54.

[57] Kapellos GE, Alexiou TS, Payatakes AC. A multiscale theoretical model for diffusive mass transfer in cellular biological media. Math Biosci 2007;210(1):177–237.

[58] Alexiou TS, Kapellos GE. Plane Couette-Poiseuille flow past a homogeneous poroelastic layer. Phys Fluids 2013;25:073605.

[59] Horn H, Reiff H, Morgenroth E. Simulation of growth and detachment in biofilm systems under defined hydro-dynamic conditions. Biotechnol Bioeng 2003;81(5):607–17.

[60] Peyton BM, Characklis WG. A statistical analysis of the effect of substrate utilization and shear stress on the kinetics of biofilm detachment. Biotechnol Bioeng 1993;41:728–35.

[61] Davies DG, Parsek MR, Pearson JP, Iglewski BH, Costerton JW, Greenberg EP. The involvement of cell-to-cell signals in the development of a bacterial biofilm. Science 1998;280:295–8.

[62] Miller MB, Bassler BL. Quorum sensing in bacteria. Annu Rev Microbiol 2001;55:165–99.

[63] Parsek MR, Greenberg EP. Sociomicrobiology: the connections between quorum sensing and biofilms. Trends Microbiol 2005;13(1):27–33.

[64] Baveye P, Valocchi A. An evaluation of mathematical models of the transport of biologically reacting solutes in saturated soils and aquifers. Water Resour Res 1989;25(6):1413–21.

[65] Dupin HJ, Kitanidis PK, McCarty PL. Pore-scale modeling of biological clogging due to aggregate expansion: a material mechanics approach. Water Resour Res 2001;37(12):2965–79.

[66] Rittmann BE. The significance of biofilms in porous media. Water Resour Res 1993;29(7):2195–202.

[67] Thullner M, Zeyer J, Kinzelbach W. Influence of microbial growth on hydraulic properties of pore networks. Transport Porous Med 2002;49:99–122.

[68] Dupin HJ, McCarty PL. Impact of colony morphologies and disinfection on biological clogging in porous media. Environ Sci Technol 2000;34:1513–20.

[69] Paulsen JE, Oppen E, Bakke R. Biofilm morphology in porous media, a study with microscopic and image techniques. Water Sci Technol 1997;36(1):1–9.

[70] Vayenas DV, Michalopoulou E, Constantinides GN, Pavlou S, Payatakes AC. Visualization experiments of biodegradation in porous media and calculation of the biodegradation rate. Adv Water Resour 2002;25:203–19.

[71] Cunningham AB, Characklis WG, Abedeen F, Crawford D. Influence of biofilm accumulation on porous media hydrodynamics. Environ Sci Technol 1991;25:1305–11.

[72] Sharp RR, Cunningham AB, Komlos J, Billmayer J. Observation of thick biofilm accumulation and structure in porous media and corresponding hydrodynamic and mass transfer effects. Water Sci Technol 1999;39 (7):195–201.

[73] Stoodley P, Dodds I, de Beer D, Lappin-Scott H, Boyle JD. Flowing biofilms as a transport mechanism for biomass through porous media under laminar and turbulent conditions in a laboratory reactor system. Biofouling 2005;21 (3/4):161–8.

[74] DeLeo PC, Baveye P, Ghiorse WC. Use of confocal laser scanning microscopy on soil thin-sections for improved characterization of microbial growth in unconsolidated soils and aquifer materials. J Microbiol Methods 1997;30:193–203.

[75] Leis AP, Schlicher S, Franke H, Strathmann M. Optically transparent porous medium for nondestructive studies of microbial biofilm architecture and transport dynamics. Appl Environ Microbiol 2005;71(8):4801–8.

[76] Seymour JD, Gage JP, Codd SL, Gerlach R. Magnetic resonance microscopy of biofouling induced scale dependent transport in porous media. Adv Water Resour 2007;30:1408–20.

[77] Kapellos G.E., Transport phenomena and dynamics of microbial biofilm growth during the biodegradation of organic pollutants in porous materials: hierarchical theoretical modeling and experimental investigation (in Greek), University of Patras, 2008.

[78] Bouwer EJ, Rijnaarts HHM, Cunningham AB, Gerlach R. Biofilms in porous media. In: Bryers JD, editor. Biofilms II: process analysis and applications. New York: Willey-Liss; 2000. p. 123–58.

[79] Scow KM, Hicks KA. Natural attenuation and enhanced bioremediation of organic contaminants in groundwater. Curr Opin Biotechnol 2005;16:246–53.

[80] McLean J, Beveridge TJ. Chromate reduction by a *Pseudomonad* isolated from a site contaminated with chromated copper arsenate. Appl Environ Microbiol 2001;67(3):1076–84.

[81] Ellis DE, Lutz EJ, Odom JM, Buchanan Jr RJ, Bartlett CL, Lee MD, et al. Bioaugmentation for accelerated in situ anaerobic bioremediation. Environ Sci Technol 2000;34:2254–60.

[82] Banat IM. Biosurfactants production and possible uses in microbial enhanced oil recovery and oil pollution remediation: a review. Bioresour Technol 1995;51:1–12.

[83] Baveye P, Vandevivere P, Hoyle BL, DeLeo PC, de Lozada DS. Environmental impact and mechanisms of the biological clogging of saturated soils and aquifer materials. Crit Rev Environ Sci Technol 1998;28 (2):123–91.

[84] Lazar I, Petrisor IG, Yen TF. Microbial enhanced oil recovery (MEOR). Petrol Sci Technol 2007;25:1353–66.

[85] Cogan NG, Guy RD. Multiphase flow models of biogels from crawling cells to bacterial biofilms. HFSP J 2010;4 (1):11–25.

[86] Kapellos GE, Alexiou TS, Payatakes AC. Theoretical modeling of fluid flow through cellular biological media: An overview. Math Biosci 2010;225(2):83–93.

[87] Eberl HJ, Parker DF, van Loosdrecht MCM. A new deterministic spatiotemporal continuum model for biofilm development. J Theor Med 2001;3(3):161–76.

[88] Dockery J, Klapper I. Finger formation in biofilm layers. J SIAM Appl Math 2001;62(3):853–69.

[89] Cogan NG, Keener JP. The role of biofilm matrix in structural development. Math Med Biol 2004;21(2):147–66.

[90] Wimpenny JWT, Colasanti R. A unifying hypothesis for the structure of microbial biofilms based on cellular automaton models. FEMS Microbiol Ecol 1997;22:1–16.

[91] Chang I, Gilbert ES, Eliashberg N, Keasling JD. A three-dimensional, stochastic simulation of biofilm growth and transport-related factors that affect structure. Microbiology 2003;149:2859–71.

[92] Kreft J-U, Booth G, Wimpenny JWT. BacSim, a simulator for individual-based modeling of bacterial colony growth. Microbiology 1998;144:3275–87.

[93] Picioreanu C, van Loosdrecht MCM, Heijnen JJ. A new combined differential-discrete cellular automaton approach for biofilm modeling: application for growth in gel beads. Biotechnol Bioeng 1998;57(6):718–31.

[94] Kreft J-U, Picioreanu C, Wimpenny JWT, van Loosdrecht MCM. Individual-based modeling of biofilms. Microbiology 2001;147:2897–912.

[95] Hermanowicz SW. A simple 2D biofilm model yields a variety of morphological features. Math Biosci 2001;169:1–14.

[96] Laspidou CS, Rittmann BE. Modeling the development of biofilm density including active bacteria, inert biomass, and extracellular polymeric substances. Water Res 2004;38:3349–61.

[97] Dupin HJ, Kitanidis PK, McCarty PL. Simulations of two-dimensional modeling of biomass aggregate growth in network models. Water Resour Res 2001;37(12):2981–94.

[98] Suchomel BJ, Chen BM, Allen MB. Network model of flow, transport and biofilm effects in porous media. Transport Porous Med 1998;30:1–23.

[99] Kapellos GE, Alexiou TS, Payatakes AC. Hierarchical simulator of biofilm growth and dynamics in granular porous materials. Adv Water Resour 2007;30:1648–67.

[100] Knutson CE, Werth CJ, Valocchi AJ. Pore-scale simulation of biomass growth along the transverse mixing zone of a model two-dimensional porous medium. Water Resour Res 2005;41:W07007.

[101] Pintelon TRR, Picioreanu C, van Loosdrecht MCM, Johns ML. The effect of biofilm permeability on bioclogging of porous media. Biotechnol Bioeng 2011;109(4):1031–42.

[102] Pintelon TRR, von der Schulenburg DAG, Johns ML. Towards optimum permeability reduction in porous media using biofilm growth simulations. Biotechnol Bioeng 2009;103(4):767–79.

[103] Kapellos GE, Alexiou TS, Payatakes AC. A multiscale theoretical model for fluid flow in cellular biological media. Int J Eng Sci 2012;51:241–71.

[104] Taylor SW, Jaffé PR. Biofilm growth and the related changes in the physical properties of a porous medium. 1. Experimental investigation. Water Resour Res 1990;26(9):2153–9.

[105] Taylor SW, Jaffé PR. Substrate and biomass transport in a porous medium. Water Resour Res 1990;26(9):2181–94.

[106] Vandevivere P, Baveye P. Effect of bacterial extracellular polymers on the saturated hydraulic conductivity of sand columns. Appl Environ Microbiol 1992;58(5):1690–8.

[107] Geesey GG, Mittelman MW, Lieu VT. Evaluation of slime-producing bacteria in oil field core flood experiments. Appl Environ Microbiol 1987;53:278–83.

[108] Shaw JC, Bramhill B, Wardlaw NC, Costerton JW. Bacterial fouling in a model core system. Appl Environ Microbiol 1985;49(3):693–701.

[109] Oberdrofer JA, Peterson FL. Waste-water injection: geochemical and biogeochemical clogging processes. Ground Water 1985;23:753–61.

[110] de Lozada SD, Vandevivere P, Baveye P, Zinder S. Decrease of the hydraulic conductivity of sand columns by *Methanosarcina barkeri*. World J Microbiol Biotechnol 1994;10:325–33.

[111] Frankenberger Jr WT, Troeh FR, Dumenil LC. Bacterial effects on hydraulic conductivity of soils. Soil Sci Soc Am J 1979;43(2):333–8.

[112] Kim D-S, Foggler HS. Biomass evolution in porous media and its effects on permeability under starvation conditions. Biotechnol Bioeng 2000;69:47–56.

Flow Through a Permeable Tube

C. Pozrikidis

University of Massachusetts, Amherst, MA, USA

9.1 INTRODUCTION

Fluid-carrying biological vessels are typically embedded in an ambient medium consisting of tissue or another interstitial material. Fluid escapes though the vessel walls to transport oxygen, nutrients, and other substances in a process that is known as extravasation. It is striking that a few dozen liters of biological fluid perfuse through the capillaries of the human body each day. The pronounced permeability of the neoplastic capillary walls of solid tumors allows fluid to penetrate the tumor interstitium and either exit the surface of the tumor or be removed by lymphatic nodes (e.g., [1]). Other biophysical applications can be found in the hydrodynamics of the proximal renal tubule [2] and in the perfusion of blood through organic and artificial hollow-fiber kidneys [3]. Industrial applications are encountered in reverse osmosis, desalination, and the flow through tubular nanostructures (e.g., [4]).

In biomechanics, fluid-carrying vessels are modeled as tubes with permeable walls occupied by membranes. A hydrodynamic or osmotic pressure difference across a membrane drives a convective transport through the wall. The mathematical modeling of the pertinent fluid flow requires a fundamental assumption regarding the coupling between the interior tube flow and the flow outside the tube in terms of an appropriate membrane law, such as that expressed by Starling's equation. If the flow outside the tube is weak, a constant or otherwise prescribed exterior pressure can be specified or assumed. If the flow outside the tube is strong, the surrounding material hosting the tube can be viewed as a porous medium, and Darcy's law can be used to describe the pressure field developing in the surroundings. The formulation results in Laplace's equation for the ambient pressure that is to be solved subject to proper boundary and far-field conditions by numerical methods even for simple geometrical configurations (e.g., [1]).

The simplest possible flow configuration consists of an infinite straight tube with a permeable wall subjected to a constant ambient pressure. Because the Reynolds number of the flow through the tube is typically small, the motion of the fluid inside the tube and across the tube wall occurs under conditions of Stokes flow. Marshall and Trowbridge [2] derived a remarkable exact solution of the unsimplified equations of the pertinent axisymmetric Stokes flow. In this chapter, their solution is rederived and generalized, and the significance of the membrane law is examined and critically assessed. A novel and noteworthy generalization of the analysis presented in this chapter pertains to the effect of the apparent wall slip at the permeable tube surface.

In Section 9.2, the description of axisymmetric Stokes flow in terms of the Stokes stream function is reviewed and solutions that vary exponentially with respect to downstream position from a designated origin are derived. The general solution derived in Section 9.2 is employed in the analysis of Section 9.3 to study flow through an infinite permeable tube in terms of an unspecified constant with dimensions of inverse length. In Section 9.4, the computation of this constant by way of Starling's equation for the pressure or normal stress is addressed. Flow through a tube with finite length is discussed in Section 9.5, the effect of wall slip is discussed in Section 9.6, and general remarks are made in Section 9.7.

9.2 AXISYMMETRIC STOKES FLOW

To describe an axisymmetric flow, we introduce cylindrical polar coordinates, (x, σ, φ), where x is the axial position, σ is the distance from the x axis, and φ is the azimuthal angle measured around the x axis with origin at the xy plane varying in the interval $[0, 2\pi]$, as shown in Figure 9.1. By assumption, neither the fluid velocity, $\mathbf{u}$, nor the fluid pressure, p, depends on the azimuthal angle, φ. The azimuthal velocity component is identically zero

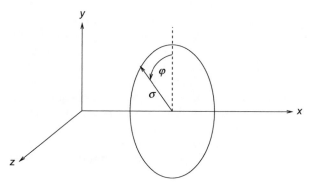

FIGURE 9.1 Definition of cylindrical polar coordinates, (x, σ, φ), with reference to companion Cartesian coordinates, (x, y, z). The azimuthal angle, φ, is measured around the x axis from the xy plane.

in the absence of swirling motion, $u_\varphi = 0$, and the vorticity, $\omega = \nabla \times \mathbf{u}$, points in the azimuthal direction at every point so that

$$\mathbf{u} = u_x \mathbf{e}_x + u_\sigma \mathbf{e}_\sigma, \quad \boldsymbol{\omega} = \omega_\varphi \mathbf{e}_\varphi, \tag{9.1}$$

where $\mathbf{e}_x$, $\mathbf{e}_\sigma$, and $\mathbf{e}_\varphi$ are unit vectors in the subscripted directions (e.g., [5]).

9.2.1 Stokes Stream Function

To expedite the analysis, we introduce the Stokes stream function, ψ, defined by the equation

$$\mathbf{u} = \nabla \times \left(\frac{1}{\sigma}\psi \mathbf{e}_\varphi\right). \tag{9.2}$$

It will be noted that the continuity equation for an incompressible fluid, $\nabla \cdot \mathbf{u} = 0$, is satisfied for any twice differentiable stream function, $\psi(x,\sigma)$. Explicitly, the axial and radial components of the velocity are given by

$$u_x = \frac{1}{\sigma}\frac{\partial \psi}{\partial \sigma}, \quad u_\sigma = -\frac{1}{\sigma}\frac{\partial \psi}{\partial x}, \tag{9.3}$$

and the azimuthal vorticity component is given by

$$\omega_\varphi = -\frac{1}{\sigma}\mathscr{E}^2\psi, \tag{9.4}$$

where

$$\mathscr{E}^2 \equiv \frac{\partial^2}{\partial x^2} + \frac{\partial^2}{\partial \sigma^2} - \frac{1}{\sigma}\frac{\partial}{\partial \sigma} \tag{9.5}$$

is a second-order differential operator (e.g., [5]). Note that this operator is different from the Laplacian operator in the (x,σ) plane, except sufficiently far from the axis of revolution.

The vorticity transport equation for axisymmetric Stokes flow requires that the Stokes stream function satisfies a fourth-order linear partial differential equation,

$$\mathscr{E}^4\psi \equiv \mathscr{E}^2\left(\mathscr{E}^2\psi\right) = 0. \tag{9.6}$$

Far from the axis of revolution, this equation reduces to the biharmonic equation describing two-dimensional flow in an azimuthal plane.

It is useful to resolve Equation (9.6) into two second-order constituent equations,

$$\mathscr{E}^2\psi = \widetilde{\psi}, \quad \mathscr{E}^2\widetilde{\psi} = 0, \tag{9.7}$$

where a tilde denotes an intermediate solution.

9.2.2 Exponential Solutions

In the case of axisymmetric flow with an anticipated exponential dependence of the velocity and pressure on the axial distance, x, we set

$$\psi(x,\sigma) = \phi_c(\sigma)\cosh kx + \phi_s(\sigma)\sinh kx, \tag{9.8}$$

where k is an unspecified constant with dimensions of inverse length, and $\phi_c(\sigma)$ and $\phi_s(\sigma)$ are profile functions associated with the hyperbolic cosine and sine. Substituting this expression into Equation (9.7), we obtain the ordinary differential equations

$$\left(\frac{d^2}{d\sigma^2} - \frac{1}{\sigma}\frac{d}{d\sigma} + k^2\right)\phi_q = \widetilde{\phi}_q \tag{9.9}$$

and

$$\left(\frac{d^2}{d\sigma^2} - \frac{1}{\sigma}\frac{d}{d\sigma} + k^2\right)\widetilde{\phi}_q = 0, \tag{9.10}$$

and q stands for c or s and a tilde denotes an intermediate function. The general solution is

$$\phi_q(\sigma) = \sigma\left[A_1^q J_1(\hat{\sigma}) + A_2^q \sigma J_0(\hat{\sigma}) + B_1^q Y_1(\hat{\sigma}) + B_2^q \sigma Y_0(\hat{\sigma})\right], \tag{9.11}$$

where J_0, J_1, Y_0, and Y_1 are Bessel functions, A_1^q, A_2^q, B_1^q, and B_2^q for $q = c, s$ are eight constants, and

$$\hat{\sigma} \equiv k\sigma \tag{9.12}$$

is a dimensionless distance from the axis of revolution. The first and third terms on the right-hand side of (9.11) satisfy (9.9) with $\widetilde{\phi}_q = 0$. The second and fourth terms represent a particular solution of (9.9) with $\widetilde{\phi}_q$ being the general solution of (9.10) involving two arbitrary constants.

Substituting the preceding expressions into (9.3) and using the following properties of the Bessel functions,

$$J_0'(z) = -J_1(z), \quad J_1'(z) = J_0(z) - \frac{1}{z}J_1(z), \quad Y_0'(z) = -Y_1(z), \quad Y_1'(z) = Y_0(z) - \frac{1}{z}Y_1(z), \tag{9.13}$$

we derive the velocity components

$$u_\beta = U_\beta^c(\sigma)\cosh kx + U_\beta^s(\sigma)\sinh kx \tag{9.14}$$

for $\beta = x, \sigma$, where

$$\begin{aligned}U_x^q(\sigma) &= \frac{1}{\sigma}\frac{d\phi_q}{d\sigma} = kA_1^q J_0(\hat{\sigma}) + A_2^q[2J_0(\hat{\sigma}) - \hat{\sigma}J_1(\hat{\sigma})] \\ &\quad + kB_1^q Y_0(\hat{\sigma}) + B_2^q[2Y_0(\hat{\sigma}) - \hat{\sigma}Y_1(\hat{\sigma})]\end{aligned} \tag{9.15}$$

for $q = c, s$,

$$U_\sigma^c(\sigma) = -\frac{k}{\sigma}\phi_s = -\left[kA_1^s J_1(\hat{\sigma}) + A_2^s \hat{\sigma}J_0(\hat{\sigma}) + kB_1^s Y_1(\hat{\sigma}) + B_2^s \hat{\sigma}Y_0(\hat{\sigma})\right], \tag{9.16}$$

and

$$U_\sigma^s(\sigma) = -\frac{k}{\sigma}\phi_c = -\left[kA_1^c J_1(\hat{\sigma}) + A_2^c \hat{\sigma}J_0(\hat{\sigma}) + kB_1^c Y_1(\hat{\sigma}) + B_2^c \hat{\sigma}Y_0(\hat{\sigma})\right]. \tag{9.17}$$

Note that U_σ^c is proportional to ϕ_s, whereas U_σ^s is proportional to ϕ_c.

The hydrodynamic pressure field due to the fluid motion can be expressed in the corresponding form

$$p = \chi_c(\sigma)\cosh kx + \chi_s(\sigma)\sinh kx + \pi_0, \tag{9.18}$$

where π_0 is a reference pressure. To compute the profile functions $\chi_c(\sigma)$ and $\chi_s(\sigma)$, we consider the x component of the Stokes equation for axisymmetric flow,

$$\frac{\partial p}{\partial x} = \mu\left[\frac{\partial^2 u_x}{\partial x^2} + \frac{1}{\sigma}\frac{\partial}{\partial\sigma}\left(\sigma\frac{\partial u_x}{\partial\sigma}\right)\right]. \tag{9.19}$$

Substituting expression (9.18) for the pressure together with the first equation in (9.14) for the velocity, we obtain

$$\chi_c(\sigma) = \mu k U_x^s + \frac{\mu}{k}\frac{1}{\sigma}\frac{d}{d\sigma}\left(\sigma\frac{dU_x^s}{d\sigma}\right) = \frac{\mu}{k}\left[\frac{1}{\sigma}\frac{d}{d\sigma}\left(\sigma\frac{dU_x^s}{d\sigma}\right) + k^2 U_x^s\right], \tag{9.20}$$

yielding

$$\chi_c(\sigma) = \frac{\mu}{k}\left(\frac{d^2 U_x^s}{d\sigma^2} + \frac{1}{\sigma}\frac{dU_x^s}{d\sigma} + k^2 U_x^s\right). \tag{9.21}$$

Substituting the expression for U_x^s given in (9.15) with $q = s$, we obtain

$$\chi_c(\sigma) = -2\mu k\left[A_2^s J_0(\hat{\sigma}) + B_2^s Y_0(\hat{\sigma})\right]. \tag{9.22}$$

Working in a similar fashion, we obtain the companion equation

$$\chi_s(\sigma) = -2\mu k\left[A_2^c J_0(\hat{\sigma}) + B_2^c Y_0(\hat{\sigma})\right]. \tag{9.23}$$

Note that only two terms contribute to the pressure field.

The components of the Newtonian stress tensor, $\boldsymbol{\sigma}$, can be computed readily from the preceding expressions for the velocity and pressure. The axial normal stress is given by

$$\sigma_{xx} = -p + 2\mu\frac{\partial u_x}{\partial x} = \Sigma_{xx}^c(\sigma)\cosh kx + \Sigma_{xx}^s(\sigma)\sinh kx - \pi_0, \tag{9.24}$$

where

$$\Sigma_{xx}^c(\sigma) = -\chi_c(\hat{\sigma}) + 2\mu k U_x^s(\hat{\sigma}), \quad \Sigma_{xx}^s(\sigma) = -\chi_s(\hat{\sigma}) + 2\mu k U_x^c(\hat{\sigma}). \tag{9.25}$$

Making substitutions, we obtain

$$\Sigma_{xx}^c(\sigma) = 2\mu k\left(kA_1^s J_0(\hat{\sigma}) + A_2^s[3J_0(\hat{\sigma}) - \hat{\sigma}J_1(\hat{\sigma})] + kB_1^s Y_0(\hat{\sigma}) + B_2^s[3Y_0(\hat{\sigma}) - \hat{\sigma}Y_1(\hat{\sigma})]\right) \tag{9.26}$$

and

$$\Sigma_{xx}^s(\sigma) = 2\mu k\left(kA_1^c J_0(\hat{\sigma}) + A_2^c[3J_0(\hat{\sigma}) - \hat{\sigma}J_1(\hat{\sigma})] + kB_1^c Y_0(\hat{\sigma}) + B_2^c[3Y_0(\hat{\sigma}) - \hat{\sigma}Y_1(\hat{\sigma})]\right). \tag{9.27}$$

The radial normal stress is given by

$$\sigma_{\sigma\sigma} = -p + 2\mu\frac{\partial u_\sigma}{\partial\sigma} = \Sigma_{\sigma\sigma}^c(\sigma)\cosh kx + \Sigma_{\sigma\sigma}^s(\sigma)\sinh kx - \pi_0, \tag{9.28}$$

where

$$\Sigma^q_{\sigma\sigma}(\sigma) = -\chi_q(\hat{\sigma}) + 2\mu\frac{\partial U^q_\sigma(\hat{\sigma})}{\partial\sigma} \tag{9.29}$$

for $q=c,s$. Making substitutions, we obtain

$$\Sigma^c_{\sigma\sigma}(\sigma) = 2\mu k\left(A^s_2 J_0(\hat{\sigma}) + B^s_2 Y_0(\hat{\sigma}) - kA^s_1 J'_1(\hat{\sigma}) - A^s_2[\hat{\sigma}J_0(\hat{\sigma})]' - kB^s_1 Y'_1(\hat{\sigma}) - B^s_2[\hat{\sigma}Y_0(\hat{\sigma})]'\right) \tag{9.30}$$

and

$$\Sigma^s_{\sigma\sigma}(\sigma) = 2\mu k\left(A^c_2 J_0(\hat{\sigma}) + B^c_2 Y_0(\hat{\sigma}) - kA^c_1 J'_1(\hat{\sigma}) - A^c_2[\hat{\sigma}J_0(\hat{\sigma})]' - kB^c_1 Y'_1(\hat{\sigma}) - B^c_2[\hat{\sigma}Y_0(\hat{\sigma})]'\right), \tag{9.31}$$

where a prime denotes a derivative with respect to $\hat{\sigma}$. Carrying out the differentiations and using the properties of the Bessel functions stated in Equation (9.13), we obtain

$$\begin{aligned}\Sigma^c_{\sigma\sigma}(\sigma) = -2\mu k&\left(kA^s_1\left[J_0(\hat{\sigma}) - \frac{1}{\hat{\sigma}}J_1(\hat{\sigma})\right] - A^s_2\hat{\sigma}J_1(\hat{\sigma})\right.\\ &\left.+ kB^s_1\left[Y_0(\hat{\sigma}) - \frac{1}{\hat{\sigma}}Y_1(\hat{\sigma})\right] - B^s_2\hat{\sigma}Y_1(\hat{\sigma})\right)\end{aligned} \tag{9.32}$$

and

$$\begin{aligned}\Sigma^s_{\sigma\sigma}(\sigma) = -2\mu k&\left(kA^c_1\left[J_0(\hat{\sigma}) - \frac{1}{\hat{\sigma}}J_1(\hat{\sigma})\right] - A^c_2\hat{\sigma}J_1(\hat{\sigma})\right.\\ &\left.+ kB^c_1\left[Y_0(\hat{\sigma}) - \frac{1}{\hat{\sigma}}Y_1(\hat{\sigma})\right] - B^c_2\hat{\sigma}Y_1(\hat{\sigma})\right).\end{aligned} \tag{9.33}$$

The shear stress is given by

$$\sigma_{x\sigma} = \sigma_{\sigma x} = \mu\left(\frac{\partial u_x}{\partial\sigma} + \frac{\partial u_\sigma}{\partial x}\right) = \overset{c}{\underset{x\sigma}{\Sigma}}(\sigma)\cosh kx + \overset{s}{\underset{x\sigma}{\Sigma}}(\sigma)\sinh kx, \tag{9.34}$$

where

$$\Sigma^c_{x\sigma}(\sigma) = \mu\left(\frac{\partial U^c_x(\hat{\sigma})}{\partial\sigma} + kU^s_\sigma(\hat{\sigma})\right) \tag{9.35}$$

and

$$\Sigma^s_{x\sigma}(\sigma) = \mu\left(\frac{\partial U^s_x(\hat{\sigma})}{\partial\sigma} + kU^c_\sigma(\hat{\sigma})\right). \tag{9.36}$$

Making substitutions, we obtain

$$\begin{aligned}\Sigma^c_{x\sigma}(\sigma) = \mu k&\left(kA^c_1 J'_0(\hat{\sigma}) + A^c_2[2J_0(\hat{\sigma}) - \hat{\sigma}J_1(\hat{\sigma})]' + kB^c_1 Y'_0(\hat{\sigma}) + B^c_2[2Y_0(\hat{\sigma}) - \hat{\sigma}Y_1(\hat{\sigma})]' - kA^c_1 J_1(\hat{\sigma})\right.\\ &\left.- A^c_2\hat{\sigma}J_0(\hat{\sigma}) - kB^c_1 Y_1(\hat{\sigma}) - B^c_2\hat{\sigma}Y_0(\hat{\sigma})\right),\end{aligned} \tag{9.37}$$

where a prime denotes a derivative with respect to $\hat{\sigma}$. Carrying out the differentiations and using properties (9.13), we obtain

$$\Sigma_{x\sigma}^{q}(\sigma) = -2\mu k\big(kA_{1}^{q}J_{1}(\hat{\sigma}) + A_{2}^{q}[J_{1}(\hat{\sigma}) + \hat{\sigma}J_{0}(\hat{\sigma})] + kB_{1}^{q}Y_{1}(\hat{\sigma}) + B_{2}^{q}[Y_{1}(\hat{\sigma}) + \hat{\sigma}Y_{0}(\hat{\sigma})]\big) \qquad (9.38)$$

for $q = c, s$.

9.2.3 General Solution

The exponential solution derived in Section 9.2.2 involves an *a priori* unknown, unspecified real or complex constant with dimensions of inverse length, k, and eight unknown coefficients. These constants are determined by boundary, far-field, and regularity conditions imposed for each specific application. A general solution can be constructed in terms of an integral with respect to k, modulated by a density distribution function. In the case of flow through a porous tube, the value of k is determined uniquely by the membrane law, as discussed in Section 9.4.

9.3 FLOW THROUGH AN INFINITE PERMEABLE TUBE

With reference to the general solution presented in Section 9.2, now we address the main problem of interest, which is Stokes flow through a cylindrical tube of radius a with a permeable wall that allows fluid to enter or escape according to a specified physical law, as shown in Figure 9.2. In this section, we consider flow through an infinite tube, and in Section 9.5, we consider flow through a tube with finite length.

To ensure regularity of velocity and pressure at the tube axis, we suppress the Bessel functions Y_0 and Y_1 by setting

$$B_{1}^{q} = 0, \quad B_{2}^{q} = 0, \qquad (9.39)$$

where q stands for c or s. The regularized solution derived in Section 9.2 involves four coefficients, $A_{1}^{q} = 0$ and $A_{2}^{q} = 0$, and an unspecified constant, k.

9.3.1 No-Slip Boundary Condition

Enforcing the no-slip boundary condition at the tube surface by requiring that

FIGURE 9.2 Illustration of Stokes flow through an infinite cylindrical tube with a permeable wall that allows fluid to enter or escape. A test section of length L is shown.

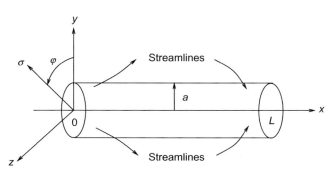

$$U_x^c(a) = 0, \quad U_x^s(a) = 0, \tag{9.40}$$

we obtain

$$kA_1^q J_0(\hat{k}) + A_2^q \left[2J_0(\hat{k}) - \hat{k}J_1(\hat{k})\right] = 0, \tag{9.41}$$

which can be rearranged into

$$kA_1^q = \left(\hat{k}\frac{J_1(\hat{k})}{J_0(\hat{k})} - 2\right)A_2^q, \tag{9.42}$$

where

$$\hat{k} \equiv ka \tag{9.43}$$

is a dimensionless constant to be computed as part of the solution.

9.3.2 Velocity Field

Substituting expression (9.42) into Equation (9.15), we obtain the axial velocity profile functions

$$U_x^q(\sigma) = A_2^q \Psi(\hat{\sigma}) \tag{9.44}$$

for $q = c, s$, where

$$\Psi(\hat{\sigma}) = \hat{k}\frac{J_1(\hat{k})}{J_0(\hat{k})}J_0(\hat{\sigma}) - \hat{\sigma}J_1(\hat{\sigma}). \tag{9.45}$$

The axial velocity distribution is

$$u_x(x, \sigma) = \left(A_2^c \cos hkx + A_2^s \sin hkx\right)\Psi(\hat{\sigma}), \tag{9.46}$$

where the coefficients A_2^c and A_2^s have physical dimensions of velocity.

Normalized profiles of the function $\Psi(\hat{\sigma})$ are shown in Figure 9.3a for several values of $\hat{k}$. The profile is parabolic for sufficiently small $\hat{k}$, approximately less than 1 (sold line), but develops undulations at higher values of $\hat{k}$ (broken lines). The significance of this behavior will be discussed later in this section.

Substituting expression (9.42) into Equations (9.16) and (9.17), we obtain the radial velocity profile functions

$$U_\sigma^c(\sigma) = -A_2^s \Phi(\hat{\sigma}), \quad U_\sigma^s(\sigma) = -A_2^c \Phi(\hat{\sigma}), \tag{9.47}$$

where

$$\Phi(\hat{\sigma}) = \left(\hat{k}\frac{J_1(\hat{k})}{J_0(\hat{k})} - 2\right)J_1(\hat{\sigma}) + \hat{\sigma}J_0(\hat{\sigma}). \tag{9.48}$$

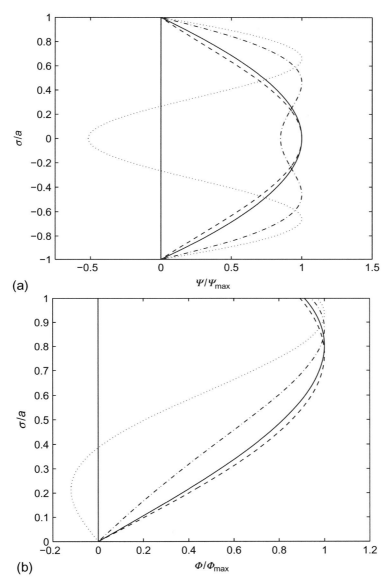

FIGURE 9.3   (a) Axial velocity profile function $\Psi(\hat{\sigma})$ and (b) radial velocity profile function $\Phi(\hat{\sigma})$ for $\hat{k} = 1.0$ (solid line), 2.5 (dashed line), 3.5 (dash-dotted line), and 4.0 (dotted line).

The radial velocity distribution is

$$u_\sigma(x, \sigma) = -\left(A_2^s \cos hkx + A_2^c \sin hkx\right)\Phi(\hat{\sigma}).\tag{9.49}$$

Note the juxtaposition of the hyperbolic functions and corresponding coefficients, A_2^q.

Normalized profiles of the function $\Phi(\hat{\sigma})$ are shown in Figure 9.3b for several values of $\hat{k}$. We observe that regions of inward radial flow may develop at high values of $\hat{k}$ (dotted line).

9.3.3 Flow Rate

Having obtained an expression for the axial velocity component, we may compute the flow rate across the tube cross section,

$$Q(x) \equiv 2\pi \int_0^a u_x(x, \sigma)\sigma d\sigma. \tag{9.50}$$

Substituting expression (9.46) for the axial velocity, we find that

$$Q(x) = \frac{2\pi a}{k}\left(A_2^c \cosh kx + A_2^s \sinh kx\right)\Phi(\hat{k}), \tag{9.51}$$

where

$$\Phi\left(\hat{k}\right) \equiv \frac{J_1(\hat{k})}{J_0(\hat{k})}\int_0^{\hat{k}} J_0(\hat{\sigma})\hat{\sigma}d\hat{\sigma} - \frac{1}{\hat{k}}\int_0^{\hat{k}} J_1(\hat{\sigma})\hat{\sigma}^2 d\hat{\sigma}. \tag{9.52}$$

Performing the integration taking into account the indefinite integrals

$$\int wJ_0(w)dw = wJ_1(w) \tag{9.53}$$

and

$$\int w^2 J_1(w)dw = 2wJ_1(w) - w^2 J_0(w), \tag{9.54}$$

we obtain

$$Q(x) = \frac{2\pi a}{k}\left(A_2^c \cosh kx + A_2^s \sinh kx\right)\Phi(\hat{k}), \tag{9.55}$$

where

$$\Phi(\hat{k}) = \left(\hat{k}\frac{J_1(\hat{k})}{J_0(\hat{k})} - 2\right)J_1\left(\hat{k}\right) + \hat{k}J_0(\hat{k}). \tag{9.56}$$

A graph of the function $\Phi(\hat{k})$ is shown in Figure 9.4. As expected, an infinite sequence of singularities occur at the zeros of J_0.

For future reference, we note that the coefficient A_2^c is related to the flow rate at $x=0$ by

$$A_2^c = \frac{k}{2\pi a}\frac{Q(0)}{\Phi(\hat{k})}. \tag{9.57}$$

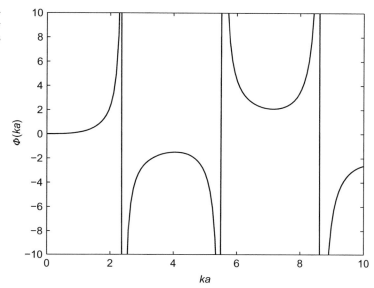

FIGURE 9.4 Graph of the function $\Phi(\hat{k})$ determining the axial flow rate. A singularity occurs when $\hat{k}$ is a zero of the Bessel function J_0.

Combining expression (9.55) for the flow rate with expression (9.46) for the axial velocity, we obtain

$$u_x(x,\sigma) = Q(x)\frac{k}{2\pi a}\frac{\Psi(\hat{\sigma})}{\Phi(\hat{k})}. \tag{9.58}$$

Making substitutions, we obtain

$$u_x(x,\sigma) = \frac{k}{2\pi a}\frac{\hat{k}J_1(\hat{k})J_0(\hat{\sigma}) - J_0(\hat{k})\hat{\sigma}J_1(\hat{\sigma})}{\left[\hat{k}J_1(\hat{k}) - 2J_0(\hat{k})\right]J_1(\hat{k}) + \hat{k}J_0^2(\hat{k})}Q(x), \tag{9.59}$$

which can be regarded as a generalized Poiseuille velocity profile expressed in terms of the flow rate.

9.3.4 Pressure Field

The general pressure profile functions derived in Section 9.2.1 simplify to

$$\chi_c(\sigma) = -2\mu k A_2^s J_0(\hat{\sigma}), \quad \chi_s(\sigma) = -2\mu k A_2^c J_0(\hat{\sigma}), \tag{9.60}$$

yielding the pressure field

$$p = -2\mu k\left(A_2^s \cos hkx + A_2^c \sin hkx\right)J_0(\hat{\sigma}) + \pi_0, \tag{9.61}$$

where π_0 is an unspecified constant.

The mean pressure over the tube cross section is defined as

$$\bar{p}(x) \equiv \frac{2}{a^2} \int_0^a p(x, \sigma) \sigma d\sigma. \tag{9.62}$$

Substituting expression (9.61), we obtain

$$\bar{p}(x) = -\frac{4\mu}{ka^2} \left(A_2^s \cosh kx + A_2^c \sinh kx \right) \int_0^{ka} J_0(\hat{\sigma}) \hat{\sigma} d\hat{\sigma} + \pi_0. \tag{9.63}$$

Performing the integration, taking into consideration the indefinite integral given in (9.53), we obtain

$$\Delta \bar{p}(x) = -\frac{4\mu}{a} \left(A_2^s \cosh kx + A_2^c \sinh kx \right) J_1(\hat{k}), \tag{9.64}$$

where

$$\Delta \bar{p}(x) \equiv \bar{p}(x) - \pi_0 \tag{9.65}$$

is a mean transmural pressure. We observe that the coefficient A_2^s is related to the mean transmural pressure at $x = 0$ by

$$A_2^s = -\frac{a}{4\mu} \frac{\Delta \bar{p}(0)}{J_1(\hat{k})}. \tag{9.66}$$

Combining (9.64) with (9.61), we find that the pressure distribution is related to the mean transmural pressure by

$$\frac{p(x, \sigma) - \pi_0}{\Delta \bar{p}(x)} = \frac{1}{2} \frac{\hat{k}}{J_1(\hat{k})} J_0(\hat{\sigma}). \tag{9.67}$$

Combining (9.64) with (9.49), we obtain the radial velocity distribution

$$u_\sigma(x, \sigma) = \frac{a \Delta \bar{p}(x)}{4\mu} \frac{\Phi(\hat{\sigma})}{J_1(\hat{k})}. \tag{9.68}$$

Making substitutions, we obtain

$$u_\sigma(x, \sigma) = \Delta \bar{p}(x) \frac{a}{4\mu} \frac{\left[\hat{k} J_1(\hat{k}) - 2 J_0(\hat{k}) \right] J_1(\hat{\sigma}) + J_0(\hat{k}) \hat{\sigma} J_0(\hat{\sigma})}{J_0(\hat{k}) J_1(\hat{k})}. \tag{9.69}$$

9.3.5 Stress Field

Substituting expression (9.42) into Equations (9.26) and (9.27), we obtain the axial normal stress profile functions

$$\Sigma_{xx}^c(\sigma) = 2\mu k A_2^s H(\hat{\sigma}), \quad \Sigma_{xx}^s(\sigma) = 2\mu k A_2^c H(\hat{\sigma}), \tag{9.70}$$

where

$$H(\hat{\sigma}) = \left(\hat{k} \frac{J_1(\hat{k})}{J_0(\hat{k})} - 2 \right) J_0(\hat{\sigma}) + 3J_0(\hat{\sigma}) - \hat{\sigma}J_1(\hat{\sigma}). \tag{9.71}$$

Substituting expression (9.42) into Equations (9.30) and (9.33), we obtain the radial normal stress profile functions

$$\Sigma_{\sigma\sigma}^c(\sigma) = 2\mu k A_2^s \Omega(\hat{\sigma}), \quad \Sigma_{\sigma\sigma}^s(\sigma) = 2\mu k A_2^c \Omega(\hat{\sigma}), \tag{9.72}$$

where

$$\Omega(\hat{\sigma}) = \left(2 - \hat{k} \frac{J_1(\hat{k})}{J_0(\hat{k})} \right) \left[J_0(\hat{\sigma}) - \frac{1}{\hat{\sigma}} J_1(\hat{\sigma}) \right] + \hat{\sigma}J_1(\hat{\sigma}) \tag{9.73}$$

or

$$\Omega(\hat{\sigma}) = J_0(\hat{\sigma}) + \frac{\Phi(\hat{\sigma})}{\hat{\sigma}} - \hat{k}J_1(\hat{k}) + \hat{\sigma}J_1(\hat{\sigma}). \tag{9.74}$$

Substituting expression (9.42) into Equation (9.38), we obtain the shear stress profile functions

$$\Sigma_{x\sigma}^q(\sigma) = 2\mu k A_2^q \Pi(\hat{\sigma}) \tag{9.75}$$

for $q = c, s$, where

$$\Pi(\hat{\sigma}) = \left(2 - \hat{k} \frac{J_1(\hat{k})}{J_0(\hat{k})} \right) J_1(\hat{\sigma}) - J_1(\hat{\sigma}) - \hat{\sigma}J_0(\hat{\sigma}). \tag{9.76}$$

9.3.6 Solution in Terms of the Flow Rate and Pressure at the Origin

The velocity and pressure fields can be expressed in terms of the axial flow rate at the designated origin of the x axis, $Q(0)$, and corresponding mean transmural pressure, $\Delta\bar{p}(0)$. Following Marshall and Trowbridge [2], we introduce the negative of the ratio of the coefficients A_2^s and A_2^c given in (9.66) and (9.57),

$$C \equiv -\frac{A_2^s}{A_2^c} = \frac{\Delta\bar{p}(0)}{Q(0)} \frac{\pi a^3}{2\mu} \frac{\Phi(\hat{k})}{\hat{k}J_1(\hat{k})}. \tag{9.77}$$

Making substitutions, we obtain

$$C = \frac{\Delta\bar{p}(0)}{Q(0)} \frac{\pi a^3}{2\mu} \left(\frac{J_1(\hat{k})}{J_0(\hat{k})} + \frac{J_0(\hat{k})}{J_1(\hat{k})} - \frac{2}{\hat{k}} \right). \tag{9.78}$$

Substituting expressions (9.57) and (9.66) for the coefficients A_2^c and A_2^s into expression (9.46) for the axial velocity, we obtain

$$u_x(x,\sigma) = A_2^c(\cosh kx - C\sinh kx)\Psi(\hat{\sigma}) \qquad (9.79)$$

or

$$u_x(x,\sigma) = \left(\frac{k}{2\pi a}\frac{Q(0)}{\Phi(\hat{k})}\cosh kx - \frac{a}{4\mu}\frac{\Delta\bar{p}(0)}{J_1(\hat{k})}\sinh kx\right)\Psi(\hat{\sigma}). \qquad (9.80)$$

Substituting expressions (9.57) and (9.66) into expression (9.49) for the radial velocity, we obtain

$$u_\sigma(x,\sigma) = A_2^c(C\cosh kx - \sinh kx)\Phi(\hat{\sigma}) \qquad (9.81)$$

or

$$u_\sigma(x,\sigma) = \left(-\frac{k}{2\pi a}\frac{Q(0)}{\Phi(\hat{k})}\sinh kx + \frac{a}{4\mu}\frac{\Delta\bar{p}(0)}{J_1(\hat{k})}\cosh kx\right)\Phi(\hat{\sigma}). \qquad (9.82)$$

Substituting expressions (9.57) and (9.66) into expression (9.64) for the averaged pressure, we obtain

$$\Delta\bar{p}(x) = \frac{4\mu}{a}A_2^c(C\cosh kx - \sinh kx)J_1(\hat{k}) \qquad (9.83)$$

or

$$\Delta\bar{p}(x) = \Delta\bar{p}(0)\cosh kx - Q(0)\frac{2\mu}{\pi a^3}\hat{k}\frac{J_1(\hat{k})}{\Phi(\hat{k})}\sinh kx. \qquad (9.84)$$

Substituting expressions (9.57) and (9.66) into expression (9.51) for the flow rate, we obtain

$$Q(x) = \frac{2\pi a}{k}A_2^c(\cosh kx - C\sinh kx)\Phi(\hat{k}) \qquad (9.85)$$

or

$$Q(x) = Q(0)\cosh kx - \Delta\bar{p}(0)\frac{\pi a^3}{2\mu}\frac{\Phi(\hat{k})}{\hat{k}J_1(\hat{k})}\sinh kx. \qquad (9.86)$$

Marshall and Trowbridge [2] discuss the structure of the flow in terms of the dimensionless coefficient C.

9.4 STARLING'S EQUATION

The expressions derived in Section 9.3 involve an unspecified constant, k, with dimensions of inverse length. To determine this constant, we stipulate that fluid enters or escapes through the walls of the tube with a wall velocity according to a sensible membrane law.

9.4.1 Starling's Equation in Terms of the Pressure

The standard Starling equation employed in all previous studies (e.g., [2]) prescribes that the radial velocity at the tube surface is given by

$$u_\sigma(x, a) = L_p[p(x, a) - p_a], \tag{9.87}$$

where $p(x, a)$ is the fluid pressure at the inner tube surface, p_a is the ambient pressure, and L_p is the wall permeability. Theoretical estimates for the permeability have been obtained for idealized membrane configurations consisting of periodic arrays of cylinders (e.g., [8]).

Substituting in Equation (9.87) the expressions derived in Section 9.3 for the radial velocity and pressure, we obtain

$$U_\sigma^c(a)\cos hkx + U_\sigma^s(a)\sin hkx = L_p(\chi_c(a)\cos hkx + \chi_s(a)\sin hkx + \pi_0 - p_a), \tag{9.88}$$

yielding

$$\pi_0 = p_a \tag{9.89}$$

and

$$\Phi(\hat{k}) = 2\mathcal{L}_p\hat{k}J_0(\hat{k}), \tag{9.90}$$

where

$$\mathcal{L}_p \equiv \frac{\mu L_p}{a} \tag{9.91}$$

is a dimensionless permeability.

Solving Equation (9.90) for $\hat{k} \equiv ka$ in terms of $\mathcal{L}_p$ by Newton's method, we obtain the solution shown with the solid line in Figure 9.5a on a log-linear scale. As $\mathcal{L}_p$ tends to infinity, $\hat{k}$ tends to an asymptotic value that is the first zero of J_0, which is approximately equal to $\hat{k}_0 \simeq 2.404$. Substituting the Taylor series of the Bessel functions about this value into (9.90) provides us with a quadratic equation for the difference $\Delta\hat{k} \equiv \hat{k} - \hat{k}_0$,

$$\hat{k}_0(2\mathcal{L}_p - 1)\Delta\hat{k}^2 - 2\Delta\hat{k} - \hat{k}_0 = 0. \tag{9.92}$$

Retaining the root with the minus sign in the standard quadratic formula provides us with the results shown with the dotted line in Figure 9.5a.

In physiological applications, the dimensionless permeability $\mathcal{L}_p$ is typically small. For example, in the case of flow in a proximal tubule, $\mathcal{L}_p$ is on the order of 10^{-7} (e.g., [2]). The numerical results plotted with the solid line in Figure 9.5a reveal that, as $\mathcal{L}_p$ tends to zero, $\hat{k}$ also tends to zero. Using the following Maclaurin series expansions of the Bessel functions,

$$J_0(z) = 1 - \frac{1}{4}z^2 + \mathcal{O}(z^4), \quad J_1(z) = \frac{1}{2}z - \frac{1}{16}z^3 + \mathcal{O}(z^5), \tag{9.93}$$

we find that

$$\Phi(\hat{k}) \simeq \frac{1}{8}\hat{k}\left(\hat{k}^2 + \frac{1}{12}\hat{k}^4\right). \tag{9.94}$$

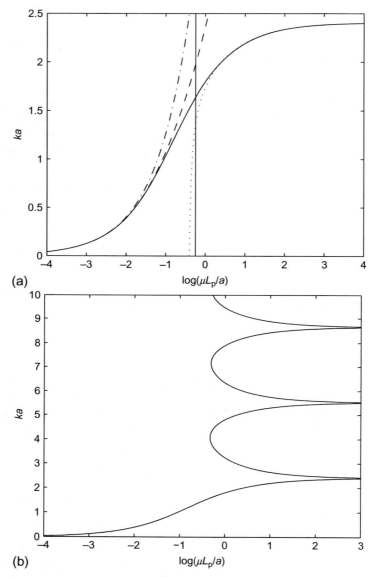

FIGURE 9.5   (a) Dependence of the coefficient $\hat{k} \equiv ka$ on the scaled wall permeability, $\mathcal{L}_p \equiv \mu L_p / a$ for Starling's equation in terms of the pressure. The dotted line represents asymptotic predictions for large scaled permeabilities, and the dashed and dot-dashed lines represent asymptotic predictions for small scaled permeabilities. The vertical solid line corresponds to dimensionless permeability $\mathcal{L}_p = 10^{-0.25} \simeq 0.5623$. (b) Complete solution space.

The transcendental Equation (9.90) reduces to a biquadratic equation,

$$\frac{1}{3}\hat{k}^4 + \hat{k}^2 \simeq 16\mathcal{L}_{\mathrm{p}}, \tag{9.95}$$

whose solution is plotted with the dashed line in Figure 9.5a. The asymptotic solutions represented by the dashed and dotted lines are approximately valid up to a lower or upper threshold, respectively, represented by the vertical solid line. These approximations can be used as initial guesses for solving the nonlinear algebraic Equation (9.90) using an iterative method (e.g., [5]). Neglecting the quartic term in (9.95) provides us with a linear equation in $\hat{k}^2$, yielding

$$\hat{k} \simeq 4\mathcal{L}_{\mathrm{p}}^{1/2}, \tag{9.96}$$

represented by the dash-dotted line in Figure 9.5a.

Using the Maclaurin expansions of the Bessel functions given in (9.93) along with (9.96), we find that the axial velocity distribution described in (9.59) is

$$u_x(x,\sigma) \simeq Q(x)\frac{1}{2\pi a^4}\left(a^2 - \sigma^2\right), \tag{9.97}$$

corresponding to quasi-unidirectional Poiseuille flow with a spatially evolving parabolic velocity profile. The radial velocity distribution described in (9.69) is given by

$$u_\sigma(x,\sigma) \simeq \Delta\bar{p}(x)\frac{a}{4\mu\frac{1}{2}\hat{k}}\left[\left(\frac{1}{2}\hat{k}^2 - 1\right)\hat{\sigma}\left(1 - \frac{1}{8}\hat{\sigma}^2\right) + \left(1 - \frac{1}{4}\hat{k}^2\right)\hat{\sigma}\left(1 - \frac{1}{4}\hat{\sigma}^2\right)\right]. \tag{9.98}$$

Simplifying, we obtain

$$u_\sigma(x,\sigma) \simeq \Delta\bar{p}(x)\frac{1}{16\mu k}\left(2\hat{k}^2\hat{\sigma} - \hat{\sigma}^3\right), \tag{9.99}$$

yielding

$$u_\sigma(x,\sigma) = \Delta\bar{p}(x)\frac{1}{16\mu a^2}\hat{k}^2\sigma\left(2a^2 - \sigma^2\right). \tag{9.100}$$

Expressing (9.96) in the form $\hat{k}^2 \simeq 16\mu L_{\mathrm{p}}/a$, we obtain

$$u_\sigma(x,\sigma) = \Delta\bar{p}(x)\frac{L_{\mathrm{p}}}{a^3}\sigma\left(2a^2 - \sigma^2\right). \tag{9.101}$$

This expression will be recovered in Section 9.5.3 in the framework of nearly unidirectional flow. The pressure profile given in (9.67) becomes

$$\frac{p(x,\sigma) - p_{\mathrm{a}}}{\Delta\bar{p}(x)} \simeq \frac{1 - \frac{1}{4}\hat{\sigma}^2}{1 - \frac{1}{8}\hat{k}^2} \simeq \left(1 - \frac{1}{4}\hat{\sigma}^2\right)\left(1 + \frac{1}{8}\hat{k}^2\right) \simeq 1 + \frac{1}{8}k^2\left(a^2 - 2\sigma^2\right), \tag{9.102}$$

yielding

$$\frac{p(x,\sigma)-p_a}{\Delta\overline{p}(x)} \simeq 1 + 2\mathcal{L}_p \frac{a^2 - 2\sigma^2}{a^2}. \tag{9.103}$$

This expression cannot be predicted working in the framework of nearly unidirectional flow.

In fact, the nonlinear Equation (9.90) has multiple solutions when $\mathcal{L}_p$ is approximately greater than 0.47, as shown in Figure 9.5b. Solution branches appear as lobes originating from the zeros of the Bessel function J_0 in the limit of infinite wall permeability. For a fixed permeability, each solution for $\hat{k}$ has a distinct corresponding velocity profile, as discussed in Section 9.3. However, the primary solution branch shown in Figure 9.5a is expected to prevail in physiological applications.

9.4.2 Starling's Equation in Terms of the Normal Stress

An alternative and seemingly more consistent implementation of Starling's equation involves the wall normal stress instead of the pressure according to the membrane law

$$u_\sigma(x,a) = -L_p[\sigma_{\sigma\sigma}(x,a) + p_a], \tag{9.104}$$

where $\sigma_{\sigma\sigma}^a = -p_a$ is the ambient normal stress. Substituting the expressions derived in Section 9.3 for the radial velocity and normal stress, we obtain

$$U_\sigma^c(a)\cos hkx + U_\sigma^s(a)\sin hkx = -L_p\left(\sigma_{\sigma\sigma}^c(a)\cos hkx + \sigma_{\sigma\sigma}^s(a)\sin hkx - \pi_0 + p_a\right), \tag{9.105}$$

yielding the familiar equation $\pi_0 = p_a$ and the new equation

$$(1 - 2\mathcal{L}_p)\Phi(\hat{k}) = 2\mathcal{L}_p\hat{k}J_0(\hat{k}), \tag{9.106}$$

which differs from (9.90) only by the presence of the factor $1 - 2\mathcal{L}_p$ on the left-hand side.

The solution space of the nonlinear algebraic Equation (9.106) is shown in Figure 9.6 on a log-linear scale. For small $\mathcal{L}_p$, we obtain a biquadratic equation,

$$\frac{1}{3}\hat{k}^4 + \hat{k}^2 \simeq 16\frac{\mathcal{L}_p}{1 - 2\mathcal{L}_p}, \tag{9.107}$$

whose solution is represented by the dashed line in Figure 9.6. Neglecting the quartic term provides us with a linear equation in $\hat{k}^2$ represented by the dash-dotted line in Figure 9.6. Real solutions exist only when $\mathcal{L}_p < 0.5$, and multiple real solutions exist when $\mathcal{L}_p$ is sufficiently close to 0.5 from lower values. A family of complex solutions whose real and imaginary parts are drawn with the solid and dotted lines in Figure 9.6 arise when $\mathcal{L}_p > 0.5$. Physically, the corresponding velocity and pressure fields exhibit exponentially modulated undulations with wavelength nearly equal to 2π. The complex solution branches are attached to the real solution branches at the zeros of J_0 and tend to well defined asymptotic values at large $\mathcal{L}_p$.

Comparing the results shown in Figures 9.5 and 9.6 demonstrates the paramount importance of the membrane perfusion law at sufficiently high wall permeabilities.

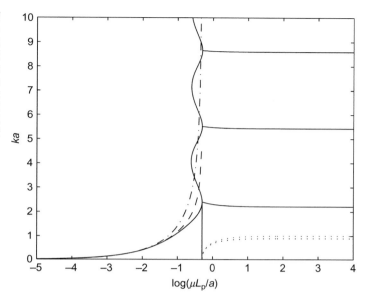

FIGURE 9.6 Solution space of Equation (9.106) corresponding to Starling's equation in terms of the normal stress. The dashed and dash-dotted lines represent asymptotic predictions for small-scaled permeability. The nearly horizontal solid lines on the left represent the real part, and the dotted lines emanating from the horizontal axis represent the imaginary part of complex solutions.

9.5 FLOW THROUGH A TUBE WITH FINITE LENGTH

The exact solution derived in Sections 9.3 and 9.4 applies for an infinite tube embedded in a constant-pressure ambient medium. In applications, we are interested in flow through a tube of finite length extending from $x=0$ to L. The corresponding solution can be derived by an adaptation of that derived in Section 9.3, neglecting entrance and exit effects.

9.5.1 Solution in Terms of Entrance and Exit Flow Rates

Assuming that the inlet and outlet flow rates, $Q(0)$ and $Q(L)$, are specified, we introduce the ratio of the flow rates at the beginning and end of the test section,

$$\chi \equiv \frac{Q(L)}{Q(0)}. \tag{9.108}$$

Referring to the general expression for the flow rate given in (9.51), we find that

$$A_2^s = \frac{\chi - \cosh kL}{\sinh kL} A_2^c. \tag{9.109}$$

Substituting this expression along with expression (9.57) for A_2^c into (9.51), we obtain

$$Q(x) = [F(x) + \chi F(L-x)]Q(0), \tag{9.110}$$

where

$$F(x) \equiv \frac{\sinh k(L-x)}{\sinh kL} \tag{9.111}$$

is the scaled hyperbolic sine, plotted with the solid lines in Figure 9.7a,b for $\mathcal{L}_{\mathrm{p}}=0.01$ and 0.10 and values of $\hat{k}$ satisfying (9.90) corresponding to Starling's equation in terms of the pressure.

For large-scaled permeability, $kL>1$, we obtain boundary layers with thickness $\delta\simeq 1/k$. For small dimensionless wall permeability, $\hat{k}\equiv ka<1$, we obtain

$$kL=\hat{k}\frac{L}{a}\simeq 4\mathcal{L}_{\mathrm{p}}^{1/2}\frac{L}{a}=\sqrt{\alpha},\tag{9.112}$$

where

$$\alpha\equiv 16\mathcal{L}_{\mathrm{p}}\left(\frac{L}{a}\right)^2=16L_{\mathrm{p}}\frac{\mu L^2}{a^3}\tag{9.113}$$

is a dimensionless permeability. The function $F(x)$ simplifies to

$$F(x)\simeq\frac{\sinh[\sqrt{\alpha}(1-\hat{x})]}{\sinh\sqrt{\alpha}},\tag{9.114}$$

where $\hat{x}=x/L$. Using (9.96), we obtain boundary layers with thickness $\delta\sim 1/k=a\hat{k}=a/\left(4\mathcal{L}_{\mathrm{p}}^{1/2}\right)$. The asymptotic solution, plotted with the dotted lines in Figure 9.7a,c, provides us with an accurate approximate approximately when $\mathcal{L}_{\mathrm{p}}<0.01$.

Substituting expressions (9.128) and (9.57) into the averaged pressure distribution given in (9.64), we obtain

$$\Delta\bar{p}(x)=-\frac{\mu}{a^3}[\chi G(x)-G(L-x)]Q(0),\tag{9.115}$$

where

$$G(x)=\frac{2\cos hkx}{\pi\sin hkL}\hat{k}\frac{J_1(\hat{k})}{\Phi(\hat{k})}\tag{9.116}$$

is a scaled hyperbolic cosine. Using Equation (9.90), we obtain the alternative expression

$$G(x)=\frac{4}{\pi\mathcal{L}_{\mathrm{p}}}\frac{\cos hkx J_1(\hat{k})}{\sin hkL J_0(\hat{k})}.\tag{9.117}$$

The function $G(x)$ is plotted with the solid lines in Figure 9.7c,d for $\mathcal{L}_{\mathrm{p}}=0.01$ and 0.10 and values of $\hat{k}$ satisfying (9.90). For small dimensionless wall permeability, $\hat{k}\equiv ka<1$, we obtain the approximation

$$G(x)\simeq\frac{8}{\pi}\frac{1}{\mathcal{L}_{\mathrm{p}}^{1/2}}\frac{\cosh\sqrt{\alpha}\hat{x}}{\sinh\sqrt{\alpha}},\tag{9.118}$$

where $\hat{x}=x/L$. The asymptotic solution is plotted with the dotted lines in Figure 9.7c,d.

From (9.115), we find that the ratio of the outlet and inlet transmural pressures is

$$\xi\equiv\frac{\Delta\bar{p}(L)}{\Delta\bar{p}(0)}=\frac{\chi G(L)-G(0)}{\chi G(0)-G(L)}=\frac{\chi\cos hkL-1}{\chi-\cos hkL}.\tag{9.119}$$

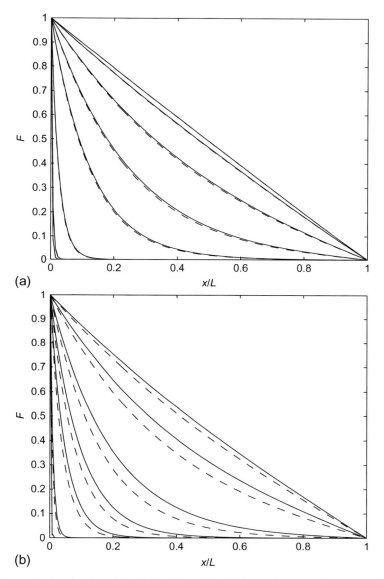

FIGURE 9.7 Dimensionless functions $F(x/L)$, $G(x/L)$, and $R(x/L)$ determining the flow rate and pressure drop for values of $\hat{k}$ satisfying (9.90) and (a, c, e) $\mathcal{L}_p = 0.01$ or (b, d, f) 0.10. The graphs correspond to tube aspect ratio $L/a = 1$, 2, 5, 10, 20, 100, 500, and 800 (lowest curves). The broken lines represent the asymptotic solution representing nearly unidirectional flow for small $\hat{k}$.

(Continued)

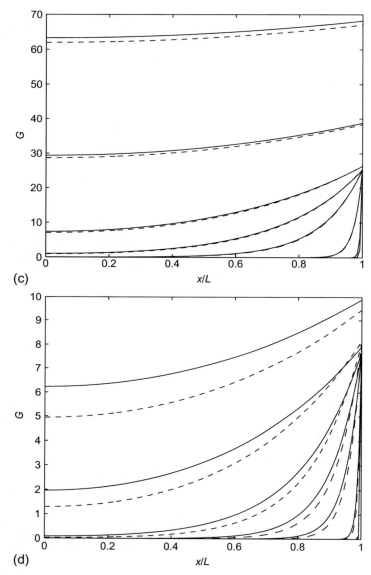

FIGURE 9.7   (Continued)

Conversely,

$$\chi = \frac{\xi \cosh kL - 1}{\xi - \cosh kL}. \tag{9.120}$$

This equation can be used to eliminate χ in favor of ξ, if so desired. For small $\hat{k}$,

$$\xi \simeq \frac{\chi \cosh \sqrt{\alpha} - 1}{\chi - \cosh \sqrt{\alpha}}, \quad \chi \simeq \frac{\xi \cosh \sqrt{\alpha} - 1}{\xi - \cosh \sqrt{\alpha}}. \tag{9.121}$$

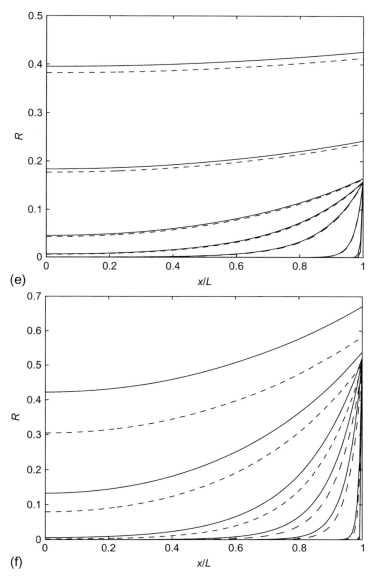

(e)

(f)

FIGURE 9.7   (Continued)

9.5.2 Solution in Terms of the Entrance and Exit Pressures

Referring to the general expression for the transmural given in (9.64), we find that

$$A_2^c = \frac{\xi - \cosh kL}{\sinh kL} A_2^s,$$ (9.122)

where ξ is the transmural pressure ratio defined in (9.119). Substituting expression (9.122) along with expression (9.66) for A_2^c into the general expression for the pressure (9.64), we obtain

$$\Delta\bar{p}(x) = [\xi F(x) + F(L-x)]\Delta\bar{p}(0), \qquad (9.123)$$

where the function $F(x)$ is defined in (9.111) and plotted in Figure 9.7.

Substituting expressions (9.66) and (9.122) into the expression for the flow rate given in (9.51), we obtain

$$Q(x) = \frac{a^3}{\mu}[R(L-x) - \xi R(x)]\Delta\bar{p}(0), \qquad (9.124)$$

where

$$R(x) \equiv \frac{\pi \cosh kx}{2 \sinh kL} \frac{1}{\hat{k}} \frac{\Phi(\hat{k})}{J_1(\hat{k})}. \qquad (9.125)$$

Using (9.90), we obtain the alternative expression

$$R(x) = \pi \mathcal{L}_p \frac{\cosh kx}{\sinh kL} \frac{J_0(\hat{k})}{J_1(\hat{k})}. \qquad (9.126)$$

The function $R(x)$ is plotted with the solid lines in Figure 9.7e,f for $\mathcal{L}_p = 0.10$ and 0.01 and values of $\hat{k}$ satisfying (9.90). For small dimensionless wall permeability, $\hat{k} < 1$, we obtain the approximation

$$R(x) \simeq \frac{\pi}{2} \mathcal{L}_p^{1/2} \frac{\cosh(\sqrt{\alpha}\hat{x})}{\sinh\sqrt{\alpha}}, \qquad (9.127)$$

where $\hat{x} = x/L$ and the dimensionless permeability α is defined in (9.113).

9.5.3 Starling's Equation in Terms of the Normal Stress

The results shown in Figure 9.7 correspond to real values of $\hat{k}$ satisfying the algebraic Equation (9.90), corresponding to Starling's equation in terms of the pressure. The real and imaginary parts of the functions $F(\hat{x})$, $G(\hat{x})$, and $R(\hat{x})$ for $\mathcal{L}_p = 1.0$ and 10.0 and complex values of $\hat{k}$ satisfying (9.106) corresponding to Starling's equation in terms of the normal stress are plotted with the solid and dashed lines in Figure 9.8. The occurrence of overshooting associated with local maxima is noteworthy.

9.5.4 Nearly Unidirectional Flow Model

Asymptotic solutions for small wall permeability were derived in Section 9.4 based on the exact solution for flow through an infinite tube. These asymptotic solutions are consistent with a simple model based on the assumption of nearly unidirectional flow with a parabolic velocity profile and uniform pressure over the tube cross section at every streamwise position, x.

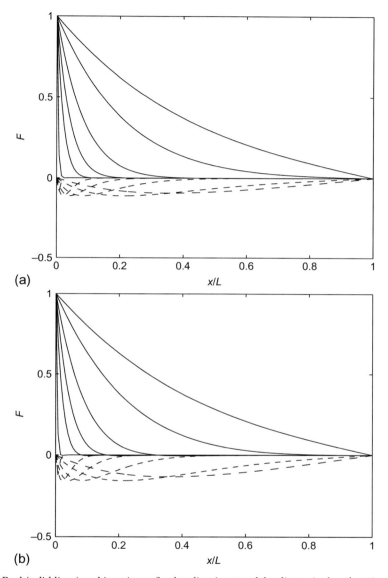

FIGURE 9.8 Real (solid lines) and imaginary (broken lines) parts of the dimensionless functions $F(x/L)$, $G(x/L)$, and $R(x/L)$ determining the flow rate and pressure drop for complex values of $\hat{k}$ satisfying (9.106) and (a, c, e) $\mathcal{L}_p = 0.01$ or (b, d, f) 0.10. The graphs correspond to tube aspect ratio $L/a = 1, 2, 5, 10, 20, 100, 500$, and 800 (lowest curves). The dotted lines represent the asymptotic solution describing nearly unidirectional flow for small $\hat{k}$.

(Continued)

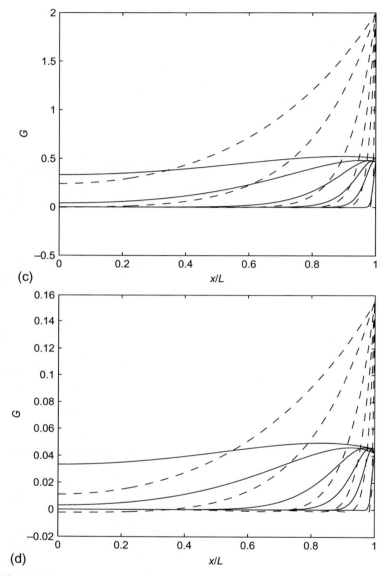

FIGURE 9.8   (Continued)

Using the Poiseuille solution for flow through a straight circular pipe (e.g., [5]), we obtain the axial velocity,

$$u_x(x, \sigma) = -\frac{dp}{dx}\frac{1}{4\mu}\left(a^2 - \sigma^2\right),$$

(9.128)

and flow rate,

$$Q(x) = -\frac{dp}{dx}\frac{\pi a^4}{8\mu},$$

(9.129)

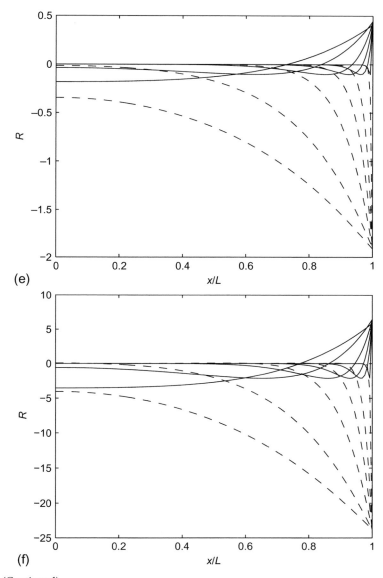

(e)

(f)

FIGURE 9.8 (Continued)

with the understanding that the pressure gradient dp/dx is a function of x. Fluid escapes through the walls of the tube with wall velocity that is given by Starling's equation,

$$u_\sigma(x, a) = L_p(p - p_a), \tag{9.130}$$

where p_a is a constant ambient pressure and L_p is the wall permeability. A mass balance requires that

$$\frac{dQ}{dx} = -2\pi u_\sigma(x,a). \tag{9.131}$$

Combining the preceding expressions, we find that the pressure inside the tube satisfies the second-order differential equation

$$\frac{d^2p}{dx^2} = \frac{16\mu L_p}{a^3}(p-p_a), \tag{9.132}$$

subject to specified end-point pressures, $p_0 = p(x=0)$ and $p_L = p(x=L)$, or specified end-point flow rates. The solution of (9.132) is consistent with that derived in Section 9.4 for small $\hat{k}$.

The radial velocity can be found by writing a mass balance over a cylindrical differential control volume of radius σ,

$$\frac{\partial}{\partial x}\int_0^\sigma 2\pi\sigma' u_x(x,\sigma')d\sigma' = -2\pi\sigma u_\sigma. \tag{9.133}$$

Substituting the expression for u_x from (9.129) and rearranging, we obtain

$$u_\sigma = \frac{d^2p}{dx^2}\frac{1}{4\mu\sigma}\int_0^\sigma \left(a^2 - \sigma'^2\right)\sigma'd\sigma'. \tag{9.134}$$

Performing the integration and using (9.132), we find that

$$u_\sigma(x,\sigma) = \frac{L_p}{a^3}(p-p_a)\sigma\left(2a^2 - \sigma^2\right), \tag{9.135}$$

which is consistent with (9.101).

9.6 EFFECT OF WALL SLIP

In the analysis presented in Sections 9.3–9.5, we have assumed that the fluid satisfies the no-slip boundary condition at the tube surface, which is an approximation. Imposing instead the slip boundary condition, we specify that

$$u_x = -\frac{\ell}{\mu}\sigma_{x\sigma} \tag{9.136}$$

at the tube surface located at $\sigma=a$, where $\sigma_{x\sigma}$ is the shear stress and ℓ is a slip length. The evaluation of the slip length, inherent in the slip velocity and associated drift velocity, has been the subject of theoretical and computational investigations (e.g., [6,7]).

Substituting expression (9.15) for the velocity and expression (9.38) for the shear stress into Equation (9.136), both with $B_1^q = 0$ and $B_2^q = 0$, we obtain

$$kA_1^q J_0(\hat{k}) + A_2^q\left[2J_0(\hat{k}) - \hat{k}J_1(\hat{k})\right] = 2\hat{\ell}\hat{k}\left(kA_1^q J_1(\hat{k}) + A_2^q\left[J_1(\hat{k}) + \hat{k}J_0(\hat{k})\right]\right) \tag{9.137}$$

for $q=c,s$, where $\hat{\ell} \equiv \ell/a$ is a dimensionless slip length. Rearranging, we obtain the counterpart of Equation (9.42),

$$kA_1^q = \left(\hat{k}S(\hat{k}) - 2\right)A_2^q,$$ (9.138)

where

$$S(\hat{k}) = \frac{J_1(\hat{k}) + 2\hat{\ell}\hat{k}J_0(\hat{k})}{J_0(\hat{k}) - 2\hat{\ell}\hat{k}J_1(\hat{k})}.$$ (9.139)

The no-slip boundary condition corresponds to $\hat{\ell} = 0$, yielding $S(\hat{k}) = J_1(\hat{k})/J_0(\hat{k})$. The axial velocity profile functions are given in (9.44), where

$$\Psi(\hat{\sigma}) = \hat{k}S(\hat{k})J_0(\hat{\sigma}) - \hat{\sigma}J_1(\hat{\sigma}).$$ (9.140)

Normalized profiles of the function $\Psi(\hat{\sigma})$ are shown in Figure 9.9a,b for $\ell/a = 0.01$ and 0.1, and for several values of $\hat{k}$. The corresponding profiles for $\ell/a = 0$ are shown in Figure 9.3a. The profile of the function Ψ is parabolic for sufficiently small $\hat{k}$, approximately less than 1 (solid line) and develops undulations at higher values of $\hat{k}$ (broken lines) for any slip coefficient.

The radial velocity profile functions are given in (9.47), where

$$\Phi(\hat{\sigma}) = \left(\hat{k}S(\hat{k}) - 2\right)J_1(\hat{\sigma}) + \hat{\sigma}J_0(\hat{\sigma}).$$ (9.141)

Normalized profiles of the function $\Phi(\hat{\sigma})$ are shown in Figure 9.9c,d for $\ell/a = 0.01$ and 0.1 and several values of $\hat{k}$. The corresponding profiles for $\ell/a = 0$ are shown in Figure 9.3b.

The axial flow rate is given in Equation (9.51), where

$$\Phi\left(\hat{k}\right) = \left(\hat{k}S(\hat{k}) - 2\right)J_1(\hat{k}) + \hat{k}J_0(\hat{k}).$$ (9.142)

Graphs of the function $\Phi\left(\hat{k}\right)$ for three values of ℓ/a are shown in Figure 9.10. An infinite sequence of singularities occur at the zeros of the denominator of the function $S(k)$, determined by the slip coefficient.

The pressure field is given by Equation (9.60) in terms of two unspecified coefficients, A_s^q for $q=c,s$. The xx component of the stress is given by Equation (9.70), where

$$H(\hat{\sigma}) = \left(\hat{k}S(\hat{k}) - 2\right)J_0(\hat{\sigma}) + 3J_0(\hat{\sigma}) - \hat{\sigma}J_1(\hat{\sigma}).$$ (9.143)

The $\sigma\sigma$ component of the stress is given by Equation (9.72), where

$$\Omega(\hat{\sigma}) = \left(2 - \hat{k}S(\hat{k})\right)\left[J_0(\hat{\sigma}) - \frac{1}{\hat{\sigma}}J_1(\hat{\sigma})\right] + \hat{\sigma}J_1(\hat{\sigma}).$$ (9.144)

The $x\sigma$ component of the stress is given by Equation (9.75), where

$$\Pi(\hat{\sigma}) = \left(2 - \hat{k}S(\hat{k})\right)J_1(\hat{\sigma}) - J_1(\hat{\sigma}) - \hat{\sigma}J_0(\hat{\sigma}).$$ (9.145)

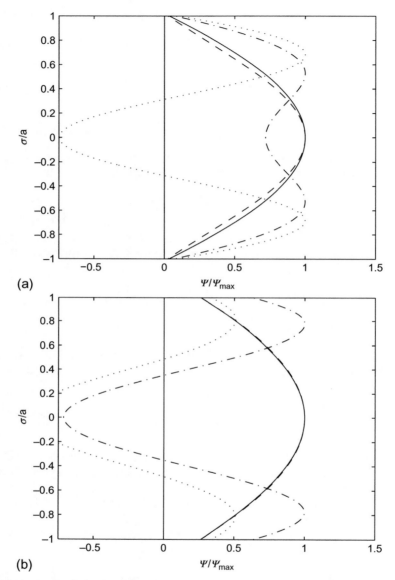

FIGURE 9.9   (a, b) Axial velocity profile function $\Psi(\hat{\sigma})$, and (c, d) radial velocity profile function $\Phi(\hat{\sigma})$, for $\hat{k} = 1.0$ (solid line), 2.5 (dashed line), 3.5 (dash-dotted line), and 4.0 (dotted line), and scaled slip length (a, c) $\ell/a = 0.01$ or (b, d) 0.10.

(Continued)

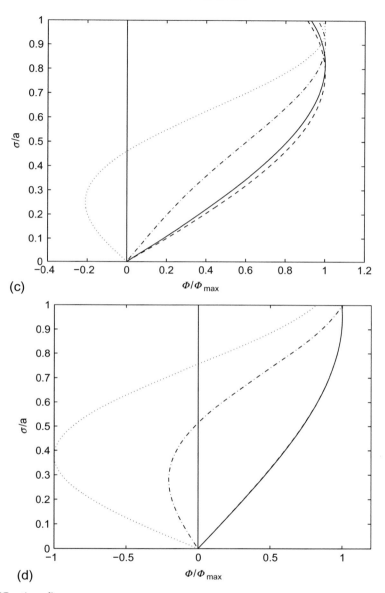

(c)

(d)

FIGURE 9.9 (Continued)

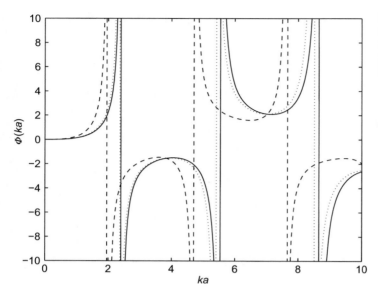

FIGURE 9.10   Graph of the function $\Phi\left(\hat{k}\right)$ determining the axial flow rate for scaled slip length $\ell/a=0$ (solid line), 0.01 (dotted line), and 0.10 (dashed line).

To derive the solution for flow through an infinite tube, we introduce Starling's equation for the pressure or normal stress.

9.6.1 Starling's Equation

Adopting Starling's equation in terms of the pressure, as discussed in Section 9.5, we derive the algebraic Equation (9.90), where the function $\Phi(\hat{k})$ now is given in Equation (9.142). Solving Equation (9.90) for $\hat{k} \equiv ka$ in terms of $\mathcal{L}_p$ by Newton's method, we obtain the solution shown in Figure 9.11a on a log-linear scale for three permeabilities. As $\mathcal{L}_p$ tends to infinity, $\hat{k}$ tends to an asymptotic value that is a zero of the denominator of the function $S(k)$. In fact, as in the case of no-slip, the nonlinear Equation (9.90) has multiple solutions when $\mathcal{L}_p$ is greater than a threshold, as shown in Figure 9.11b. Solution branches appear as lobes originating from the zeros of the Bessel function $S(\hat{k})$ in the limit of infinite wall permeability.

Adopting instead Starling's equation in terms of the normal stress, as discussed in Section 9.6, we derive the algebraic Equation (9.106), where the function $\Phi(\hat{k})$ now is given in (9.93). The solution space is shown in Figure 9.12 on a log-linear scale. Real solutions exist only when $\mathcal{L}_p$ is less than a threshold that depends on the slip length, and multiple real solutions exist when $\mathcal{L}_p$ is slightly less than this threshold. A family of complex solutions arise for sufficiently high values of $\mathcal{L}_p$.

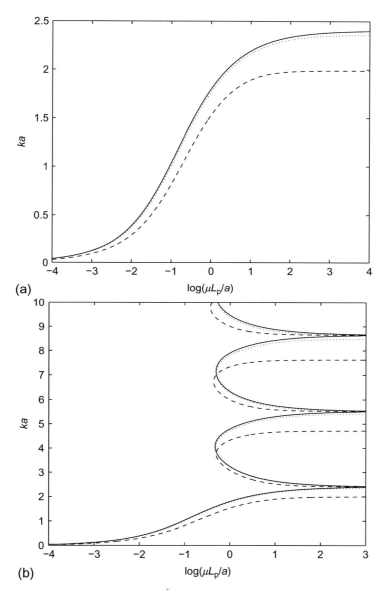

FIGURE 9.11 (a) Dependence of the coefficient $\hat{k} \equiv ka$ on the scaled wall permeability, $\mathcal{L}_p \equiv \mu L_p/a$ for scaled slip coefficient $\ell/a = 0$ (solid line-no slip), 0.01 (dotted line), and 0.10 (dashed line) corresponding to Starling's equation in terms of the pressure. (b) Complete solution space.

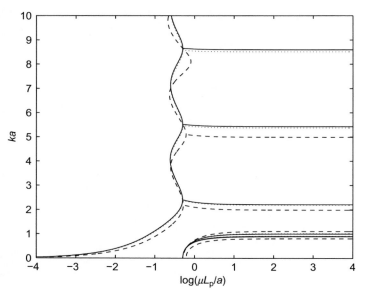

FIGURE 9.12 Solution space of Equation (9.106) corresponding to Starling's equation in terms of the normal stress for scaled slip coefficient $\ell/a=0$ (solid line), 0.01 (dotted line), and 0.10 (dashed line). The nearly horizontal lines on the left represent the real part, and the lines emanating from the horizontal axis represent the imaginary part of complex solutions.

9.7 SUMMARY

In this chapter, we have presented a detailed derivation of the equations describing Stokes flow through an infinite tube subject to a uniform ambient pressure. The expressions for the velocity and pressure involve a scalar constant determined by the wall permeability according to Starling's equation for the pressure or normal stress. We have seen that the precise form of Starling's equation has a significant impact on the solution at sufficiently high membrane permeabilities, but a minor influence at small and moderate permeabilities encountered in physiological and biomedical applications. Likewise, moderate wall slip has a small influence on the structure of the flow.

We envision that the formulas derived in this chapter will be useful in the engineering design and analysis of biomedical and physiological systems involving fluid flow through conduits with permeable walls embedded in constant pressure surroundings.

References

[1] Pozrikidis C. Leakage through a permeable capillary tube into a poroelastic tumor interstitium. Eng Anal Bound Elem 2013;37:728–37.
[2] Marshall EA, Trowbridge EA. Flow of a Newtonian fluid through a permeable tube: the application to the proximal renal tubule. Bull Math Biol 1974;36:457–76.
[3] Ross SM. A mathematical model of mass transport in a long permeable tube with radial convection. J Fluid Mech 1974;63:157–75.
[4] Probstein R. Desalination: some fluid mechanical problems. Trans ASME J Basic Eng 1972;94:286–313.
[5] Pozrikidis C. Introduction to theoretical and computational fluid dynamics. 2nd ed. New York: Oxford University Press; 2011.
[6] Pozrikidis C. Effect of membrane thickness on the slip and drift velocity in parallel shear flow. J Fluids Struct 2005;20:177–87.
[7] Pozrikidis C. Slip velocity over a perforated or patchy surface. J Fluid Mech 2010;643:471–7.
[8] Wang CY. Stokes slip flow through a grid of circular cylinders. Phys Fluids 2002;14:3358–60.

Transdermal Drug Delivery and Percutaneous Absorption: Mathematical Modeling Perspectives

Filippo de Monte[a], *Giuseppe Pontrelli*[b], *Sid M. Becker*[c]

[a]University of L'Aquila, L'Aquila, Italy
[b]Institute for Applied Mathematics—CNR, Rome, Italy
[c]University of Canterbury, Christchurch, New Zealand

Nomenclature

c concentration (kg m^{-3})
C uniform initial concentration (kg m^{-3})
D effective diffusivity of the drug in the vehicle or skin (m^2 s^{-1})
J mass flux due to a concentration gradient (kg s^{-1} m^{-2})
k partition coefficient (dimensionless)
k_P tissue permeability (m s^{-1})
K skin/capillary clearance coefficient (m s^{-1})
l thickness (m)
M mass per unit of area (kg m^{-2})
P mass transfer coefficient (m s^{-1})
t time (s)

x Cartesian space coordinate (m)
X eigenfunction

Greek Symbols

β binding/unbinding rate constant (s^{-1})
δ unbinding/binding rate constant (s^{-1})
λ eigenvalue

Acronyms

PDE partial differential equation
SC stratum corneum
TDD transdermal drug delivery
TH theophylline

Subscripts

0 unbound (free) drug in the vehicle
1 unbound (free) drug in the skin
b bound drug in the skin
e bound drug in the vehicle

Superscripts

k integer for series

10.1 INTRODUCTION

Systemic delivery of drugs by percutaneous permeation (transdermal drug delivery, TDD) offers several advantages compared to oral release or hypodermic injection. Because TDD's controlled release rate can provide a constant concentration for a long period of time and improved patient compliance, TDD has been shown to be an attractive alternative to oral administration [1]. The most advanced delivery systems (electroporation and cavitational ultrasound) enhance transdermal delivery through a strong and reversible disruption of the stratum corneum (SC), without damaging the deeper tissues.

Drugs can be delivered across the skin in order to have an effect on the tissues adjacent to the site of application (topical delivery) or to be effective after distribution through the circulatory system (systemic delivery). While there are many advantages to delivering drugs through the skin, the skin's barrier properties provide a significant challenge. To this aim, it is important to understand the mechanism of drug permeation from the delivery device (or vehicle, typically a transdermal patch or medicated plaster, as in Figure 10.1, or from novel delivery devices such as arrays of microneedles [2–5]).

Mathematical modeling for TDD constitutes a powerful predictive tool to arrive at a better understanding of the fundamental physics underlying biotransport processes. In the absence of experiments, many studies have used mathematical models and numerical simulations to research TDD efficacy and the optimal design of TDD devices [3,6,7]. The transdermal release

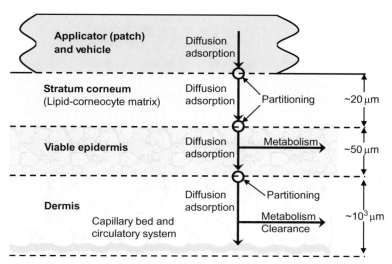

FIGURE 10.1   The composite representation of the skin with an applicator for transdermal delivery. The layers are not drawn to scale and actual skin layer thicknesses can vary depending on the body site and on the individual. Transport mechanisms are listed as well.

of a drug must be carefully tailored to achieve the optimal therapeutic effect and to deliver the correct dose in the required time [8]. The pharmacological effects of the drug (the tissue accumulation, duration, and distribution) all have an effect on its efficacy. Hence, a delicate balance between an adequate amount of drug delivered over an extended period of time and the minimal local toxicity should be found [9]. Although a large number of mathematical models are available for drug dynamics in the skin, there is a limited effort to explain the drug delivery mechanism from the vehicle platform. This is a very important issue indeed, because the polymer matrix acts as a drug reservoir, and a strategic design of its microstructural characteristics would improve the release performances [10]. It is noteworthy to emphasize that the drug elution depends on the properties of the "vehicle-skin" system, taken as a whole, and can be modeled as a coupled two-layered system. In it, coupled to diffusive effects, drug binding and unbinding phenomena are considered. In both layers, these effects play an important role.

In Section 10.4 of this chapter a "vehicle-skin" coupled model is presented and a semianalytical form is given for drug concentration and mass within the vehicle and the skin at various times. Our mathematical approach is similar to that used to describe mass dynamics from a drug-eluting stent in an arterial wall and is similarly based on a two-layer diffusion model [11]. The simulations, aimed at the design of technologically advanced vehicles, can be used to provide valuable insights into local TDD and to assess experimental procedures to evaluate drug efficacy. A major issue in modeling drug penetration is the assessment of the key parameters defining skin permeability, diffusion coefficients, and partition coefficients. A big challenge is the large number of parameters required for an advanced modeling, which are often not readily available in the literature. With this in mind, we begin this chapter with a discussion of the physiological environment of the skin and its effect on the kinetics of drug transport.

10.2 PHYSIOLOGICAL DESCRIPTION AND DRUG TRANSPORT MODELS

This introductory section is written with a particular audience in mind—engineers and mathematicians unfamiliar with the terminology and basics of the field of dermal pharmacology. We have written this introduction with the intent of providing insight into the various parameters and constants that are so commonly used in the related mathematical models. The engineer or mathematician will find comfort in the familiar approach to the solutions of the transient conservative equations. However, when it comes to interpreting the results of a parametric analysis, this audience will be at a loss if they do not have a basic understanding of the meanings and the sources of the values representing the parametric constants. With this in mind, we begin this section with a general physiological description of the body's greatest organ, the skin. This is followed by a brief discussion of the considerations that must be made when approaching the modeling of transdermal transport. The descriptions and terminology are presented with the mathematical modeling perspective in mind so that the reader may more easily make the connection between mathematics, physics, and physiology.

10.2.1 The Skin as a Composite

The human skin is not a homogeneous medium as it is made of multiple constituent composite layers, each providing a specific function with varying thicknesses. The reader unfamiliar with the physiology of the skin may find the following texts helpful: the very comprehensive introductory book by Millington and Wilkinson [12] and the shorter, but highly informative synopsis by Wood and Bladon [13].

The outer skin layer, the epidermis (0.05-1.5 mm), is without vasculature and acts as a protective barrier preventing molecular transport. Below the epidermis is the highly vascular inner skin layer, the dermis (0.3-3 mm). An important characteristic of the dermis is the large network of capillaries with high blood flow rates exceeding several times that of metabolic requirements. The primary reason for such high perfusion rates in the dermis is to regulate body temperature: perfusion rates decrease in order to conserve body heat and increase in order to cool the body. The dermis is also characterized by its high collagen content, which provides the skin with its structural support. The skin, at most sites on the body, is perforated by appendages pathways in the form of sweat glands and hair follicles. Although such shunt routes occupy less than 0.1% of the lateral surface area of the skin, in some instances, they may contribute to electrically assisted diffusion transport [14]. In any model of *in vivo* skin (living, as in skin attached to a living body) that focuses on the kinetic response of a drug to the physiology of the skin, it is important to consider the adequacy of the model's composite layer and its physiological description of the skin.

10.2.2 The SC, Its Corneocytes, and the Lipid Matrix

The thin (10-50 μm) outermost layer of the epidermis is called the *stratum corneum*. This is the most resistive layer to transport through the skin. Although the SC's barrier function is vitally important to healthy skin (by keeping harmful molecules from passing into the skin and providing an initial defense against infection), it is this high resistance to permeability that presents a major obstacle for successful transdermal delivery. Thus, in TDD modeling, a great deal of research is focused on the structure and the resulting barrier behavior of the SC.

At the sub-membrane scale, the SC is often described as a medium that is composed of two primary components: corneocytes, which are essentially flat, dead, keratinized cells and a lamellar network of lipid bilayers. The SC is composed of 15-20 layers of corneocytes which are interconnected by a lipid lamellar bilayer structure in a crystalline-gel phase [15,16]. The nature in which individual corneocytes are set within the lamellar matrix of lipid bilayers has inspired researchers to conceptualize a brick-and-mortar microstructure of the SC in which the corneocytes are represented by isolated bricks and the lipid structure is represented by a continuous mortar space encapsulating the bricks [15]. Individual SC corneocytes are 10-40 μm in diameter, and they may differ in their thickness depending on the body site and their relative location within the SC. The corneocyte thickness may also be influenced by their degree of hydration, which varies from 10% to 30% bound water. Excellent descriptions of the understanding of the SC's microstructure are provided in Refs. [15,17]. The corneocytes and lipid structures have a strong contrast in their chemical behavior: the former are regarded as hydrophilic while the surrounding extracellular lipid matrix is lipophilic (water is lipophobic). With respect to TDD, a great deal of attention is focused upon the lipid regions of the SC. This is primarily because the drugs and drug vehicles have been designed such that they are lipophilic [15]. Lipophilic drugs will naturally prefer the lipid-filled spaces of the SC to the hydrophilic corneocytes, and thus transport circumventing the barrier function of the SC is primarily associated to occur within the lamellar lipid structure of the SC [16].

The molecular structure of the lipid phase has been well researched and publicized in various works by Bouwstra et al. [16,18,19] over the past two decades. The microstructure of the lipid phase is such that the lipid bilayers (heads and tails) are organized in a series of repeating periodic parallel sheets. The long periodicity phase has a periodicity of ~13 nm and is believed to play a major role in the barrier function of the SC [18].

10.2.3 The Drug and the Vehicle

In describing the delivery of drugs, it is common to focus on the local concentration of the drug itself. However, in practice, the drug is not delivered to the skin in isolation. A transporting agent is used to administer the drug and to enhance the efficacy of delivery. Such an agent (termed the "vehicle") may serve several purposes in the transdermal delivery system. The vehicle in general is inert, having by itself no real therapeutic value. Transdermal delivery vehicles come in many forms such as colloids, creams, gels, lotions, ointments, liquids, and powders. When the vehicle is placed directly onto the skin surface, it acts as a medium to give the drug bulk and to provide the drug a contact time. In some applications, the vehicle can serve as a solvent in which the drug is dissolved. In more advanced cases, the vehicle is designed such that it encapsulates the drug on the journey through the SC [20]. Some such vehicles are designed with the chemical microstructure of the skin in mind so that the vehicle is more easily able to pass through the skin compared to the drug alone. An excellent review of penetration enhancers is provided by Williams and Barry [21]. With regards to the modeling of TDD, it is important to note that the drug's interaction with the vehicle must be considered. In the following section, some simple concepts will be described that allow for the drug vehicle interaction to be taken into account.

10.2.4 Diffusion–Transport Considerations

The structures of the skin (and of biological media in general) pose a number of unique phenomena that influence passive transport. The following is provided as a synopsis of the more detailed descriptions provided in Ref. [22] and which is summarized in the excellent recent reviews by Mitragotri et al. [3] and by Naegel et al. [23] and in the books of collected chapters by Roberts et al. [24,25]. For further insight, the reader is encouraged to consider the excellent historical review and fundamental description that is provided in the seminal work by Scheuplein and Blank [26]. The rate of distribution of a drug into the skin is dependent upon a number of critical factors associated with the skin's tissue layers. These include the tissue microstructure, the tendency for the drug or the drug's vehicle to become bound within this local microstructure, the drug's lipid or water solubility, the rates of metabolism within the tissue, and the rate of blood perfusion in the dermis.

10.2.4.1 Diffusion

In the absence of advection and electrokinetic effects, passive diffusion is responsible for the transport of a drug (solute) through the vehicle (solvent) and, then, through the skin layers. In its most general representation, for a mixture of two components (for example, drug and vehicle), the one-dimensional flux J per unit area (along x) of drug transport by diffusion through the solvent follows Fick's law (also known as Fick's first law). Hence, it is directly proportional to the gradient in concentration c:

$$J = -D\frac{\partial c}{\partial x} \qquad (10.1)$$

where D is the diffusion coefficient of the drug in the binary mixture, for example drug/vehicle.

When a steady-state condition is reached, the Fick's first law reduces to $J_{SS} = \Delta c/(l/D)$. It is important to recognize that steady state can only be reached after the lag time for solute diffusion has passed. The lag time for diffusion across a homogeneous membrane is given by $l^2/(6D)$. It is noteworthy to mention that Higuchi [27] expressed Fick's first law more appropriately in terms of thermodynamic activity rather than the more widely used concentration approximation. Thermodynamics activity for any given solute is generally defined by the fractional solubility of the solute in the medium.

It is important to consider that a drug will have a different diffusion coefficient depending upon which vehicle or layer of skin that the drug is in. For example, the drug will have one associated diffusion coefficient in the SC and a different diffusion coefficient in the neighboring epidermis. Furthermore, at the macroscale within a specific layer, the tissue may be considered to be homogeneous, while at the microscale the medium is only periodically homogeneous. For example, at the macroscale, the SC will seem homogeneous, while, the SC is depicted as a matrix of corneocytes and lamellar structures of lipid bilayers.

Thus, the diffusion coefficient of Equation (10.1) is often an "effective" or "apparent" diffusion coefficient that represents the diffusional behavior of a drug in the tissue at a macroscale perspective. In order to apply a particular effective diffusion coefficient to a solute within a particular medium, the condition must be met that the time for equilibrium to be achieved at the microscale must be much shorter than the time required for transport across the entire layer. This is not the case when the diffusion coefficient varies chemically (with concentration) [28] or if there are slow kinetic processes related to binding and unbinding [9]. This is further discussed in Section *10.2.5*.

By taking conservation of drug mass (see Section 3.3 in de Monte et al. [29]), Equation (10.1) leads to the well-known one-dimensional, transient, diffusion equation (also known as Fick's second law):

$$\frac{\partial c}{\partial t} = D \frac{\partial^2 c}{\partial x^2} \tag{10.2}$$

where the calculation of the concentration distribution $c = c(x,t)$ requires that the starting concentration as well as the conditions for concentration or flux at the boundaries (i.e., at the outermost and innermost surfaces) be specified: it requires the knowledge of the boundary and initial conditions. Therefore, different concentration solutions are obtained for different starting values of concentration and the boundary conditions.

10.2.4.2 *Evaluation of the Diffusion Coefficient*

The evaluation of apparent diffusion coefficient values of various solutes in various vehicles and in the different skin layers (or in the skin as a whole) has been heavily researched with obvious motivation. However, it seems that the coefficient's value depends highly on the conditions used to approximate it. The evaluation of the diffusion coefficient requires multiscale and multiphysics considerations. Because the SC is the most rate-limiting layer to transdermal delivery, a great deal of attention has been given to evaluating the effective diffusivity within this layer alone.

There are two primary methods that have been used to estimate the diffusion coefficient that is related to steady-state conditions. One involves experimentally-obtained data of steady-state fluxes across a medium. A steady-state permeability, k_P, of the drug or carrier across a layer of thickness, l, is determined from the experimentally-measured steady-state flux, J_{SS}, and the difference in steady concentrations across the layer, Δc, as:

$$k_P = \frac{J_{SS}}{\Delta c} \tag{10.3}$$

Then, the diffusion coefficient may be determined from the relation:

$$D = k_P l \tag{10.4}$$

In the event that the drug is carried by a vehicle, the permeability and diffusion coefficient are related not only to the translayer diffusion of the drug itself, but are related also to the drug within the vehicle [25] by the relations:

$$k_P = \frac{J_{SS}}{\Delta c_V} \tag{10.5}$$

where c_V corresponds to the concentration of the solute within the vehicle.

Then, the diffusion coefficient may be determined from the relation:

$$D = k_{SV} k_P l \tag{10.6}$$

where k_{SV} corresponds to the partition coefficient of the solute between the skin and the vehicle (partitioning is discussed in the following Section 10.2.4.3).

There are limitations to the experimental approach. This is because the diffusion coefficient derived from the steady-state flux may not always accurately represent the behavior of transient diffusion. A second limitation to this method of evaluating the diffusion coefficient is that the conditions of any model that uses this value must have conditions that precisely match the experimental conditions from which the diffusion coefficient was evaluated.

This is the motivation of the second method of diffusion coefficient estimation, which involves model-based approximations. For example, the Potts-Guy model [30] describes the diffusion coefficient within the SC that is treated as an essentially homogeneous membrane. The diffusion coefficient is related to solute molecular weight in an exponentially decaying relationship [31]:

$$\frac{D_{SC}}{l_{SC}} = 10^{-6.3} e^{-0.0061\,MW} \text{ cm s}^{-1} \tag{10.7}$$

where MW is the solute molecular weight. The predominant feature of this model (that even later, more complicated microstructure models retain) is the model's exponential dependence upon the solute volume (or molecular weight).

10.2.4.3 *Partitioning*

Partitioning is used to describe the behavior of a compound that is added to a binary mixture. When a solute is added to an isolated single-component system, it will arrange itself in a uniform concentration throughout the system. However, when this same solute is added into an isolated two-component system, the solute will arrange itself preferentially to one of the system components. To put this into context, consider a lipophilic solute that diffuses into a compartment that is filled with two components (A and B). If component A is lipophilic and component B is hydrophilic, the drug will prefer the environment occupied by lipophilic material A over that of hydrophilic material B. Thus, at steady state, a concentration will arise in which, taken separately and individually, A and B each have distinct and uniform concentrations of the drug, for which lipophilic section A will have a higher concentration than hydrophilic component B.

This concept is referred to as partitioning and can be derived from first principles perspectives as in Chapter 4 of Ref. [25] and in Chapter 5 of Ref. [32]. Following the derivations of Ref. [32], the partition coefficient may be defined as:

$$k_{AB} \equiv \frac{c_A}{c_B} \tag{10.8}$$

where c_A is the steady-state concentration of the drug in A and c_B is the steady-state concentration of the drug in B.

Equation (10.8) requires that the system be at thermal equilibrium (that the system be at a constant temperature). At the interface between component A and component $B(x=x_{AB})$, at any time t, the partition coefficient is employed into an interface condition such that:

$$c_B(x_{AB}, t)k_{AB} = c_A(x_{AB}, t) \tag{10.9}$$

Partitioning in TDD is concerned with a drug's diffusive behavior at the interface between two different tissue types or microscopic regions. At the cellular level (microscale), partitioning occurs within the SC at the interface between the lipids and the corneocytes, and between the drug's vehicle and the SC lipids. However, at the membrane level (macroscale), partitioning occurs at the layer interfaces (at the interfaces between the donor and the SC, between the SC and viable epidermis, between the epidermis and the dermis, etc.).

10.2.4.4 *Evaluation of the Partition Coefficient*

The dermal partition coefficients and dermal effective diffusion coefficients of 26 different compounds in mammalian dermis were examined and reported in an excellent work by Kretsos et al. [33]. Evaluating the partition coefficient between the SC and a vehicle is a very difficult task. The recent review of Mitragotri et al. [3] provides an excellent overview of the current methods that are employed to represent the partition coefficient within the SC. This review explains that the partitioning of the solute into the lipid bilayers is influenced by a chemical factor and a physical factor. The chemical factor accounts for the fact that the environment in the lipid bilayers is much more hydrophobic than its surroundings. The physical factor accounts for the actual molecular structure of the SC's arrangement of the lipid bilayers.

The field takes the view that "it is reasonable to assume that the partition coefficient of a solute from water into SC lipids is comparable to that into an isotropic solvent that reasonably mimics the SC lipid environment" [3]. For this reason, the reader will find that the partition coefficient of solutes in the literature is often reported in terms of the readily available octanol-water partition coefficient [34].

10.2.5 Adsorption

The concept of adsorption (sometimes referred to as binding/unbinding) of a drug or carrier into the local molecular microstructure has been extensively studied both experimentally and theoretically in various studies by Anissimov and Roberts [9,24,28,35,36]. In short, the concept can be conventionalized as follows.

On its route through the skin, a drug may encounter localized pockets (within the microstructure) that act to restrain the drug from transport. This is most often and most easily described within the SC. Consider the simple diffusion of water through the SC. Recalling that the lipid-filled spaces are lipophilic and the individual corneocytes are hydrophilic, it is not difficult to imagine that as a water molecule diffuses through the SC, it could become adsorbed into one of the corneocytes. Because the corneocyte shell provides a restrictive envelope through which the water is unable to freely diffuse, once the water is inside the corneocyte, it could be considered to be "bound" to remain within the space occupied by this individual corneocyte.

The adsorption is not restricted to occur within the corneocyte: defects within the SC lipid lamellar structures sometimes create small hydrophilic pockets within this otherwise lipophilic environment. At the molecular scale, individual drug molecules may bind to the corneocyte envelope. Furthermore, adsorption can take place outside of the SC. Consider the comprehensive discussion on the parameters associated within the dermis by Kretsos et al. [33]. There it is postulated that, in some instances, the solute can bind to large, relatively immobile macromolecules, such as proteins in the dermis. This, too, would constitute a binding process.

Mathematically, the binding and unbinding of a species (as it follows the path through the skin's layers) may be considered by categorizing the species into two subspecies: one representing the concentration of the drug in its unbound state, c_u, and the other representing the concentration of the drug in its bound state, c_b. The generalized description of adsorption, presented in a recent study by Nitsche and Frasch [37], comprehensively addresses the interpretation of this phenomenon from a macroscopic perspective, distinguishing between slow binding and fast binding.

10.2.5.1 Slow Binding

The development of the system of coupled partial differential equations (PDE's) representing this process was detailed by Anissimov and Roberts [9] in the context of the diffusion of water through the SC. The model provides a simple linear coupling between the bound (c_b) and unbound (c_1) states. The conservation of the solute mass in the unbound state is represented by:

$$\frac{\partial c_1}{\partial t} = D_1 \frac{\partial^2 c_1}{\partial x^2} - \beta_1 c_1 + \delta_1 c_b \tag{10.10}$$

Because it is not bound, it is free to diffuse spatially. Here the binding rate constant, β_1, provides a representation of the rate at which the solute leaves the free state and enters the bound state. The rate of release of bound drug into its unbound state is implied by the unbinding rate constant δ_1. This linear representation of the exchange between the two states couples the concentration of the unbound state to that of the drug in its bound state:

$$\frac{\partial c_b}{\partial t} = \beta_1 c_1 - \delta_1 c_b \tag{10.11}$$

The binding and unbinding rate constants, that appear in Equations (10.10) and (10.11), have units of inverse time: their magnitudes are inversely proportional to the time associated with the binding/unbinding process. Equations (10.10) and (10.11) are used in cases of what is termed *slow binding*, that is, when the time scale associated with the binding process is not negligible when compared to the time scale associated with transport by diffusion. Thus, the rate constants associated with slow binding are small.

For water penetration through human SC, it was shown in Ref. [9] that $\beta_1 \approx \delta_1 \approx 0.05$ min^{-1} ($\approx 10^{-3}$ s^{-1}). According to Anissimov et al. [35], "It is reasonable to expect that, for larger molecules, binding constants will be much smaller."

The binding rates of theophylline (TH) in human skin were also evaluated experimentally in Ref. [38] and then explained by an extended partition-diffusion model with reversible binding. In the case of TH binding to keratin, the above study finds: $\beta_1 = (0.466 \pm 0.147)$h^{-1}, i.e., $\beta_1 = (1.3 \pm 0.4) \times 10^{-4}$ s^{-1}, and $\beta_1/\delta_1 = 1.401$. These values were subsequently used in the theoretical development study of Nitsche and Frasch [37].

10.2.5.2 Fast Binding

While the problem and solutions developed later in this chapter (Section 10.4) are related primarily to adsorption with slow binding/unbinding rates, for completeness, the development of the equations governing the fast binding rate case is provided as well. In the event that the binding rate is very fast, that is, rate constants very large, the left hand side (LHS) of Equation (10.11) is negligible and so that the fast binding rate relation may be made:

$$\beta_1 c_1 = \delta_1 c_b \qquad (10.12)$$

This directly relates the concentration of the bound state to the unbound state. Now consider that the total concentration $\bar{c}_1$ may be defined as a sum of bound and unbound concentrations:

$$\bar{c}_1 = c_1 + c_b \qquad (10.13)$$

and, furthermore, the conservation of the total solute mass is governed by:

$$\frac{\partial \bar{c}_1}{\partial t} = D_1 \frac{\partial^2 c_1}{\partial x^2} \qquad (10.14)$$

When the relation Equation (10.12) is substituted into the definition Equation (10.13) and this in turn into the governing Equation (10.14), a single species representation of the transient diffusion may be made:

$$\frac{\partial}{\partial t}\left[\left(1 + \frac{\beta_1}{\delta_1}\right)c_1\right] = D_1 \frac{\partial^2 c_1}{\partial x^2} \qquad (10.15)$$

This simple representation of the fast binding adsorption processes has been developed and modified to capture nonlinear effects as well [39]. While the fast binding rate approximation allows for a simpler set of equations, it has been noted that the slow binding rate equations may be more representative of the local physics [23].

10.2.6 Metabolism and Clearance

The metabolic activity within the viable epidermis and the dermis may strongly influence the rate and delivery. Obviously, in some instances, the metabolism of the drug by the cells is, in itself, the motivation of the delivery: the targeted delivery cells are actively metabolizing the drug. The metabolic rate of the consumption of solute, having c as concentration, is generally represented by a simple sink term in a first-order reaction [23,40–42], that is $-\mu_{ms}c$, where μ_{ms} is the metabolic rate constant (s^{-1}).

The values of the rate constant are specific to the drug metabolized and the location within the skin. The works of Yamaguchi et al. provide values of the metabolic rate constants within the dermis for several specific drugs [40–42].

Clearance is similar to metabolism in that the effect of clearance is to remove solute from the tissue. This is most notably associated with the bulk removal of solute via the microcapillary system in the dermis. Recall that the dermis is highly perfused; it has rates of blood flow exceeding those of metabolic requirements. For a systemic delivery, the drug will be adsorbed into the dermal vasculature. The actual process of removal of solute from the dermis via the capillary system is complicated. It involves the diffusion of the solute from within the dermal interstitial space, through the capillary walls, and then by hydrodynamic means into the blood stream. An excellent description of this process is provided in the comprehensive review by Kretsos and Kasting [43] who include a compendium of experimentally derived data involving the parameters concerning capillary geometry and kinetics. In their 2004 work, Kretsos et al. [44] reduce the complexity of the representation of clearance within the dermis by providing a simple term that (from a macroscale perspective) acts as a sink, so that within the dermis, the transient transport is represented by:

- slow binding (extension of Eq. (10.10))

$$\frac{\partial c_1}{\partial t} = D_1 \frac{\partial^2 c_1}{\partial x^2} - \beta_1 c_1 + \delta_1 c_b - \kappa_1 c_1 \qquad (10.16a)$$

- fast binding (extension of Eq. (10.15))

$$\frac{\partial}{\partial t}\left[\left(1 + \frac{\beta_1}{\delta_1}\right)c_1\right] = D_1 \frac{\partial^2 c_1}{\partial x^2} - \kappa_1 c_1 \qquad (10.16b)$$

where the κ_1 parameter (s^{-1}) is the clearance coefficient, whose effects represent the removal of drug by the microvasculature system. In this study, experimental data of salicylic acid in de-epidermized rat skin is presented in order to develop an analytic approximation of the associated clearance. They report a clearance coefficient $\kappa_1 = 9.1 \times 10^{-4}$ s^{-1}.

While reducing the complexities of the dermal microvasculature to a single sink term may seem an oversimplification, it does provide information on the spatial behavior of the drug within the lower skin layers. Alternately, many reputable studies provide a further simplification of this phenomenon at the cost of this spatial information. In those studies, the dermal clearance behavior is represented by a boundary condition of the third kind at the dermis-epidermis interface:

$$K_1 c_1 + D_1 \frac{\partial c_1}{\partial x} = 0 \qquad (10.17)$$

where K_1 is the skin-capillary clearance having dimensions of velocity (m s^{-1}) and related to the κ_1 coefficient stated before by $K_1 = \kappa_1 l_1$, where l_1 is the skin thickness.

10.2.7 TDD Models

Mathematical models of TDD and percutaneous absorption are highly relevant to the development of a fundamental understanding of biotransport processes as well as to the assessment of dermal exposure to industrial and environmental hazards. The foundations of predictive modeling of transdermal and topical delivery were laid in the 1940-1970s. During this time, it was recognized that partitioning and solubility were important factors that determine skin penetration.

This review summarizes the key developments in predictive simulation of skin permeation and related solution methods over the last 50 years and also looks to the future so that such approaches are effectively harnessed for the development of better topical and transdermal formulations and for improved assessment of skin exposure to toxic chemicals.

10.2.7.1 Fickian Models

Compartment models, also called pharmacokinetic (PK) models of skin, are often used to study the fate of chemicals entering and leaving the body. The PK models treat the skin and also the body as one or several well-stirred compartments of uniform (average) concentration that act as reactors and/or reservoirs of chemical storage with transfer between the compartments depicted by first order rate constant expressions. While permeation across the skin can be described using Equation (10.3), it is often represented in a PK model as either a series of compartments to mimic the partitioning and diffusion processes in the SC or as one compartment and two compartments that separately distinguish the lipophilic SC and hydrophilic viable epidermis layers of the skin [36].

While most attention in the field of modeling of skin permeation has been focused on describing diffusion processes in the SC, it has been recognized that additional processes including binding and metabolism [45] also play an important role in determining drug uptake. Binding is especially significant because many substances bind to keratin, which significantly influences their permeation across the SC.

The effect of binding on transdermal transport in the context of the epidermal penetration has been discussed by Roberts et al. [24] where the kinetics associated with the reservoir effect of the SC was considered. It was assumed in this work that binding is instantaneous, that is equilibration between bound and unbound states is fast compared to diffusion. The advantage of such an approach is that the modeling in this case is relatively simple with the diffusion coefficient D in the diffusion equation being replaced by an effective diffusion coefficient D_{eff}, where $D_{eff}=f_u D$ and f_u is the fraction of unbound solute. As the fraction unbound is less than unity, binding leads to slower diffusion, and therefore longer lag times. If binding/partitioning is not fast compared to diffusion, the single diffusion equation has to be replaced by coupled PDE's [9], as shown in Section 10.2.5.1.

10.2.7.2 Non-Fickian Models

Often experimental results show that the predicted drug concentration distribution in the vehicle and in the skin by the Fick's model does not agree with experimental data. Recently, a non-Fickian mathematical model for the percutaneous absorption problem was proposed by Barbeiro and Ferreira [46].

In this new model, the Fick's law for the flux is modified by introducing a non-Fickian contribution defined with a relaxation parameter τ_J related to the properties of the components. This parameter is similar to the relaxation time τ_q of the heat flux for the analogous heat wave diffusion problem [47]. We have

$$J = -D\frac{\partial c}{\partial x} - \tau_J\frac{\partial J}{\partial t} \qquad (10.18)$$

Combining the flux equation with the mass conservation law, a system of integro-differential equations was established with a compatibility condition on the boundary between the two components of the physical model. Alternatively, a hyperbolic PDE can be derived in place of the above integro-differential one, as done by Haji-Sheikh et al. [47]. In order to solve the mathematical model, its discrete version was introduced by Barbeiro and Ferreira [46] and the demanding stability and convergence properties of the discrete system were studied by the analysis of numerical experiments.

10.3 REVIEW OF MATHEMATICAL METHODS

The expression of transport of a solute across a skin barrier membrane involves a number of steps and phases in a space and time variant process. The formal description of this process as a single equation is not straightforward, other than as one or more approximations in definition of the transport conditions or in presentation of the solutions. Here, we begin with the conventional Laplace transform analytical approach used to solve diffusion equations, move to numerical methods that allow variations in space and time in the transport process and various complexities to be better addressed.

10.3.1 Laplace Transform

Laplace transform is an integral transformation that is used to solve ordinary and partial differential equations. Its application for solving diffusion problems has been described in the well-known book by Crank [48] and by Carslaw and Jaeger [49] and Özişik [50] for the analogous heat conduction problems.

The popularity of the Laplace transform in the skin literature has increased since the availability of scientific software (e.g. Scientist, MicroMath Scientific software, Matlab, Mathematica) that can invert from the Laplace domain to the time domain and allow regression to experimental data without the extra work of first deriving a functional representation of the Laplace solution inverted into the time domain. With this type of software, having the Laplace solution is virtually as good as having a solution written in terms of time. Anissimov and Roberts [9,27,51,52] have used the numerical inversion of Laplace transform solutions to the diffusion equation for simulations and data analysis of skin transport experiments.

One of the useful properties of the Laplace transform is that it can be used directly (without inversion to the time domain) to determine some parameters. In transport through skin for the case of a constant donor concentration, such parameters are the steady-state flux and the lag time [51,52].

While Laplace transforms offer numerous advantages in solving diffusion equations, they also suffer from certain limitations. Most notably, to be solvable by the Laplace transform, the partial differential equations must have concentration-independent coefficients. Also, the coefficients in the differential equation (e.g., the diffusivity in Equation (10.3)) have to be independent of time (e.g., constants or functions of space only) for the Laplace transform to convert the partial differential equation into an ordinary differential equation of x only. This excludes important classes of problems in skin transport that involve the diffusion coefficient changing with concentration (nonlinear equation) or with time; for example, co-diffusion with a penetration enhancer or a diffusion coefficient that changes due to skin drying.

10.3.2 Finite Difference Method

The finite difference approach to solving a differential equation or a system thereof involves replacing the differential equation with a set of difference equations that cover the requisite space and time [50, Chapter 12]. There are many variations to this theme, the sophistication of which depends upon the problem to be solved. The most common difference approximations are centered differences, equations centered in space at the location where the approximation is made; similarly, for the time variable. But backward and forward differences are possible, too.

Finite difference methods are particularly advantageous for potentially nonlinear systems with either simple geometry or periodic geometry. Much of the efficiency is lost for disordered structures. Relative to finite element methods (FEMs), finite difference methods can be much more efficient on periodic problems such as a regular brick-and-mortar SC structure. However, considerable skill is required to construct accurate approximations at boundaries and to implement an efficient variable mesh scheme. Relative to Laplace transform methods, the biggest advantages of finite differences are the ability to handle nonlinear problems and more complex boundary conditions; for instance, mixed boundary conditions. Both call for considerable skill by the operator.

10.3.3 Finite Element Method

The FEM is related to the finite difference method in that both offer approximate numerical solutions to linear or nonlinear PDE's. The FEM is able to handle domains with regular or irregular geometries and boundaries, including moving boundaries. The primary basis for the FEM is the discretization of a continuous domain of interest—here the skin—into a discrete set of connected subdomains. The resulting mesh of triangles or higher order polygons, referred to as elements, creates a finite-dimensional linear problem whose solution can be implemented on a computer. In general, the density of the mesh varies across the domain, with greater density over those areas where greater precision in the solution is required.

An example might be the regions in the SC near a boundary between corneocyte and lipid domains. Owing to the complexity of the meshing and solution procedures, the FEM is frequently implemented using commercial software packages. Rim et al. [10] developed a finite element model consisting of two isotropic materials with different diffusion and partition coefficients, connected by an interfacial flux. The two materials are intended to represent a dermal patch or reservoir containing a drug of interest, and the skin. Addition of a permeation enhancer creates a coupled two-component system with concentration-dependent diffusivities to account for interactions between drug and enhancer.

Frasch and Barbero [53] analyzed a finite element model of the SC lipid pathway to investigate effective path length and diffusional lag times in this path compared with a homogeneous membrane of the same thickness. This research group also presented a transcellular pathway model, whereby permeants are granted access to the corneocytes via a corneocyte-lipid partition coefficient and separate diffusivity within corneocytes compared with lipids. Results pointed to a transcellular pathway with preferential corneocyte partitioning as the likely diffusional path for hydrophiles.

A secondary result from these investigations was the observation that the complex disordered geometric representation of the SC could be reduced to a simple, rectangular, brick-and-mortar geometry with very similar results [54]. Furthermore, for many realistic combinations of corneocyte/lipid partitioning and diffusivity, the short vertical connections between bricks can be ignored and the problem can be reduced to a two-layer lipid-corneocyte laminate model. This configuration is a good representation for the transcellular path with preferential corneocyte partitioning. Thus, for many purposes, the complex geometrical arrangement of the SC can be reduced to a much simpler geometry for which simpler numerical algorithms, such as the finite difference method, can be applied. In fact, analytical solutions for steady-state flux and lag time have been published for the multilayer laminate model [48–50].

10.3.4 Finite Volume Method

Heisig et al. [55] used a related method, finite volumes, to solve both one-dimensional transient and steady-state transport of drugs through a biphasic brick-and-mortar model of SC with isotropic lipids and permeable, isotropic corneocytes. This work demonstrated the contributions of corneocyte alignment, relative phase diffusivity, and phase partitioning in the barrier properties of the SC. Subsequent extensions in both two-dimensional skin models have been described in [56,57], and the group has explored the role of drug binding to corneocyte elements on skin transport [58].

10.4 MODELING TDD THROUGH A TWO-LAYERED SYSTEM

In this section, we develop the governing equations and semianalytic solution to the problem of transport from a receiver vehicle into the skin. This approach attempts to capture the kinetics of the drug behavior in a one-dimensional transient domain. The various physiological considerations that were discussed in Section 10.2 are used to introduce the terms and parameters making up the associated PDE's. The general overview of the two-layer domain is such that the vehicle is represented by one layer (Layer 0) and all the constituent parts of the skin are lumped together into a second layer (Layer 1), as suggested by Kubota et al. [59] and by Simon and Loney [60]. This model will then consider partitioning, adsorption (binding), and diffusion of a drug as it passes from the vehicle (Layer 0) into the skin (Layer 1) and experiences clearance from the skin layer via the advection of the drug through the skin's microcirculatory system. It is important for the reader to recognize that while the parameters used to represent the skin correspond in this model do not correspond to any individual layer of the skin, the physics that these parameters represent are those explained in Section 10.2.

10.4.1 Mathematical Formulation

Let us consider the two-layered delivery system. Layer 0 represents the vehicle by which the drug is administered for therapeutic purposes. The vehicle could be a transdermal patch or the film of an ointment placed directly onto the skin. Layer 1 represents the skin. In this

FIGURE 10.2   Cross-section of the vehicle and the skin layers. Due to an initial difference of unbound concentrations c_0 and c_1, a mass flux is established at the interface $x=0$ and drug diffuses through the skin. At $x=1$, the skin-receptor (capillary) is set. Figure not to scale.

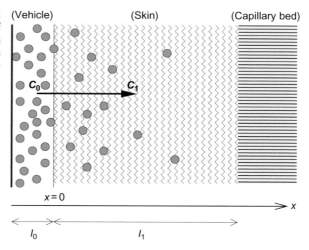

case, the skin of Layer 1 is representative of all the composite layers described in Section 10.2, the SC, the epidermis, and the dermis, and includes the dermal capillary bed. This is conceptualized in Figure 10.2. Because the lengths associated with the area of skin that is covered by the vehicle are very large compared to the lengths representing the skin and vehicle thicknesses, most of the mass dynamics occurs along the direction normal to the flat skin surface; so, we restrict our study to a simplified one-dimensional model. In particular, we consider the x-axis as normal to the skin surface and pointing outwards.

Without loss of generality, let $x_0=0$ be the vehicle-skin interface; and l_0 and l_1 the thicknesses of these layers, respectively (Figure 10.2). Hence, both vehicle and skin are treated macroscopically as two homogeneous media. In this model, the governing equations of each layer (the vehicle and the skin) are developed such that the effects of binding and unbinding of each layer are considered. This means that in each layer (Layer 0 for vehicle and Layer 1 for skin) the drug can exist in a bound state and an unbound state. Thus, there are two equations for each layer that address the possible states of the drug.

The vehicle acts as a drug reservoir made of a thin substrate (generally, a polymer or a gel) containing a therapeutic drug to be delivered. Here, we will consider that initially within the vehicle, the entire mass of the drug exists in a bound state. This would be anticipated if the drug is encapsulated at maximum concentration in a solid phase of, for example, nanoparticles. In such a state, the drug cannot be delivered by diffusion into the skin, so it is considered "bound" and we denote the concentration of the drug within the vehicle that is bound by the symbol c_e. The vehicle is designed to release the drug from its solid bound state once it is applied to the skin. As the vehicle system starts the release process, a fraction of the drug mass is first transferred, in a finite time, to an unbound—free, biologically available—phase. In this model, the unbound drug within the vehicle is denoted c_0. The drug will then be available to diffuse into the skin. The drug enters the skin in an unbound state and, in this model, the concentration of the drug in Layer 1 (that is in the unbound state) is denoted c_1. Section 10.2.5 explained at the molecular level its route through the skin and the drug experience binding due to the local tissue microstructure. This model will consider that the drug may exist in the skin layer in a bound state as well. This is denoted by the symbol c_b.

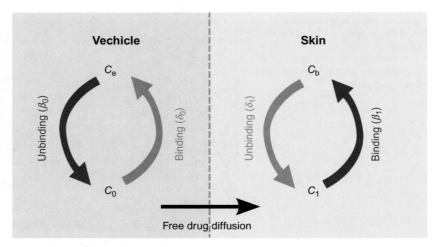

FIGURE 10.3 Schematic of drug delivery and percutaneous absorption in the vehicle-skin system. An unbinding (resp. binding) reaction occurs in the vehicle (resp. in the skin). Reverse reactions are possible in both layers. Diffusion occurs only in the free unbound phases c_0 and c_1.

Hence, the drug delivery process starts from the vehicle and ends at the skin receptors, with a phase change in a cascade sequence, as schematically represented in Figure 10.3. Bidirectional drug binding and unbinding phenomena play a key role in TDD, with characteristic times comparable (slow binding—Section 10.2.5.1) or faster (fast binding—Section 10.2.5.2) than those of diffusion.

Here we deal only with slow binding and will use the methods described in Section 10.2.5.1 in order to represent the kinetics of the drug in its bound state and its unbound state in both layers.

In Layer 0, the bound drug is governed by:

$$\frac{\partial c_e}{\partial t} = -\beta_0 c_e + \delta_0 c_0 \quad (-l_0 < x < 0; \, t > 0) \tag{10.19a}$$

where the paramters $\beta_0 \geq 0$ and $\delta_0 \geq 0$ are the unbinding and binding rate constants in the vehicle, respectively.

The unbound drug in the vehicle is governed by the coupled equation:

$$\frac{\partial c_0}{\partial t} = D_0 \frac{\partial^2 c_0}{\partial x^2} + \beta_0 c_e - \delta_0 c_0 \quad (-l_0 < x < 0; \, t > 0) \tag{10.19b}$$

where D_0 is the effective diffusion coefficient of the unbound solute within the vehicle.

The drug's behavior within the skin is also represented by a diffusion process that includes the effects of slow binding. The drug in the skin that is in the unbound state is governed by:

$$\frac{\partial c_1}{\partial t} = D_1 \frac{\partial^2 c_1}{\partial x^2} - \beta_1 c_1 + \delta_1 c_b \quad (0 < x < l_1; \, t > 0) \tag{10.20a}$$

where D_1 is the effective diffusivity coefficient of the unbound solute, $\beta_1 \geq 0$ and $\delta_1 \geq 0$ are the binding and unbinding rate constants in the skin, respectively.

$$\frac{\partial c_b}{\partial t} = \beta_1 c_1 - \delta_1 c_b \quad (0 < x < l_1; \, t > 0) \tag{10.20b}$$

To close the previous two-layered mass transfer system of Equations (10.19a), (10.19b), (10.20a), and (10.20b), a flux continuity condition and a jump in concentration (due to partitioning) must be assigned at the vehicle-skin interface:

$$D_0 \left(\frac{\partial c_0}{\partial x} \right)_{x=0} = D_1 \left(\frac{\partial c_1}{\partial x} \right)_{x=0} \quad (t > 0) \tag{10.21a}$$

$$c_0(0, t) = k_{01} c_1(0, t) \quad (t > 0) \tag{10.21b}$$

where k_{01} is the partition coefficient defined in Section *10.2.4* between Layer 0 and Layer 1.

However, as the interface condition Equation (10.21b) would be rigorously valid only for steady-state conditions, in the current treatment we prefer not to use it (contrary to what is generally done in the TDD field), but to deal with the following interface equation:

$$P[c_0(0, t) - c_1(0, t)] = -D_1 \left(\frac{\partial c_1}{\partial x} \right)_{x=0} \quad (t > 0) \tag{10.21c}$$

where P is a mass transfer coefficient which accounts for partitioning of the drug at the interface vehicle-skin.

In such a way, the concentration ratio $c_0(0, t)/c_1(0, t)$ is not constant and equal to k_{01} at any time, as indicated by Equation (10.21b), but can change with the time. Also, P (whose value is unknown) can be related to the partition coefficient k_{01} (whose value can be taken experimentally, as described in Section *10.2.4*) when a steady state is reached. In fact, Equation (10.21c) gives ($t \to \infty$):

$$c_0(0, \infty) = c_1(0, \infty) \underbrace{\left[1 + \frac{J_{SS}}{P c_1(0, \infty)} \right]}_{k_{01}} = k_{01} c_1(0, \infty) \tag{10.22a}$$

where J_{SS} is the steady-state mass flux. Also, bearing in mind Equations (10.5) and (10.6), we can write

$$P = \frac{D_1}{k_{01}(k_{01} - 1)l_1} \tag{10.22b}$$

Then, no mass flux passes between the vehicle and the surrounding, and we impose a no-flux condition:

$$D_0 \left(\frac{\partial c_0}{\partial x} \right)_{x=-l_0} = 0 \quad (t > 0) \tag{10.23}$$

Also, a boundary condition has to be imposed at the skin-receptor (capillary bed) surface. At this point, the elimination of the drug by the capillary system follows a first-order kinetics (see Section *10.2.6*):

$$-D_1 \left(\frac{\partial c_1}{\partial x} \right)_{x=l_1} = K_1 c_1(l_1, t) \quad (t > 0) \tag{10.24}$$

where K_1 is the skin-capillary clearance per unit area.

Finally, the initial conditions are

$$c_e(x, 0) = C_e, \qquad c_0(x, 0) = 0 \qquad (-l_0 < x < 0) \tag{10.25a}$$

$$c_1(x, 0) = 0, \qquad c_b(x, 0) = 0 \qquad (0 < x < l_1) \tag{10.25b}$$

10.4.1.1 Dimensionless Equations

To get easily computable quantities, all the variables and the parameters appearing in the governing equations listed previously are normalized as follows:

$$\tilde{x} = \frac{x}{l_1}, \quad \tilde{l}_0 = \frac{l_0}{l_1}, \quad \tilde{t} = \frac{D_1 t}{l_1^2}, \quad \gamma_0 = \frac{D_0}{D_1}, \quad \phi = \frac{P l_1}{D_1}$$

$$K = \frac{K_1 l_1}{D_1}, \quad \tilde{c} = \frac{c}{C_e}, \quad \tilde{\beta} = \frac{\beta l_1^2}{D_1}, \quad \tilde{\delta} = \frac{\delta l_1^2}{D_1} \tag{10.26}$$

By omitting the tilde for sake of simplicity, the mass transfer problem of the two-layered system of Figure 10.2 governed by Equations (10.19a)–(10.21a), (10.21c), (10.23)–(10.25b), can be rewritten in a dimensionless form as:

$$\frac{\partial c_e}{\partial t} = -\beta_0 c_e + \delta_0 c_0 \quad (-l_0 < x < 0; t > 0) \tag{10.27a}$$

$$\frac{\partial c_0}{\partial t} = \gamma_0 \frac{\partial^2 c_0}{\partial x^2} + \beta_0 c_e - \delta_0 c_0 \quad (-l_0 < x < 0; t > 0) \tag{10.27b}$$

$$\frac{\partial c_1}{\partial t} = \frac{\partial^2 c_1}{\partial x^2} - \beta_1 c_1 + \delta_1 c_b \quad (0 < x < 1; t > 0) \tag{10.27c}$$

$$\frac{\partial c_b}{\partial t} = \beta_1 c_1 - \delta_1 c_b \quad (0 < x < 1; t > 0) \tag{10.27d}$$

with the following inner and outer BCs:

$$\left(\frac{\partial c_0}{\partial x} \right)_{x=-l_0} = 0 \quad (t > 0) \tag{10.28a}$$

$$\gamma_0 \left(\frac{\partial c_0}{\partial x} \right)_{x=0} = \left(\frac{\partial c_1}{\partial x} \right)_{x=0} \quad (t > 0) \tag{10.28b}$$

$$\phi[c_0(0, t) - c_1(0, t)] = -\left(\frac{\partial c_1}{\partial x} \right)_{x=0} \quad (t > 0) \tag{10.28c}$$

$$\left(\frac{\partial c_1}{\partial x} \right)_{x=1} + K c_1(1, t) = 0 \quad (t > 0) \tag{10.28d}$$

supplemented with the initial conditions:

$$c_e(x, 0) = 1, \quad c_0(x, 0) = 0 \quad (-l_0 < x < 0) \tag{10.29a}$$

$$c_1(x, 0) = 0, \quad c_b(x, 0) = 0 \quad (0 < x < 1) \tag{10.29b}$$

10.4.2 Method of Solution

Preliminarily, by using the first of the two equations (10.29a), we note that the solution of the linear homogeneous ordinary differential equation (ODE) (10.27a) is

$$c_e(x, t) = \underbrace{\exp(-\beta_0 t)}_{c_e^*(t)} + \underbrace{\delta_0 \exp(-\beta_0 t) \int_0^t c_0(x, \tau) \exp(\beta_0 \tau) d\tau}_{c_e^{**}(x, t)} \qquad (10.30)$$

Hence, it turns out that c_e can be expressed as a function of c_0 and can be considered in two parts. The first part on the right hand side (RHS) of Equation (10.30) depends only on the time (exponentially) and is due to the initial drug concentration other than zero of the bound state within the vehicle. The other part depends on both space and time and is influenced by the boundary conditions (10.28a)–(10.28d) through c_0.

Similarly, from Equation (10.27d), by using the second of the two equations (10.29b), c_b can be expressed as a function of c_1 as:

$$c_b(x, t) = \underbrace{\beta_1 \exp(-\delta_1 t) \int_0^t c_1(x, \tau) \exp(\delta_1 \tau) d\tau}_{c_b^{**}(x, t)} \qquad (10.31)$$

where the part depending only on the time is absent due to the zero initial concentration of the unbound state within the vehicle.

Let us now find a solution for c_0 and c_1 by the separation-of-variables (SOV) method

$$c_0(x, t) = X_0(x)G(t), \quad c_1(x, t) = X_1(x)G(t) \qquad (10.32)$$

As a consequence of Equations (10.30) and (10.31), the part of c_e and c_b depending on both space and time, c_e^{**} and c_b^{**}, respectively, can be separated by the same eigenvector set as:

$$c_e^{**}(x, t) = X_0(x)G_e(t), \quad c_b^{**}(x, t) = X_1(x)G_b(t) \qquad (10.33)$$

Therefore, Equations (10.30) and (10.31) become

$$c_e(x, t) = \exp(-\beta_0 t) + X_0(x)G_e(t), \quad c_b(x, t) = X_1(x)G_b(t) \qquad (10.34)$$

The time-dependent exponential term, appearing in the first of the two equations (10.34) and due to the only initial drug concentration other than zero, does not allow the SOV method to be applied. Then, for purposes of computation of the functions $X_0(x)$, $X_1(x)$, $G_0(t)$, $G_1(t)$, $G_e(t)$, and $G_b(t)$, we first neglect the initial drug concentration of the bound state and then, by means of appropriate constants, we will again account for it.

Thus, substituting Equations (10.32) and (10.34) in Equations (10.27a)–(10.27d) and neglecting the exponential term as said above gives

$$\frac{dG_e}{dt} = -\beta_0 G_e + \delta_0 G_0 \qquad (10.35a)$$

$$\frac{1}{\gamma_0 G_0}\left[\frac{dG_0}{dt} - (\beta_0 G_e - \delta_0 G_0)\right] = \frac{1}{X_0}\frac{d^2 X_0}{dx^2} = -\lambda_0^2 \qquad (10.35b)$$

$$\frac{1}{G_1}\left[\frac{dG_1}{dt} - (\delta_1 G_b - \beta_1 G_1)\right] = \frac{1}{X_1}\frac{d^2 X_1}{dx^2} = -\lambda_1^2 \tag{10.35c}$$

$$\frac{dG_b}{dt} = -\delta_1 G_b + \beta_1 G_1 \tag{10.35d}$$

where λ_0 and λ_1 are separation constants. In the following, they will be denoted as the eigenvalues of the vehicle and skin, respectively.

10.4.2.1 Time-Dependent Solution

Two decoupled systems, each of two ordinary differential equations defined in the time domain, may be derived from Equations (10.35a)–(10.35d). In a matrix form, we have

$$\frac{d}{dt}\begin{pmatrix} G_e \\ G_0 \end{pmatrix} = \begin{pmatrix} -\beta_0 & \delta_0 \\ \beta_0 & -(\delta_0 + \gamma_0 \lambda_0^2) \end{pmatrix}\begin{pmatrix} G_e \\ G_0 \end{pmatrix} \quad (t > 0) \tag{10.36}$$

$$\frac{d}{dt}\begin{pmatrix} G_1 \\ G_b \end{pmatrix} = \begin{pmatrix} -(\beta_1 + \lambda_1^2) & \delta_1 \\ \beta_1 & -\delta_1 \end{pmatrix}\begin{pmatrix} G_1 \\ G_b \end{pmatrix} \quad (t > 0) \tag{10.37}$$

The general solution of the previous two systems is

$$G_e(t) = A_\mu^+\left(\frac{\delta_0}{\beta_0 + \mu_+}\right)\exp(\mu_+ t) + A_\mu^-\left(\frac{\delta_0}{\beta_0 + \mu_-}\right)\exp(\mu_- t) \tag{10.38a}$$

$$G_0(t) = A_\mu^+ \exp(\mu_+ t) + A_\mu^- \exp(\mu_- t) \tag{10.38b}$$

$$G_1(t) = A_\nu^+ \exp(\nu_+ t) + A_\nu^- \exp(\nu_- t) \tag{10.39a}$$

$$G_b(t) = A_\nu^+\left(\frac{\beta_1}{\delta_1 + \nu_+}\right)\exp(\nu_+ t) + A_\nu^-\left(\frac{\beta_1}{\delta_1 + \nu_-}\right)\exp(\nu_- t) \tag{10.39b}$$

where $\mu_\pm$ and $\nu_\pm$ may be taken as

$$\mu_\pm = \frac{-(\beta_0 + \delta_0 + \gamma_0 \lambda_0^2) \pm \sqrt{(\beta_0 + \delta_0 + \gamma_0 \lambda_0^2)^2 - 4\gamma_0 \beta_0 \lambda_0^2}}{2} \tag{10.40}$$

$$\nu_\pm = \frac{-(\beta_1 + \delta_1 + \lambda_1^2) \pm \sqrt{(\beta_1 + \delta_1 + \lambda_1^2)^2 - 4\delta_1 \lambda_1^2}}{2} \tag{10.41}$$

It is easily seen that $\mu_\pm$ and $\nu_\pm$ are both real and negative. In order to satisfy the interface conditions, Equations (10.28b) and (10.28c), we should have $G_0(t) = G_1(t)$, that is $\mu_\pm = \nu_\pm$ and $A_\mu^\pm = A_\nu^\pm$.

10.4.2.2 Space-Dependent Solution: The Eigenvalue Problem

Two ordinary differential equations defined in the space domain may be derived from Equations (10.35b) and (10.35c) as:

$$\frac{d^2 X_0}{dx^2} + \lambda_0^2 X_0 = 0 \quad (-l_0 < x < 0) \tag{10.42a}$$

$$\frac{d^2X_1}{dx^2} + \lambda_1^2 X_1 = 0 \quad (0 < x < 1) \tag{10.42b}$$

They are coupled through the interface conditions, Equations (10.28b) and (10.28c). Then, substituting Equations (10.32) and (10.34) in Equations (10.28a)–(10.28d) and neglecting the exponential term appearing in the first of the two equations (10.34), as done in Subsection 10.4.2, we have

$$\left(\frac{dX_0}{dx}\right)_{x=-l_0} = 0 \tag{10.42c}$$

$$\gamma_0 \left(\frac{dX_0}{dx}\right)_{x=0} = \left(\frac{dX_1}{dx}\right)_{x=0} \tag{10.42d}$$

$$\phi[X_0(0,t) - X_1(0,t)] = -\left(\frac{dX_1}{dx}\right)_{x=0} \tag{10.42e}$$

$$\left(\frac{dX_1}{dx}\right)_{x=1} + KX_1(1,t) = 0 \tag{10.42f}$$

Equations (10.42a)–(10.42f) represent a Surm-Liouville (eigenvalue) problem with discontinuous coefficients whose solution is

$$X_0(x) = a_0 \cos(\lambda_0 x) + b_0 \sin(\lambda_0 x) \quad (-l_0 < x < 0) \tag{10.43a}$$

$$X_1(x) = a_1 \cos(\lambda_1 x) + b_1 \sin(\lambda_1 x) \quad (0 < x < 1) \tag{10.43b}$$

with

$$a_0 = -\left[\frac{\lambda_1}{\phi} + \frac{K \tan(\lambda_1) + \lambda_1}{K - \lambda_1 \tan(\lambda_1)}\right] \tag{10.44a}$$

$$b_0 = \frac{1}{\gamma_0} \frac{\lambda_1}{\lambda_0} \tag{10.44b}$$

$$a_1 = -\frac{K \tan(\lambda_1) + \lambda_1}{K - \lambda_1 \tan(\lambda_1)} \tag{10.44c}$$

For $K \to \infty$, we have a boundary condition of the first kind at $x = 1$ and Equations (10.44a)–(10.44c) reduce to the same equations as given in de Monte et al. [29, p. 100].

The eigencondition for computing the eigenvalues is $a_0 \tan(\lambda_0 l_0) + b_0 = 0$. Substituting the coefficients a_0 and b_0 listed before gives

$$\underbrace{\left[\frac{\lambda_1}{\phi} + \frac{K \tan(\lambda_1) + \lambda_1}{K - \lambda_1 \tan(\lambda_1)}\right]}_{-a_0} \tan(\lambda_0 l_0) - \underbrace{\frac{1}{\sqrt{\gamma_0}} \frac{\lambda_1}{\sqrt{\lambda_1^2 + (\beta_1 - \delta_0)}}}_{b_0} = 0 \tag{10.45}$$

where the eigenvalues λ_0 and λ_1 are related through the constraint $\mu_\pm = \nu_\pm$, as shown in the previous paragraph, where $\mu_\pm$ and $\nu_\pm$ are given by Equations (10.40) and (10.41), respectively.

10.4.2.3 Concentration Solution

By numerically solving the transcendental equation (10.45) along with the condition $\mu_\pm = \nu_\pm$, an infinite set of real and distinct eigenvalues is computed, that is, λ_0^k and λ_1^k, with $k = 1, 2, \ldots$. Correspondingly, we will have a countable set of eigenfunctions X_0^k and X_1^k, defined through Equations (10.43a) and (10.43b), and of time-dependent functions $G_e^k(t)$, $G_0^k(t)$, $G_1^k(t)$, and $G_b^k(t)$, defined through Equations (10.38a) and (10.39b), respectively, with $\mu_\pm^k = \nu_\pm^k$.

Finally, the general solution will be given by the linear superposition of the fundamental solution in the form:

$$c_e(x, t) = \sum_{k=1}^{\infty} X_0^k(x) \left[A_k \left(\frac{\delta_0}{\beta_0 + \mu_+^k} \right) \exp\left(\mu_+^k t \right) + B_k \left(\frac{\delta_0}{\beta_0 + \mu_-^k} \right) \exp\left(\mu_-^k t \right) \right] \tag{10.46a}$$

$$c_0(x, t) = \sum_{k=1}^{\infty} X_0^k(x) \left[A_k \exp\left(\mu_+^k t \right) + B_k \exp\left(\mu_-^k t \right) \right] \tag{10.46b}$$

$$c_1(x, t) = \sum_{k=1}^{\infty} X_1^k(x) \left[A_k \exp\left(\mu_+^k t \right) + B_k \exp\left(\mu_-^k t \right) \right] \tag{10.46c}$$

$$c_b(x, t) = \sum_{k=1}^{\infty} X_1^k(x) \left[A_k \left(\frac{\beta_1}{\delta_1 + \mu_+^k} \right) \exp\left(\mu_+^k t \right) + B_k \left(\frac{\beta_1}{\delta_1 + \mu_-^k} \right) \exp\left(\mu_-^k t \right) \right] \tag{10.46d}$$

Bearing in mind the initial conditions $c_0(x,0) = 0$ and $c_1(x,0) = 0$, it follows that $B_k = -A_k$, where A_k may be computed by using the remaining initial conditions, that is, $c_e(x,0) = 1$ and $c_b(x,0) = 0$. In detail, we have

$$\sum_{k=1}^{\infty} A_k X_0^k(x) \left(\frac{\delta_0}{\beta_0 + \mu_+^k} - \frac{\delta_0}{\beta_0 + \mu_-^k} \right) = 1 \quad (-l_0 < x < 0) \tag{10.47a}$$

$$\sum_{k=1}^{\infty} A_k X_1^k(x) \left(\frac{\beta_1}{\delta_1 + \mu_+^k} - \frac{\beta_1}{\delta_1 + \mu_-^k} \right) = 0 \quad (0 < x < 1) \tag{10.47b}$$

By truncating the above two series to N terms and, then, by collocating in K points, the system of algebraic equations (10.47a) and (10.47b) is solved to get the constants A_k, with $k = 1, 2, \ldots, N$.

Then, the analytical form of the solution given by Equations (10.46a)–(10.46d) allows us an easy computation of the dimensionless drug mass (per unit of area), $\widetilde{M} = M/(C_e l_1) \to M$, as an integral of the concentration over the correspondent layer, that is,

$$M(t) = \int c(x, t) \mathrm{d}x \quad (0 < x < 1) \tag{10.48}$$

10.4.3 Numerical Simulation and Results

A common difficulty in modeling physiological processes is the identification of reliable estimates of the model parameters. Experiments of TDD are prohibitively expensive or impossible *in vivo* and the only available sources are data from literature. The physical problem depends on a large number of parameters; each of them may vary in a finite range, with a variety of combinations and limiting cases. Also, we develop the solution for a particular combination of parameters, that is $\beta_0 = \delta_1$ and $\beta_1 = \delta_0$, being the general case addressed in Ref. [61]. Consequently, as $\mu_\pm^k = \nu_\pm^k$, we have: $\gamma_0 (\lambda_0^k)^2 = (\lambda_1^k)^2$.

The physical parameter-values used are related to a beta-adrenoceptor blocking agent, timolol, released from the acrylic copolymer to local circulation through the skin. These parameter-values are given by [59,60]

- vehicle

$$D_0 = 2.7 \times 10^{-9} \, \text{cm}^2 \, \text{s}^{-1}, \quad l_0 = 40 \, \mu\text{m} \tag{10.49a}$$

- skin

$$D_1 = 7.8 \times 10^{-10} \, \text{cm}^2 \, \text{s}^{-1}, \quad l_1 = 125 \, \mu\text{m}, \quad K_1 = 3.5 \times 10^{-3} \, \text{cm} \, \text{s}^{-1} \tag{10.49b}$$

- vehicle-skin interface

$$k_{01} = 1 \Rightarrow P \to \infty \tag{10.49c}$$

Equation (10.49c) states that there is no concentration jump at the interface as the partition coefficient is equal to 1. Therefore, the boundary condition expressed by Equation (10.21c) simply reduces to $c_0(0,t) = c_1(0,t)$ for any $t > 0$.

As the binding/unbinding processes are not considered in Refs. [59,60], the corresponding reaction rates $\beta_1 = \delta_0$ and $\beta_0 = \delta_1$ are unavailable. However, as the characteristic reaction times are smaller than the diffusion times, in the current case (timolol in human skin) we can write (in dimensional form): $\beta_1, \delta_1 > D_1/l_1^2 = 5 \times 10^{-6} \, \text{s}^{-1}$. Also, according to Anissimov et al. [35] it is reasonable to expect that, for larger molecules (such as timolol), binding/unbinding constants are smaller than the ones for water penetration through human SC where $\beta_1 \approx \delta_1 \approx 10^{-3} \, \text{s}^{-1}$ [9], as already shown in Section 10.2.5.1. Therefore, we have the following constraint for timolol through human skin:

$$5 \times 10^{-6} \, \text{s}^{-1} < \beta_1, \quad \delta_1 < 10^{-3} \, \text{s}^{-1} \tag{10.50}$$

In all numerical experiments, we have assumed: $\beta_1 = \delta_0 = 10^{-4} \, \text{s}^{-1}$ (the same order of TH binding to keratin) and three different values for $\beta_0 = \delta_1$. In detail: $8 \times 10^{-5} \, \text{s}^{-1}$, $2 \times 10^{-4} \, \text{s}^{-1}$, and $10^{-3} \, \text{s}^{-1}$, as shown in Figures 10.4–10.7 to follow.

Then, the limit of the skin layer (l_1) was estimated by the following considerations. Strictly speaking, in a diffusion-reaction problem the concentration vanishes asymptotically at infinite distance. However, for computational purposes, the concentration is damped out (within a given tolerance) over a finite distance at a given time. Such a distance, known as "penetration distance" d_p [62], may be defined as the distance from the perturbed region at which

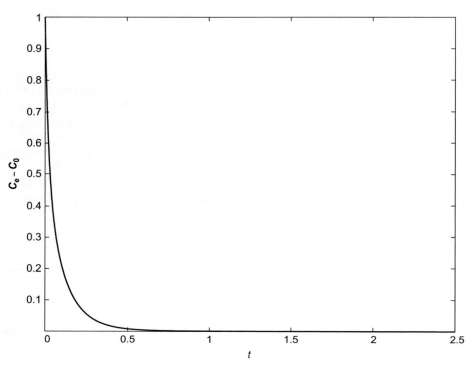

FIGURE 10.4 Difference between bound and free concentrations, $c_e - c_0$, in the vehicle as a function of time for various locations.

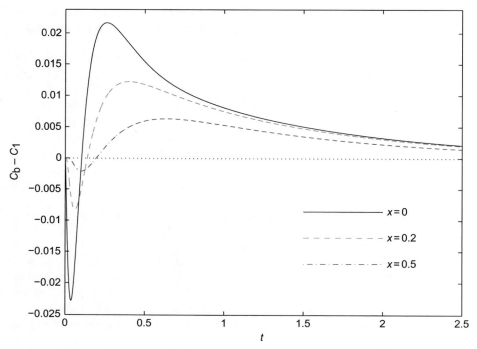

FIGURE 10.5 Difference between bound and free concentrations, $c_b - c_1$, in the skin versus time with location as a parameter.

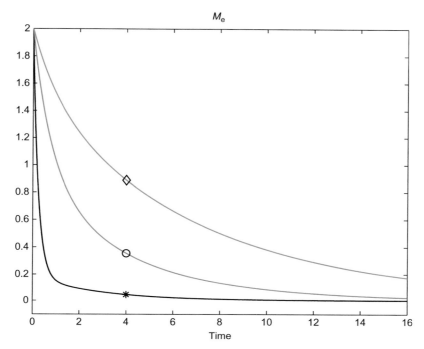

FIGURE 10.6 Bound drug mass M_e in the vehicle versus time for $\delta_0(=\beta_1)=10^{-4}\,\text{s}^{-1}$ and three different values of the unbinding rate $\beta_0=\delta_1(\text{s}^{-1})$.

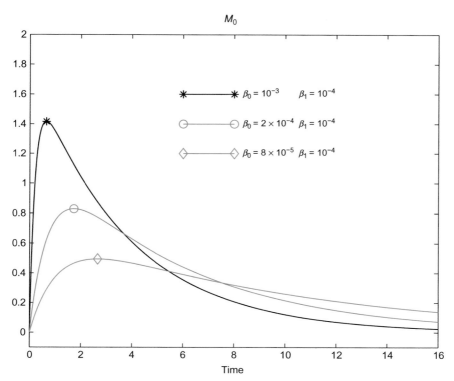

FIGURE 10.7 Unbound drug mass M_0 in the vehicle as a function of time for $\delta_0(=\beta_1)=10^{-4}\,\text{s}^{-1}$ with $\beta_0=\delta_1(\text{s}^{-1})$ as a parameter.

concentration and mass flux are just affected (errors less than 10^{-n}; $n=1,2,...10$) at a given time t by this perturbation (in the current case, the initial condition $c_e(x,0)=1$). It critically depends on the diffusive properties of both layers; in particular, it is related to the ratio $\gamma_0 = D_0/D_1$ as follows [63]

$$d_p \cong \begin{cases} \sqrt{10n\gamma_0 t}, & 0 \le t \le 0.1\, l_0^2/(n\gamma_0) \\ \sqrt{10nt} + l_0\left(\dfrac{\sqrt{\gamma_0}-1}{\sqrt{\gamma_0}}\right), & t \ge 0.1\, l_0^2/(n\gamma_0) \end{cases} \tag{10.51}$$

where d_p is dimensionless $\left(d_p \to \tilde{d}_p = d_p/l_1\right)$

Therefore, the outer boundary condition Equation (10.28d) may be replaced with

$$\begin{cases} c_1(x=0,t)=0, & 0 \le t \le 0.1\, l_0^2/(n\gamma_0) \\ c_1(x=d_p,t)=0, & t \ge 0.1\, l_0^2/(n\gamma_0) \end{cases} \tag{10.52}$$

where the first of the two prior equations states that, at early times $0 \le t \le 0.1\, l_0^2/(n\gamma_0)$, the two-layer (vehicle/SC) system reduces to only one single layer (vehicle) slab with a homogeneous boundary condition of the first kind, that is, $c_1(x=0,t)=0$, with errors less than 10^{-n} ($n=1$, $2,...,10$).

The concentration is decreasing inside each layer, and vanishes at a distance that is within the SC, at all times. Due to the relatively large value of D_0 and to the small value of l_0, the concentration profiles are almost flat in the vehicle, with levels reduced in time, and have a decreasing behavior in the skin layer. In particular, a fast decaying phase transfer is evidenced in the vehicle (Figure 10.4), whereas a fast phase change of drug occurs at early times within the skin, that is more evident at points close to the interface $x=0$, and continues at later times (Figure 10.5).

The mass $M_e(t)$ exponentially decreases in the vehicle, as shown in Figure 10.6, while the mass $M_0(t)$ first increases up to some upper bound and then decays asymptotically, as shown in Figure 10.7.

The relative size of $\beta_0 = \delta_1$ and $\beta_1 = \delta_0$ affects the binding/unbinding transfer processes, thus influencing the mechanism of the whole dynamics. The occurrence and the magnitude of the drug peak depend on the combination and the relative extent of the diffusive and reaction parameters. The outcome of the simulation provides valuable indicators to assess whether drug reaches target tissue and, hence, to optimize the dose capacity in the vehicle. For example, Figures 10.8 and 10.9 show that a lower value of the unbinding parameter $\beta_0 = \delta_1$ guarantees a more prolonged and uniform release. For the other way around, a large value of $\beta_0 = \delta_1$ is responsible for a localized peaked distribution followed by a faster decay.

The present TDD model constitutes a simple tool that can help in designing and in manufacturing new vehicle platforms that guarantee the optimal release for an extended period of time.

10.5 CONCLUSIONS

In the last decades, transdermal delivery has emerged as an attractive alternative and an efficient route for drug administration. After a general phenomenological description of the skin and its parameters, and a brief review of the main predictive modeling techniques and

M_1

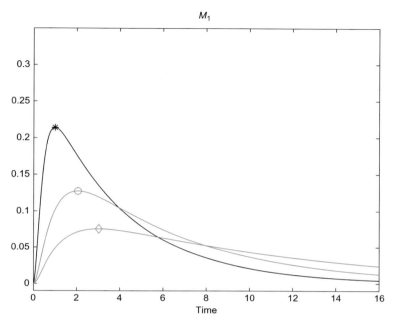

FIGURE 10.8 Time history of the unbound drug mass M_1 in the skin for $\beta_1(=\delta_0)=10^{-4}\,\mathrm{s}^{-1}$ and three values of the unbinding rate $\delta_1=\beta_0(\mathrm{s}^{-1})$.

M_b

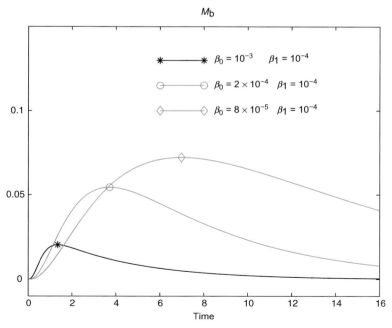

FIGURE 10.9 Time plot of the bound drug mass M_b in the skin for $\beta_1(=\delta_0)=10^{-4}\,\mathrm{s}^{-1}$ with $\delta_1=\beta_0(\mathrm{s}^{-1})$ as a parameter.

related analytical and numerical solutions, a comprehensive mathematical model of drug delivery by percutaneous permeation is presented in this chapter. To account for diffusion and reaction aspects of drug dynamics from the vehicle across the skin, a multiphase, two-layered model is developed and an eigenvalue-based semianalytic solution for drug concentration is proposed.

The model incorporates the reversible binding process and can be employed to study the effects of the various parameters that control the vehicle-skin delivery system. This can be of interest in the design of smarter devices in order to get the optimal therapeutic effect by releasing the correct dose in the required time. Although limited to a simple one-dimensional case, the results of the numerical simulations can offer a useful tool to estimate the performance of the drug delivery systems.

References

[1] Chien YW. Novel drug delivery systems. New York: Marcel Dekker; 1992.
[2] Cevc G, Vierl U. Nanotechnology and the transdermal route: a state of the art review and critical appraisal. J Control Release 2010;141:277–99.
[3] Mitragotri S, Anissimov YG, Bunge AL, Frasch HF, Guy RH, Hadgraft J, et al. Mathematical models of skin permeability: an overview. Int J Pharm 2011;418:115–29.
[4] George K, Kubota K, Twizell E. A two-dimensional mathematical model of percutaneous drug absorption. BioMed Eng 2004;3:567–78.
[5] Barry BW. Novel mechanisms and devices to enable successful transdermal drug delivery. Eur J Pharm Sci 2001;14:101–14.
[6] Manitz R, Lucht W, Strehmel K, Weiner R, Neubert R. On mathematical modeling of dermal and transdermal drug delivery. J Pharm Sci 1998;87:873–9.
[7] Addick W, Flynn G, Weiner N, Curl R. A mathematical model to describe drug release from thin topical applications. Int J Pharm 1989;56:243–8.
[8] Prausniz M, Langer R. Transdermal drug delivery. Nat Biotechnol 2008;26:1261–8.
[9] Anissimov YG, Roberts MS. Diffusion modelling of percutaneous absorption kinetics: 4. Effects of a slow equilibration process within stratum corneum on absorption and desorption kinetics. J Pharm Sci 2009;98:772–81.
[10] Rim JE, Pinsky P, van Osdol W. Finite element modeling of coupled diffusion with partitioning in transdermal drug delivery. Ann Biomed Eng 2005;33:1422–38.
[11] Pontrelli G, de Monte F. Mass diffusion through two-layer porous media: an application to the drug-eluting stent. Int J Het Mass Tran 2007;50:3658–69.
[12] Millington PF, Wilkinson R. Skin. Cambridge: Cambridge University Press; 2009.
[13] Wood EJ, Bladon PT. The human skin. London: Edward Arnold; 1985.
[14] Barry BW. Drug delivery routes in skin: a novel approach. Adv Drug Deliv Rev 2002;54(Suppl. 1):S31–40.
[15] Menon GK. New insights into skin structure: scratching the surface. Adv Drug Deliv Rev 2002;54(Suppl. 1):S3–S17.
[16] Bouwstra JA, Gooris GS. The lipid organization in human stratum corneum and model systems. Open Derm J 2010;4:10–3.
[17] Madison KC. Barrier function of the skin: "La raison d'etre" of the epidermis. J Invest Dermatol 2003;121:231–41.
[18] Bouwstra JA, Gooris GS, van der Spek JA, Bras W. Structural investigations of human stratum corneum by small-angle X-ray scattering. J Invest Dermatol 1991;97:1005–12.
[19] Bouwstra JA, Dubbelaar F, Gooris GS, Ponec M. The lipid organisation in the skin barrier. Acta Derm Venereol 2000;208:23–30.
[20] Han IH, Choi SU, Nam DY, Park YM, Kang MJ, Kang KH, et al. Identification and assessment of permeability enhancing vehicles for transdermal delivery of glucosamine hydrochloride. Arch Pharm Res 2010;33:293–9.
[21] Williams AC, Barry BW. Penetration enhancers. Adv Drug Deliv Rev 2004;56:603–18.
[22] Barry BW. Modern methods of promoting drug absorption through the skin. Mol Aspects Med 1991;12:195–241.

[23] Naegel A, Heisig M, Wittum G. Detailed modeling of skin penetration—an overview. Adv Drug Deliv Rev 2013;65:191–207.

[24] Roberts MS, Pellett MA, Cross SE. Basic mathematical principles in skin permeation. In: Keith RB, Adam CW, editors. Dermatological and transdermal formulations. Boca Raton: CRC Press; 2002.

[25] Roberts MS, Pellett MA, Cross SE. Skin transport. In: Walters KA, editor. Dermatological and transdermal formulations. Boca Raton: CRC Press; 2002.

[26] Scheuplein RJ, Blank IH. Permeability of the skin. Physiol Rev 1971;51:702–47.

[27] Higuchi T. Physical chemical analysis of percutaneous absorption process from creams and ointments. J Soc Cosmet Chem 1960;11:85–97.

[28] Anissimov YG, Roberts MS. Diffusion modeling of percutaneous absorption kinetics: 3. Variable diffusion and partition coefficients, consequences for stratum corneum depth profiles and desorption kinetics. J Pharm Sci 2004;93:470–87.

[29] de Monte F, Pontrelli G, Becker SM. Drug release in biological tissues. In: Becker SM, Kuznetsov AV, editors. Transport in biological media. 1st ed. New York: Elsevier Publishing; 2013. p. 59–118 [chapter 3].

[30] Potts RO, Guy RH. Predicting skin permeability. Pharm Res 1992;9:663–9.

[31] Kumins CA, Kwei TK. Free volume and other theories. In: Crank J, Park GS, editors. Diffusion in polymers. New York: Academic Press; 1968. p. 107–25.

[32] Streng WH. Partition coefficient. In: Streng WH, editor. Characterization of compunds in solution (Theory and Practice. USA: Springer; 2001. p. 47–60.

[33] Kretsos K, Miller MA, Zamora-Estrada G, Kasting GB. Partitioning, diffusivity and clearance of skin permeants in mammalian dermis. Int J Pharm 2008;346:64–79.

[34] Johnson ME, Berk DA, Blankschtein D, Golan DE, Jain RK, Langer RS. Lateral diffusion of small compounds in human stratum corneum and model lipid bilayer systems. Biophys J 1996;71:2656–68.

[35] Anissimov YG, Jepps OG, Dancik Y, Roberts MS. Mathematical and pharmacokinetic modelling of epidermal and dermal transport processes. Adv Drug Deliv Rev 2013;65:169–90.

[36] Roberts MS, Anissimov YG. Mathematical models in percutaneous absorption. In: Bronaugh RL, Maibach HI, editors. Percutaneous absorption drugs–cosmetics–mechanisms–methodology. 4th ed. New York: Marcel Dekker; 2005. p. 1–44.

[37] Nitsche JM, Frasch HF. Dynamics of diffusion with reversible binding in microscopically heterogeneous membranes: general theory and applications to dermal penetration. Chem Eng Sci 2011;66:2019–41.

[38] Frasch HF, Barbero AM, Hettick JM, Nitsche JM. Tissue binding affects the kinetics of theophylline diffusion through the stratum corneum barrier layer of skin. J Pharm Sci 2011;100:2989–95.

[39] Kubota K, Twizell EH. A nonlinear numerical model of percutaneous drug absorption. Math Biosci 1992;108:157–78.

[40] Yamaguchi K, Mitsui T, Yamamoto T, Shiokawa R, Nomiyama Y, Ohishi N, et al. Analysis of in vitro skin permeation of 22-oxacalcitriol having a complicated metabolic pathway. Pharm Res 2006;23:680–8.

[41] Yamaguchi K, Mitsui T, Aso Y, Sugibayashi K. Analysis of in vitro skin permeation of 22-oxacalcitriol from ointments based on a two- or three-layer diffusion model considering diffusivity in a vehicle. Int J Pharm 2007;336:310–8.

[42] Yamaguchi K, Morita K, Mitsui T, Aso Y, Sugibayashi K. Skin permeation and metabolism of a new antipsoriatic vitamin D3 analogue of structure 16-en-22-oxa-24-carboalkoxide with low calcemic effect. Int J Pharm 2008;353:105–12.

[43] Kretsos K, Kasting GB. Dermal capillary clearance: physiology and modeling. Skin Pharmacol Physiol 2005;18:55–74.

[44] Kretsos K, Kasting GB, Nitsche JM. Distributed diffusion-clearance model for transient drug distribution within the skin. J Pharm Sci 2004;93:2820–35.

[45] Liu P, Higuchi WI, Ghanem AH, Good WR. Transport of beta-estradiol in freshly excised human skin in vitro: diffusion and metabolism in each skin layer. Pharm Res 1994;11:1777–84.

[46] Barbeiro S, Ferreira JA. Coupled vehicle-skin models for drug release. Comput Method Appl M 2009;198:2078–86.

[47] Haji-Sheikh A, de Monte F, Beck JV. Temperature solutions in thin films using thermal wave Green's function solution equation. Int J Heat Mass Tran 2013;62:78–86.

[48] Crank J. The mathematics of diffusion. 2nd ed. Oxford: Clarendon Press; 1979.

[49] Carslaw HS, Jaeger JC. Conduction of heat in solids. 2nd ed. Oxford: Oxford University Press; 1959.

[50] Özişik MN. Heat conduction. 2nd ed. New York: John Wiley & Sons; 1993.

[51] Anissimov YG, Roberts MS. Diffusion modeling of percutaneous absorption kinetics: 1. Effects of flow rate, receptor sampling rate and viable epidermal resistance for a constant donor concentration. J Pharm Sci 1999;88:1201–9.

[52] Anissimov YG, Roberts MS. Diffusion modeling of percutaneous absorption kinetics: 2. Finite vehicle volume and solvent deposited solids. J Pharm Sci 2001;90:504–20.

[53] Frasch HF, Barbero AM. Steady-state flux and lag time in the stratum corneum lipid pathway: results from finite element models. J Pharm Sci 2003;92:2196–207.

[54] Barbero AM, Frasch HF. Modeling of diffusion with partitioning in stratum corneum using a finite element model. Ann Biomed Eng 2005;33:1281–92.

[55] Heisig M, Lieckfeldt R, Wittum G, Mazurkevich G, Lee G. Non steady-state descriptions of drug permeation through stratum corneum. I. The biphasic brick-and-mortar mode. Pharm Res 1996;13:421–6.

[56] Naegel A, Hansen S, Newmann D, Lehr CM, Schaefer UF, Wittum G, et al. Insilico model of skin penetration based on experimentally determined input parameters. Part II: mathematical modelling of in vitro diffusion experiments, identification of critical input parameters. Eur J Pharm Biopharm 2008;68:368–79.

[57] Naegel A, Heisig M, Wittum G. A comparison of two- and three-dimensional models for the simulation of the permeability of human stratum corneum. Eur J Pharm Biopharm 2009;72:332–8.

[58] Hansen S, Naegel A, Heisig M, Wittum G, Neumann D, Kostka KH, et al. The role of corneocytes in skin transport revised—a combined computational and experimental approach. Pharm Res 2009;26:1379–97.

[59] Kubota K, Dey F, Matar S, Twizell E. A repeated dose model of percutaneous drug absorption. Appl Math Model 2002;26:529–44.

[60] Simon L, Loney N. An analytical solution for percutaneous drug absorption: application and removal of the vehicle. Math Biosci 2005;197:119–39.

[61] Pontrelli G, de Monte F. A two-phase two-layer model for transdermal drug delivery and percutaneous absorption. Math Biosci 2014 (February) http://dx.doi.org/10.1016/j.mbs.2014.05.001 [available online 14.05.14], in press.

[62] de Monte F, Beck JV, Amos DE. Diffusion of thermal disturbances in two-dimensional Cartesian transient heat conduction. Int J Heat Mass Tran 2008;51:5931–41.

[63] Pontrelli G, de Monte F. A multi-layer porous wall model for coronary drug-eluting stents. Int J Heat Mass Tran 2010;53:629–3637.

Mechanical Stress Induced Blood Trauma

Katharine Fraser

Imperial College London, London, UK

11.1 INTRODUCTION

The essential functions of blood are to deliver oxygen and nutrients to all tissues, to remove waste products, and to defend the body against infection. Whole blood is a concentrated suspension of formed cellular elements including red blood cells (erythrocytes), white blood cells (leukocytes), and platelets, suspended in plasma, an aqueous solution of proteins and other substances.

In the normal circulation of blood: from the chambers of the heart, through the heart valves to the arteries, capillaries, veins, and on return back to the heart, the blood is subjected to a range of mechanical stresses. Specifically, pressures from 0 to 150 mmHg, shear stresses up to 15 Pa and elongational flows which occur as the blood enters narrower arteries. Blood functions well under these conditions. However, under pathological conditions, created for example by an atherosclerotic stenosis, or under conditions of mechanical life support systems, blood is exposed to a considerably wider range of mechanical stresses. These conditions have a host of consequences on the blood constituents. Numerical models now play an important role in the understanding of these consequences, and are becoming vital in the development of novel blood contacting medical devices. This chapter begins with an analysis of the range of mechanical stresses experienced by blood in different cardiovascular devices. The current literature on the effects of mechanical stresses on different blood elements and functions is then reviewed. This is followed by a comprehensive description of state-of-the-art numerical models for analyzing blood trauma. Finally, some of the experimental and numerical issues which remain to be addressed are highlighted.

11.2 MECHANICAL STRESSES EXPERIENCED BY BLOOD

Mechanical stress on blood is conveyed via fluid dynamical forces which, at every location, can be described by the viscous stress tensor (σ):

$$\boldsymbol{\sigma} = \begin{pmatrix} \sigma_{xx} & \sigma_{xy} & \sigma_{xz} \\ \sigma_{yx} & \sigma_{yy} & \sigma_{yz} \\ \sigma_{zx} & \sigma_{zy} & \sigma_{zz} \end{pmatrix} \tag{11.1}$$

The stress tensor can be split into a pressure term (P) which gives a change in volume, and a deviatoric tensor, which gives a change in shape:

$$\boldsymbol{\sigma} = -P + \begin{pmatrix} \tau_{xx} & \tau_{xy} & \tau_{xz} \\ \tau_{yx} & \tau_{yy} & \tau_{yz} \\ \tau_{zx} & \tau_{zy} & \tau_{zz} \end{pmatrix} \tag{11.2}$$

Where:

$$\sigma_{ii} = -P + \tau_{ii} \quad \text{and} \quad \sigma_{ij} = \tau_{ij} \tag{11.3}$$

The definition of a fluid is that it deforms continuously when acted upon by a shearing stress, and so deformation is an important property of the fluid. The deformation can be described by the deformation tensor which is composed of the velocity gradients:

$$D = \begin{pmatrix} \dfrac{\partial u}{\partial x} & \dfrac{\partial u}{\partial y} & \dfrac{\partial u}{\partial z} \\[2mm] \dfrac{\partial v}{\partial x} & \dfrac{\partial v}{\partial y} & \dfrac{\partial v}{\partial z} \\[2mm] \dfrac{\partial w}{\partial x} & \dfrac{\partial w}{\partial y} & \dfrac{\partial w}{\partial z} \end{pmatrix} = \dfrac{\partial u_i}{\partial x_j} \tag{11.4}$$

Like any second order tensor, the deformation tensor can be decomposed into a symmetric part and an antisymmetric part:

$$D = \frac{\partial u_i}{\partial x_j} = \frac{1}{2}\left(\frac{\partial u_i}{\partial x_j} + \frac{\partial u_j}{\partial x_i}\right) + \frac{1}{2}\left(\frac{\partial u_i}{\partial x_j} - \frac{\partial u_j}{\partial x_i}\right) \tag{11.5}$$

The symmetric part is the rate of strain tensor:

$$\varepsilon = \frac{1}{2}\left(\frac{\partial u_i}{\partial x_j} + \frac{\partial u_j}{\partial x_i}\right) = \begin{pmatrix} \frac{\partial u}{\partial x} & \frac{1}{2}\left(\frac{\partial u}{\partial y} + \frac{\partial v}{\partial x}\right) & \frac{1}{2}\left(\frac{\partial u}{\partial z} + \frac{\partial w}{\partial x}\right) \\ \frac{1}{2}\left(\frac{\partial v}{\partial x} + \frac{\partial u}{\partial y}\right) & \frac{\partial v}{\partial y} & \frac{1}{2}\left(\frac{\partial v}{\partial z} + \frac{\partial w}{\partial y}\right) \\ \frac{1}{2}\left(\frac{\partial w}{\partial x} + \frac{\partial u}{\partial z}\right) & \frac{1}{2}\left(\frac{\partial w}{\partial y} + \frac{\partial v}{\partial z}\right) & \frac{\partial w}{\partial z} \end{pmatrix} \tag{11.6}$$

And, the antisymmetric part is the vorticity tensor ($\omega \equiv \nabla \times u$):

$$\omega = \frac{1}{2}\left(\frac{\partial u_i}{\partial x_j} - \frac{\partial u_j}{\partial x_i}\right) = \begin{pmatrix} 0 & \frac{1}{2}\left(\frac{\partial u}{\partial y} - \frac{\partial v}{\partial x}\right) & \frac{1}{2}\left(\frac{\partial u}{\partial z} - \frac{\partial w}{\partial x}\right) \\ \frac{1}{2}\left(\frac{\partial v}{\partial x} - \frac{\partial u}{\partial y}\right) & 0 & \frac{1}{2}\left(\frac{\partial v}{\partial z} - \frac{\partial w}{\partial y}\right) \\ \frac{1}{2}\left(\frac{\partial w}{\partial x} - \frac{\partial u}{\partial z}\right) & \frac{1}{2}\left(\frac{\partial w}{\partial y} - \frac{\partial v}{\partial z}\right) & 0 \end{pmatrix} \tag{11.7}$$

11.2.1 Couette Flow

In the case of simple shear flow, or Couette flow, named after the French physicist Maurice Marie Alfred Couette (1858-1943), the velocity field is steady, fully developed, and unidirectional. The magnitude of the velocity depends linearly on the spatial component with the axis perpendicular to the direction of flow. Simple shear flow can be generated between parallel plates with constant separation where one plate is moving relative to the other (Figure 11.1). In practice, Couette flow is obtained in rotating concentric cylinders (Figure 11.2), which approximate real Couette provided the flow regime is below the critical Taylor number: above this, the flow transitions to a set of toroidal vortices which become unstable.

11.2.2 Elongational Flow

In contrast to simple shear flow, simple elongational, or extensional, flow is generated when there is a velocity gradient in the same direction as the flow. Extensional flows are generated during pipe constrictions or nozzles. In reality, in the complex flow fields of cardiovascular devices, the shear stresses are a combination of shear and extension.

FIGURE 11.1 Couette flow between parallel plates.

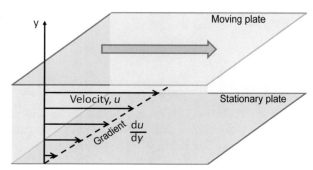

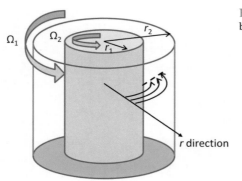

FIGURE 11.2   Taylor-Couette flow between rotating cylinders below the critical Taylor number.

11.2.3 Wall Shear Stress

Wall shear stress (WSS) is the value of shear stress at the surface, be it an artery wall, or the surface of a medical device. The WSS is important in vascular biology, because this is the stress exerted on the endothelial cells of the vasculature, but it is also important in the study of mechanical stresses on the blood, as in many cases it is the largest value of the shear stress. For example, steady flow of a Newtonian fluid through a long cylindrical tube will have a parabolic velocity profile, and the shear rate is then proportional to the distance from the tube axis.

11.2.4 Scalar Shear Stress

In an elastic material which is exposed to a general set of loads, an inhomogeneous, anisotropic stress field develops. A yield criterion is used to determine whether the material will fail, and is a way of relating what happens in three dimensions to the results of uniaxial tests. One commonly used yield criterion is the von Mises criterion which states that if the von Mises stress exceeds the yield stress then the material will fail. The von Mises stress is given by a combination of the components of the stress tensor. If a viscoelastic fluid is gradually sheared between two plates, a yield criterion can be used to determine when the fluid will flow. When the shear stress field is more complicated, an equivalent stress is again required for the yield criterion. Investigations of blood trauma have again made use of an equivalent stress. According to the work of Bludszuweit, the equivalent stress used is termed the Scalar Shear Stress and is analogous to the von Mises stress of solid mechanics [1]:

$$\sigma_s = \left(\frac{1}{6} \sum \left(\sigma_{ii} - \sigma_{jj} \right)^2 + \sum \sigma_{ij}^2 \right)^{1/2} \tag{11.8}$$

11.2.5 Fluid Dynamic Stresses in the Circulation

In the circulation, the highest stress on the blood is the wall shear stress (WSS). Flow in the arteries is pulsatile, which means WSS is also pulsatile, whereas flow in the capillaries and veins is steady. The mean WSS in the aorta is around 1 Pa with systolic WSS peaking around 4 Pa (Figure 11.3). The WSS remains similar in the conduit arteries (carotid, femoral, and brachial) but increases significantly in the arterioles: the smallest arterioles have WSS up to 15 Pa.

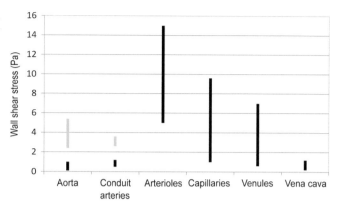

FIGURE 11.3 Mean WSS (black) and peak systolic WSS (grey) in the normal circulation. *Data from [106–114].*

On the venous side of the circulation, the flow is steady and the mean WSS decreases as the size of the vessels increases. During the pathological case of arterial stenosis in the conduit arteries, WSS may increase by an order of magnitude from around 1 Pa to over 10 Pa [2].

11.2.6 Fluid Dynamic Stresses in Blood Contacting Devices

11.2.6.1 Needles, Catheters, Cannulae

In needles, catheters, and cannulae, as in other medical devices, three types of forces act on the blood: pressure gradients, elongational forces, and shear forces. Needles of size 21 gauge are commonly used for drawing blood for testing, while 16 or 17 gauge needles are most commonly used for blood donation. So the range of internal diameters is 0.5 to 1.2 mm. Flow rates for blood donation are in the region of 60-75 mL/min, meaning the flow is laminar and therefore has a parabolic Poiseuille flow profile; then the wall shear stress in the needle is around 0.015 Pa. Based on the flow rate and a needle length of 40 mm, the time in the needle is 0.035 s. This shear stress is comparable with that experienced by blood in the normal circulation and should not cause any trauma. However, blood may be more sensitive to the extensional stress at the entrance to the needle. In microflow channels reminiscent of small needles Down et al. [3] found that damage to red blood cells was related not to the shear stress within the channel but to the shape of the entrance, and suggested the extensional stress may be the reason. Extensional stresses occur when there is an acceleration in the velocity due to the geometry. Rounding the edges results in lower extensional forces but sharp edges give short regions of high acceleration.

11.2.6.2 Rotary Ventricular Assist Devices

The rotating component of a rotary ventricular assist device (VAD) is responsible for huge shear stresses on the blood. Shear stresses in a range of VADs were investigated using computational fluid dynamics [4]. With a flow rate of 3 L/min and pressure head of 100 mmHg, around 4.0-5.5% of the volume of the axial VADs investigated experienced scalar shear stress (SSS) greater than 50 Pa for 4-7 ms. The same volume in the centrifugal VADs investigated was a little smaller: 1.0-3.0% for 4-5 ms. The volume experiencing SSS greater than 150 Pa was smaller in all the VADs studied: 0.75-1.25% for 1.5 ms in the axial VADs and less than 0.25% for under 0.5 ms in the centrifugal VADs. Even smaller regions, "hot spots," of the

VADs experience significantly higher SSS: up to 1,250-3,000 Pa in the axial VADs and 300-1,100 Pa in the centrifugal VADs.

11.2.6.3 *Displacement Ventricular Assist Devices*

In displacement VADs, a large vortex is established during diastole and keeps the blood moving during systole [5]; however, flow stasis is still a concern. Hochareon et al. [6] experimentally measured velocities in the 50 cc Penn State Artificial Heart and found early diastolic velocities up to 80 cm/s. However, due to the nonuniform flow across the chamber, a region with low wall shear rate, below 250 s^{-1}, persists throughout the cardiac cycle and this correlated with clot formation. In a poorly designed displacement VAD, blood can remain in the chamber for more than three cardiac cycles [7].

11.2.6.4 *Mechanical Heart Valves*

The leaflets of mechanical heart valves create unphysiological turbulent vortices in their wake in which the blood is exposed to increased shear stresses up to 15 Pa [8]. The hinges of the valves, and leakage between the leaflets when the valve is in the closed position, result in very high velocity flows (up to 208 cm/s [9]) which cause the high shear stresses and turbulence. Malfunctioning heart valves, which open only partially, exert still higher stresses which have been found to activate platelets by a factor of eight times that of properly opening valves [10].

11.2.6.5 *Membrane Oxygenators*

Blood velocities inside the hollow fibre bundles typical of membrane oxygenators are usually slow with consequentially low shear stresses exerted on the blood. Zhang et al. [11] calculated the velocities in a model of a small portion of an oxygenator to be up to 33 cm/s which resulted in peak wall shear stress (WSS) on the fibres up to 52 Pa. The average WSS depended on both the flow rate and the geometry of the fibre bundle, with the square array having the lowest area weighted average WSS and the diagonal array the highest. With relatively slow flows and long residence times necessary for diffusion of oxygen into the blood, and carbon dioxide out of the blood, the potential for blood stagnation is a concern [12]. Jones et al. [13] used X-ray imaging of a contrast agent to look at flow through a radial oxygenator; their results showed that blood would typically spend around 2.5 s in the fibre bundle. The radial design, also used in some artificial lungs [14], ensures a symmetrical flow field that minimizes the risk of stagnation zones.

11.3 FLUID DYNAMIC EFFECTS ON BLOOD CONSTITUENTS

11.3.1 Red Blood Cells

Blood rheology is a large field and here it is sufficient to simply outline the information pertinent to blood trauma. Erythrocytes make up between 36% and 54% of the blood volume [15] and are largely responsible for its non-Newtonian behavior. Blood is a shear thinning, viscoelastic, thixotropic fluid, which may have a yield stress.

At low shear rates ($< \sim 50$ s^{-1}) and in the presence of fibrinogen and other plasma proteins, the erythrocytes aggregate in rod-shaped stack structures of individual cells called rouleaux. At very low shear rates, the rouleaux align perpendicular to each other creating a three dimensional branching secondary structure. These structures of rouleaux are responsible for the higher viscosity at low shear rates. Erythrocyte aggregation is a reversible dynamic phenomenon, so as the shear rate is increased the rouleaux come apart. Faraheus demonstrated rouleaux formation at low shear rates in both physiological and pathological conditions [16]. In most blood vessels, the shear rates are too high for rouleaux formation. However, they are likely to form in some veins and venules, and some diseases, for example pneumonia, cause a tendency for rouleaux. As the shear rate is increased, the rouleaux break up; the higher the shear rate, the smaller the structures. When the shear rate is reduced again, large structures reform. However, it takes time to form large rouleaux, which leads to a time dependent change in viscosity.

At higher shear rates, several different types of individual red blood cell motions have been reported: the cells have been observed to progress through different types of motion as the shear stress increases. These different types of motion include flipping, tumbling, rolling like a wheel, and tank-treading, a state in which the membrane of the cell exhibits precession [17].

11.3.1.1 Deformation of RBCs

At still higher shear stresses, the red blood cells undergo volume and area preserving shape changes and deform into ellipsoids [18] (Figure 11.4). This deformation is like that of immersed fluid droplets [19] and is a consequence of the tensile and compressive forces the cell experiences in alternate quadrants. In long exposure time experiments ($\sim$5 mins) the elongation is a linear function of shear stress up to 250 Pa when the elongation is 100% [18]. The ellipsoidal cells align according to the direction of the flow, with the major axis of their ellipsoidal shape at a small angle to the flow direction, with the angle depending on the shear stress [20–22]. Their membrane performs a "tank-treading" motion in which it rotates about the axis perpendicular to both its velocity and the velocity gradient. This tank-treading motion has a frequency which also depends on the shear stress.

Deformation of RBCs is the cause of blood's shear thinning viscosity at shear rates above those affecting rouleaux formation. The deformation of RBCs into ellipsoids with their major axis aligned with the flow leads to reduced viscous hindrance. More importantly, tank-treading transfers the fluid stress from the immersion medium to the inside of the cell so that the cell participates in the flow rather than purely distorting it [19].

The stress in extensional or elongational flow also causes deformation of the RBC. Indeed, recent data suggest that RBC deformation may actually be more susceptible to extensional than pure shear flow [3,23]. However, there are still far fewer studies on the relationship between RBC deformation and extensional stress.

Another important consideration is the reversibility of RBC deformation. When the shear stress is removed, the RBC returns to its biconcave disc shape with a relaxation time of around 100-300 ms [24]. However, while cells sheared at 200 Pa for 5 mins returned to their normal resting biconcave shape afterwards, those sheared at 350 Pa and left to recover were crenellated and fragmented, suggesting RBC recovery is limited [18].

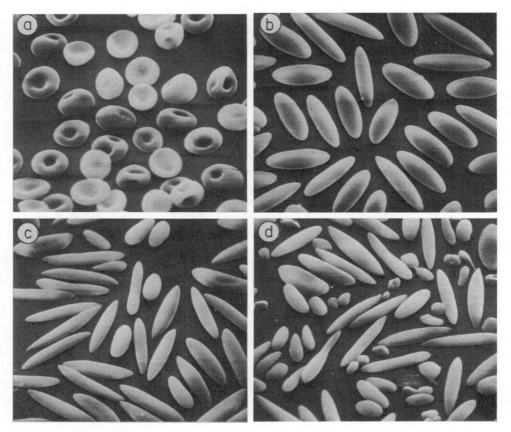

FIGURE 11.4   Human red blood cells fixed while immersed in shear flow in a concentric cylinder viscometer. Applied shearing stress was (a) 10, (b) 200, (c) 350, (d) 450 Pa for 4 min at 37 °C. *(Reproduced, with permission, from [18].)*

11.3.1.2 *Hemolysis*

Hemolysis is the best-studied aspect of mechanically induced blood damage and is defined as the release of hemoglobin into the plasma due to mechanical compromise of the erythrocyte membrane. An RBC in a shear flow deforms, as described earlier, at first undergoing volume and surface area preservation. Above a critical shear stress the deformation no longer maintains the cell's surface area and the membrane is forced to stretch. Then the mechanism for the release of hemoglobin may be by complete rupture of the cell [25] or through pores appearing in the viscoelastic cell membrane during high stress [26].

As discussed by Leverett et al. [27] RBC damage may be influenced by a number of factors in addition to fluid shear stress which include: solid surface interactions, air interfaces, and cell-cell interactions. Leverett et al. [27] used a shear stress-exposure time plot to summarize the results of studies, including their own and those of previous authors, on the critical shear stress threshold required for hemolysis. With this plot, they showed that, below the critical threshold, surface effects dominate, whereas shear stress effects are important above it. In their own study, they varied the surface area to volume ratio of their Couette viscometer and showed that hemolysis as a function of shear stress was unaffected by this ratio. Significant hemolysis was found in the region of the viscometer close to the air interface. By varying the hematocrit, they showed that cell-cell interactions were not an important mechanism for hemolysis, at least in the Couette viscometer.

A series of investigations in throughflow Couette devices by different groups have shown that hemolysis occurs as a power-law function of the shear stress acting on the cells, τ, and the time of the exposure to that shear stress, t (Figure 11.5). The index of hemolysis, IH, is the percentage difference in plasma free hemoglobin, Hb, after a single pass through the device:

$$IH = \frac{\Delta Hb}{Hb} \times 100 = C\tau^a t^b \tag{11.9}$$

Wurzinger et al. [28,29] and Giersiepen et al. [30] were the first to use such a device with resuspended human RBCs under the conditions $\tau < 255$ Pa and $t < 700$ ms. Heuser and Opitz [31] used the same device with porcine blood at comparable exposure times ($t < 700$ ms), and higher shear stresses ($\tau < 700$ Pa). Problems with the fluid seal in this device are thought to have led to increased hemolysis. Paul et al. [32] developed a device with an improved seal and studied porcine blood at $t < 1200$ ms and shear stresses $\tau < 450$ Pa. They found that measureable hemolysis only occurred with $\tau < 425$ Pa and $t < 620$ ms. Most recently, Zhang et al. [33] designed novel throughflow Couette devices based on rotary VAD designs. One of these

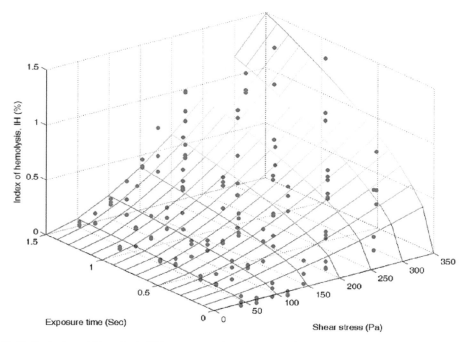

FIGURE 11.5   Index of Hemolysis (IH) is a power-law function of shear stress magnitude and exposure time. *(Reproduced, with permission, from [33].)*

had pin bearings while the other was magnetically levitated to completely eliminate damage due to mechanical contact. The constants were then $C = 1.228 \times 10^{-5}$, $a = 1.9918$, and $b = 0.6606$. The characteristics of the hemolysis curve were then quite different from those found by Paul et al. [32] as the hemolysis increased gradually from the lowest values of t and τ measured. This difference is likely to be due to the difference in characteristics of ovine versus porcine blood.

The susceptibility of RBCs to shear stress has been shown to depend on the species. In turbulent jets, the critical speed for hemolysis increases as the size of the RBCs decreases [34] which is possibly due to the size of the Kolmogorov vortices which decreases as the jet speed increases. It is thought that for turbulent vortices to damage cells the fluctuations must be of the same order of magnitude as the cells themselves. Human RBCs, with a volume of 103 μm^3, hemolyzed with a jet speed of 1,930 cm/s whereas bovine and ovine RBCs, 32 and 31 μm^3, hemolyzed at 3850 and 3880 cm/s, respectively. In contrast, there must exist another mechanism for the differences in fragility of blood cells in laminar shear stress. Jikuya et al. [35] using a concavo-convex rheometer at 30 Pa for exposure times up to 15 min, found that ovine RBCs produced 1.8 times the plasma free hemoglobin of human RBCs, while bovine RBCs produced only 0.5 times the plasma free hemoglobin.

Older RBCs have more viscous, stiffer membranes and a smaller surface area [36]. The amount of damage done by shear stress is thought to depend on the current level of damage and damage accumulates over the lifetime of the cells. When they are sufficiently damaged, the RBCs are removed from the circulation by the spleen.

11.3.2 White Blood Cells

Infection rates for cardiovascular devices vary substantially; while infections related to vena cava filters and coronary artery or peripheral vascular stents are rare, incidence of infection in pacemakers, defibrillators, LVADs, ventriculoarterial shunts, vascular grafts, aortic balloon pumps, angioplasty, angiography, and arterial patches is significant [37]. The latest INTERMACS report shows that for mechanical circulatory support devices in clinical use in the United States, there are 8.01 infections/100 patient months. The overall rate for cardiovascular devices is 3% [38] and poses a significant medical challenge with most cases requiring surgical intervention; there is a 20-40% associated mortality rate. There are three considerations in the pathogenesis of cardiovascular device related infections: (1) pathogen virulence factors, (2) the physical and chemical characteristics of the medical device, and (3) host response to the presence of an artificial device. Here we focus on the last of these, with an emphasis on the response of the white blood cells, the key players in the immune response. However, it is important to remember that the interplay between the different factors in the host response will determine the course of the infection.

Of the different types of leukocytes, neutrophils and monocytes are the most important when considering device related infection. Neutrophils defend against bacterial and fungal infections. They are the body's first response and are the most common cell type seen in the early stages of acute inflammation. They make up 60-70% of the leukocyte count in human blood and have a life span of about five days. Neutrophils destroy bacteria by phagocytosis. They are unable to renew their lysosomes, which are used in digesting microbes, and so die after phagocytosing a few bacterial cells. Moncytes are less prolific, making up around 5% of the leukocyte count, but they have a longer life span. They have two main roles: they present pieces of pathogens to T-cells to start an antibody response and to ensure that the pathogen will be remembered in future; and they assist the neutrophils with phagocytosis. These tasks are performed in the bloodstream, but moncytes may leave the blood by passing through the vascular wall to the underlying tissue. There they differentiate, becoming macrophages that phagocytose dead cell debris and microorganisms.

The ability of the body to defend itself against infection is, therefore, determined by the phagocytic ability of the neutrophils, monocytes, and macrophages. The first step in phagocytosis is for the receptors on the leukocyte surface to bind with ligands on the particle. This triggers the leukocyte to deform, surrounding and engulfing the particle. So deformability is key to phagocytic ability. Furthermore, to destroy bacteria within tissue, as opposed to within the blood, the monocytes must first penetrate the vascular wall. This requires them to squeeze through small openings in the wall, so again deformability is required.

An artificial surface placed in the body is a substrate for a host of molecules including fibronectin, vitronectin, fibrinogen, glycoproteins, proteoglycans, polysaccharides, and lipids which interact to form a biofilm coating the surface [39]. The properties of the artificial material such as its chemistry, charge, microarchitecture, and affinity for water determine the constitution of the biofilm. Bacteria such as *Staphylococcus epidermidis* and *Staphylococcus aureus* can adhere either directly to the artificial surface or through one or more components of the biofilm. The biofilm makes it more difficult for the host defense system: the presence of the film reduces the phagocytic ability of macrophages and neutrophils.

The processes by which leukocytes arrive at a surface are influenced by the hemodynamics [40]. Leukocytes roll along the normal vascular wall becoming firmly adhered when they are recruited to sites of infection. During periods of chronic infection, monocytes may adhere to endothelial cells or artificial surfaces for hours or days and therefore the accumulated shear stress—exposure time product—is likely to be significant. Rosenson-Schloss et al. [41] examined adherent leukocytes under shear stresses in the vascular range (0.1-2.5 Pa). At 0.3 Pa, cells exhibited contracted lamellipodia and ruffled membranes. At 1 Pa, cells were rounded and most had completely retracted lamellipodia. At the highest shear stress studied, 2.5 Pa, leukocytes were elongated and aligned with the direction of flow. The phagocytic ability of the cells changed with shear stress: under static control conditions leukocytes ingested 3-4 particles/cell whereas above 1 Pa phagocytosis increased up to 2.5 Pa when the phagocytic rate was 7 particles/cell.

Neutrophils adherent to the surface of a cardiovascular device are subjected to significantly higher shear stress than in the normal circulation. Shive et al. [38] showed that when neutrophils are adherent to polyurethane, a material commonly used in cardiovascular devices, physiological levels of shear stress lead to apoptosis, programmed cell death. The effect of shear stress was rapid; apoptosis was 100% after 60 min exposure to 0.6 Pa. Neutrophils adherent in low shear (0-0.2 Pa) areas efficiently cleared adherent bacteria whereas those in high shear (>0.6 Pa) areas were incapable of interacting with the bacteria.

In contrast, higher levels of shear stress are required to alter the morphology of neutrophils in suspension. Carter et al. [42] sheared leukocytes by passing a suspension of the cells through capillary tubes. They found that, with an exposure time of 90 ms, a WSS of 13.4 Pa produced a decrease in phagocytic ability of 10%. Increasing the wall shear stress decreased the phagocytic ability up to a plateau at 764 Pa when the phagocytic ability was around 60%. There was no significant change in the leukocyte count with increasing shear stress.

Chan et al. [43] investigated leukocyte morphology in rotary ventricular assist devices *in vitro* by recirculating the blood flow for 25 hours (Figure 11.6). They compared the ROTA-FLOW and VentrAssist at the same pressure head and flow rate, estimating the maximum shear stress in the two devices as 12 and 284 Pa, respectively. After 25 hours, the total percentage decrease in leukocytes was 37% in ROTAFLOW and 63% in VentrAssist, compared with 23% in static blood. The neutrophils were the most susceptible to the shear stress in the VentrAssist with a decrease of 59% compared with 6% in both ROTAFLOW and static blood. Damage to the leukocytes was assessed using flow cytometry which showed larger numbers of both physically damaged and necrotic leukocytes in blood from the VentrAssist, as compared with either static blood or blood from the ROTAFLOW. In contrast to the work of Shive et al. [38], which investigated adherent neutrophils under shear stress < 6 Pa and found that the cells underwent apoptosis, at the considerably higher shear stresses investigated by Chan et al. [43] the leukocytes are physically broken down. In this process, parts of the cell membrane break off to form leukocyte microparticles (LMPs); increased levels of LMPs have been found in VAD patients [44].

11.3.3 Platelets

Platelets are small (2-3 microns diameter) cell fragments and are prominent in both physiological hemostasis and pathological thrombosis [45]. In normal hemostasis, when platelets interact with a thrombogenic surface, for example exposed collagen in the injured vessel wall, the platelets then become activated, release procoagulant agonists, and bind more tightly. The released agonists induce further activation of nearby platelets in a positive feedback mechanism and the clot grows. The clot seals the injured vessel preventing further blood loss. So, clotting is central to hemostasis and requires three main steps: adhesion of platelets, platelet activation, and the coagulation cascade. However, under different mechanical conditions there are important differences in these steps. Different mechanical conditions are found in the healthy cardiovascular system, for example in arteries versus veins or in the center of an artery versus the wall, but are exaggerated by pathological conditions such as atherosclerosis, and medical devices such as heart valves or VADs, and, because much of the evidence has been obtained from nonphysiological *in vitro* experiments, we also know about these.

11.3.3.1 Adhesion

Under low shear conditions ($<1,000$ s^{-1}) platelets move slowly relative to their surroundings including surfaces, other platelets, and proteins dissolved in the plasma. The slow movement allows time for their integrin $\alpha 2\beta 1$ receptors to interact and bind to thrombogenic surfaces. Activated integrin $\alpha 2\beta 1$ then binds to dissolved fibrinogen which bridges neighboring activated platelets.

At higher shear rates (1,000-10,000 s^{-1}) initial adhesion of platelets is via von Willebrand factor (vWf) [46]. A thrombogenic surface, which could be an exposed subendothelial layer,

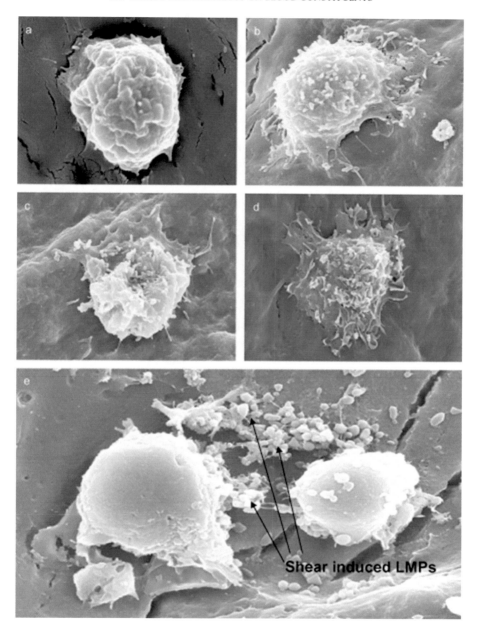

FIGURE 11.6   Scanning electron microscope images of leukocytes exposed to shear stress. Blood was recirculated in a flow loop through the ROTAFLOW CP: (a) after 5 mins the leukocyte shows early signs of damage, (b) after 360 min a cell shows increased damage, (c) and (d) are cells after 22 hours showing further damage and collapse, respectively, and (e) after 22 hours shear induced leukocyte microparticles can be seen. *(Reproduced, with permission, from [43].)*

the membrane of a second platelet, or an artificial material, has molecules of vWf immobilized on it. These immobilized vWf molecules bind to the transmembrane protein GPIbα on platelets. This adhesion is rapid, but likewise the interaction is rapidly reversible. The platelets slide over the thrombogenic surface and are slowed down by their interaction with vWf. When platelets slide across vWf, thin tethers are pulled from their membranes which act as anchors to slow their progress [47]. The number of platelets forming tethers, tether lifetime, and tether growth rate are all shear rate dependent. As the tethers form and then detach, the platelet translates with a stop-start motion. The interactions of vWf and GPIbα cause transient intracellular calcium concentration spikes which activate integrin αIIbβ3. Subsequent interaction of activated integrin αIIbβ3 with vWf triggers sustain calcium oscillations and further integrin αIIbβ3 activation [48]. Interaction via the integrin family receptors is only possible at sufficiently low platelet speeds because this interaction is slow. So, when the platelet has been slowed down enough, the activated integrin α2β1 receptors bind to vWf, collagen, and fibrinogen which stabilizes the clot [49].

At still higher shear rates ($>10,000$ s^{-1}) both the initial adhesion of platelets and stabilization of the aggregate involves vWf [50]. The vWf immobilized on the platelet surface rapidly engages with GPIbα. At these pathologically high shear rates, platelet adhesion and aggregation depends on soluble as well as surface bound vWf. There is no requirement for platelet activation in order for the platelets to form a short-term stable aggregate. However, in the longer term, activation will result in a stable thrombus.

The combination of alternating periods of high shear stress and low shear stress has been found to induce significantly higher levels of platelet aggregation compared with either pure high shear stress or pure low shear stress [51,52]. This is likely to be because the period of high shear stress promotes initial interactions between vWf and GPIbα and the subsequent low shear period enhances the decelerating property of the GPIbα -vWf interaction to allow more time for the binding of αIIbβ3 to fibrinogen and P-selectin to its ligand.

11.3.3.2 Activation

When the platelets adhere to the surface, or to each other (aggregate), they are held together via vWf, connected to their GPIbα receptor (usually under high shear conditions), or collagen connected to their integrin α2β1 receptors (usually under low shear conditions). Platelets adhered to collagen via the integrin α2β1 are held close to the collagen long enough that they can interact with the lower affinity receptor GPVI. GPVI stimulates platelet activation.

On activation, P-selectin translocates rapidly to the cell surface and platelets change from a discoid shape to a sphere with protrusions called pseudopods. Activated platelets form membrane blebs or microparticles when part of the cell membrane detaches [53]. However, it is thought that not all the small particles seen after platelet exposure to high shear stress are true microparticles: Slack et al. [54] found that after exposure to high shear stress (10 Pa for 60 s) the number of particles smaller than platelets increased even when the platelets had previously been exposed to the anti-integrin αIIbβ3 monoclonal antibody, 7E3, which prevents platelet activation. The authors thought that the shear stress mechanically broke up the platelet membrane.

Upon activation, platelets also release alpha granules which contain a vast number of biologically active molecules. These include: adhesive proteins (such as fibrinogen, vWf, thrombospondin), plasma proteins (such as albumin), cellular mitogens (such as platelet derived growth factor and TGF beta), coagulation factors (such as Factor V), and protease inhibitors [55]. Activated platelets express phosphatidylserine on their surface membrane and this lipid is essential to the coagulation cascade. In particular, phosphatidylserine is involved in the conversion of prothrombin to thrombin and activation of factor X.

Fibrinogen from the plasma binds to the activated integrin $\alpha IIb\beta 3$ on more than one platelet, stabilizing the platelet aggregate. Aggregation of activated platelets into a plug which stops the bleeding completes primary hemostasis. Secondary hemostasis is the cascade of coagulation factor reactions which leads to fibrin formation and permanent binding of the clot.

Hellums [56] showed that the locus of points on the shear stress-exposure time plane at which the platelets become activated follow a consistent curve over a wide range of conditions (Figure 11.7). One method for measuring levels of platelet activation is the platelet activation state (PAS) assay [10,57]. Acetylated prothrombin is incubated in the presence of factor Xa and Ca^{2+} with the platelet sample. The amount of acetylated thrombin generated is related to the activation state of the platelets in the sample because the phosphatidylserine on the surface of activated platelets is required for the prothrombin to thrombin conversion. Because acetylated thrombin cannot itself activate platelets, there is no feedback and the relationship between acetylated thrombin and PAS is linear. The PAS assay has been used by Bluestein and colleagues to investigate platelet activation by defective heart valves [10].

In an *in vitro* study, Bakir et al. [58] found a significantly reduced platelet count at moderate shear stress (6.5 Pa for 15 mins) while significant hemolysis was only seen at higher shear stresses (around 13 Pa for 15 mins). In contrast, in an *in vivo* study, Loffler et al. [59] found no evidence that patients with the Jarvik 2000 LVAD had platelet activation levels above those of normal healthy patients, as determined by measurements of P-selectin and activated integrin $\alpha IIb\beta 3$, despite increased hemolysis.

In addition to activation, high shear stress can also trigger platelet apoptosis. Leytin et al. [60] found that applying shear stress in the range 11.4 to 38.8 Pa for 90 s at 37 °C caused caspase 3 activation and mitochondrial transmembrane potential depolarization, which are both signs of apoptosis.

FIGURE 11.7 Hellums' shear stress-exposure time threshold required to activate platelets, with additional data. *(Reproduced, with permission, from [4].)*

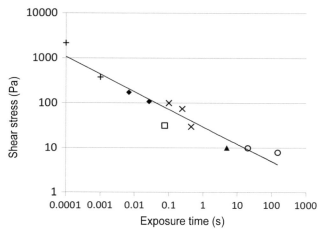

11.3.4 von Willebrand factor (vWf)

von Willebrand factor (vWf) is a glycoprotein secreted by endothelial cells (and megakaryocytes) and crucial to hemostasis through its interaction with platelets and the coagulation cascade (as described in Section 11.3.3). Some vWf is secreted immediately after synthesis and the rest is stored and released upon stimulation as a very large multimer [61]: the longest are over 20,000 kDa and are composed of 500 kDa dimers. These high molecular weight (HMW) vWf multimers are broken down to a series of different sized multimers via proteolysis. The spectrum of vWf multimers controls the clotting ability of blood, because the larger multimers have significantly more binding potency. In von Willebrand disease (vWd) type 1 and type 2A, a deficiency in total vWf, or HMW multimers, respectively, causes increased bleeding risk. ADAMTS13, released by endothelial cells, cleaves the multimers at the A2 domain [62,63] and so exerts control on the spectrum of multimers. Shear stress in the circulation unravels the multimers to reveal the cleavage site accelerating this decomposition. Cleavage of vWf by ADAMTS13 was found to be enhanced by the presence of platelets [64]. In thrombotic thrombocytopenic purpura (TTP), inhibition or deficiency of ADAMTS13 results in excess HMW vWfs causing microscopic clotting. So in the normal circulation, ADAMTS13 influences the balance between prohemorrhagic and prothrombotic states. Under conditions of increased fluid shear stresses, such as those found in blood contacting devices, the balance may be pushed more towards prohemorrhagic.

Patients implanted with rotary VADs have reduced levels of HMW vWf which is known as Acquired Von Willebrand Syndrome (AVWS) [65–67]. In one study [65] all patients with a HeartMateII (a rotary VAD), both with and without gastrointestinal bleeding, had reduced HMW vWf. Reduction in the HMW vWf is detectable from the first postoperative day [66,68]. In contrast, pulsatile Total Artificial Heart (TAH) patients did not have reduced HMW vWf [66].

The following theory describes how fluid shear stress may influence vWf multimer lengths. In low stress conditions, the attractive static forces in the vWf multimer result in a coiled up, globular configuration. When exposed to fluid dynamic stresses, the vWf globule unravels, with the unravelled portion extended by the flow [69,70]. Unravelling permits elongational force loading on the A2 domain. At the molecular level, the domains are constantly unfolding and refolding, but under extensional force the lifetime of the unfolded A2 domain is much longer (140 s compared with 1.9 s) [69]. A2 can only be cleaved by ADAMTS13 in its unfolded state. The observation that platelets catalyze the cleavage of vWf by ADAMTS13 could be the result of larger fluid dynamic forces on the platelets attached to vWf, which more rapidly stretch the vWf molecules [64].

11.3.5 Thrombosis and Emboli

Once activated, platelets accumulate, coagulate, and adhere to the highly thrombogenic surfaces of medical devices. Blood clotting has long been associated with three factors which are collectively known as Virchow's triad: the nature of the surface, the condition of the blood, and the local flow conditions. Slow, stagnant, or recirculating flow and low shear stresses promote thrombosis [6,71].

11.4 NUMERICAL MODELS OF DAMAGE TO THE BLOOD CONSTITUENTS

11.4.1 Red Blood Cells

Our primary concern is a universal model for hemolysis in blood contacting devices. As discussed, the range of hemodynamic stresses is known and the more detailed stress field can be solved analytically, or numerically, or measured experimentally. Given that we know these stresses, we want to know what damage is done.

Fundamentally, there are two approaches to modeling hemolysis. In the first approach, which we call stress-based, hemolysis calculation begins with experimental measurements of hemolysis in a uniform shear stress field. From the experimental measurements, an empirical relationship between stress and hemolysis is determined. While this approach is conceptually straightforward, there are a number of challenges associated with applying it to the spatially and temporally inhomogeneous stress fields of real blood contacting devices: these challenges are discussed below. In the second approach, which we call strain-based, we again begin with an experimental measurement of hemolysis in a uniform stress field, but in addition we measure the deformation, or strain, of the RBC. The strain of the RBC in an inhomogeneous stress field is then found using an analytical model of the RBC. So, in the strain-based approach hemolysis is related to stress via strain, rather than directly. This approach is technically more demanding that the stress-based approach, but has the potential to be applicable across all devices.

11.4.1.1 Stress-Based

A model for hemolysis, which can be applied to all devices, is desirable for VAD design. Many groups have attempted to calculate hemolysis in various blood wetted devices. As described in Section 11.3.1.2, hemolysis in a uniform shear stress field can be described by a power-law equation. The power-law equation has been implemented in computational fluid dynamics (CFD) calculations using postprocessing techniques in Eulerian and Lagrangian coordinates, as well as by solving a transport equation for plasma free hemoglobin. In the Eulerian approach, a damage index, H, is integrated over all the cells in the numerical space, for example Garon and Farinas [72] use:

$$H = \left(\frac{1}{Q} \int_V C^{1/b} \tau^a dV \right)^b \qquad (11.10)$$

where Q is the flow rate and V the volume. Garon and Farinas [72] compared their numerical predictions with published hemolysis data for cannulae [73] and with their own measurements in VADs and found good agreement. Arvand et al. [74] also used an Eulerian approach but incorporated pressure head and a high stress fraction and found their constants experimentally. This simple Eulerian approach is the easiest to implement but cannot show the distribution of hemolysis production and fails to account for differing residence times in different regions of the flow field. An improvement is to solve a scalar transport equation for the flow of plasma free hemoglobin, Hb, through the fluid domain with a source term accounting for the release of hemoglobin from the cells. In practice, because the release of plasma free hemoglobin is governed by a power-law we solve a transport equation for Hb' where: $Hb' = Hb^{1/b}$

$$\frac{d(Hb')}{dt} = S + \nabla \cdot (D\nabla Hb') - \nabla \cdot (vHb') \tag{11.11}$$

with

$$S = C^{1/b}\tau^{a/b}$$

The source term, S, is a function of shear stress, τ, which is usually the SSS. The diffusivity, D, for plasma free hemoglobin is very small, so the diffusion term may be omitted. This scalar transport approach has been used to calculate hemolysis in a range of ventricular assist devices [4,75].

In the Lagrangian formulation, the damage index is integrated along tracer particle streaklines [76] or post processed pathlines [77–80]. After solving the governing flow equations, pathlines are calculated starting from the flow inlet using forward Euler integration. The lines are then discretized and the total damage calculated by summing the individual damage portions. Different authors have used various damage functions, for example Mitoh et al. [76] and Yano et al. [77], based their model on Bludszuweit's idea [1] of linear damage accumulation and defined the damage for a section i of the pathline as:

$$H_i = H_{i-1} + (1 - H_{i-1})C\tau_i^a \Delta t_i^b \tag{11.12}$$

Several authors [73,80–82], define the infinitesimal damage simply by using a mini power law:

$$dH = C\tau^a dt^b \tag{11.13}$$

and integrate along the streamline. The temporal derivative has also been suggested as an infinitesimal damage [83]:

$$dH = Cb\tau^a t^{b-1} dt \tag{11.14}$$

and this does at least predict the same total blood damage under uniform shear stress as the power-law equation. Grigioni et al. [83,84] showed that the mini power law form of the infinitesimal damage did not reproduce the power law result for the total damage under constant shear stress, and that while the temporal derivative does reproduce the total damage given by the power-law equation, the infinitesimal damage in the temporal derivative form does not depend on the damage history. They formulate a new infinitesimal damage based on the idea of a mechanical dose, and show that it depends not only on the current shear stress and exposure time but also on the shear stress-exposure time history:

$$dH = Cb\left(\int_{t_0}^{t} \tau(\varepsilon)^{a/b} d\varepsilon + H(t_0)\right)^{b-1} \tau(t)^{a/b} dt \tag{11.15}$$

Goubergrits and Affeld [85,86] also developed a similar infinitesimal damage accounting for damage history although theirs was formulated in a different way.

The Lagrangian pathline approach has had some success in predicting hemolysis [76–78]. However, obtaining pathlines has computational problems at near zero velocities and in recirculation regions, and does not necessarily take into account the entire flow field. Carswell et al. [87] showed that, at least in the VAD they studied, more than 20,000 pathlines were required to achieve a converged hemolysis prediction. The scalar transport and Lagrangian pathline approaches are compared by Taskin et al. [88], who conclude that the scalar transport approach can predict relative hemolysis levels and is simpler to implement.

11.4.1.2 Strain-Based

As mentioned earlier, in the strain-based approach, an empirical relationship between hemolysis rate and strain is required. This has been found by several authors and is already discussed in Section 11.3.1.1. Also required for the strain-based approach is an analytical model for the strain of the RBC in a stress field. These analytical models are the primary focus of this section.

If the resting shape of the RBC is simplified to a sphere, then droplet theory can be used to describe the shape change in flow. The phenomenological model of Maffetone and Minale [89] was used by Arora et al. [78] and Pauli et al. [90]. In this work, the droplet is assumed to remain ellipsoidal at all times, until the droplet breaks up, at which time the model fails. The droplet shape can be described by a symmetric, positive-definite, second rank tensor S with eigenvalues representing the square semi-axes of the ellipsoid. There are two competing forces on the droplet: the interfacial tension which tries to recover the undeformed spherical shape; and the drag exerted by the surrounding fluid. Because the droplet is assumed to be incompressible, its volume must be preserved. The evolution equation for S is then:

$$\frac{ds}{dt} - \omega \cdot S + S \cdot \omega = -\frac{f_1}{t_c}[S - g(S)I] = f_2(\varepsilon \cdot S + S \cdot \varepsilon) \qquad 11.16$$

where $t_c = \sigma_t / \mu R$ is a characteristic time; I is the unit tensor; ε and ω are the deformation rate and vorticity tensors (see Section 11.2) σ_t is the interfacial tension, μ is the viscosity of the immersion fluid, and R is the radius of the spherical droplet. The left hand side is a Jaumann derivative (which accounts for translation with the fluid, like the substantial derivative, but also includes rotation with the fluid). The first term on the right hand side accounts for the interfacial tension, acting to recover the undistorted spherical shape, represented by the unit tensor I. The function $g(S)$ preserves the droplet volume, and is $g(S) = 3^{III_S}/_{II_S}$ where II_S and III_S are the second and third invariants of S, respectively. In the general case, f_1 and f_2 can be shown to be functions of the ratio of the viscosities of the droplet and immersion fluids; f_1 is a relaxation time constant and in the specific case of modeling RBCs, the value used by Arora et al. [79] was 5.0 s^{-1}; f_2 was found from the excess surface area of the RBC compared to a sphere with the same volume, and the areal strain an RBC can experience before hemolyzing.

The shape deformation model then needs to be coupled to the fluid shear stress field. Arora et al. [79] use a pathline approach for this in which the changing droplet shape is calculated along pathlines traced from the inlet to the outlet. The droplet shape is then known at all locations along a large number of routes through the device. In a later paper, the same group account for the real device shear stress field by solving the shape equation for the whole domain with a least squares finite element method (LSFEM) [90].

As well as the shape, the hemolysis rate as a function of the shape is also required. To find this hemolysis rate Arora et al. [79] make use of an effective shear stress. At any given point, the cell in the nonuniform shear stress field has a shape: the effective shear stress is defined to be the uniform shear stress that would give the same shape. They then use experimental hemolysis data, which is reported as a function of uniform shear stress fields, to obtain the varying hemolysis rate. They then integrate this along their pathlines to find the total hemolysis for the device. In contrast, Pauli et al. [90] solve a scalar transport equation for plasma free hemoglobin with a source term based on the hemolysis rate.

One advantage of the strain-based model, compared with the stress-based model, is that it accounts for viscoelastic deformation of the RBCs. Experimental data comes from observations under constant uniform stress fields. In these experiments, the cells will have reached their maximum steady state deformed shape. In contrast, in the nonuniform stress fields present in real devices, the stresses experienced by the cells will be changing such that the cell is unlikely to reach its steady state deformation. Because the cells are less deformed, they will have smaller hemolysis rates. Hence, if viscoelastic deformation is not accounted for, we are likely to overestimate hemolysis.

A second advantage of the strain-based model is that it can account for the tank-treading motion of the RBCs (see Section 11.3.1.1). In stress-based modeling, a fundamental assumption is that hemolysis is a function of shear stress magnitude and exposure time. However, in reality, the relationship also depends on the viscosity of the RBC suspension fluid. This is due to the tank-treading motion which helps to dissipate some of the energy, which in turn means the cell deforms less than it might otherwise do. The tank-treading motion depends on the ratio of the viscosity inside and outside the cell, is independent of the shear rate, [21] and cannot be incorporated into stress-based models.

It is well known that RBCs from different species have different sizes and different susceptibility to mechanical stress and these differences cause differences in hemolysis rates between species (see Section 11.3.1.2). The interspecies differences are one of the factors leading to confusion over what parameters to use for stress-based models and they make it hard to compare results from different studies. The properties of the species-specific RBC could, in principle, be incorporated into the strain-based model via the parameters R, σ, f_1, and f_2, allowing us to make sense of data from multiple animal species and different studies.

Although the features discussed so far represent important advantages of strain-based over stress-based models, arguably their most important advantage is the ability to account for all components of the fluid stress tensor (Equation 11.2). This is important because a model which properly incorporates the effects of all components has the potential to be used as a universal hemolysis model suitable for all types of blood contacting devices. In stress-based modeling, the components of the shear stress tensor are incorporated via the scalar shear stress; therefore, any potential differences in the way different components of the tensor hemolyze RBCs cannot be accounted for. As discussed in Section 11.3.1.1, at the same shear stress level, RBCs deform more in extensional flow compared to shear flow. This suggests extensional components of the shear stress tensor cause more hemolysis than their pure shear counterparts. This difference is included naturally in the strain-based approach because the shape depends explicitly on the whole strain rate tensor.

11.4.1.3 *Conclusion*

To date, a great deal of experimental data has been collected and this is complemented by excellent developments in theoretical models of hemolysis. If current modeling approaches can be extended, a universal hemolysis model seems within reach. This would allow hemolysis in all types of devices to be estimated, including rotary blood pumps, in which shear stresses likely dominate hemolysis production, and needles, in which extensional flows likely dominate.

11.4.2 Platelets

As discussed in Sections 11.3.3.1 and 11.3.3.2, the first two stages in the formation of a blood clot are platelet aggregation and activation. Once activated, platelets release chemicals which in turn activate more platelets in a positive feedback loop. Because of these complexities, before beginning to model platelet activation it is important to understand the question we would like to answer. For example, when designing a blood contacting device we might want to know how many platelets are activated directly by mechanical shear stress in the first round of platelet activation. However, if we want to understand clot formation, for example, then we will need to know the concentrations of clotting factors: in this case, our model will need to account for the feedback loop in the activation cascade. In this section, we will consider numerical models of the direct effect of mechanical shear stress on platelet aggregation and activation. The modeling of the chemical processes and feedback will be discussed later in Section 11.4.4.

Models featuring individual cells may capture the complex interactions involved in platelet aggregation. Fogelson's microscale approach [91] can model a collection of individual platelets and their interactions with the surrounding fluid, each other, and the vessel walls. They use an immersed boundary method to simulate the passage of the Lagrangian platelet models through an Eulerian grid of the flow domain. Each platelet has a membrane with properties influencing the platelet's deformation. Blyth and Pozrikidis [92] modeled a single platelet adhering to a surface using a boundary-element method with adhesion bond dynamics. This type of model lends itself to incorporation of the interactions between vWf and platelets which is required to make the modeling more realistic and account for the shear dependence of platelet aggregation. However, to the best of our knowledge, this is yet to be done. Currently, these cellular models are unsuitable for studying platelet activation and adhesion in real devices because they are computationally too demanding.

Work on numerical models for device-induced platelet aggregation and activation has focused on developing mathematical formulae capable of predicting various indicators of aggregation or activation. These formulae are usually implemented using a Lagrangian pathline approach (analogous to the methods for hemolysis previously). As discussed by Grigioni et al. [84] , these formulae should fulfill a number of conditions:

Condition 0: The principle of causality applies so that damage levels always increase, and a decreasing shear stress does not produce a reduction in damage.
Condition 1: The equation should be mathematically consistent so that the sum of the damage along the pathline in a uniform shear stress field should result in the original equation for damage in a uniform shear stress field.

Condition 2: The equation should account for damage history. That is, the damage done during the current timestep is influenced by the total damage that has been accumulated to that point.

Platelet activation (A) has been modeled assuming a hemolysis-type power-law [86] , an infinitesimal damage based on Grigioni et al.'s [83] hemolysis model [93] and using a transcendental equation [94]:

$$\dot{A}(t) = \left(\frac{\tau}{\tau_0}\right)^r \frac{1}{(1 - D(t))^k} \tag{11.17}$$

where k is negative to give a positive first derivative and, therefore, ensure damage always accumulates. Alemu and Bluestein applied this transcendental equation to platelet activation by mechanical heart valves [94].

Although the platelet activation equations described here are based on sound physical principles, they do not attempt to incorporate biological interactions; the high shear interaction between platelets and vWf, the reduction in speed of a platelet relative to a surface or another platelet, the activation of integrin receptors, the formation of platelet aggregates, and finally the activation of platelets. To make progress in the development of numerical models of platelet activation for medical devices, a hybrid model incorporating the mathematical formulae for mechanical influences and the modeling of biological consequences.

11.4.3 Blood Proteins Including von Willebrand Factor

Von Willebrand factor has been modeled using techniques from polmer physics by Alexander-Katz and colleagues [95–97]. They use Brownian dynamics to study the motion of vWf in a flow field, taking account of hydrodynamic interactions between the vWf dimers. Each vWf dimer is represented as a sphere connected to the adjacent dimer by a spring. These springs represent the self association between the dimers and act to collapse the vWf multimer into a globule. The flow field acts on the vWf such that shear flows and elongational flow stretch the multimer. Sing and Alexander-Katz used the model to show that lower elongational force magnitudes unravel the vWf compared with shear forces [98]. To date these models have not been applied to blood flow through cardiovascular devices.

11.4.4 Thrombosis and Emboli

Clotting within the device compromises its efficiency and can lead to device failure, while embolization of clots from the device can cause neurological events and strokes, among other problems. At its simplest, modeling of these effects can be the prediction of low shear or low velocity flow regions [6]. Fraser et al. [71] showed that stagnation regions with an LVAD drainage cannula correspond with regions of clot formation in an animal model (Figure 11.8).

Continuum models for platelet aggregation which include chemical activation have been formulated by Fogelson et al. [91,99–101], Sorensen et al. [102,103] and Goodman et al [104] and these models are fundamentally similar. A convection-diffusion-reaction equation is solved for each of the species involved: resting, and activated platelets, platelet released

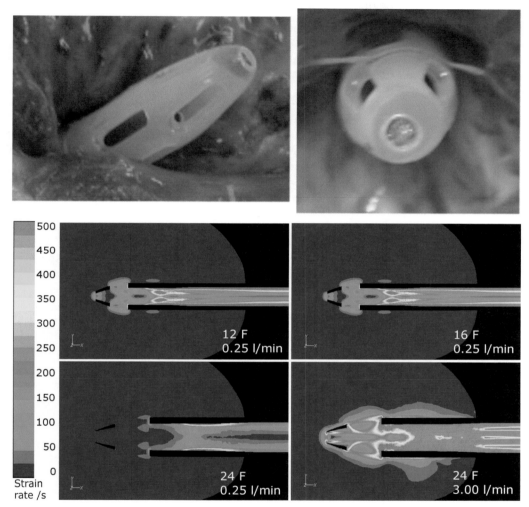

FIGURE 11.8   Spatial correlation between blood clots and low shear rate in the tip of a ventricular assist device drainage cannula. *(Reproduced, with permission, from [71].)*

agonists (such as ADP), platelet synthesized agonists (such as thrombin and thromboxane A_2), and prothrombin and antithrombin III which interact with thrombin. Platelet-platelet and platelet-surface adhesion are modeled via specified surface flux boundary conditions such that, as the number of adhered platelets increases and the available free surface area decreases, the probability of surface adhesion decreases. Surface activation can be included [102,103], or alternatively shear activation of adhered cells may be modeled [104]. Red blood cell enhanced platelet and large molecule diffusion can be incorporated either by specifying a diffusion constant, which is proportional to the wall shear rate [102,103], or by allowing the diffusion constant to vary with the local shear stress [104]. The unknown parameters such as platelet adhesion rates and platelet diffusivity were found by fitting to experimental data from the literature or by performing experiments in straight tubes. Sorensen et al [103] tested their model by comparison with experimental results from the literature and platelet deposition was in good agreement with one anticoagulant (but not another). One of the main limitations of their model is that it does not account for mechanical interactions between the flow and the thrombus.

The model by Fogelson et al. [91,99–101] uses an elastic link function to describe the interplatelet bonds and their evolution, and predicts a phase transition which can be interpreted as platelet aggregation. A forcing term ensures fluid flows around the formed thrombus, and the model can predict embolization. They used their model to show how clot development in atherosclerosis depends on the plaque rupture site. Goodman et al. [104] incorporate thrombus growth in their model using a simpler approach: when the volume of adhered platelets within a mesh element exceeds the volume of that element, the element becomes thrombus. The viscosity is increased 100,000 times, the boundary geometry is updated, and the surface flux conditions are applied at the new boundary. Fluid forces on the growing thrombus are then tracked and the entire thrombus embolizes (and is removed from the grid) when the forces exceed a given characteristic adhesion strength. Their model was validated with experiments in constricted tubes. Video microscopy was used to measure formation, growth, and embolization of thrombi, while scanning electron microscopy was used to measure adhered platelets. Good agreement was found for the location and quantity of adherent platelets, as well as for thrombus growth rates and embolization frequency.

Xu et al [105] have developed a thrombosis model which uses convection-diffusion-reaction equations for the biochemistry of the plasma with a discrete cellular Potts model of the thrombus involving platelets and RBCs.

11.5 SUMMARY

Mechanical stress-induced blood trauma may be caused by many blood contacting devices: it is a primary factor in device-induced complications such as neurological symptoms, infection, bleeding, and anemia. Understanding the effects of mechanical stresses on the various blood components will help us to design less traumatic devices which induce fewer complications.

In this chapter, the types of fluid dynamic forces induced by blood contacting devices have been described and the effects of these on blood components, including red blood cells, white blood cells, platelets, and blood proteins have been outlined. The numerical models which have been developed to date for predicting these effects have been discussed in detail.

With an understanding of the numerical models available, the reader should now be in a position to choose the most appropriate of the existing modeling methods to apply to their own research.

However, there remain a number of challenges still to be addressed. We need a universal hemolysis model that accounts for extensional as well as shear stresses. There are no useful models for white blood cell damage. A model for platelet activation and aggregation, which accounts both for mechanical effects and biological consequences and can be applied to whole devices is needed. And, finally, models for damage to blood proteins such as von Willebrand factor would be useful. Each of these areas represents a challenging problem. In addition, the current lack of experimental data in some areas means that close collaboration with experimentalists will be essential. Another challenge is to bring all of the available experimental data into current and future models, thereby making more widely applicable models.

In spite of these challenges, important progress is being made, and we can be confident that numerical modeling will contribute to improved blood contacting devices and, therefore, better cardiovascular health solutions.

References

[1] Bludszuweit C. Model for a general mechanical blood damage prediction. Artif Organs 1995;19:583–98.
[2] Kaazempur-Mofrad MR, Isasi AG, Younis HF, Chan RC, Hinton DP, Sukhova G, et al. Characterization of the atherosclerotic carotid bifurcation using MRI, finite element modeling and histology. Ann Biomed Eng 2004;32:932–46.
[3] Down LA, Papavassiliou DV, O'Rear EA. Significance of extensional stresses to Red blood cell lysis in a shearing flow. Annals Biomed Eng 2011;39:1632–42.
[4] Fraser KH, Zhang T, Taskin ME, Griffith BP, Wu ZJ. A quantitative comparison of mechanical blood damage parameters in rotary ventricular assist devices: shear stress, exposure time and hemolysis index. J Biomech Eng 2012;134:081002.
[5] Nanna JC, Wivholm JA, Deutsch S, Manning KB. Flow field study comparing design iterations of a 50 cc left ventricular assist device. ASAIO J 2011;57:349–57.
[6] Hochareon P, Manning KB, Fontaine AA, Tarbell JM, Deutsch S. Correlation of in vivo clot deposition with the flow characteristics in the 50 cc penn state artificial heart: a preliminary study. ASAIO J 2004;50:537–42.
[7] Konig CS, Clark C, Mokhtarzadeh-Dehghan MR. Comparison of flow in numerical and physical models of a ventricular assist device using low- and high-viscosity fluids. Proc IMechE Part H: J Engineering in Medicine 1999;213:423–32.
[8] Ge L, Dasi LP, Sotiropoulos F, Yoganathan AP. Characterization of hemodynamic forces induced by mechanical heart valves: reynolds vs. Viscous stresses. Ann Biomed Eng 2008;36:276–97.
[9] Leo H-L, Simon HA, Dasi LP, Yoganathan AP. Effect of hinge Gap width on the microflow structures in 27-mm bileaflet mechanical heart valves. J Heart Valve Dis 2006;15:800–8.
[10] Bluestein D, Yin W, Affeld K, Jesty J. Flow-induced platelet activation in mechanical heart valves. The Journal of Heart Valve Disease 2004;13:501–8.
[11] Zhang J, Chen X, Ding J, Fraser KH, Taskin ME, Griffith BP, et al. Computational study of the blood flow in three types of 3D hollow fiber membrane bundles. J Biomech Eng 2013;135:121009.
[12] Graefe R, Borchardt R, Arens J, Schlanstein P, Schmitz-Rode T, Steinseifer U. Improving oxygenator performance using computational simulation and flow field-based parameters. Artif Organs 2010;34:930–6.
[13] Jones CC, McDonough JM, Capasso P, Wang D, Rosenstein KS, Zwischenberger JB. Improved computational fluid dynamic simulations of blood flow in membrane oxygenators from X-ray imaging. Ann Biomed Eng 2013;41:2088–98.
[14] Wu ZJ, Taskin E, Zhang T, Fraser KH, Griffith BP. Computational model-based design of a wearable artificial pump-lung for cardiopulmonary/respiratory support. Artif Organs 2012;36:387–99.
[15] Billett HH. Hemoglobin and hematocrit. In: Walker HK, Hall WD, Hurst JW, editors. Clinical methods: the history, physical, and laboratory examinations. 3rd ed. Boston: Butterworths; 1990. p. 718–9, Chapter 15.
[16] Fahraeus R. The suspension stability of the blood. Physiol Rev 1929;2:241–74.
[17] Dupire J, Socol M, Viallat A. Full dynamics of a red blood cell in shear flow. PNAS 2012;109:20808–13.
[18] Sutera SP, Mehrjardi MH. Deformation and fragmentation of human Red blood cells in turbulent shear flow. Biophys J 1975;15:1–10.
[19] Schmid-Schonbein H, Wells R. Fluid drop-like transition of erythrocytes under shear. Science 1969;165:288–91.
[20] Fischer TM. Tank-tread frequency of the Red blood cell membrane: dependence on the viscosity of the suspending medium. Biophys J 2007;93:2553–61.
[21] Keller SR, Skalak R. Motion of a tank-treading ellipsoidal particle in a shear flow. J Fluid Mech 1982;120:27–47.

[22] Bitbol M. Red blood cell orientation in orbit C = 0. Biophys J 1986;49:1055–68.

[23] Lee SS, Yim Y, Ahn KH, Lee SJ. Extensional flow-based assessment of red blood cell deformability using hyperbolic converging microchannel. Biomed Microdevices 2009;11(5):1021–7.

[24] Bronkhorst PJH, Streekstra GJ, Grimbergen J, Nijhof EJ, Sixma JJ, Brakenhoss GJ. A New method to study shape recovery of Red blood cells using multiple optical trapping. Biophys J 1995;69:1666–73.

[25] Rand RP. The Red cell membrane II. Viscoelastic breakdown of the membrane. Biophys J 1964;4:303–16.

[26] Zhoa R, Antaki JF, Naik T, Bachman TN, Kameneva MV, Wu ZJ. Microscopic investigation of erythrocyte deformation dynamics. Biorheology 2006;43:747–65.

[27] Leverett LB, Hellums JD, Alfrey CP, Lynch EC. Red blood cell damage by shear stress. Biophys J 1972;12:257–72.

[28] Wurzinger LJ, Opitz R, Wol M, Schmid-Schonbein H. Shear induced platelet activation: a critical reappraisal. Biorheology 1985;22:399–413.

[29] Wurzinger LJ, Opitz R, Eckstein H. Mechanical blood trauma: an overview. Angeiologie 1986;38:81–97.

[30] Giersiepen M, Wurzinger LJ, Opitz R, Reul H. Estimation of shear stress-related blood damage in heart valve prosthesis-in vitro comparison of 25 aortic valves. Int J Artif Organs 1990;13:300–6.

[31] Heuser G, Opitz R. A couette viscometer for short time shearing of blood. Biorheology 1980;17:17–24.

[32] Paul R, Apel J, Klaus S, Schugner F, Schwindke P, Reul H. Shear stress related blood damage in laminar couette flow. Artif Organs 2003;27:517–29.

[33] Zhang T, Taskin ME, Fang H-B, Jarvik R, Griffith BP, Wu ZJ. Study of flow-induced hemolysis using couette-type blood shearing devices. Artif Organs 2011;35:1180–6.

[34] Blackshear PL, Blackshear GL. Mechanical hemolysis. In: Skalak R, Chien S, editors. Handbook of bioengineering. New York: McGraw-Hill Book Company; 1987. p. 15.1–15.19.

[35] Jikuya Y, TsuTsui T, Shigeta O, Sankai Y, Mitsui T. Species differences in erythrocyte mechanical fragility: comparison of human, bovine and ovine cells. ASAIO J 1998;44:M452–5.

[36] Sutera SP, Gardner RA, Boylan CW, Carroll GL, Chang KC, Marvel JS, et al. Age-related changes in deformability of human erythrocytes. Blood 1985;2:275–82.

[37] Baddour LM, Bettmann MA, Bolger AF, Epstein AE, Ferrieri P, Gerger MA, et al. Nonvalvular cardiovascular device-related infections. Circulation 2003;108:2015–31.

[38] Shive MS, Salloum ML, Anderson JM. Shear stress-induced apoptosis of adherent neutrophils: a mechanism for persistence of cardiovascular device infections. PNAS 2000;97:6710–5.

[39] Padera RF. Infection in ventricular assist devices: the role of biofilm. Cardiovasc Pathol 2006;15:264–70.

[40] Schmid-Schonbein WG. Analysis of inflammation. Ann Rev Biomed Eng 2006;8:93–151.

[41] Rosenson-Schloss RS, Vitolo JL, Moghe PV. Flow-mediated cell stress induction in adherent leukocytes is accompanied by modulation of morphology and phagocytic function. Med Biol Eng Comput 1999;37:257–63.

[42] Carter J, Hristova K, Harasaki H, Smith WA. Short exposure time sensitivity of white cells to shear stress. ASAIO J 2003;49:687–91.

[43] Chan CHH, Hilton A, Foster G, Hawkins KM, Badiei N, Thornton CA. The evaluation of leukocytes in response to the in vitro testing of ventricular assist devices. Artif Organs 2013;37:793–801.

[44] Diehl P, Aleker M, Helbing T, Sossong V, Beyersdorf F, Olschewski M, et al. Enhanced microparticles in ventricular assist device patients predict platelet, leukocyte and endothelial cell activation. Interactive CardioVascular and Thoracic Surgery 2010;11:133–7.

[45] Stassen JM, Arnout J, Deckmyn H. The hemostatic system. Curr Med Chem 2004;11:2245–60.

[46] Varga-Szabo D, Pleines I, Nieswandt B. Cell adhesion mechanisms in platelets. Arterioscler Thromb Vasc Biol 2008;28:403–12.

[47] Dopheide SM, Maxwell MJ, Jackson SP. Shear-dependent tether formation during platelet translocation on von Willebrand factor. Blood 2002;99:159–67.

[48] Nesbitt WS, Kulkarni S, Giuliano S, Goncalves I, Dopheide SM, Yap CL, et al. Distinct Glycoprotein Ib/V/IX and Integrin alpha IIb beta3-dependent Calcium Signals Cooperatively Regulate Platelet Adhesion under Flow. J Biol Chem 2002;25:2965–72.

[49] Maxwell MJ, Westein E, Nesbitt WS, Giuliano S, Dopheide SM, Jackson SP. Identification of a 2-stage platelet aggregation process mediating shear-dependent thrombus formation. Blood 2007;109:566–76.

[50] Ruggeri ZM, Mendolicchio GL. Adhesion mechanisms in platelet function. Circ Res 2007;100:1673–85.

[51] Merten M, Chow T, Hellums D, Thiagarajan P. A New role for P-selectin in shear-induced platelet aggregation. Circulation 2000;2045–2050:102.

[52] Sheriff J, Bluestein D, Girdhar G, Jesty J. High-shear stress sensitizes platelets to subsequent Low-shear conditions. Annals Biomed Eng 2010;38:1442–50.

[53] Miyazaki Y, Nomura S, Miyake T, Kagawa H, Kitada C, Taniguchi H, et al. High shear stress can initiate both platelet aggregation and shedding of procoagulant containing microparticles. Blood 1996;88:3456–64.

[54] Slack SM, Jennings LK, Turitto VT. Platelet size distribution measurements as indicators of shear stress-induced platelet aggregation. Annals Biomed Eng 1994;22:653–9.

[55] Harrison P, Cramer EM. Platelet alpha-granules. Blood Rev 1993;7:52–62.

[56] Hellums JD. Whitaker lecture: biorheology in thrombosis research. Ann Biomed Eng 1993;1994(22):445–55.

[57] Jesty J, Yin W, Perrotta P, Bluestein D. Platelet activation in a circulating flow loop: combined effects of shear stress and exposure time. Platelets 2003;14:143–9.

[58] Bakir I, Hoylaerts MF, Kink T, Foubert L, Luyten P, Van kerckhoven S, et al. Mechanical stress activates platelets at a subhemolysis level: an in vitro study. Artif Organs 2007;31:316–23.

[59] Loffler C, Straub A, Bassler N, Pemice K, Beyersdorf F, Bode C, et al. Evaluation of platelet activation in patients supported by the jarvik 2000* high-rotational speed impeller ventricular assist device. J Thoracic Cardiovascular Surg 2009;137:736–41.

[60] Leytin V, Allen D, Mykhaylov S, Mis L, Lyubimov EV, Garvey B, et al. Pathologic high shear stress induces apoptosis events in human platelets. Biochem Biophys Res Commun 2004;320:303–10.

[61] Sporn LA, Marder VJ, Wagner DD. Inducible secretion of large biologically potent von Willebrand factor multimers. Cell 1986;46:185–90.

[62] Luken BM. Extracellular control of VWF multimer size and thiol-disulfide exchange. J Thromb Haemost 2008;6:1131–4.

[63] Xie L, Chesterman CN, Hogg PJH. Reduction of von Willebrand factor by endothelial cells. Thromb Haemost 2000;84:506–13.

[64] Shim K, Andersen PJ, Tuley EA, Wiswall E, Sadler JE. Platelet-VWF complexes are preferred substrates of ADAMTS13 under fluid shear stress. Blood 2008;111:651–7.

[65] Uriel N, Pak S-W, Jorde UP, Jude B, Susen S, Vincentelli A, et al. Acquired von Willebrand syndrome after continuous-flow mechanical device support contributes to a high prevalence of bleeding during long-term support and at the time of transplantation. J Am Coll Cardiol 2010;56:1207–13.

[66] Heilmann C, Geisen U, Beyersdorf F, Nakamura L, Benk C, Berchtold-Herz M, et al. Acquired von Willebrand syndrome in patients with ventricular assist device or total artificial heart. Thromb Haemostasis 2010;103:962–7.

[67] Meyer AL, Malehsa D, Bara C, Budde U. Acquired von Willebrand syndrome in patients with an axial flow left ventricular assist device. Circ Heart Fail 2010;3:675–81.

[68] Heilmann C, Geisen U, Beyersdorf F, Lea Nakamura L, Georg Trummer G, Berchtold-Herz M, et al. Acquired Von Willebrand syndrome is an early-onset problem in ventricular assist device patients. Eur J Card Thorac Surg 2011;40:1328–33.

[69] Zhang X, Halvorsen K, Zhang C-Z, Wong WP, Springer TA. Mechanoenzymatic cleavage of the ultralarge vascular protein von Willebrand factor. Science 2009;324:1330–4.

[70] Siedlecki CA, Lestini BJ, Kottke-Marchant K, Eppell SJ, Wilson DL, Marchant RE. Shear-dependent changes in the three-dimensional structure of human von Willebrand factor. Blood 1996;88:2939–50.

[71] Fraser KH, Zhang T, Taskin ME, Griffith BP, Wu ZJ. Computational fluid dynamics analysis of thrombosis potential in ventricular assist device drainage cannulae. ASAIO J 2010;56:157–63.

[72] Garon A, Farinas M-I. Fast three-dimensional numerical hemolysis approximation. Artif Organs 2004;28:1016–25.

[73] De Wachter D, Verdonck P. Numerical calculation of hemolysis levels in peripheral hemodialysis cannulas. Artif Organs 2002;26:576–82.

[74] Arvand A, Hormes M, Reul H. A validated computational fluid dynamics model to estimate hemolysis in a rotary blood pump. Artif Organs 2005;29:531–40.

[75] Taskin ME, Zhang T, Gellman B, Fleischli A, Dasse KA, Griffith BP, et al. Computational characterization of flow and hemolytic performance of the UltraMag blood pump for circulatory support. Artif Organs 2010;34:1099–113.

[76] Mitoh A, Yano T, Sekine K, Mitamura Y, Okamato E, Kim D-W, et al. Computational fluid dynamics analysis of an intra-cardiac axial flow pump. Artif Organs 2003;27:34–40.

[77] Yano T, Sekine K, Mitoh A, Mitamura Y, Okamoto E, Kim D-W, et al. An estimation method of hemolysis within an axial flow blood pump by computational fluid dynamics analysis. Artif Organs 2003;2003:920–5.

[78] Arora D, Behr M, Pasquali M. Hemolysis estimation in a centrifugal blood pump using a tensor-based measure. Artif Organs 2006;30:539–47.

[79] Arora D, Behr M, Pasquali M. A tensor-based measure for estimating blood damage. Artif Organs 2004;28:1002–15.

[80] Song X, Throckmorton AL, Wood HG, Antaki JF, Olsen DB. Computational fluid dynamics prediction of blood damage in a centrifugal pump. Artif Organs 2003;27:938–41.

[81] Chan K, Wong YW, Ding Y, Chua LP, Yu SCM. Investigation of the effect of blade geometry on blood trauma in a centrifugal blood pump. Artif Organs 2002;26:785–93.

[82] Throckmorton AL, Untaroiu A. CFD analysis of a Mag-Lev ventricular assist device for infants and children: fourth generation design. ASAIO J 2008;54:423–31.

[83] Grigioni M, Morbiducci U, D'Avenio G, Di Benedetto G, Del Gaudio C. A novel formulation for blood trauma prediction by a modified power-law mathematical model. Biomechanical Modeling and Mechanobiology 2005;4:249–60.

[84] Grigioni M, Daniele C, Morbiducci U, D'Avenio G, Di Benedetto G, Barbaro V. The power-law mathematical model for blood damage prediction: analytical developments and physical inconsistancies. Artif Organs 2004;28:467–75.

[85] Goubergrits L, Affeld K. Numerical estimation of blood damage in artificial organs. Artif Organs 2004;28:499–507.

[86] Goubergrits L. Numerical modeling of blood damage: current status, challenges and future prospects. Expert Reviews on Medical Devices 2006;3:527–31.

[87] Carswell D, McBride D, Croft TN, Slone AK, Cross M, Foster G. A CFD model for the prediction of haemolysis in micro axial left ventricular assist devices. Appl Math Model 2013;37:4199–207.

[88] Taskin ME, Fraser KH, Zhang T, Griffith BP, Wu ZJ. Evaluation of eulerian and lagrangian models for hemolysis estimation. ASAIO J 2012;58:363–72.

[89] Maffettone PL, Minale M. Equation of change for ellipsoidal drops in viscous flow. J Non-Newtonian Fluid Mech 1998;78:227–41.

[90] Pauli L, Nam J, Pasquali M, Behr M. Transient stress-based and strain-based hemolysis estimation in a simplified blood pump. Int J Numer Meth Biomed Eng 2013;29:1148–60.

[91] Fogelson A, Guy RD. Immersed-boundary-type models of intravascular platelet aggregation. Comput Methods Appl Mech Eng 2008;197:2087–104.

[92] Blyth MG, Pozrikidis C. Adhesion of a blood platelet to injured tissue. Engineering Analysis with Boundary Elements 2009;33:695–703.

[93] Nobili M, Sheriff J, Morbiducci U, Redaelli A, Bluestein D. Platelet Activation Due to Hemodynamic Shear Stresses: Damage Accumulation Model and Comparison to In Vitro Measurements. ASAIO J 2008;54:64–72.

[94] Alemu Y, Bluestein D. Flow induced platelet activation and damage accumulation in a mechanical heart valve: numerical studies. Artif Organs 2007;31:677–88.

[95] Alexander-Katz A, Schneider MF, Schneider SW, Wixforth A, Netz RR. Shear-flow-induced unfolding of polymeric globules. Phys Rev Lett 2006;97:138101.

[96] Alexander-Katz A, Netz RR. Dynamics and instabilities of collapsed polymers in shear flow. Macromolecules 2008;41:3363–74.

[97] Sing CE, Selvidge JG, Alexander-Katz A Von. Willlebrand adhesion to surfaces at high shear rates is controlled. Biophys J 2013;105:1475–81.

[98] Sing CE, Alexander-Katz A. Elongational flow induces the unfolding of von Willebrand factor at physiological flow rates. Biophys J 2010;98:L35–7.

[99] Fogelson AL. Continuum models of platelet aggregation: formulation and mechanical properties. SIAM J Appl Math 1992;52:1089–110.

[100] Fogelson AL, Tania N. Coagulation under flow: the influence of flow-mediated transport on the initiation and inhibition of coagulation. Pathophysiol Haemost Thromb 2005;34:91–108.

[101] Fogelson AL, Guy RD. Platelet-wall interaction in continuum models of platelet thrombosis: formulation and numerical solution. Mathematical Medicine and Biology 2004;21:293–334.

[102] Sorensen EN, Burgreen GW, Wagner WR, Antaki JF. Computational Simulation of Platelet Deposition and Activation: I. Model Developement and Properties. Ann Biomed Eng 1999;27:436–48.

[103] Sorensen EN, Burgreen GW, Wagner WR, Antaki JF. Computational Simulation of Platelet Deposition and Activation: II. Results for Poiseuille Flow over Collagen. Ann Biomed Eng 1999;27:449–58.

[104] Goodman PD, Barlow ET, Crapo PM, Mohammed SF, Solen KA. Computational model of device-induced thrombosis and thromboembolism. Ann Biomed Eng 2005;33:780–97.

[105] Xu Z, Chen N, Kamocka MM, Rosen ED, Alber M. A multiscale model of thrombus development. J R Soc Interface 2008;5:705–22.

[106] Pries AR, Secomb TW, Gaehtgens P. Design principles of vascular beds. Circ Res 1995;77:1017–23.

[107] Oyre S, Pedersen EM, Ringgaard S, Boesiger P, Paaske WP. In vivo wall shear stress measured by magnetic resonance velocity mapping in the normal human abdominal aorta. Eur J Vasc Endovasc Surg 1997;13:263–71.

[108] Oyre S, Ringgaard S, Kozerke S, Paaske WP, Erlandsen M, Boesiger P, et al. Accurate noninvasive quantification of blood flow, cross-sectional lumen vessel area and wall shear stress by three-dimensional paraboloid modeling of magnetic resonance imaging velocity data. JACC 1998;32:128–34.

[109] Cheng CP, Herfkens RJ, Taylor CA. Inferior vena caval hemodynamics quantified in vivo at rest and during cycling exercise using magnetic resonance imaging. Am J Physiol Heart Circ Physiol 2008;284:H1161–7.

[110] Ping S, Ringgaard S, Oyre S, Hansen MS, Rasmus S, Pedersen EM. Wall shear rates differ between the normal carotid, femoral and brachial arteries: an in vivo MRI study. J Magnetic Resonance Imaging 2004;19:188–93.

[111] Nagaoka T, Yoshida A. Noninvasive evaluation of wall shear stress on retinal microcirculation in humans. Invest Ophthalmol Vis Sci 2006;47:1113–9.

[112] Lipowsky HH, Kovalcheck S, Zweifach BW. The distribution of blood rheological parameters in the microvasculature of cat mesentery. Circ Res 1978;43:738–49.

[113] Koutsiaris AG, Tachmitzi SV, Batis N, Kotoula MG, Karabatsas CH, Tsironi E, et al. Volume flow and wall shear stress quantification in the human conjunctival capillaries and post-capillary venules in vivo. Biorheology 2007;44:375–86.

[114] Oshinski JN, Ku DN, Mukundan SJ, Loth F, Pettigrew RI. Determination of wall shear stress in the aorta with the use of MR phase velocity mapping. J Magn Reson Imaging 1995;5:640–7.

Modeling of Blood Flow in Stented Coronary Arteries

Claudio Chiastra, Francesco Migliavacca

Politecnico di Milano, Milano, Italy

OUTLINE

Abbreviations

2D two-dimensional
3D three-dimensional
CAD computer-aided design
CCA conventional coronary angiography
CFD computational fluid dynamics
CHD coronary heart disease
CT computed tomography
CTA computed tomography angiography
FKB final kissing balloon
FSI fluid-structure interaction
FSP flow separation parameter
ICAM-1 intercellular adhesion molecule-1
ISR in-stent restenosis

LAD left anterior descending coronary artery
LNH local normalized helicity
MB main branch
MET mean exposure time
micro-CT micro-computed tomography
NH neointimal hyperplasia
OCT optical coherence tomography
OSI oscillatory shear index
PCI percutaneous coronary intervention
PIV particle image velocimetry
PSB provisional side branch
RCA right coronary artery
RRT relative residence time
SB side branch
TAWSS time-averaged wall shear stress
VCAM-1 vascular cell adhesion molecule-1
WSS wall shear stress
WSSG wall shear stress gradient

12.1 INTRODUCTION

Coronary heart disease (CHD) is one of the major causes of death and premature disability in developed societies. The American Heart Association estimates 15.4 million adults are affected by CHD in the United States alone [1].

CHD is caused by atherosclerotic lesions that reduce arterial lumen size through plaque formation and arterial thickening, decreasing blood flow to the heart and frequently leading to severe complications like myocardial infarction or angina pectoris [1,2].

CHD treatments can be medical or surgical, including percutaneous coronary intervention (PCI) or coronary artery bypass grafting. PCI is performed under local anesthesia and requires only a short hospitalization, decreasing recovery time, and costs compared to coronary bypass surgery. Since PCI introduction in the late 1970s, the use of this mini-invasive procedure for coronary revascularization has rapidly expanded [3]. Today, about 1 million procedures are performed annually in the United States, representing twice the annual number of coronary bypass operations [1,4].

PCI consists of balloon angioplasty usually followed by stenting (Fig. 12.1): the lumen of the diseased vessel is restored by balloon inflation and, subsequently, wire mesh tubular structures, known as stents, are deployed in order to hold open the newly expanded artery.

Although PCI with stenting is the most widely performed procedure for the treatment of CHD, it is still associated with serious clinical complications such as in-stent restenosis (ISR) [5]. ISR is the reduction of the lumen size as a result of neointimal hyperplasia (NH), an excessive growth of tissue inside the stented vessel. The phenomenon of ISR has been partially attenuated by the introduction, in 2004, of drug eluting stents, which are able to release antiproliferative drugs with programmed pharmacokinetics into the arterial wall. However, both clinical and histologic studies on drug eluting stents have demonstrated evidence of continuous neointimal growth during long-term follow-up [6,7].

The mechanisms and causes of ISR are not fully understood. In addition to stent design and vascular injury caused by device implantation, hemodynamic alterations induced by stent

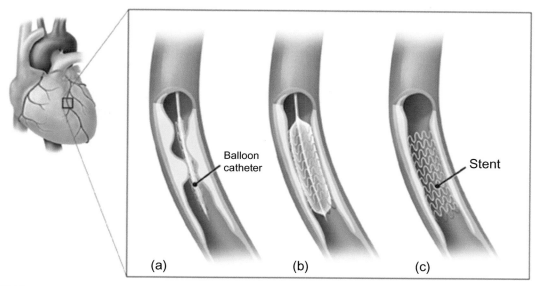

FIGURE 12.1   Procedure of stent deployment: (a) A balloon catheter with a crimped stent is positioned in the atherosclerotic lesion; (b) The balloon is inflated and the stent expands compressing the plaques; (c) The balloon is deflated and removed, leaving behind the stent which holds open the newly expanded artery. (© 2009 *University of Ottawa Heart Institute.*)

presence can be associated with NH [8]. Therefore, the study of the fluid dynamics of stented coronary arteries is of extreme importance for a better comprehension of the mechanisms involved in ISR.

In this chapter a review of works on the hemodynamics of stented coronary arteries is given, focusing in particular on computational fluid dynamics (CFD) models. Indeed, CFD allows the investigation of local hemodynamics at a level of detail not always accessible with experimental techniques, calculating fluid flow variables (e.g., wall shear stress) that can be used as indicators to predict sites where NH is excessive [8]. Applications of CFD to idealized and patient-specific coronary stented models are outlined as well.

In Section 12.2, the mathematical definitions of the most investigated hemodynamic quantities in this research field and their relation with ISR are introduced. In Section 12.3, the fluid dynamic models of idealized stented coronary arteries are discussed, focusing on coronary bifurcations. In Section 12.4, the fluid dynamic models based on geometries reconstructed through imaging techniques are presented. In Section 12.5, the most significant limitations of the CFD studies reviewed in the previous sections are discussed. Suggestions for the improvement of future models are also provided. Finally, in Section 12.6, conclusive observations are listed. These indicate that, while the results of CFD are very useful for studying the ISR phenomenon, fluid dynamic results need to be integrated with other modeling information as ISR is not driven purely by hemodynamic factors. Structural simulations of stent deployment for the calculation of the stress state in the arterial wall and transport simulations accounting for the drug release need to be considered as well.

12.2 HEMODYNAMIC QUANTITIES OF INTEREST

12.2.1 Introduction

Disturbed flow patterns, such as flow separation and reattachment, recirculation zones, significant secondary flow velocities, and stagnation point regions, play a key role in the onset and progression of atherosclerosis and NH [9]. These altered flow patterns are due to the flow-input waveforms and to the geometrical features of the vessel (e.g., sudden expansions, sharp bends, bifurcations, and the presence of the stent struts).

Several quantities are considered in the study of the disturbed flow, including the hemodynamic alterations induced by the stent presence, to identify sites in the coronary artery that could be prone to ISR. In this section, for each of the widely used quantities, the mathematical definition and the relation between them and the ISR is provided. The quantities are classified as near-wall quantities, quantities for the measure of flow stasis and flow separation, and bulk flow quantities.

12.2.2 Near-Wall Quantities

The main investigated quantity is the wall shear stress (WSS) which plays a fundamental role in the atherosclerotic processes [10]. WSS is defined by the following relation:

$$\vec{\tau}_w = \vec{n} \cdot \vec{\tau}_{ij} \tag{12.1}$$

where $\vec{n}$ is the normal vector to the arterial wall surface and $\vec{\tau}_{ij}$ is the fluid viscous stress tensor. The magnitude of WSS vector is equal to the viscous stress on the surface and the direction of WSS is the direction of the viscous stress acting on the surface.

When time-dependent flows are studied, the time-averaged WSS (TAWSS) is introduced. It is calculated as

$$TAWSS = \frac{1}{T} \int_0^T |\vec{\tau}_w| \, dt \tag{12.2}$$

where T is the duration of a cardiac cycle.

From a biological standpoint, endothelial cell morphology and orientation are altered according to WSS. Endothelial cells subjected to WSS greater than 1 Pa have been shown to elongate and align themselves in the direction of flow, while those experiencing WSS lower than 0.4 Pa or oscillatory WSS were circular in shape, without showing any preferred alignment pattern [11]. WSS changes may alter the endothelial cells in shape as well as function. This alteration could increase cell permeability to serum substrates, either through the intercellular junction or by vesicular transport to the subendothelium, as in the case of atherosclerotic lesions [12,13]. It also has been highlighted through animal experiments that tissue regrowth in a stented coronary artery is prominent at the sites of low WSS [14].

Analyses based only on the magnitude of WSS alone are not sufficient because variations in shear stress in both time and space have an influence on ISR phenomenon [10]. The oscillatory nature of shear stress induced by pulsatile flow is quantified using the oscillatory shear index (OSI), which was defined by Ku and colleagues [15] as

$$\text{OSI} = \frac{1}{2}\left(1 - \frac{\left|\int_0^T \vec{\tau}_w \, dt\right|}{\int_0^T |\vec{\tau}_w| \, dt}\right) \tag{12.3}$$

where T is the period of the cardiac cycle, and $\vec{\tau}_w$ is the WSS vector. OSI values range between 0, when there is no oscillatory WSS, and 0.5 when there is the maximum oscillatory WSS. Regions characterized by high oscillatory WSS (OSI > 0.1 [16]) have shown a greater risk of arterial narrowing [17]. High OSI have been associated with an increase of the endothelial permeability to blood borne particles [18] and with a greater production of the gene endothelin-1 mRNA which increases cell proliferation [19]. In addition, it has been shown that low and oscillatory WSS increase the expression of inflammatory markers, including the intercellular adhesion molecular-1 (ICAM-1) and the vascular cell adhesion molecule-1 (VCAM-1) [10,20], thus indicating cell activation. In the experimental study by Yin et al. [21] it was found that cell surface ICAM-1 expression is significantly enhanced when coronary endothelial cells are exposed to low pulsatile shear stresses, indicating endothelial cell activation. These results, in agreement with the findings from other works [22,23], suggested that this hemodynamic condition is atherogenic.

In order to combine the information provided by WSS and OSI, the relative residence time (RRT) was recently introduced [18]:

$$\text{RRT} = \frac{1}{(1 - 2 \cdot \text{OSI}) \cdot \text{TAWSS}} \tag{12.4}$$

High RRT is recognized as critical for the problem of atherogenesis and ISR [24]. RRT is also associated with the residence time of the particles near the wall [18].

Another quantity of interest is the spatial wall shear stress gradient (WSSG). To mathematically define it, a local *m-n-l* coordinate system has to be introduced, where *m* is the WSS direction tangential to the endothelial surface (temporal mean WSS direction for pulsatile flows), *n* is the direction tangential to the surface and normal to *m*, and *l* is the endothelial surface normal direction. Using this coordinate system, the WSSG can be obtained, calculating spatial derivatives of WSS, which results in a nine-component tensor:

$$\nabla\vec{\tau}_w = \begin{bmatrix} \dfrac{\partial\tau_{w,m}}{\partial m} & \dfrac{\partial\tau_{w,m}}{\partial n} & \dfrac{\partial\tau_{w,m}}{\partial l} \\[2mm] \dfrac{\partial\tau_{w,n}}{\partial m} & \dfrac{\partial\tau_{w,n}}{\partial n} & \dfrac{\partial\tau_{w,n}}{\partial l} \\[2mm] \dfrac{\partial\tau_{w,l}}{\partial m} & \dfrac{\partial\tau_{w,l}}{\partial n} & \dfrac{\partial\tau_{w,l}}{\partial l} \end{bmatrix} \tag{12.5}$$

where $\nabla = \hat{m}\frac{\partial}{\partial m} + \hat{n}\frac{\partial}{\partial n} + \hat{l}\frac{\partial}{\partial l}$ and $\hat{m}$, $\hat{n}$, and $\hat{l}$ are the unit vectors in their respective direction. As suggested by Lei et al. [25], the components $\frac{\partial\tau_{w,m}}{\partial m}$ and $\frac{\partial\tau_{w,n}}{\partial n}$ are the most important ones for intimal thickening due to atherosclerosis or hyperplasia. In fact, these components generate intracellular tension, which causes widening and shrinking of the cellular gaps. Therefore, the WSSG can be written as:

$$\text{WSSG} = \left[\left(\frac{\partial \tau_{w,m}}{\partial m} \right)^2 + \left(\frac{\partial \tau_{w,n}}{\partial n} \right)^2 \right]^{1/2} \tag{12.6}$$

Results of *in vitro* and numerical studies [26,27] showed that a large spatial WSSG induce morphological and functional changes in the endothelial cells, which contribute to an increase in the wall permeability, and hence possible atherosclerotic lesions. Endothelial cells have been shown to migrate downstream of an area with high WSSG [27]. This phenomenon could affect the process of growth of a new layer of endothelial cells after the damage provoked by stent implantation [28]. Moreover, regions that are susceptible to NH have been correlated with sites where WSSG has been predicted to exceed 200 N/m^3 in an end-to-side anastomosis model [29] and a rabbit iliac model [14].

12.2.3 Flow Stasis Quantities

In order to quantify the flow stasis, the mean exposure time (MET) can be calculated. By releasing a high concentration of fluid particles at the inlets of an artery model, MET measures how long each particle resides within each element of the mesh [30,31]. For each element e, MET is defined as

$$\text{MET}_e = \frac{1}{N_e V_e^{1/3}} \sum_{p=1}^{N_t} H_e^p(t) \mathrm{d}t \tag{12.7}$$

where N_e is the number of times a particle passes through the element, V_e is the volume of the element, N_t is the total number of particles released, and $H_e^p(t)$ is equal to 1 when a particle p is located inside the element at time t and is equal to 0 otherwise. In stented models, highly anisotropic meshes with finer elements close to the struts are needed. Because the MET index depends on the element size, these meshes are not suitable. Therefore, an auxiliary isotropic mesh has to be used during the postprocessing step for the MET calculation.

Because the duration that a particle resides within an element is normalized by the N_e, the MET is able to distinguish between recirculating particles and stagnant particles [30]. For example, a particle that passes twice through an element, each pass spanning one time unit, does not contribute as much to MET as a particle that passes once for two time units [31]. In this particular case, the first particle is probably recirculating while the second one is stagnant.

Other quantities have been used in the study of stented coronary artery models to analyze stagnation and recirculation regions. For example, the flow separation parameter (FSP) quantifies the fraction of time with respect to one cardiac cycle where the flow at some location on the arterial wall is separated from the mainstream flow due to stagnation or recirculation [32]. Mathematically, FSP was calculated by He and colleagues [32] as:

$$\text{FSP} = \frac{T_S}{T} \tag{12.8}$$

where T_S is the amount of time that the flow is separated from the mainstream, and T the period of the cardiac cycle. FSP only varies spatially. It is defined as occurring when the wall shear has the opposite sign from the mainstream flow. A value of 0 implies no flow separation while a value of 1 implies constant flow separation throughout the entire flow period (recirculation or stagnation).

Equation (12.8) is valid only for a high flow condition (e.g., Reynolds number $=240$, WSS $=1\pm0.5$ Pa) while, for a low flow condition (e.g., Reynolds number $=50$, WSS $=0.2\pm1$ Pa), it has to be modified to account for the natural shear stress oscillation as follows [32]:

$$\text{FSP} = \frac{T_P\left(\phi_{\text{pos}}\right) + T_N\left(\phi_{\text{neg}}\right)}{T} \tag{12.9}$$

where ϕ_{pos} is the FSP during the time of forward mainstream flow T_P, ϕ_{neg} is the FSP during the time of reverse mainstream flow T_N, and T is the period of the cardiac cycle ($T=T_P+T_N$).

FSP is large in regions where stent strut spacing is small, as demonstrated by Berry et al. [33] comparing *in vitro* and CFD results.

12.2.4 Bulk Flow Quantities

The effect of stenting procedures on the bulk flow can be evaluated through the analysis of the helicity [34]. By definition, the helicity of a fluid flow confined to a domain D of the Euclidean space R^3 is given by the integral value of the kinetic helicity density H_k, defined as [35]:

$$H_k = \left(\nabla \times \vec{v}\right) \cdot \vec{v} \tag{12.10}$$

where $\vec{v}$ is the velocity vector and $\left(\nabla \times \vec{v}\right)$ is the vorticity. To better visualize the topological features of the flow field, the kinetic helicity density can be normalized with the velocity and vorticity magnitude resulting in the local normalized helicity (LNH):

$$\text{LNH} = \frac{\left(\nabla \times \vec{v}\right) \cdot \vec{v}}{\left|\left(\nabla \times \vec{v}\right)\right|\left|\vec{v}\right|} = \cos\theta \tag{12.11}$$

where θ is the angle between $\vec{v}$ and $\left(\nabla \times \vec{v}\right)$. LNH is a nondimensional quantity that ranges between -1 and 1. Physically, this quantity describes the arrangement of fluid streams into spiral patterns as they evolve within conduits [36]. In fact, it is a measure of the alignment/misalignment of the local velocity and vorticity vectors and its sign is a useful indicator of the direction of rotation of helical structures: positive LNH values indicate left-handed rotating fluid structures, while negative LNH values indicate right-handed rotating structures, when viewed in the direction of the forward movement.

The helical flow is a peculiar feature of natural blood flow present in arteries [37,38] and it has been recently found to be essential in suppressing flow disturbances in healthy vessels [39] and in stented arteries and bypass grafts [36,40,41].

12.3 FLUID DYNAMIC MODELS OF IDEALIZED STENTED GEOMETRIES

12.3.1 Introduction

In recent years, numerous CFD studies on stented coronary arteries have been proposed in the literature. Initially, the models were characterized by two-dimensional (2D) highly simplified geometries. For example, in the study conducted by Berry et al. [33] only a cross-section of the region very close to the arterial wall, with eight struts, was investigated. With the gradual increase of the computational resources, the analysis of three-dimensional (3D) stented geometries became possible. Complex models with or without bifurcations and with the presence of multiple stents were investigated.

As proposed by Murphy and Boyle [28], the works on stented coronary arteries can be classified in two categories:

− Effect-based studies: two or more models with a well-defined difference are compared from the fluid dynamic point of view. This category includes the comparison of straight and curved coronary arteries, Newtonian versus non-Newtonian blood flow, different stent sizing, and different stenting techniques.
− Design-based studies: the impact of the geometric configuration itself on hemodynamic quantities is investigated. This category includes the comparison between different strut designs, strut spacing, or commercial stent designs.

In this section the main works on stented coronary bifurcations based on geometries that were not reconstructed through imaging techniques (idealized geometries) are presented. Indeed, the studies on idealized single-vessel geometries are discussed in detail in the review by Murphy and Boyle [28].

12.3.2 Coronary Bifurcation Models

The treatment of coronary bifurcation lesions is a challenging area in interventional cardiology because of a lower rate of procedural success and a higher rate of restenosis compared to nonbifurcation interventions [42]. It is estimated that the rate of lesions involving bifurcations is 15–20% of all the PCI procedures [43]. Bifurcations are characterized by recirculation and stagnation zones that cause low WSS, making these regions very prone to developing atherosclerosis. No single strategy exists for the treatment of atherosclerotic lesions because of the variability of bifurcations in anatomy (plaque burden, location of plaque, angle between branches, diameter of branches, bifurcation site) and in the dynamic changes in anatomy during treatment (plaque shift, dissection) [44]. Many stenting techniques that involve the implantation of one or two stents, standard or dedicated to bifurcation, have been proposed in the literature (Fig. 12.2) [45]. However, each technique is associated with some limitations that make the choice of the best-fitting treatment uncertain.

The CFD studies presented here all belong to the effect-based category. Different stenting strategies for the treatment of bifurcation lesions are compared. The first CFD study on stented coronary bifurcations was carried out by Deplano and coworkers in 2004 [46]. These researchers studied the hemodynamic changes induced by the presence of two stents in a 90°

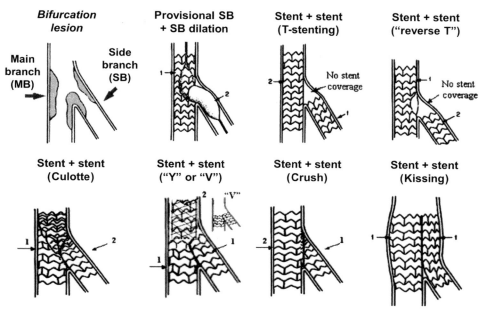

FIGURE 12.2 Main stenting techniques used for the treatment of coronary bifurcation lesions. (*Adapted with permission from Sharma et al. [43]. © 2010 Elsevier.*)

bifurcated coronary (Fig. 12.3a). They investigated six different cases, simulating different post-implantation configurations (Fig. 12.3b). Case 1 was modeled as a healthy coronary bifurcation without any stent in order to provide a reference case; Case 2 was representative of a double implantation not followed by an enlargement of the side branch (SB) cell, while Case 3 was followed by an optimum enlargement; Cases 4–6 modeled an imperfect enlargement with a partial opening of the stent struts in the main branch (MB) lumen. The Palmaz® P308 stent (Cordis Corporation, Bridgewater, NJ, USA), which belongs to an old generation of slotted tube stent, was considered in this study. A flow-rate curve recorded in a human left coronary artery [47] was imposed at the inlet of the models. A constant flow-rate repartition, based on the diameters of the two daughter arteries, was applied at the outlets. The no-slip condition was imposed along the walls. Blood was modeled as a Newtonian fluid characterized by a density of 1060 kg/m^3 and a dynamic viscosity of 3.6 cP. Because the Reynolds number based on the inlet diameter was ~200 at the peak of flow rate, an order of magnitude smaller than the Reynolds number for transition to turbulence (2300) [48], the flow was assumed to be laminar. The commercial software ANSYS Fluent (ANSYS Inc., Canonsburg, PA, USA), which is based on the finite volume method, was used to solve the Navier-Stokes equations. Results demonstrated that, behind the protruding stent struts located near the inner and the outer walls of the SB, some flow stasis and recirculation areas develop, causing low WSS values. Figure 12.3c shows the flow characteristic in the longitudinal plane ($z=0$) at the first time of acceleration. Cases 1 and 3 (without stent or completely opened) are similar without stasis areas. For Cases 2 and 4–6, the two lateral

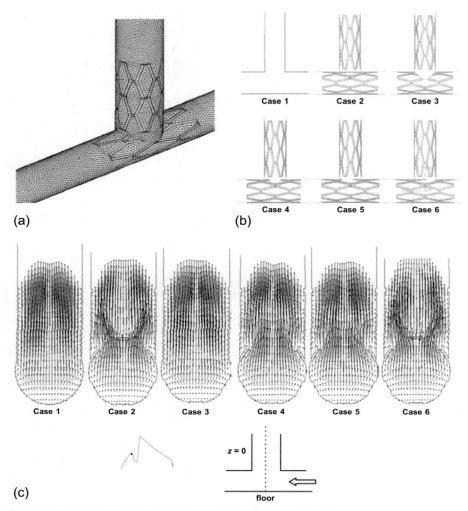

(a)

(b)

(c)

FIGURE 12.3 (a) 90° bifurcated coronary artery model with two implanted stents used by Deplano et al. [46]. The surface mesh is shown. (b) Projection views of the six investigated geometrical configurations. (c) Flow evolution in the longitudinal plane ($z=0$) at the first time of acceleration. *(Adapted with permission from Deplano et al. [46]. © 2004 Springer.)*

stent struts act as a convergence pipe that orientates and accelerates the flow at the center. Moreover, for Cases 2 and 6, two lateral jets are created, with stasis areas downstream from the luminal struts. The WSS at the stent wall were also investigated. High WSS values of about 20 Pa, which could stimulate platelet activation potentially leading to thromboembolic complications, were induced by the struts protruding into the SB. The authors concluded that, in terms of fluid dynamics, the best situation is obtained when the stent struts were ideally removed from the SB (Case 3).

Williams et al. [16] quantified the altered hemodynamics caused by MB stenting and subsequent SB angioplasty that removed stent struts from the ostium of a representative coronary bifurcation. Four different scenarios were compared: (1) pre-stenting, (2) post MB stenting with the best stent orientation, (3) post MB stenting with the worst stent orientation, and (4) post SB balloon angioplasty. An idealized bifurcation model was created with a typical bifurcation angle (46°) and Finet's law for the inlet and outlet diameter values [49], using a computer-aided design (CAD) software. A Boolean intersection command was implemented to subtract the geometrical model of the stent from the lumen, obtaining the fluid domain for the CFD simulation. The Taxus® Express 2™ stent (Boston Scientific, Natick, MA, USA) was considered in this work. As boundary conditions, resting and hyperemic inflow waveforms (Reynolds numbers ∼90 and ∼240, respectively), obtained from a canine left anterior descending coronary artery (LAD), were applied at the inlet cross-section as temporally varying Womersley velocity profiles. In order to estimate the behavior of the downstream vasculature, a three-element Windkessel model was imposed at the outlets using a multidomain approach [50]. The three-element Windkessel model is defined by three parameters with physiologic meaning: R_c, C, and R_d. R_c is the characteristic impedance representing the resistance, compliance, and inertance of the proximal artery of interest, C is the arterial capacitance and describes the collective compliance of all arteries beyond a model outlet, and R_d represents the total distal resistance beyond a given outlet.

Blood was assumed to behave as a Newtonian fluid with a density of 1060 kg/m^3 and a dynamic viscosity of 4 cP. TAWSS and OSI were quantified. In Fig. 12.4a, the contour maps of TAWSS are depicted. The nondiseased bifurcation model was not characterized by TAWSS lower than 0.4 Pa. Stenting introduced areas of low TAWSS next to the struts both under resting and hyperemic flow conditions. Moreover, eccentric regions of low TAWSS were present along the lateral wall of the MB after MB stenting. Virtual SB angioplasty resulted in a more concentric area of low TAWSS in the distal MB and along the lateral SB. Despite the regional variation in the location of low TAWSS, the luminal surface exposed to low TAWSS was similar before and after virtual SB angioplasty (rest: 43% vs. 41%; hyperemia: 18% vs. 21%). High OSI (>0.1) were absent in the nondiseased model (Fig. 12.4b). Considering the models after MB stenting, only 4% and 8% of the luminal surface was exposed to elevated OSI during rest and hyperemia, respectively. Virtual SB angioplasty reduced the percentage area exposed to high OSI at rest (1%) but had little impact under hyperemic conditions (7%). These results demonstrated that SB angioplasty intervention could not be a real benefit from a fluid dynamic point of view.

The method for obtaining the fluid domain used in the previous works [16,46] is exclusively based on geometrical/anatomic information that neglects both the mechanical behavior of the stent and the biological tissues, as well as their interaction during stenting procedures. To overcome this limitation, Morlacchi et al. [51] proposed a sequential structural and fluid dynamic approach that allowed the researchers to consider the artery deformation during stenting procedure for the creation of the fluid domain.

Firstly, the provisional side branch (PSB) technique, which is the preferred coronary bifurcation stenting procedure [52], was simulated by means of structural simulations. PSB consists of stenting of the MB with a conventional stent and it is frequently concluded by the final kissing balloon (FKB) inflation, in other words, the simultaneous expansion of two balloons in both the branches of the bifurcation to free the access to the SB from the struts protruding into

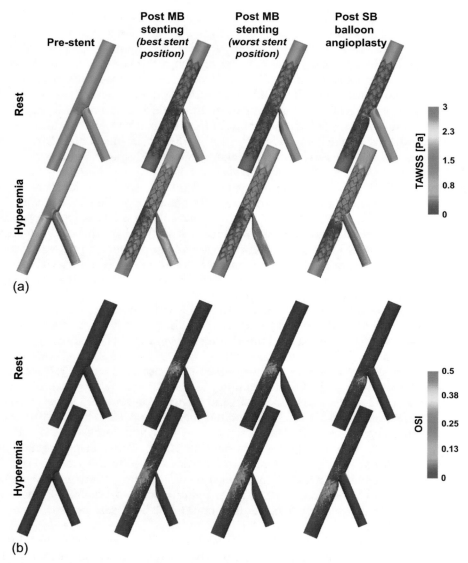

FIGURE 12.4   Contour maps of (a) time-averaged wall shear stress (TAWSS) and (b) oscillatory shear index (OSI) obtained by Williams and colleagues [16] for the four different scenarios: pre-stenting, post MB stenting with the best stent orientation, post MB stenting with the worst stent orientation, and post SB balloon angioplasty. Results are shown under resting and hyperemic conditions. *(Adapted with permission from Williams et al. [16]. © 2010 American Physiological Society.)*

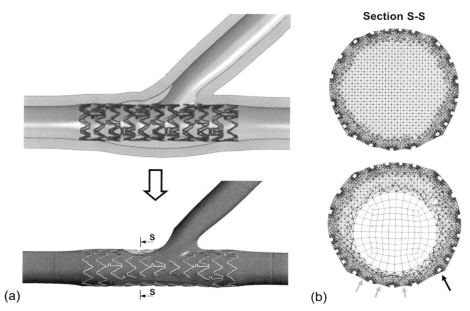

FIGURE 12.5   (a) Creation of the fluid domain from the final geometrical configuration obtained through structural simulations. (b) Example of a fully tetrahedral mesh (top) and a hybrid mesh (bottom) of the same cross-section. The hybrid mesh is characterized by an internal cylinder discretized with hexahedral elements and the region between the arterial wall and the cylinder meshed with tetrahedral elements. Tissue prolapse among stent struts (gray arrows) and malapposed struts (black arrow) can be detected by this model. *(Adapted with permission from Morlacchi et al. [51]. © 2011 ASME.)*

the lumen. Secondly, the final geometrical configurations obtained through structural simulations were used as fluid domains to perform transient CFD analyses (Fig. 12.5a).

The geometrical model of an LAD with its first diagonal branch and the Multi-Link Vision® stent (Abbott Laboratories, Abbott Park, IL, USA) were considered in this study. Three different cases were compared: (1) bifurcation after MB stenting (MBst), (2) bifurcation after FKB was performed with standard balloons (FKBstd), (3) bifurcation after "modified FKB" (FKBmod) using a new tapered balloon proposed to limit the structural damage induced to the arterial wall and to enhance the fluid dynamic performance. The 3D geometries were highly complex with stent struts both in contact and not in contact with the arterial wall (malapposed struts). As a consequence, extremely fine computational grids were needed resulting in high computational resources required to solve the CFD models. In order to obtain accurate results in the shortest time compatible with computational resources available, these authors applied a hybrid meshing method that uses both hexahedral and tetrahedral elements [53] (Fig. 12.5b). Hexahedral elements should be preferred because of higher accuracy and reduced number of elements. However, hexahedral elements are difficult to use in complex geometries as the intersections between arterial wall and struts. In these regions tetrahedral elements become necessary. To obtain the hybrid mesh, internal cylinders

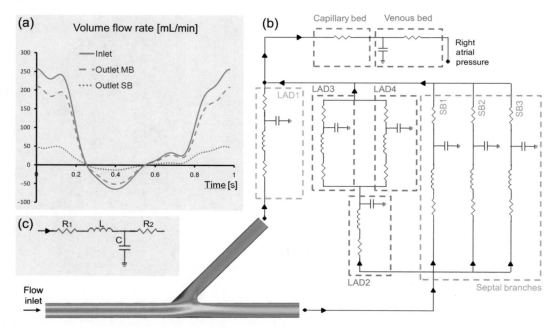

FIGURE 12.6   Multidomain model of an LAD proposed by Morlacchi et al. [51]. (a) Flow waveform used as inlet boundary condition (solid line) [54], and flow tracings obtained at MB (dashed line) and SB (dotted line) outlets. (b) 3D model of the coronary bifurcation coupled with the lumped-parameter scheme proposed by Pietrabissa et al. [55] representing the downstream coronary tree. (c) Example of the lumped-parameter scheme used to model a segment of a coronary vessel composed of two resistances, a capacitance and an inductance. *(Reprinted with permission from Morlacchi et al. [51]. © 2011 ASME.)*

were created along the MB and SB and discretized using hexahedral elements. Then the region between the cylinders and the arterial wall was meshed with tetrahedral elements obtaining the final grid.

At the inlet cross-section of the models, a pulsatile flow tracing [54] was applied as a paraboloid-shaped velocity profile (Fig. 12.6a). As previously done by Williams et al. [16], a multidomain approach was used to define the boundary condition at the outlets. In particular, preliminary simulations were performed for the bifurcation model without and with plaques (without the presence of the stent), considering the lumped-parameter scheme proposed by Pietrabissa et al. [55] (Fig 12.6b and c). Results in terms of flow rates at the outlets were identical for both the two limit-cases, highlighting that the downstream districts play a dominant role in establishing the blood flow distribution to downstream tissues. Because the resistance induced by the struts in the SB area were not influencing blood flow distribution, flow tracings at boundaries were maintained constant in the investigated cases, which were solved as stand-alone models. At the MB outlet, the flow curve resulting from the lumped-parameter model of the coronary tree was imposed while, at the SB outlet, a reference zero pressure was applied. Blood was assumed to be an incompressible, non-Newtonian fluid with a density equal to 1060 kg/m^3 and the viscosity varying according to the Carreau model [56]. CFD simulations were carried out by means of ANSYS Fluent. The hemodynamic forces

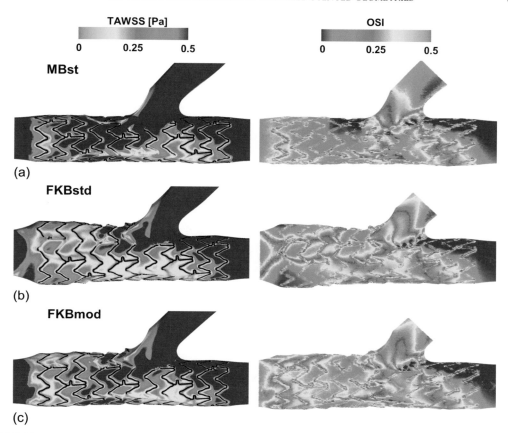

FIGURE 12.7   Results in terms of contour maps of time-averaged wall shear stress (TAWSS) and oscillatory shear index (OSI) obtained by Morlacchi et al. [51] for the following cases: (a) Stenting of MB (MBst); (b) Standard FKB inflation (FKBstd) performed with two cylindrical balloons (3.00 mm in the MB and 2.00 mm in the SB); (c) Modified FKB inflation (FKBmod) performed with a cylindrical 3.00 mm balloon in the MB and a dedicated conical balloon in the SB. *(Reprinted with permission from Morlacchi et al. [51]. © 2011 ASME.)*

acting on the arterial wall were analyzed in terms of TAWSS and OSI (Fig. 12.7). The FKB removes the struts from the blood flow in the bifurcation region, freeing the access to the SB and lowering the hemodynamic disturbances downstream of the bifurcation. However, FKB creates a larger area characterized by low WSS if compared to MB stenting only (MBst case). In fact, the percentage area with TAWSS lower than 0.5 Pa was 79.0% and 62.3%, respectively, for FKBstd and MBst cases. This was caused by the overexpansion in the proximal part of the MB during the FKB procedure. The use of the tapered balloon partially improves the results, with a smaller area exposed to low TAWSS (71.3%) and decreasing the zone with low and oscillating WSS.

A further work based on the same sequential approach was proposed by Chiastra et al. [53]. The authors examined the different hemodynamic scenarios provoked by FKB performed (1) with a proximal or (2) a distal access to the SB. CFD models similar to those

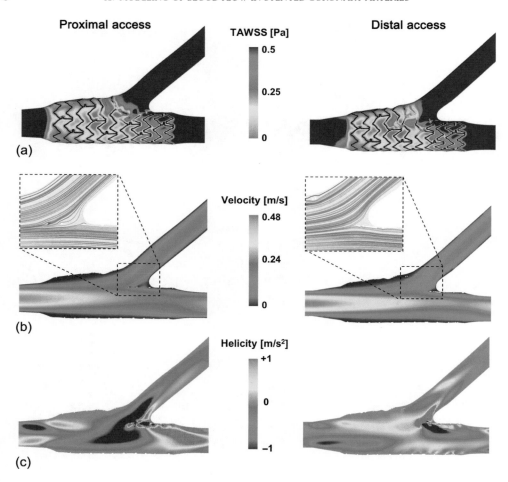

Proximal access   **Distal access**

(a)

(b)

(c)

FIGURE 12.8   Comparison of the proximal (on the left) and distal (on the right) accesses within the PSB approach: (a) Time-averaged wall shear stress (TAWSS) contour maps; (b) Velocity contours in the middle plane and velocity streamlines (in the boxes); (c) Helicity field in the middle plane. *(Reprinted with permission from Chiastra et al. [53]. © 2012 Elsevier.)*

investigated in Morlacchi et al. [51] were implemented. Different boundary conditions were considered. At the inlet, the velocity curve measured in a human LAD by Davies et al. [57] was imposed as a paraboloid-shaped profile. At the outlets, a typical flow split for 45° coronary bifurcations (70% MB, 30% SB) was imposed. Results showed that the distal access led to a smaller area exposed to TAWSS lower than 0.5 Pa if compared to the proximal access (84.7% vs. 88.0%, respectively for distal and proximal access cases) (Fig. 12.8a). Low velocity zones were present in the flow divider region due to the presence of the struts inside the vessel (Fig. 12.8b). These struts represent an obstacle to the SB access, modifying the flow patterns. A lower flow disturb can be detected in the distal access case where the opened struts are moved towards the external side of the SB wall. This evidence supports the clinical experience which shows that opening the struts through the most distal strut available improves the blood flow distribution across the coronary bifurcation. The flow disturbance also can be qualitatively observed in Fig. 12.8c where the helicity field on the middle plane of the models is depicted, giving a description of the bulk flow behavior. The helicity was higher in the proximal access case than in the distal one, resulting in a more disturbed flow condition. The results obtained by these researchers gave a quantitative support to the clinical experience that suggests it is better to perform the distal access instead of the proximal one if the PSB stenting strategy followed by FKB is chosen.

Also in the work by Katritsis et al. [58], a comparison of different stenting techniques was made from the fluid dynamic perspective. The authors investigated both single and double stenting procedures, calculating the near-wall quantities (TAWSS, OSI, and RRT). In particular they compared six cases: (1) stenting of the MB only, (2) stenting of the MB followed by balloon angioplasty of the SB, (3) balloon angioplasty of the SB followed by stenting of the MB, (4) Culotte technique [59], (5) Crush technique [59], and (6) T-stenting [59]. A geometrical model of an idealized human LAD and its diagonal branch was considered with a bifurcation angle of 50° following the Finet's law [49]. A CAD software was used to draw a single stent or two stents inside the bifurcation, with a procedure similar to Williams et al. [16], obtaining the six different investigated configurations. As the stent design, the PROMUS Element™ stent (Boston Scientific) was considered. Blood was modeled as a Newtonian fluid with dynamic viscosity 3.5 cP and density 1060 kg/m³. At the inlet, a pulsatile blood flow curve was prescribed [47] while, at the outlet boundaries, the flow was assumed to split proportionally to the (3/2) power of the outlet diameters. CFD simulations were performed using ANSYS Fluent. Figure 12.9 shows the contour maps of RRT for the six investigated cases. Larger areas exposed to high RRT are located in the cases with two implanted stents with respect to a single

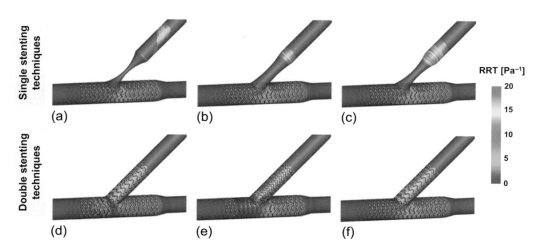

FIGURE 12.9   Contour maps of relative residence time (RRT) for the six cases investigated by Katritsis et al. [58]: (a) stenting of MB only, (b) stenting of MB followed by balloon angioplasty of SB, (c) balloon angioplasty of SB followed by stenting of MB, (d) Culotte technique, (e) Crush technique, and (f) T-stenting. *(Adapted with permission from Katritsis et al. [58]. © 2012 American Heart Association.)*

implanted stent. This result indicated that stenting of the MB with or without balloon angio-plasty of the SB offers hemodynamic advantages over double stenting. This finding supported the results of large clinical trials that documented that stenting of the MB only is preferable in the majority of bifurcation lesions [52,60]. Considering double stenting procedures, crush technique could result in better hemodynamics if compared to culotte or T-stenting.

In the end, Chen et al. [61] investigated the role of SB diameter, angle, and lesion on hemodynamics in case of MB stenting. Three dimensional geometrical models of coronary bifurcation with a generic coronary stent inside were created using a CAD software. Correct-sized as well as 5% and 10% undersized stent models were considered. Bifurcation angles of 30° and 70° as well as bifurcations with diameter ratios of SB/MB $=1/2$ and 3/4 were studied. A mild stenosis at the SB (40% in area) was also introduced in the analyses. Blood was modeled as a non-Newtonian Carreau fluid with a density of 1060 kg/m^3. At the inlet, a pul-satile velocity curve based on a human LAD was applied. At the outlets, a traction-free bound-ary condition was imposed. The results highlighted that the stent undersizing decreased the WSS and increased the WSSG and OSI values. Stenting of the MB in bifurcations with larger SB/MB ratios or smaller SB angle resulted in lower WSS and higher WSSG and OSI. Stenosis at the SB lowered WSS and increased WSSG and OSI. Considering these results, MB stenting could not be optimal in bifurcations with large SB diameter, small angle, and SB stenosis.

12.4 FLUID DYNAMIC MODELS OF IMAGE-BASED STENTED GEOMETRIES

12.4.1 Introduction

The CFD studies on stented coronary arteries mentioned in Section 12.3 are based on ide-alized geometries. Although in two of these works [51,53] a sequential structural and fluid dynamic approach was used to consider the effects of stenting procedure on the geometry of the model, realistic fluid domains obtained through imaging techniques are needed for a more reliable evaluation of stent performance and hemodynamic alterations induced by stenting, and also for a better comprehension of the ISR phenomenon.

In this section, CFD studies based on geometries reconstructed through imaging tech-niques are reviewed. These studies are classified in three categories according to the origin of the images used for the creation of the fluid domain: (1) *in vitro* models, (2) animal models, and (3) patients.

12.4.2 CFD Studies from *In Vitro* Model Images

Foin and colleagues [62] evaluated post-dilatation strategies after MB stenting. The follow-ing scenarios were compared: (1) MB stenting only, (2) stenting of the MB followed by FKB, and (3) stenting of the MB concluded by 2-step sequential post-dilation. A series of drug elut-ing stents was implanted in silicone models of coronary bifurcations. Micro-computed tomography (micro-CT) scans of the stents were used to create the fluid volumes for subse-quent CFD simulations (unfortunately, no detailed information was provided by the authors about this step). As boundary conditions, a flow waveform recorded in a human LAD was applied at the inlet cross-section of each model. A flow split of 70% and 30%, respectively for the MB and the SB, was imposed at the outlets. Blood was assumed to be a Newtonian fluid. The analyses were performed with the commercial software ANSYS CFX. The results of the work in terms of flow patterns and shear rates are depicted in Fig. 12.10. Flow distribution

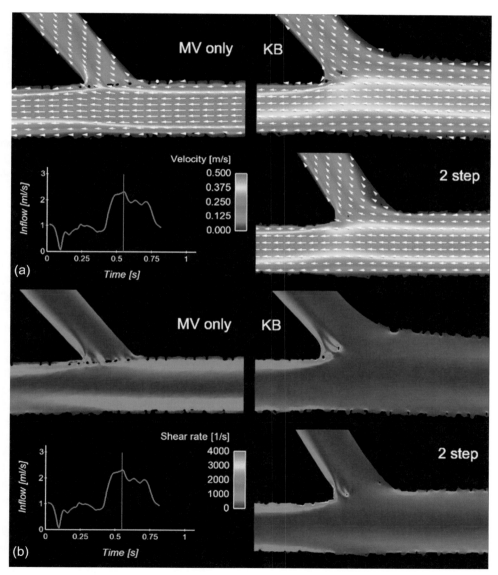

FIGURE 12.10   Velocity field (a) and shear rate contour maps (b) for the different stenting techniques: MB stenting only (MV only), stenting of the MB followed by FKB (KB), and stenting of the MB concluded by 2-step sequential post-dilation (2 step). *(Reprinted with permission from Foin et al. [62]. © 2012 American College of Cardiology Foundation.)*

toward the SB showed higher velocity near the carina and lower velocity opposite the carina. This difference was more evident in the case of MB stenting because the flow directed to the SB was impaired by the stent struts covering the SB ostium. Moreover, a large area of high shear rate was located around these struts. Flow patterns in FKB and 2-step cases were more stable with a less disturbed SB flow if compared to the MB stenting only. High shear rates were situated in correspondence of the malapposed struts near the carina. The fluid dynamic analyses represent a part of the work of Foin et al. [62]. The rate of strut malapposition was calculated analyzing the reconstructed geometrical models of the bifurcations. FKB and 2-step post-dilation techniques significantly reduced the rate of strut malapposition with respect to MV stenting only. However, FKB led to a higher risk of incomplete stent apposition at the proximal stent edge. Considering all the results, the authors concluded that sequential 2-step post-dilation might offer a simpler and more efficient alternative to FKB.

CFD analyses were performed also in a similar study by the same research group [63]. The Crush technique followed by FKB inflation was investigated. Extensive layers of malapposed struts were observed in the lumen. The unapposed struts in the neocarina caused severe flow disturbances with high shear rate that might increase the risk of platelet adhesion and stent thrombosis.

12.4.3 CFD Studies from Animal Models

Morlacchi et al. [64] investigated blood flow patterns in a realistic stented porcine coronary artery geometry using a model that combines data from *in vivo* experiments, 3D imaging techniques, and CFD simulations. Twelve BiodivYsio[TM] stents (Biocompatibles International Plc, Farnham, UK) were implanted individually into the right coronary artery (RCA) and the LAD of six healthy Yorkshire White pigs. The stents were intentionally over-expanded in order to damage the vessels and, consequently, produce a significant growth of neointimal tissue. The animals were sacrificed at different times, from 6 h to 28 days post-implantation. The stented segments were scanned using micro-CT and then embedded and cut with a high-speed precision saw in order to obtain sections for histological analyses. One case was selected for numerical analyses (stent deployed in RCA, explanted after 14 days). Starting from micro-CT images, the 3D geometry of the stent was reconstructed. Because the arterial wall could not be extracted from micro-CT images, a structural simulation was performed (Fig. 12.11a): the model of the artery was expanded under displacement control until its internal diameter was greater than the stent diameter; then the artery retracted due to elastic recoil coming into contact with the undeformable structure of the stent. Fluid domain was extracted with a similar procedure used by the same research group [51]. In order to reduce the high computational cost, only the first two repeating units of the proximal end of the stent were considered in the CFD analyses. As boundary conditions, at the inlet a velocity waveform of a porcine RCA taken from literature [66] was imposed with a paraboloid-shaped profile. At the outlet a relative pressure of 0 Pa was set. The artery and the stent were considered as wall boundaries. Blood was considered as a non-Newtonian Carreau fluid with a density of $1060 \, kg/m^3$.

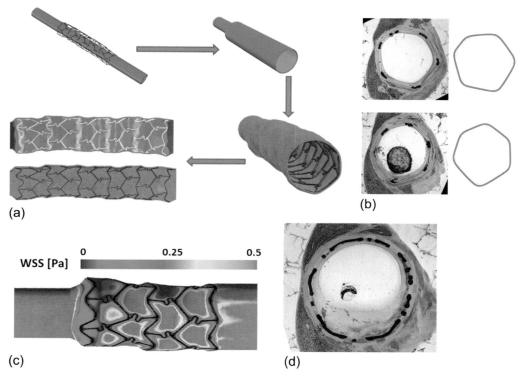

FIGURE 12.11   (a) Structural model implemented to identify the geometrical configuration of the stented arteries of *in vivo* porcine coronary models: CAD model of the undeformed artery and of the stent reconstructed from micro-CT images, expansion of the artery through a cylindrical surface dilatation, stent-artery coupling obtained after the recoil of the artery, and longitudinal section of the final configuration. (b) Comparison between two histological images and the corresponding sections obtained with the structural simulation (on the left) and the numerical geometrical configuration (on the right). (c) WSS spatial distribution along the arterial wall. (d) Correlation between areas characterized by low WSS (orange lines) and ISR phenomenon after 14 days in a proximal section of the stented artery. *(Reprinted with permission from Morlacchi and Migliavacca [65]. © 2012 Biomedical Engineering Society.)*

Simulations were carried out using ANSYS Fluent. The realistic geometry obtained from micro-CT images and structural simulation was characterized by proximal overexpansion and asymmetric stent deployment leading to a nonuniform distribution of WSS values (Fig. 12.11b and c). In particular, the regions associatedwith the highest risk of NH (WSS < 0.5 Pa) were located at the proximal end of the stent where the abrupt enlargement of the artery was observed. A good correlation was found between the computed hemodynamic parameters and the asymmetric neointimal growth evaluated by means of histomorphometric analyses of the explanted vessels (Fig. 12.11d).

Rikhtegar et al. [67] studied the hemodynamics of stented coronary arteries of *ex vivo* porcine hearts. Absorbable metal stents (Biotronik AG, Bülach, Switzerland) were implanted by an interventional cardiologist in the left coronary artery of *ex vivo* porcine hearts. Vascular corrosion casting was performed by injection of a radio-opaque resin into the stented coronary vascular tree under physiological pressure of 90 mmHg. Then, the stented coronary artery cast was scanned using micro-CT in order to obtain the 3D geometrical model for subsequent CFD analyses. Both steady-state and transient CFD simulations were carried out with ANSYS CFX. A blood inflow rate of 0.95 mL/s was applied at the ostium for steady-state calculation and time-dependent flow rate replicating systolic and diastolic LCA blood flow for the transient analysis [47]. Zero relative pressure was imposed at the outlet with the largest diameter, and outflow rates were imposed at the remaining outlet according to Murray's law [68]. A no-slip condition was prescribed at the vessel wall. Accurate local flow information was obtained thanks to the method used for the creation of the fluid domain. In fact, this method ensured anatomic fidelity, capturing arterial tissue prolapse, radial and axial arterial deformation, and stent malapposition, otherwise difficult to include in an idealized model. In Fig. 12.12a, an example of malapposed struts is shown. The presence of these struts provoked tunneling of blood flow between the strut and the arterial wall, leading to perturbation of the local flow field. Low WSS were located at bifurcations and next to the struts. Low WSS were not only present close to struts arranged perpendicular to the flow direction, but also occurred in the vicinity of inter-strut connectors arranged parallel to the flow direction. The distribution of low (<0.5 Pa), intermediate (0.5–2.5 Pa), and high (>2.5 Pa) WSS is shown in Fig. 12.12b relative to the surface area of selected arterial segments in a vessel with two stents. In the stented sections, more than 40% of the wall surface area was exposed to low WSS, indicating these segments of artery as potentially atheroprone regions.

In a further study by the same research group [69], the hemodynamics in *ex vivo* porcine left coronary arteries with overlapping stents was investigated. Six cases with partially overlapping stents were compared to five cases with two nonoverlapping stents. The same methodology as used in their previous study [67] was adopted. Their results showed that the area exposed to WSS lower than 0.5 Pa is higher in the overlapping stent segments than in the regions without overlap of the same samples and in the nonoverlapping stents. Moreover, the configuration of the overlapping stent struts relative to each other influenced the size of the low WSS area. The positioning of the struts in the same axial location resulted in larger low WSS areas compared to alternating struts.

12.4.4 CFD Studies from Patient Images

Gundert et al. [30] developed a method to virtually implant a stent in patient-specific coronary artery models for CFD analyses. This method was based on the following three steps: (1) the creation of a 3D geometrical model of a human LAD starting from computed tomography (CT) scans (Fig. 12.13a), (2) the preparation of an idealized model of a thick stent matching arterial geometry (Fig. 12.13b), and (3) the production of the fluid domain by subtracting the stent volume from the coronary solid model (Fig. 12.13c). Two different stent designs were compared from the fluid dynamic perspective: an open-cell ring-and-link stent (stent A) and a closed-cell slotted tube prototype stent (stent B). As in a previous work by the same research group [16], discussed in Section 12.3.2, an LAD blood flow waveform at rest was applied at

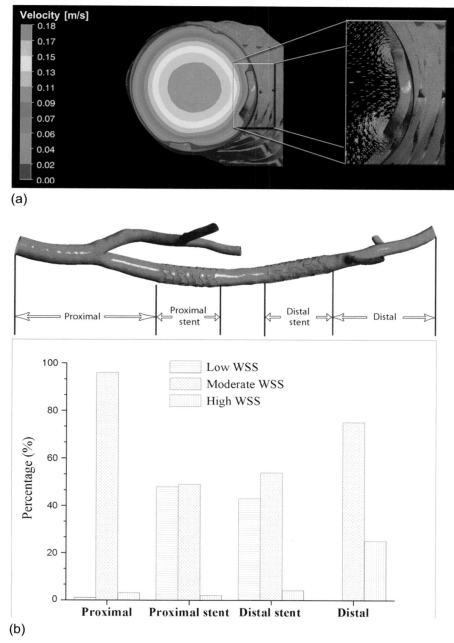

FIGURE 12.12   (a) Contour map of velocity magnitude in an axial cross-section of the stented artery near a malapposed strut. In the magnification box, the velocity vector plot near the malapposed strut is depicted showing the presence of vortices. (b) Distribution of WSS relative to the surface area of selected segments of a porcine left coronary artery model with two stents. (© 2013 Rikhtegar et al. [67].)

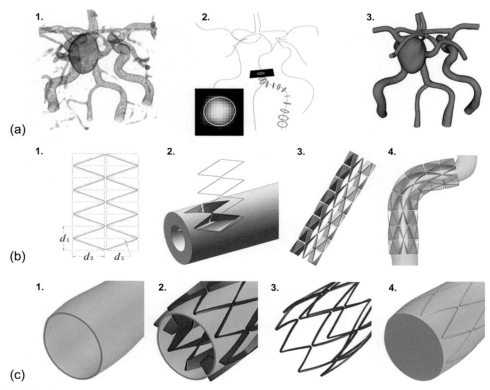

FIGURE 12.13   Methodology developed by Gundert et al. [30] of patient-specific geometrical model creation for CFD analyses of stented coronary arteries: (a) Workflow from imaging data to 3D solid model of a patient-specific artery tree; (b) Preparation of an idealized model of a thick stent matching arterial geometry; (c) Production of the fluid domain by subtracting the stent volume from the coronary solid model. *(Reprinted with permission from Gundert et al. [30]. © 2012 Biomedical Engineering Society.)*

the inlet while a three-element Windkessel model representing the downstream vasculature was imposed at the outlets using a multidomain approach. Blood was assumed to be a Newtonian fluid with a density of 1060 kg/m^3 and a dynamic viscosity of 4 cP. CFD simulations were performed by means of the open-source software Simvascular (https://simtk.org/). Regions of low TAWSS were localized close to the struts and more prevalent distal to the bifurcation in both the cases. The total intrastrut area of the lumen that was exposed to TAWSS lower than 0.4 Pa was shown to be higher for stent A (75.6%) than for stent B (59.3%). In Fig. 12.14, cross-sections of MET at the bifurcation are shown for the model without stent and the two stented models. The presence of the stent increased the region of high MET near the arterial wall in the MB (Fig. 12.14, bifurcation plane and cross-section A). In the SB, just distal to the stent (Fig. 12.14, cross-section B), a pronounced difference in MET between the models can be appreciated. The stent design, placement, and number of struts

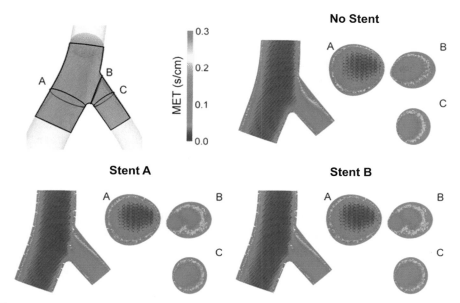

FIGURE 12.14 Mean exposure time (MET) at the coronary bifurcation for the model without stent and the two stented models. MET is visualized in a plane parallel to the bifurcation and three planes perpendicular to the vessel. *(Adapted with permission from Gundert et al. [30]. © 2011 Biomedical Engineering Society.)*

in the SB influenced the exposure time of the particles. The lowest values of MET were found near the carina of the bifurcation for all the models, corresponding to an increased velocity in this zone.

In a further study by the same research group [70], the hemodynamics of a patient-specific stented left circumflex artery with a thrombus was investigated. The geometrical model of the vessel was reconstructed combining CT and optical coherence tomography (OCT) images acquired immediately post-stenting and after a 6-month follow-up period. In order to obtain the fluid domains, the OCT guide-wire pathway was determined by applying a shortest path algorithm; then, segments from OCT images were registered orthogonal to the wire pathway using rotational orientation consistent with geometry estimated by CT. The stent was drawn inside the artery applying the previously developed method [30]. CFD models were similar to those of Gundert et al. [30]. The same boundary conditions were used. Simulations were performed with the commercial flow solver LesLib (Altair Engineering Inc., Troy, MI, USA). Considering TAWSS results, the percentage area of the vessel exposed to TAWSS lower than 0.4 Pa was 11% in the post-stent model and only 3% in the follow-up model, limited to areas of localized curvature. The areas of stent-induced low WSS returned to physiological levels at follow-up and a good correlation was found with the measures of neointimal thickness in OCT images. Finally, high OSI values were present next to struts in the post-stent model and in the areas of curvature of the follow-up model.

Chiastra et al. [34] studied the hemodynamics of two cases (A and B) of pathologic LAD with their bifurcations treated with a provisional T-stenting technique without FKB. In case A, a Xience Prime™ stent (Abbott Laboratories) was implanted while, in case B, two Endeavor® Resolute stents (Medtronic, Minneapolis, MN, USA) were deployed. Preoperative computed tomography angiography (CTA) and conventional coronary angiography (CCA) were used to reconstruct the internal surface of the 3D prestenting geometries [71]. CTA gave the 3D trajectories followed by the arteries and CCA gave accurate lumen radius estimation. The external walls were obtained by lofting circumferential cross-sections perpendicular to the centerlines of the models. Then, structural simulations that replicated all the implantation steps followed by the clinicians were performed [72]. Following the same approach of Morlacchi et al. [51], the final geometrical configurations obtained through these simulations were used as fluid domains to carry out CFD analyses. At the inlet cross-section a pulsatile flow curve which is representative of a human LAD [57] was applied as a paraboloid-shaped velocity profile. At the outlets, the flow distribution was calculated using the relations defined by van der Giessen et al. [73], based on human coronary bifurcation data. The following flow splits were imposed on the models: case A, 72.8% for the MB and 27.2% for the SB; case B, 57.6% for the MB, 32.9% for the proximal SB, and 9.5% for the distal SB. The arterial wall and the stent struts were assumed to be rigid and defined with a no-slip condition. The blood density was considered constant with a value of 1060 kg/m^3. The non-Newtonian nature of the blood was taken into account using the Carreau model. CFD simulations were performed using ANSYS Fluent. Both near-wall and bulk-flow quantities were investigated, thus providing a comprehensive study of the hemodynamics of the two cases. Figure 12.15a shows the contour maps of TAWSS along the arterial wall. Both cases were characterized by low WSS next to the stent struts. Moreover, a wider area with low WSS was present in the region of the bifurcations and, for case B, in the overlapping zone between the two stents. The regions of the vessels outside the stent have WSS higher than 0.4 Pa except the proximal part of case A where the presence of a stagnation zone resulted in low WSS. Quantitatively, the percentage area of the stented region exposed to values of TAWSS lower than 0.4 Pa was significant: 35.0% for case A and 38.4% for case B. The results obtained from the two models for TAWSS were confirmed by the contour maps of RRT. In order to visualize peculiar topological features in the bulk flow, the mutual orientation of velocity and vorticity vectors, given by LNH, was used. Helical flow patterns originated in the region of the vessel upstream from the stent, where helicity was mainly driven by the shape of the unstented segment proximal to the stented one, and in the bifurcations (Fig. 12.15b). The straightening of the vessel induced by stent implantation determined the gradual disappearance of the helical structures. Smaller helical structures generated as a consequence of the presence of the stent struts protruding into the lumen of the vessel are also present (Fig. 12.15b). Results on helicity are preliminary. The role of the small helical structures on ISR (either beneficial or detrimental) have not been fully clarified by this study yet and additional investigations are needed. More cases should also be investigated to find a relationship between the surface area exposed to disturbed shear and helical fluid structures as it was previously done in other vascular districts such as carotid arteries [39].

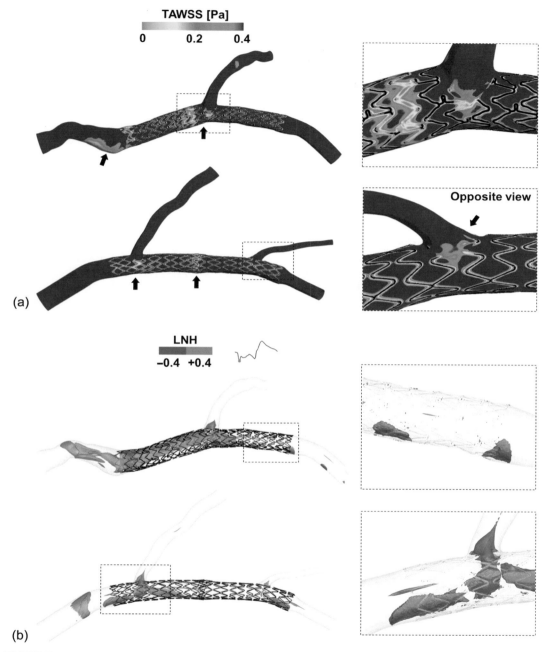

FIGURE 12.15 (a) Contour maps of time-averaged wall shear stress (TAWSS) for case A (top) and case B (bottom). Low WSS regions are indicated in red. Regions where the stent struts are in contact with the arterial wall are colored black. Black arrows point out the widest area with low WSS. (b) Example of isosurfaces of LNH at the systolic phase for case A (top) and case B (bottom). In the magnification boxes small helical fluid structures generated by the struts are visible. *(Adapted with permission from Chiastra et al. [34]. © 2011 The Royal Society.)*

12.5 LIMITATIONS OF THE CURRENT CFD MODELS AND FUTURE REMARKS

12.5.1 Introduction

The CFD studies on stented coronary arteries introduced in Sections 12.3 and 12.4 are subjected to some assumptions and limitations that need to be considered in the interpretation of the results, especially when clinical consideration is made. In this section the main limitations of the proposed models are discussed.

12.5.2 Heart Motion

The motion of coronary arteries is complex because of the presence of the beating heart and can be described by overall vessel translation, stretching, bending, and twisting and, to a minor degree, by radial expansion and axial torsion [74]. In the CFD studies reviewed in this chapter, the cardiac-induced motion was always neglected. However, its effects on hemodynamics are still a subject of study. Several works have been proposed in the literature with conflicting results. These studies do not analyze stented coronary arteries, but unstented vessels. In particular, Zeng et al. [75] studied the fluid dynamics of an RCA model under physiologically realistic heart motion. The authors concluded that the motion effects were small compared to flow pulsation effects. On the contrary, Ramaswamy et al. [74] found that heart motion substantially affects the hemodynamics in LAD. Prosi et al. [76] considered a curved model of LAD with its first diagonal branch by attaching it to the surface of a sphere with time-varying radius based on experimental dynamic curvature data. Their results indicated that the effect of curvature dynamics on the flow field were negligible. Theodorakakos et al. [77] studied the effect of cardiac motion on the flow field and WSS distribution of an image-based human LAD and its main branches in the presence of a stenosis. Their results showed that hemodynamics was considerably affected by the pulsatile nature of the flow and myocardial motion had only a minor effect on flow patterns within the arterial tree. Although the absolute values of WSS were different, the spatial distribution of WSS was very similar between the stationary and the moving coronary tree. Hasan et al. [78] investigated the effects of cyclic motion (i.e., bending and stretching) on blood flow in a 3D model of a segment of the LAD, which was created on the basis of anatomical studies. Their results highlighted that, although the motion of the coronary artery could significantly affect blood particle trajectory, it had slight effect on velocity and WSS.

12.5.3 Rigid Walls

In all the studies discussed in this chapter the arterial wall and the stents are assumed to be rigid and fixed. This assumption might produce different local hemodynamic results. Fluid-structure interaction (FSI) simulations of coronary artery models were performed in a limited number of studies [79–83], without considering the presence of the stents.

In order to understand the effects of the wall compliance on hemodynamics quantities and to show that FSI can now be carried out, we performed a preliminary FSI study on a straight coronary artery with a common cobalt-chromium, open-cell stent deployed. The results of the

FSI simulation were compared to a rigid-wall fluid dynamic simulation which was carried out using the same initial model. In the remainder of this section, a summary of this study is outlined.

The geometry of the artery was created with realistic dimensions (Fig. 12.16a) using a CAD software. The stent was expanded inside the vessel through a structural simulation [84]. The final geometrical configuration after the elastic recoil was exported as triangulated surfaces to discretize the model [51]. A mixed tetrahedral and prismatic mesh of about 700,000

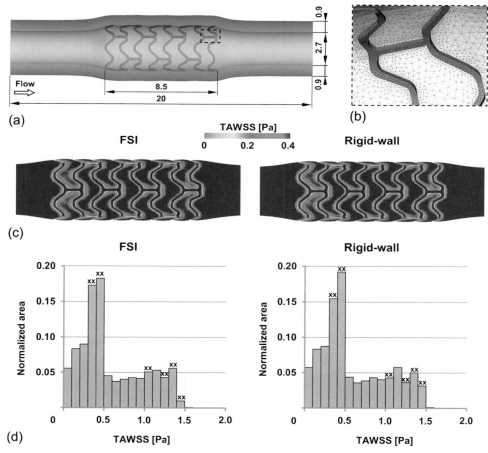

FIGURE 12.16 (a) Geometrical model of the straight coronary artery for FSI simulations with a typical deployed open-cell stent. The fluid domain is visualized in light gray while the solid domain is shown in dark gray. Dimensions are in millimeters. (b) Particular of the fluid flow grid obtained at the fluid-structure interface characterized by a refinement in correspondence of the struts. (c) Contour maps of TAWSS along the fluid-structure interface for the FSI (left) and rigid-wall (right) model. (d) TAWSS distributions for the FSI (left) and rigid-wall (right) model. Each bar of the histograms represents the amount of normalized area with a defined range of TAWSS. Bar widths are 0.1 Pa. The bars with an absolute difference greater than 0.005 Pa between the FSI and rigid-wall model are indicated by the symbol "xx".

tetrahedral elements (190,000 nodes) was used for the arterial wall and stent. The fluid flow grid, which considers a refinement around the device (Fig. 12.16b), was composed of 900,000 tetrahedral elements (222,500 nodes).

The arterial wall was modeled as a hyperelastic, incompressible, isotropic and homogeneous material using a Mooney-Rivlin model [85]. The following mechanical properties were assigned to the stent: Young's modulus $= 233$ GPa and Poisson's ratio $= 0.35$. The blood was modeled as incompressible Newtonian fluid, with a density of 1060 kg/m^3 and a dynamic viscosity of 0.0035 Pa s [67]. The extremities of the solid model were constrained by preventing axial and transaxial motion. The physiological curves of flow and pressure measured *in vivo* by Davies et al. [57] were imposed at the inlet and at the outlet of the model, respectively.

A qualitative comparison of the contour maps of TAWSS along the fluid-structure interface for the FSI and rigid-wall case (Fig. 12.16c) did not point out any significant difference between the two. In Fig. 12.16d, the area distribution of TAWSS in the stented region is presented for the two analyzed cases. To better visualize the differences between the histograms, the bars with an absolute difference greater than 0.005 Pa between FSI and rigid-wall model are indicated by symbols. Differences between the bars are small. The maximum absolute difference is 0.022 Pa (interval 1.4–1.5 Pa). The percentage of area exposed to TAWSS lower than 0.4 Pa in the stented region is similar between the FSI and rigid-wall case (40.22% and 38.65%, respectively).

The results of this preliminary work suggest that, for idealized models of a stented coronary artery, the rigid-wall assumption for fluid dynamic simulations is adequate when the aim of the study is the analysis of near-wall quantities such as WSS.

12.5.4 Boundary Conditions

The choice of boundary conditions in the CFD models is fundamental for obtaining reliable results. For all the numerical models introduced in this chapter the inlet boundary condition was similar and consisted in a pulsatile blood flow waveform taken from the literature applied as either a paraboloid-shaped (e.g., Ref. [34]) velocity profile or a Womersley-like velocity profile (e.g. Ref. [16]). For the outlet boundary conditions, different choices were proposed: traction-free boundary condition [61], flow split based on empirical laws [34,67], or the multidomain approach that considers a lumped-parameter model representing the downstream vasculature coupled to the 3D domain [16,30,51,70].

Because *in vivo* measurements of flow-rate and pressure in coronary arteries are highly invasive and not always required in a clinical routine exam, boundary conditions used for CFD simulations of patient-specific geometries (Section 12.4.4) were mainly based on literature data.

12.5.5 Accuracy of Three-Dimensional Geometrical Models

The results of a fluid dynamic simulation are more accurate when the numerical analysis is performed with a geometrical model closer to the reality. Considering the CFD models of patient-specific geometries (Section 12.4.4), only the arterial vessel was obtained from imaging data while the stent was created using a CAD software.

Currently, OCT is a promising imaging technique for coronary arteries, due to its high resolution (10–20 µm) and the possibility to detect both the stent and the vessel wall [86]. Thanks to these characteristics, OCT will be a useful tool to reconstruct 3D geometries of stented coronary arteries.

12.5.6 Model Validation

The validation of the numerical model is an important aspect in order to guarantee the validity of the results. It is defined as the process of determining whether the CFD model is an accurate representation of the real system, for the particular objectives of each study [87].The small dimensions of the coronary vessels (2–4 mm in diameter) and the stents (80–100 µm for the strut size) make *in vivo* local measurements of velocities and velocity gradients very difficult, resulting in the impossibility of mapping the shear stress distribution at the arterial wall. Also, the *in vitro* quantification of the local fluid dynamics of stented coronary artery models is very complex to realize, mainly because of the small dimensions.

In the works introduced in Sections 12.3 and 12.4, validation was not directly done. However, in an older study [33] both CFD simulations and *in vitro* experiments were performed. The hemodynamic effects of two different stent geometries were investigated using CFD and dye injection flow visualization. A simplified 2D CFD model of the region very close to the arterial wall was considered. Numerical results were qualitative in agreement with the experiments. Morlacchi [88] compared particle image velocimetry (PIV) flow measurements in *in vitro* stented and nonstented coronary bifurcation models with the fluid dynamic results of the corresponding numerical replica of the experimental cases. Despite the intrinsic differences and modeling assumptions of the two approaches, the results were qualitatively in agreement. Both PIV and CFD analyses were able to capture the main features of the fluid flows, highlighting the influence of different bifurcation angles and stenting procedures.

Validation remains a key aspect of 3D numerical models of stented coronary arteries. In the future, the improvement of experimental techniques for fluid dynamic measurements such as PIV [54] or 3D particle tracking velocimetry [89] will allow a quantitative comparison between CFD and *in vitro* models of stented coronary arteries.

12.6 CONCLUSIONS

In this chapter, the main computational studies on the fluid dynamics of stented coronary arteries have been reviewed. Both the idealized bifurcation and the image-based models have been considered.

CFD analyses have been demonstrated to be a useful tool to study the local fluid dynamics within stented arteries. They allow the calculation of hemodynamic quantities like WSS that have been demonstrated to be correlated to the process of ISR. However, the indications obtained from these simulations should be carefully evaluated because the altered fluid dynamics provoked by the presence of stents is not the only factor involved in the ISR process. The vascular injury caused by device implantation, the stent design (e.g., bare-metal or drug eluting stent) and hypoxia should also be taken into account.

Several works focused on structural simulations of stent implantation have been proposed in the literature [65,90]. The results of these kinds of simulations provide better insight on the changes of the mechanical environment due to the stent expansion.

Virtual models that simultaneously take into account the fluid dynamics and the drug release would also be useful to better predict ISR regions, making models more predictive. Some multiphysics works have already been proposed in the literature coupling fluid dynamics and drug release analysis [91]. However, these works consider simplified vessel geometries [92] or approximate the stent as a line and not as a 3D structure [93,94], with the impossibility of an accurate study of the local fluid dynamics.

In the end, the correlation between ISR and hypoxia is another important aspect to be considered. Recent indications of this correlation have been elucidated through *in vivo* [95,96] and computational [97,98] studies.

Acknowledgments

The authors would like to acknowledge Prof. Mauro Malvè (Universidad Pública de Navarra, Departamento de Ingeniería Mecánica, Energética y de Materiales, Pamplona, Spain) for the collaboration on the fluid-structure interaction model of a straight stented coronary artery introduced in Section 12.5.3.

References

[1] Go AS, Mozaffarian D, Roger VL, et al. Heart disease and stroke statistics—2013 update: a report from the American Heart Association. Circulation 2013;127:e6–e245.

[2] Libbi P. The pathogenesis, prevention, and treatment of atherosclerosis. In: Fauci AS, Kasper DL, Longo DL, Braunwald E, Hauser SL, Jameson JL, Loscalzo J, editors. Harrison's principles of internal medicine. 17th ed. Newyork: McGraw-Hill Publ. Comp; 2008, part 9, section 235.

[3] Frye RL, Alderman EL, Andrews K, Bost J, Bourassa M, Chaitman BR, et al. Comparison of coronary bypass surgery with angioplasty in patients with multivessel disease. The Bypass Angioplasty Revascularization Investigation (BARI) Investigators. N Engl J Med 1996;335:217–25.

[4] Baim DS. Percutaneous coronary intervention. In: Fauci AS, Kasper DL, Longo DL, Braunwald E, Hauser SL, Jameson JL, Loscalzo J, editors. Harrison's principles of internal medicine. 17th ed. Newyork: McGraw-Hill Publ.Comp; 2008, part 9, section 240.

[5] Park S-J, Kang S-J, Virmani R, Nakano M, Ueda Y. In-stent neoatherosclerosis: a final common pathway of late stent failure. J Am Coll Cardiol 2012;59:2051–7.

[6] Grube E, Dawkins K, Guagliumi G, Banning A, Zmudka K, Colombo A, et al. TAXUS VI final 5-year results: a multicentre, randomised trial comparing polymer-based moderate-release paclitaxel-eluting stent with a bare metal stent for treatment of long, complex coronary artery lesions. Eurointervention 2009;4:572–7.

[7] Nakazawa G, Finn AV, Vorpahl M, Ladich ER, Kolodgie FD, Virmani R. Coronary responses and differential mechanisms of late stent thrombosis attributed to first-generation sirolimus- and paclitaxel-eluting stents. J Am Coll Cardiol 2011;57:390–8.

[8] Wentzel JJ, Gijsen FJH, Schuurbiers JCH, van der Steen AFW, Serruys PW. The influence of shear stress on in-stent restenosis and thrombosis. Eurointervention 2008;4(Suppl. C):C27–32.

[9] Kleinstreuer C, Hyun S, Buchanan Jr. JR, Longest PW, Archie Jr. JP, Truskey GA. Hemodynamic parameters and early intimal thickening in branching blood vessels. Crit Rev Biomed Eng 2001;29:1–64.

[10] Chatzizisis YS, Coskun AU, Jonas M, Edelman ER, Feldman CL, Stone PH. Role of endothelial shear stress in the natural history of coronary atherosclerosis and vascular remodeling: molecular, cellular, and vascular behavior. J Am Coll Cardiol 2007;49:2379–93.

[11] Malek AM, Alper SL, Izumo S. Hemodynamic shear stress and its role in atherosclerosis. J Am Med Assoc 1999;282:2035–42.

[12] Gerlach H, Esposito C, Stern DM. Modulation of endothelial hemostatic properties: an active role in the host response. Annu Rev Med 1990;41:5–24.

[13] Okano M, Yoshida Y. Influence of shear stress on endothelial cell shapes and junction complexes at flow dividers of aortic bifurcations in cholesterol-fed rabbits. Front Med Biol Eng 1993;5:95–120.

[14] LaDisa Jr. JF, Olson LE, Molthen RC, Hettrick DA, Pratt PF, Hardel MD, et al. Alterations in wall shear stress predict sites of neointimal hyperplasia after stent implantation in rabbit iliac arteries. Am J Physiol Heart Circ Physiol 2005;288:H2465–75.

[15] Ku DN, Giddens DP, Zarins CK, Glagov S. Pulsatile flow and atherosclerosis in the human carotid bifurcation. Positive correlation between plaque location and low and oscillating shear stress. Arteriosclerosis 1985; 5:293–302.

[16] Williams AR, Koo B, Gundert TJ, Fitzgerald PJ, LaDisa Jr. JF. Local hemodynamic changes caused by main branch stent implantation and subsequent virtual side branch balloon angioplasty in a representative coronary bifurcation. J Appl Physiol 2010;109:532–40.

[17] Zarins CK, Giddens DP, Bharadvaj BK, Sottiurai VS, Mabon RF, Glagov S. Carotid bifurcation atherosclerosis. Quantitative correlation of plaque localization with flow velocity profiles and wall shear stress. Circ Res 1983;53:502–14.

[18] Himburg HA, Grzybowski DM, Hazel AL, LaMack JA, Li XM, Friedman MH. Spatial comparison between wall shear stress measures and porcine arterial endothelial permeability. Am J Physiol Heart Circ Physiol 2004;286: H1916–22.

[19] Malek AM, Izumo S. Physiological fluid shear stress causes downregulation of endothelin-1 mRNA in bovine aortic endothelium. Am J Physiol 1992;263:C389–96.

[20] Chiu JJ, Chien S. Effects of disturbed flow on vascular endothelium: pathophysiological basis and clinical perspectives. Physiol Rev 2011;91:327–87.

[21] Yin W, Shanmugavelayudam SK, Rubenstein DA. The effect of physiologically relevant dynamic shear stress on platelet and endothelial cell activation. Thromb Res 2011;127:235–41.

[22] Dardik A, Chen L, Frattini J, Asada H, Aziz F, Kudo FA, et al. Differential effects of orbital and laminar shear stress on endothelial cells. J Vasc Surg 2005;41:869–80.

[23] Dai G, Kaazempur-Mofrad MR, Natarajan S, Zhang Y, Vaughn S, Blackman BR, et al. Distinct endothelial phenotypes evoked by arterial waveforms derived from atherosclerosis-susceptible and -resistant regions of human vasculature. Proc Natl Acad Sci U S A 2004;101:14871–6.

[24] Hoi Y, Zhou Y, Zhang X, Henkelman X, Steinman DA. Correlation between local hemodynamics and lesion distribution in a novel aortic regurgitation murine model of atherosclerosis. Ann Biomed Eng 2011;39:1414–22.

[25] Lei M, Kleinstreuer C, Truskey GA. A focal stress gradient-dependent mass transfer mechanism for atherogenesis in branching arteries. Med Eng Phys 1996;18:326–32.

[26] Truskey GA, Barber KM, Robey TC, Olivier LA, Combs MP. Characterization of a sudden expansion flow chamber to study the response of endothelium to flow recirculation. J Biomech Eng 1995;117:203–10.

[27] DePaola N, Gimbrone Jr MA, Davies PF, Dewey Jr CF. Vascular endothelium responds to fluid shear stress gradients. Arterioscler Thromb 1992;12:1254–7.

[28] Murphy JB, Boyle FJ. Predicting neointimal hyperplasia in stented arteries using time-dependent computational fluid dynamics: a review. Comput Biol Med 2010;40:408–18.

[29] Ojha M. Spatial and temporal variations of wall shear stress within an end-to-side arterial anastomosis model. J Biomech 1993;26:1377–88.

[30] Gundert TJ, Shadden SC, Williams AR, Koo B, Feinstein JA, LaDisa Jr. JF. Rapid and computationally inexpensive method to virtually implant current and next-generation stents into subject-specific computational fluid dynamics models. Ann Biomed Eng 2011;39:1423–36.

[31] Lonyai A, Dubin AM, Feinstein JA, Taylor CA, Shadden SC. New insights into pacemaker lead-induced venous occlusion: simulation-based investigation of alterations in venous biomechanics. Cardiovasc Eng 2010;10:84–90.

[32] He Y, Duraiswamy N, Frank AO, Moore Jr JE. Blood flow in stented arteries: a parametric comparison of strut design patterns in three dimensions. J Biomech Eng 2005;127:637–47.

[33] Berry JL, Santamarina A, Moore Jr JE, Roychowdhury S, Routh WD. Experimental and computational flow evaluation of coronary stents. Ann Biomed Eng 2000;28:386–98.

[34] Chiastra C, Morlacchi S, Gallo D, Morbiducci U, Cárdenes R, Larrabide I, et al. Computational fluid dynamic simulations of image-based stented coronary bifurcation models. J R Soc Interface 2013;10:20130193.

[35] Moffatt HK, Tsinober A. Helicity in laminar and turbulent flow. Annu Rev Fluid Mech 1992;24:281–312.

[36] Morbiducci U, Ponzini R, Grigioni M, Redaelli A. Helical flow as fluid dynamic signature for atherogenesis in aortocoronary bypass A numeric study. J Biomech 2007;40:519–34.

[37] Morbiducci U, Ponzini R, Rizzo G, Cadioli M, Esposito A, Montevecchi FM, et al. Mechanistic insight into the physiological relevance of helical blood flow in the human aorta: an in vivo study. Biomech Model Mechanobiol 2011;10:339–55.

[38] Kilner PJ, Yang GZ, Mohiaddin RH, Firmin DN, Longmore DB. Helical and retrograde secondary flow patterns in the aortic arch studied by three-directional magnetic resonance velocity mapping. Circulation 1993;88:2235–47.

[39] Gallo D, Steinman DA, Bijari PB, Morbiducci U. Helical flow in carotid bifurcation as surrogate marker of exposure to disturbed shear. J Biomech 2012;45:2398–404.

[40] Zheng T, Wen J, Diang W, Deng X, Fan Y. Numerical investigation of oxygen mass transfer in a helical-type artery bypass graft. Comput Methods Biomech Biomed Engin 2012;17(5):549–59.

[41] Chen Z, Fan Y, Deng X, Xu Z. A new way to reduce flow disturbance in endovascular stents: a numerical study. Artif Organs 2011;35:392–7.

[42] Louvard Y, Thomas M, Dzavik V, et al. Classification of coronary artery bifurcation lesions and treatments: time for a consensus! Catheter Cardiovasc Interv 2008;71:175–83.

[43] Sharma SK, Sweeny J, Kini AS. Coronary bifurcation lesions: a current update. Cardiol Clin 2010;28:55–70.

[44] Yazdani SK, Nakano M, Otsuka F, Kolodgie FD, Virmani R. Atheroma and coronary bifurcations: before and after stenting. Eurointervention 2010;6:J24–30.

[45] Sheiban I, Omedè P, Biondi-Zoccai G, Moretti C, Sciuto F, Trevi GP. Update on dedicated bifurcation stents. J Interv Cardiol 2009;22:150–5.

[46] Deplano V, Bertolotti C, Barragan P. Three-dimensional numerical simulations of physiological flows in a stented coronary bifurcation. Med Biol Eng Comput 2004;42:650–9.

[47] Berne RM, Levy MN. Cardiovascular physiology. St. Louis, MO, U.S.A.: Mosby; 1967.

[48] Spurk JH, Aksel N. Fluid mechanics. 2nd ed. Berlin Heidelberg: Springer-Verlag; 2008.

[49] Finet G, Gilard M, Perrenot B, Rioufol G, Motreff P, Gavit L, et al. Fractal geometry of arterial coronary bifurcations: a quantitative coronary angiography and intravascular ultrasound analysis. Eurointervention 2008;3:490–8.

[50] Vignon-Clementel IE, Figueroa AC, Jansen KE, Taylor CA. Outflow boundary conditions for three-dimensional finite element modeling of blood flow and pressure in arteries. Comput Methods Appl Mech Eng 2006;195:3776–96.

[51] Morlacchi S, Chiastra C, Gastaldi D, Pennati G, Dubini G, Migliavacca F. Sequential structural and fluid dynamic numerical simulations of a stented bifurcated coronary artery. J Biomech Eng 2011;133:121010.

[52] Behan MW, Holm NR, Curzen NP, et al. Simple or complex stenting for bifurcation coronary lesions: a patient-level pooled-analysis of the Nordic bifurcation study and the British bifurcation coronary study. Circ Cardiovasc Interv 2011;4:57–64.

[53] Chiastra C, Morlacchi S, Pereira S, Dubini G, Migliavacca F. Fluid dynamics of stented coronary bifurcations studied with a hybrid discretization method. Eur J Mech B/Fluids 2012;35:76–84.

[54] Charonko J, Karri S, Schmieg J, Prabhu S, Vlachos P. In vitro, time-resolved PIV comparison of the effect of stent design on wall shear stress. Ann Biomed Eng 2009;37:1310–21.

[55] Pietrabissa R, Mantero S, Marotta T, Menicanti L. A lumped parameter model to evaluate the fluid dynamics of different coronary bypasses. Med Eng Phys 1996;18:477–84.

[56] Gijsen F. Modeling of wall shear stress in large arteries. PhD thesis. Eindhoven University of Technology, Eindhoven; 1998.

[57] Davies JE, Whinnett ZI, Francis DP, Manisty CH, Aguado-Sierra J, Willson K, et al. Evidence of dominant backward-propagating suction wave responsible for diastolic coronary filling in humans, attenuated in left ventricular hypertrophy. Circulation 2006;113:1768–78.

[58] Katritsis DG, Theodorakakos A, Pantos I, Gavaises M, Karcanias N, Efstathopoulos EP. Flow patterns at stented coronary bifurcations: computational fluid dynamics analysis. Circulation Cardiovascular Interventions 2012;5:530–9.

[59] Iakovou I, Ge L, Colombo A. Contemporary stent treatment of coronary bifurcations. J Am Coll Cardiol 2005;46:1446–55.

[60] Katritsis DG, Siontis GCM, Ioannidis JPA. Double versus single stenting for coronary bifurcation lesions: a meta-analysis. Circ Cardiovasc Interv 2009;2:409–15.

[61] Chen HY, Moussa ID, Davidson C, Kassab GS. Impact of main branch stenting on endothelial shear stress: role of side branch diameter, angle and lesion. J R Soc Interface 2012;9:1187–893.

[62] Foin N, Torii R, Mortier P, De Beule M, Viceconte N, Chan PH, et al. Kissing balloon or sequential dilation of the side branch and main vessel for provisional stenting of bifurcations: lessons from micro-computed tomography and computational simulations. JACC Cardiovasc Interv 2012;5:47–56.

[63] Foin N, Alegria-Barrero E, Torii R, Chan PH, Viceconte N, Davies JE, et al. Crush, culotte, T and protrusion: which 2-stent technique for treatment of true bifurcation lesions?—insights from in vitro experiments and micro-computed tomography. Circ J 2012;77:73–80.

[64] Morlacchi S, Keller B, Arcangeli P, Balzan M, Migliavacca F, Dubini G, et al. Hemodynamics and in-stent restenosis: micro-CT images, histology, and computer simulations. Ann Biomed Eng 2011;39:2615–26.

[65] Morlacchi S, Migliavacca F. Modeling stented coronary arteries: where we are, where to go. Ann Biomed Eng 2013;41:1428–44.

[66] Huo Y, Choy JS, Svendsen M, Sinha AK, Kassab GS. Effects of vessel compliance on flow pattern in porcine epicardial right coronary arterial tree. J Biomech 2009;42:594–602.

[67] Rikhtegar F, Pacheco F, Wyss C, Stok KS, Ge H, Choo RJ, et al. Compound ex vivo and in silico method for hemodynamic analysis of stented arteries. PLoS One 2013;8:e58147.

[68] Murray CD. The physiological principle of minimum work: I. The vascular system and the cost of blood volume. Proc Nat Acad Sci USA 1926;12:207–14.

[69] Rikhtegar F, Wyss C, Stok KS, Poulikakos D, Müller R, Kurtcuoglu V. Hemodynamics in coronary arteries with overlapping stents. J Biomech 2014;47:505–11.

[70] Ellwein LM, Otake H, Gundert TJ, Koo B, Shinke T, Honda Y, et al. Optical coherence tomography for patient-specific 3D artery reconstruction and evaluation of wall shear stress in a left circumflex coronary artery. Cardiovasc Eng Technol 2011;2:212–27.

[71] Cárdenes R, Díez JL, Duchateau N, Pashaei A, Frangi AF. Model generation of coronary artery bifurcations from CTA and single plane angiography. Med Phys 2013;40:013701.

[72] Morlacchi S, Colleoni SG, Cárdenes R, Chiastra C, Diez JL, Larrabide I, et al. Patient-specific simulations of stenting procedures in coronary bifurcations: two clinical cases. Med Eng Phys 2013;35(9):1272–81.

[73] van der Giessen AG, Groen HC, Doriot P, de Feyter PJ, van der Steen AFW, van de Vosse FN, et al. The influence of boundary conditions on wall shear stress distribution in patients specific coronary trees. J Biomech 2011;44:1089–95.

[74] Ramaswamy SD, Vigmostad SC, Wahle A, Lai Y-G, Olszeski ME, Braddy KC, et al. Fluid dynamic analysis in a human left anterior descending coronary artery with arterial motion. Ann Biomed Eng 2004;32:1628–41.

[75] Zeng D, Ding Z, Friedman MH, Ethier CR. Effects of cardiac motion on right coronary artery hemodynamics. Ann Biomed Eng 2003;31:420–9.

[76] Prosi M, Perktold K, Ding Z, Friedman MH. Influence of curvature dynamics on pulsatile coronary artery flow in a realistic bifurcation model. J Biomech 2004;37:1767–75.

[77] Theodorakakos A, Gavaises M, Andriotis A, Zifan A, Liatsis P, Pantos I, et al. Simulation of cardiac motion on non-Newtonian, pulsating flow development in the human left anterior descending coronary artery. Phys Med Biol 2008;53:4875–92.

[78] Hasan M, Rubenstein D, Yin W. Effects of cyclic motion on coronary blood flow. J Biomech Eng 2013;135:121002.

[79] Asanuma T, Higashikuni Y, Yamashita H, Nagai R, Hisada T, Sugiura S. Discordance of the areas of peak wall shear stress and tissue stress in coronary artery plaques as revealed by fluid-structure interaction finite element analysis: a case study. Int Heart J 2013;54:54–8.

[80] Malvè M, García A, Ohayon J, Martínez MA. Blood flow and mass transfer of a human left coronary artery bifurcation: FSI vs. CFD. Int Comm Heat Mass Transfer 2012;39:745–51.

[81] Torii R, Wood NB, Hadjiloizou N, Dowsey AW, Wright AR, Hughes AD, et al. Fluid-structure interaction analysis of a patient-specific right coronary artery with physiological velocity and pressure waveforms. Commun Numer Methods Eng 2009;25:565–80.

[82] Yang C, Bach RG, Zheng J, Naqa IE, Woodard PK, Teng Z, et al. In vivo IVUS-based 3-D fluid-structure interaction models with cyclic bending and anisotropic vessel properties for human atherosclerotic coronary plaque mechanical analysis. IEEE Trans Biomed Eng 2009;56:2420–8.

[83] Koshiba N, Ando J, Chen X, Hisada T. Multiphysics simulation of blood flow and LDL transport in a porohyperelastic arterial wall model. J Biomech Eng 2007;129:374–85.

[84] Gastaldi D, Morlacchi S, Nichetti R, Capelli C, Dubini G, Petrini L, et al. Modeling the provisional side-branch stenting approach for the treatment of atherosclerotic coronary bifurcations: effects of stent positioning. Biomech Model Mechanobiol 2010;9:551–61.

[85] Malvè M, Chiastra C, Morlacchi S, Martínez MÁ, Migliavacca F. Comparison between fluid-structure interaction and fluid dynamic simulations of stented coronary arteries. Proceedings 23rd International Congress of Theoretical and Applied Mechanics, Beijing, China, 19-24 August 2012.

[86] Farooq MU, Khasnis A, Majid A, Kassab MY. The role of optical coherence tomography in vascular medicine. Vasc Med 2009;14:63–71.

[87] Law AM, McComas MG. How to build valid and credible simulation models. In: Peters BA, Smith JS, Medeiros DJ, Rohrer MW, editors. Proceedings of the winter simulation conference. VA, USA: Arlington; 2001. p. 22–9.

[88] Morlacchi S. Structural and fluid dynamic assessment of stenting procedures for coronary bifurcations. PhD thesis. Politecnico di Milano, Italy; 2013.

[89] Gülan U, Lüthi B, Holzner M, Liberzon A, Tsinober A, Kinzelbach W. Experimental study of aortic flow in the ascending aorta via particle tracking velocimetry. Exp Fluids 2012;53:1469–85.

[90] Martin D, Boyle FJ. Computational structural modelling of coronary stent deployment: a review. Comput Methods Biomech Biomed Engin 2011;14:331–48.

[91] O'Connell BM, McGloughlin TM, Walsh MT. Factors that affect mass transport from drug eluting stents into the artery wall. Biomed Eng Online 2010;9:15.

[92] Kolachalama VB, Levine EG, Edelman ER. Luminal flow amplifies stent-based drug deposition in arterial bifurcations. PLoS One 2009;4:e8105.

[93] Cutrì E, Zunino P, Morlacchi S, Chiastra C, Migliavacca F. Drug delivery patterns for different stenting techniques in coronary bifurcations: a comparative computational study. Biomech Model Mechanobiol 2013;12:657–69.

[94] D'Angelo C, Zunino P, Porpora A, Morlacchi S, Migliavacca F. Model reduction strategies enable computational analysis of controlled drug release from cardiovascular stents. SIAM J Appl Math 2011;71:2312–33.

[95] Santilli SM, Tretinyak AS, Lee ES. Transarterial wall oxygen gradients at the deployment site of an intra-arterial stent in the rabbit. Am J Physiol Heart Circ Physiol 2000;279:H1518–25.

[96] Sanada J-I, Matsui O, Yoshikawa J, Matsuoka T. An experimental study of endovascular stenting with special reference to the effects on the aortic vasa vasorum. Cardiovasc Intervent Radiol 1998;21:45–9.

[97] Caputo M, Chiastra C, Cianciolo C, Cutrì E, Dubini G, Gunn J, et al. Simulation of oxygen transfer in stented arteries and correlation with in-stent restenosis. Int J Numer Method Biomed Eng 2013;29:1373–87.

[98] Coppola G, Caro C. Arterial geometry, flow pattern, wall shear and mass transport: potential physiological significance. J R Soc Interface 2009;6:519–28.

Hemodynamics in the Developing Cardiovascular System

C. Poelma[a], *B.P. Hierck*[b]

[a]Delft University of Technology, Delft, The Netherlands
[b]Leiden University Medical Center, Leiden, The Netherlands

13.1 INTRODUCTION

The cardiovascular system of most organisms already starts functioning before its development is complete. As soon as the heart starts beating, blood flow is induced. This blood flow not only provides mass transfer but also has an important secondary role in the further development of the cardiovascular system: through hemodynamic forces, flow acts as an epigenetic factor that guides development, complementing the genetic regulation [1–3]. Breakthrough experiments by Hove et al. [4] showed that hemodynamic forces are essential for the morphogenesis of the zebrafish heart. Disturbing normal flow patterns can trigger anomalous cardiac development, as was shown by Hogers et al. [5] in chicken embryos. In the same model system, manipulation of flow in vascular networks has been observed to cause structural changes [6]. Similar observations are reported for mice, where hemodynamic forces were found to be essential for remodeling in the yolk sac [7].

While the fact that flow plays an important role is well established, the exact mechanisms of how flow supplements genetic control are not fully understood. In this chapter, we give an overview of the research into the role of hemodynamics in the developing cardiovascular system. The focus is on experimental work in the chicken embryo, a well-documented and accessible model system for human development. In Section 13.2, we give an introduction to this model system. Section 13.3 describes the relevant flow regimes as they occur during development, showing that the flow is dominated by viscous forces. In Section 13.4, we summarize findings from the literature and describe the various tools that are available. In the second half of this chapter, we focus on microscopic aspects of the interaction between flow and tissue. We briefly look at the mechanisms used by cells to sense hemodynamic forces (Section 13.5). At these scales, it is also essential to consider the nature of blood as a complex fluid, which is discussed in Section 13.6.

13.2 THE CHICKEN EMBRYO MODEL SYSTEM

The chicken embryo has been the focus of research for at least two millennia: from Aristotle's observations, via the seminal work of Harvey and Malpighi in the seventeenth century, it evolved into one of the most important model systems for cardiovascular studies currently in use [8]. Low cost and ease of handling are significant advantages compared to mammalian systems, while it retains, for instance, the complexity of a four-chambered heart (as in humans). The system offers excellent visualization and measurement opportunities: either the embryo can be transferred to a petri dish [9] or experiments can be done "in ovo" [10]. In the latter method, a window is created in the shell, which allows for observation and manipulation with minimal disturbance of the development. The window can be resealed and eggs can be placed back into the incubator until the next relevant developmental stage has been reached. Finally, chicken embryos are not considered to be test animals in most jurisdictions. They may generally be used without regulation during the first half of the total gestation time. Among other reasons, this is due to their inability to experience pain in that stage of development.

13.2.1 Key Stages in Cardiovascular Development

In describing the progression of embryonic development, relying on a classification based on the total time of incubation has the potential to lead to significant variation between embryos of the same "age." To overcome this, a standardized system based on anatomical features was introduced by Hamburger and Hamilton [11], which has become the *de facto* standard. We will refer to these "HH" stages throughout, with approximate times of incubation for convenience.

At stage HH12 (~45-49 h of incubation), the embryonic heart—the first organ to be formed—starts beating. Until this moment, the embryo was small enough to rely solely on diffusion for mass transport. To grow further, flow (i.e., advection) is needed for efficient transport. However, there is a short period where the circulation appears to be redundant: restricting the cardiac output from HH14 to HH25 (day 3-4) does not significantly affect short-term growth [12].

At the onset of the heartbeat, the heart is a curved tube; it is in the early stage of the process of "cardiac looping," in which the initially symmetric tube curls up [13]. This process starts at HH9 (29-33 h) and by HH18 the heart resembles an S-shaped tube. The first stages of looping occur before onset of the heartbeat and, indeed, it has been shown that this stage of looping occurs independent of flow [14]. Later stages, however, do rely on flow for proper development [5]. Note that septation, which leads to atria and ventricles, does not occur until later (completed around HH34, ~8 days). Valves also form in a later stage and originate from endocardial cushions, see also Figure 13.8.

Despite its simplicity during these early stages, the heart soon creates a unidirectional flow. A circulatory arc is present within the embryo but also extends to the yolk sac that surrounds the embryo [16]. The blood vessels that form in the yolk sac constitute the vitelline circulation, which ensures the transport of nutrients to the embryo and disposes waste products [17]. The blood vessels in the vitelline circulation originate from migrating blood islands and initially form an unstructured primary vascular plexus [18,19]. At around 33 h, this plexus is connected to the embryo (and thus the heart) via the omphalomesenteric veins [16]. From this moment on, blood also fills the intraembryonic vessels. At the onset of circulation, remodeling of the vitelline circulation starts. This remodeling occurs due to a combination of genetic and hemodynamic cues [20]. In contrast to the blood vessels inside the embryo, the extraembryonic vessels remain relatively unobscured during development. This makes the vitelline arteries and veins excellent candidates for studying vascular remodeling. In Figure 13.1, two embryos are shown after 2 and 3 days of incubation. Notice the difference in the structure of the vitelline circulation: in the younger embryo, blood leaves the embryo near the presumptive navel (denoted by the asterisk) and flows through the network of blood vessels. Eventually, it reaches the sinus terminalis (the circular vessel in the corners of the image), which collects the fluid and brings it back to the embryo; it re-enters the embryo either at the head or near the tail. In contrast, in the older embryo the vitelline circulation consists of (paired) arteries and veins: blood still leaves near the presumptive navel, but after flowing through the yolk sac it returns via a hierarchical series of veins. This hierarchy (larger arteries feeding

FIGURE 13.1 Two-day and three-day old chicken embryos. *(Left hand source image adapted from Kloosterman et al. [21]; Right hand provided by Erasmus MC.)*

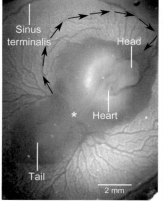
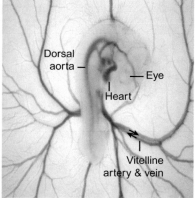

systematically smaller vessels) is also much more pronounced in the latter stage embryo compared to the more unstructured nature of the younger vitelline circulation. It is generally assumed that remodeling leads to an "efficient" network, which is able to achieve its function (transport of nutrients and waste) with minimal effort (work performed by the heart, homeostasis). The hierarchical structure of the cardiovascular system that is observed during both the embryonic and adult stages in most animals is a prime example: large blood vessels ensure efficient transport over greater distances with minimum energy loss, while small vessels are efficient for nutrients exchange due to their large cumulative exchange area. Theoretical work by Murray [22] predicts a 1:1.26 ratio in diameter at bifurcating vessels, which is indeed generally observed [21,23].

As mentioned earlier, the vitelline arteries and veins remain largely unobstructed for observation. In contrast, the dorsal aorta is already difficult to distinguish in these bright field images. An important advantage of the vitelline circulation as an observation area is that it remains relatively two-dimensional. The network of vessels floats on top of the yolk, and typical observation area dimensions are small compared to the radius of curvature (order of centimeters). However, in later stages arteries and veins may occasionally cross over or under each other, which can locally introduce three-dimensional effects.

Besides optical access, the *in ovo* model system also allows for relatively straightforward intervention possibilities. These include (local) mechanical approaches, such as ligation by means of a clip [5,6], incision [6], or suture [12]. Furthermore, chemical agents can be added relatively easily [6].

Later on during development (starting around day 4-5 of development), a third circulatory arc is formed (together with the intraembryonic and vitelline arcs) in the allantois: the chorioallantoic membrane or CAM [16]. This serves as a respiratory system for the embryo until day 19 [24]. The CAM is a heavily vascularized structure and is similarly suited for *in vivo* observation and experimentation, but it only develops relatively late during the development. This may conflict with the aforementioned ethical and/or legal regulations with respect to animal tests.

13.3 RELEVANT FLUID MECHANIC REGIMES

The embryonic circulation is a pulsatile flow of a non-Newtonian suspension in a very complex, compliant geometry. This would suggest that modeling this flow would be nearly impossible for practical purposes. Fortunately, a number of simplifications can be made that will lead to a less complex description without losing essential details. These simplifications are based on scaling arguments that consider the relevance of each of the terms (i.e., forces) in the equations describing fluid motion, the Navier-Stokes equations [25]. To evaluate the relative importance of the various forces, their influences are usually compared in the form of a relevant dimensionless number. The most well-known—and important—example of this approach is the Reynolds number:

$$Re \equiv \frac{UD\rho}{\eta} \qquad (13.1)$$

In this equation, U and D are a representative velocity and length scale of the flow; conventionally, the mean blood flow velocity and the blood vessel diameter are used. The density ρ and dynamic viscosity η describe the relevant fluid properties. The density of blood is generally assumed to be slightly higher than that of water, $\rho = 1035\,\text{kg/m}^3$. The viscosity of blood is more complicated (see Section 13.6), but in general it is assumed that it is around 0.004 Pa s [26].

13.3.1 Viscous Effects Dominate in the Developing Circulation

The Reynolds number (Equation 13.1) describes the ratio of inertial forces ($\sim \rho U^2/D$) to viscous forces ($\sim \eta U/D^2$). For low values of Re, viscous effects are dominant over inertial forces. This has a number of important consequences. First of all, disturbances in the flow are damped and the flow will remain laminar. For flows in circular geometries, the threshold is typically around $Re = 2000\text{-}2300$; for higher values, any perturbation (e.g., due to wall roughness or motion) can lead to turbulence. Although for the laminar case some exact solutions are available, the turbulent flow can only be described using empirical relations that depend on the particular conditions [27]. Alternatively, one can use numerical simulations, which generally are computationally demanding.

Laminar flow represents the conditions in which fluid is transported with the least amount of energy expenditure and thus seems the optimal regime for biological systems, generally speaking. For the steady, fully developed, pressure-driven laminar flow in a cylindrical geometry, the exact solution is the famous Hagen-Poiseuille result; extensive derivations are available in introductory fluid mechanics text books (e.g., White [25]). Table 13.1 summarizes the relevant results.

Another implication of a low Reynolds number, in particular when it is of order unity or lower, is that the flow very rapidly approaches the fully developed situation, represented by a parabolic (Poiseuille) velocity profile. Once the flow is fully developed, there is no longer any variation of velocity (and thus also WSS) in the downstream/axial direction. With a

TABLE 13.1   Overview of Useful Poiseuille Flow Relations

$u(r) = u_c \left(1 - \dfrac{4r^2}{D^2}\right)$	$u(r)$ streamwise velocity component	(m/s)
	u_c centerline velocity	(m/s)
	r radial position	(m)
	D vessel diameter	(m)
$U = u_c/2$	U mean velocity	(m/s)
$\Delta P = \dfrac{128\eta L Q}{\pi D^4}$	ΔP pressure drop	(Pa)
	η dynamic viscosity	(Pa s)
	L length of vessel segment	(m)
$Q = \dfrac{\pi}{4}D^2 U$	Q flow rate	(m^3/s)
$\tau_w = \dfrac{32\mu Q}{\pi D^3} = \dfrac{8\mu U}{D}$	τ_w wall shear stress (WSS)	(Pa)

decreasing Reynolds number, the effect of the viscous forces increases and any deviation from the parabolic profile is more rapidly negated. Such deviations can arise from, e.g., vessel bifurcations or upstream curvature of the blood vessel. The distance required to return to the fully developed parabolic profile is the so-called entrance length (or inlet length) and is directly proportional to the Reynolds number [25]:

$$L/D \approx 0.08Re + 0.7 \qquad (13.2)$$

The constant term is included to match the observations that for Re approaching zero, L/D approaches a constant value of approximately 0.7 D. Note that some sources give slightly lower values as a result of different criteria that are used to define fully developed flow (e.g., $L/D \approx 0.06Re$ in many engineering text books), but it is the order of magnitude that is relevant here.

For larger arteries in an adult human, where the Reynolds number is of order $\mathcal{O}(10^3)$ [26], this means that it takes 50-100 blood vessel diameters before the flow is fully developed. In the case of the embryonic circulation, however, typical diameters are less than 1 mm and the flow rarely exceeds a few mm/s. This means that the Reynolds number is of order unity (Equation 13.1) and that the flow will thus "develop" within one diameter (Equation 13.2). Only at the actual site of bifurcations or in strongly curved regions will the flow deviate from the parabolic flow profile; in the bulk of the embryonic circulation, the flow can be assumed to match the Poiseuille solution.

13.3.2 Curvature Effects Are Minimal, Except in the Embryonic Heart

When blood flows through a strongly curved vessel, additional forces act on the fluid and velocity profile will no longer remain axisymmetric. The effect of curvature depends on the magnitude of the centrifugal forces compared to the viscous and inertial forces. It can, therefore, be anticipated that the effect of curvature will depend on the Reynolds number and a parameter which describes the relative curvature. Indeed, curvature effects are described by the product of the relative curvature ($D/2r_c$, with r_c the radius of curvature) and the Reynolds number; this is usually referred to as the Dean number and written in the following form [28,29]:

$$De \equiv \sqrt{\frac{D}{2r_c}}Re \qquad (13.3)$$

This dimensionless number is named after W.D. Dean, whose pioneering work provided an analytical solution for the case of laminar flow through a weakly curved geometry, in other words, $D/2 \ll r_c$ [30]. The pressure gradient due to centrifugal forces distorts the initially parabolic flow, skewing it to the outer wall (see Figure 13.2, left-hand side). As fluid is pushed toward the outer wall, a secondary motion is created in the form of a counter-rotating vortex pair (Dean vortices), see the bottom inset of Figure 13.2. The theoretical framework and terminology for the weakly curved case is also used to describe cases where the radius of curvature is much smaller and Dean's theoretical model is no longer valid, such as, e.g., the aortic arch [31].

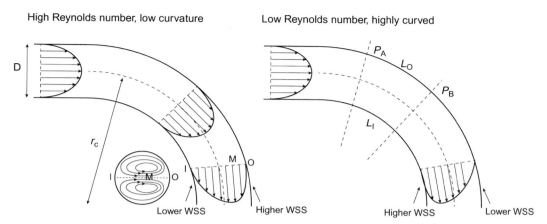

FIGURE 13.2   The influence of curvature on the velocity profile depends on the Reynolds number.

The formation of the secondary flow pattern is a result of the inertia of the fluid. The curved geometry implies that fluid elements must change direction. The centrifugal force that accomplishes this increases with the local radius of curvature ($\sim u/r^2$) so that forces that act on the fluid elements in the inner curve are larger than the forces acting on the fluid near the outer wall [32]. Flow near the outer wall, therefore, retains its original trajectory (i.e., inertia) longer than the flow near the inner wall, leading to the observed asymmetric flow profile. Both an increase in curvature (a larger value of $D/2r_c$) or an increase in the Reynolds number will lead to a more skewed velocity profile (cf. Equation 13.3).

The previous discussion raises the question what would occur in a similar flow, yet with negligible inertial forces; in other words, a flow with a Reynolds number of unity or lower. Surprisingly, the velocity profile of the flow will still be skewed, but the maximum will shift to the *inner* wall of the curve. This becomes evident when the curvature is very strong ($\frac{1}{4} < D/2r_c < \frac{1}{2}$). A schematic representation of this case is shown in Figure 13.2 (right-hand side). Viscous forces dominate and we can assume that the pressure variation in the radial direction is small compared to the pressure drop in the direction of the flow. For a very strongly curved geometry, the path length of the outer wall is much longer than the path length of the inner wall (see L_I and L_O, Figure 13.2). A fluid element near the inner wall will experience the same pressure difference ($p_A - p_B$) but only has to travel a short distance compared to elements near the outer wall. This implies that the fluid elements will travel faster near the inner curve. This phenomenon has been described theoretically by Ward-Smith [33] and has been observed in several studies [34,35].

Figure 13.3 shows this important distinction between the low- and high-Reynolds number cases. The two computational fluid dynamic studies show the flow in the same 180° curved tube with a relative curvature of $D/2r_c = 1/2$. In the top figure, the Reynolds number is chosen to be 100. This leads to a Dean number of 71. The parabolic inflow can be seen to develop into the characteristic horse-shoe shape that is caused by the strong Dean vortex pair (indicated by the vectors that show the in-plane velocity field in the outlet). In the case at the bottom, the Reynolds number is reduced to unity. This gives a Dean number of 0.71. The parabolic inlet

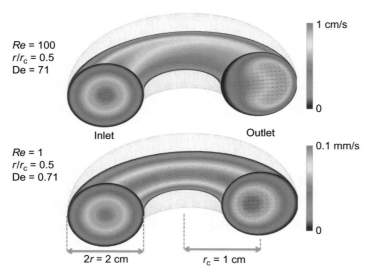

$Re = 100$
$r/r_c = 0.5$
$De = 71$

1 cm/s

0

Inlet   Outlet

$Re = 1$
$r/r_c = 0.5$
$De = 0.71$

0.1 mm/s

0

$2r = 2$ cm $r_c = 1$ cm

FIGURE 13.3   CFD simulations of the flow through a strongly curved tube. Two cases are shown: moderate Reynolds number (top, $Re=100$) and low Reynolds number (bottom, $Re=1$). The geometry is described by a relative curvature $r/r_c=1/2$. Each figure shows the inlet, outlet, and the mid-plane, with color representing the velocity magnitude.

profile gradually skews toward the inner curve by the time the flow reaches the outflow plane. The transition between these two cases is around $Re=20$ [36].

The secondary flow patterns due to curvature have three consequences: (i) the pressure drop will be (slightly) higher than the equivalent straight geometry, (ii) mixing is enhanced, as the counter-rotating vortex pair leads to spiraling streamlines, and (iii) the wall shear stress (WSS) distribution becomes nonuniform. The WSS is the normalized tangential shear force that is experienced by the vessel wall. As the cells of the vessel wall can respond to variation in the WSS, this nonuniform distribution can have a profound effect. The WSS is defined as the tangential velocity gradient at the wall multiplied by the dynamic viscosity:

$$\tau_w = \eta \frac{\partial u}{\partial n}\Big|_{wall} \tag{13.4}$$

In this definition, n represents the unit vector in the wall-normal direction; for a cylindrical geometry, this can be replaced by the radial coordinate r. For a constant viscosity, the WSS directly follows from the gradient in the velocity near the wall. As a brief example, consider the gradients of the velocity at the wall in Figure 13.2: the velocity profiles are skewed so that in the case on the left the gradient on the inner wall is lower. This also implies a lower WSS than at the outer wall. For the right-hand side case, the situation is reversed. Examples of a nonuniform WSS due to curvature are abundant in theoretical, numerical, and experimental studies, see, e.g., Siggers and Waters [29], Alastruey et al. [37], and Poelma et al. [38]. In Section 13.5, the response of endothelial cells to WSS and the consequences for development are discussed in more detail.

In the developing cardiovascular network, the curvature of the blood vessels is minimal. A recent study by Kloosterman et al. [21] showed that the tortuosity[1] is approximately 10% for

---

[1] The tortuosity is defined as the length of a vessel segment divided by the distance between the beginning and end point of the vessel segment, minus one. A tortuosity of 0 would thus be a straight line.

stages HH 13-HH 17. Combined with the low Reynolds number (as discussed in the previous section), this leads to very low Dean numbers (De ≪ 1) and secondary flows due to curvature can thus be neglected. This does *not* hold for the embryonic heart [34] and aortic arches [39]. While the Reynolds number is also relatively small, the very strong curvature will induce skewed velocity and WSS profiles.

To conclude this section on curvature effects, it should be remarked that flow separation in strongly curved geometries at large *Re* is beyond the scope of this chapter and not likely to occur in the developing cardiovascular system.

13.3.3 Pulsatile Effects Can Be Ignored in the Embryonic Phase

All of the preceding discussions tacitly assumed steady flow. In reality, blood flow is pulsatile. During development, the heart rate of a chicken embryo is not constant, it rises from 103 beats per minute bpm at HH12 to 146 bpm at HH18 and 208 bpm at HH29 (~6 days) [40,41]. This gives rise to a nonstationary driving pressure for the circulation, $\Delta P(t)$.

To assess the relevance of the pulsatile nature, we can again consider the relevance of the forces in the Navier-Stokes equations. In this case, the *transient* inertial terms are compared to the viscous forces. The transient inertial term scales as $\rho U/T$, with T the duration of a cardiac cycle. As before, the viscous terms scale as $\eta U/D^2$. The ratio of these forces is, therefore, [32]:

$$\alpha^2 = \frac{\rho D^2}{\eta T} \rightarrow \text{Wo} = D\sqrt{\frac{2\pi f}{\nu}} \tag{13.5}$$

The most common form is obtained by taking the square root of this dimensionless number, which is usually referred to as the Womersley number, named after the British mathematician.[2] Also introduced in this definition are the frequency $f \equiv 1/T$ and the kinematic viscosity $\nu = \eta/\rho$. Notice that the velocity, U, cancels out in the ratio. Also, assuming that the viscosity of blood and the heart beat frequency is fairly constant, it can be predicted that the Womersley number will generally be dependent on the diameter D.

For very large Womersley numbers (Wo → ∞), viscous terms can be neglected completely. Solving the equations of motion will lead to a uniform velocity distribution (i.e., a flat velocity profile, "plug flow") that oscillates in response to the driving pressure $\Delta P(t)$. This regime is expected, for instance, in the adult human aorta, where the Womersley number can reach a value of 20 [32].

Alternatively, for very low Womersley numbers (Wo ≪ 1), the viscous forces dominate. Solving the equations of motion leads to a parabolic velocity profile, with a maximum value modulated by the driving pressure $\Delta P(t)$. In this case, the maximum of the velocity profile (u_c) is in phase with the driving pressure (i.e., $u_c \sim \Delta P(t)$). This is not the case in the inertial regime, where a phase lag of up to 90° will occur.

---

[2] The squared value (α^2) is sometimes referred to as the Stokes number (Fung 1997), as it arises in his analytical solution of the flow field near an oscillating plate [126]; obviously this problem is very closely related to pulsatile pipe flow.

In the intermediate range (Wo $\sim$ 1 - 10), the flow is a combination of the two limiting cases. For a fully developed, straight tube, an analytical solution[3] was presented by Womersley [42]. These "Womersley profiles" are characterized by a flattened profile compared to the viscous, parabolic case. This situation is relevant for the larger arteries in human adults, for example. In early blood flow measurements in small vessels, blunted velocity profiles were reported despite being in the viscous range [43]. This has been clarified as being a measurement artifact [44] and should not be confused with inertial effects.

We can now estimate in which regime the flow in the developing (chicken) cardiovascular system will be by evaluating the Womersley number. Using the typical heart rates mentioned at the beginning of this section, we choose a typical frequency of 2 Hz for the first week of development. With a kinematic viscosity $\nu = \eta/\rho \approx 4 \times 10^{-6}\,\mathrm{m^2/s}$, we then find Wo $\approx 1800\,\mathrm{D}$. In other words, blood vessels with a diameter less than 0.5 mm will be in a regime where viscosity is dominant over inertial forces (Wo $<$ 1). This is expected to hold for all blood vessels, as well as the heart, for at least the first week of development (see, e.g., the dimensions of the aorta as reported by Hu and Clark [41]). The low Womersley number thus indicates that the velocity profiles in the developing cardiovascular system are (nearly) parabolic. This has indeed been confirmed in several studies [45–47].

13.3.4 Local Flow Can Be Described Using Two Parameters Only

As shown in the previous sections, the flow in the embryonic circulation is dominated by viscous effects. This means that effects due to curvature, flow separation/turbulence, and pulsatility—which are all due to inertia—can safely be ignored. This is illustrated in Figure 13.4. The left-hand side of this figure shows a snapshot at systole of the velocity field in a bifurcation of a chicken vitelline artery [47]. The figure actually shows two separate measurements superimposed, one of the plasma velocity and one of red blood cells; the purpose of this experiment was mainly to show that these are identical in these conditions. The top right figure shows the velocity profile along location A as a function of the phase within the cardiac cycle Φ (with $\Phi = 0$ corresponding to peak systole). As can be seen in this figure, the velocity profile is parabolic throughout the cycle. Notice that there is a strong variation of the centerline velocity and even some retrograde flow (e.g., at $\Phi = 0.6$, see also the bottom figure), but no flattening due to pulsatile effects can be observed. The Womersley number is approximately 0.27 in this flow. As the blood vessel bifurcates into two smaller vessels, the flow profile develops nearly instantaneously: at locations B and C, the velocity is again nearly parabolic. This directly follows from the fact that the Reynolds number is very low, approximately 0.07 based on the peak velocity. Any changes in the profile are due to variations in the local geometry and not due to inertia.

Based on the scaling arguments—and the experimental evidence—it is possible to characterize the local flow in a blood vessel with a strongly reduced number of parameters. In fact, just two parameters are sufficient to describe the parabolic flow: a diameter and the centerline

---

[3] Note that the analytical solution assumes a sinusoidal variation of the driving pressure. As long as the flow is fully developed, the governing equations are linear and we can make use of the superposition principle; realistic physiological driving pressures can be constructed by a Fourier series.

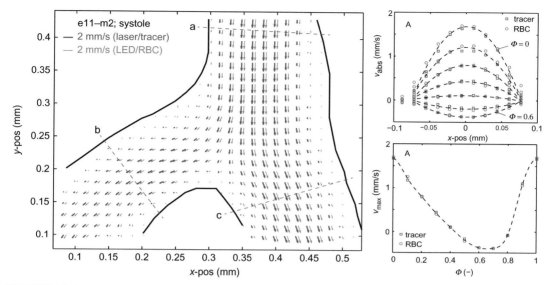

FIGURE 13.4   The flow in the vitelline circulation is dominated by viscosity and effects due to curvature and pulsatility can be ignored. (Left) Snapshot of the velocity field. (Top right) Velocity profile at location A during the cardiac cycle. (Bottom right) Centerline velocity at location A during the cardiac cycle. (*Reproduced from Poelma et al. [47].*)

velocity. All other parameters (such as WSS, flow rate, or mean velocity) can be derived from these two. This implicitly assumes that the diameter remains constant throughout the cardiac cycle. Poelma et al. [48] report a variation of an vitelline artery of approximately 1 μm, similar to the accuracy of the measurement. Note that the proximal arteries experience the largest pressure variations [32]. The diameters were determined by extrapolating the velocity profiles to zero (cf. the dashed lines in the top right of Figure 13.4). In the embryonic heart, the diameter variations are naturally larger, as described by Männer et al. [49], Poelma et al. [38], and Lee et al. [50]. Typical ranges of the outflow tract diameter of 100-250 μm are reported in the latter two studies.

The fact that blood vessel segments can be described using only two parameters dramatically reduces the amount of data needed to describe large vascular networks [21]. Very large vector fields describing local velocities (e.g., Figure 13.9, bottom) are much more difficult to interpret than, e.g., graph representations, which can highlight network topology and functioning [51]. An example of such a graph representation is given in Figure 13.5, taken from Kloosterman et al. [21]. Reducing the local hemodynamics of each certain blood vessel can be done without significant loss of detail. Changes in hemodynamics can then be studied in a much more efficient way, by analyzing the changes in, e.g., diameter distribution and number of vessels of large sample sizes.

Note that the simplification of describing blood flow with only two parameters is not suitable for parts of the cardiovascular system that are strongly curved (e.g., heart, aortic arch). Here, such a description will lead to the loss of essential information. However, it is still possible to reduce the complexity of such data. An example is a parameterization of the flow, in

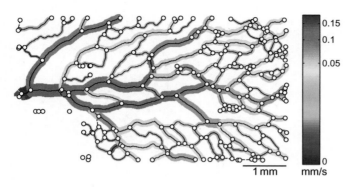

0.15

0.1

0.05

0

1 mm   mm/s

FIGURE 13.5   An example of a graph representation of a vascular network. Each blood vessel segment is characterized by a diameter (line thickness) and mean velocity (line color). Branch points are indicated by open circles. Note that the color scaling is nonlinear. *(Reproduced from Kloosterman et al. [21].)*

other words, to describe it as a function of the position along the centerline of the vessel. This can be done with a two-dimensional polynomial that describes the velocity field in the plane perpendicular to the local direction of the centerline. Depending on the required complexity, an appropriate function can be chosen that captures, e.g., the WSS distribution with sufficient accuracy. An example of this approach is given in Poelma et al. [52].

In conclusion, we can state that the circulation during embryonic development is fortunately much simpler than that of, e.g., the adult arterial system. Effects due to curvature, pulsatility, wall compliance, and turbulence are largely absent. This means that flow can be described by the local diameter and mean centerline velocity, and all other parameters can be derived. However, for the heart and the aortic arch a more comprehensive description is needed.

13.4 EXPERIMENTAL STUDIES

Experimental work using the chicken embryo—and other model systems—has seen a shift from qualitative descriptions to very detailed quantitative measurements of multiple hemodynamic parameters. In this section, we briefly introduce some techniques that can be used to study hemodynamics during embryonic development. As an example of the very rapid technical progress in the last decade, Figure 13.6 shows two similar studies of the flow patterns in the embryonic heart before and after ligation of an vitelline artery. On the left-hand side, the drawings show qualitative results based on observation of streamlines in the flow, obtained by injected dye [5]. On the right-hand side, quantitative results based on microscopic particle image velocimetry are shown: the velocity field is represented by the vectors, the wall is color coded with WSS [54].

The level of detail needed obviously depends on the biological questions that are addressed in the study. It can range from straightforward observation of the topology of the vasculature and general flow directions [6] or dye visualizations [5] to very complex instrumentation for combined flow and functional imaging data [55]. A compromise usually has to be made for the spatial resolution, with measurements spanning either entire networks [21,56] or detailing cell-level phenomena [57,58]. It is assumed here that the rheological behavior is known (see Section 13.6) so that the hemodynamics can be described in terms

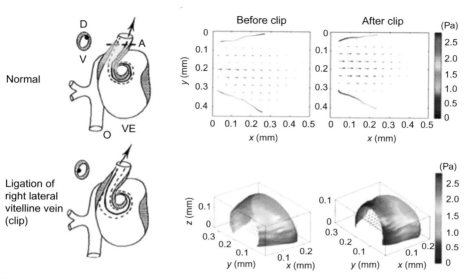

FIGURE 13.6   (Left) Dye visualizations showing the effect of clipping the vitelline artery on the flow pattern in the embryonic heart. (Right) PIV results showing change in flow patterns in the outflow tract due to clip. *(Left: Images adapted from Hogers et al. [5]. Right: Images adapted from Egorova et al. [53,54].)*

of pressure and velocities. Before discussing measurement of these parameters, it is useful to consider the influence of temperature. This influence serves as an example of how many important parameters that serve as boundary conditions for the flow are linked.

13.4.1 Influence of Temperature

The temperature of the model system is usually carefully kept at appropriate values (37 °C), unless temperature effects are the specific purpose of the study [50]. Increasing the temperature has a twofold effect[4]: the heart rate increases and the cardiac output changes. The heart rate increase is approximately linear, from 2.25 Hz at 35 °C to 3.5 Hz at 41 °C for embryos at HH18 [59]; normal values are in the range of 2.4-2.7 Hz or 144-162 bpm at 37 °C during that stage [41,59].

Besides this increase in heart beat rate, the "overall performance" of the heart also changes. The latter, generally expressed as the cardiac output, is much less straightforward to measure. The cardiac output is the product of heart beat rate and the stroke volume. The stroke volume (the amount of blood pumped from the heart into the arterial system in one cardiac cycle) follows from integration (over one cycle) of the flow through a certain cross section. Both flow and cross section change dramatically during the cycle, as is demonstrated in Figure 13.7. This figure, taken from Lee et al. [50], shows the centerline velocity (panel a) and inner diameter (b) of the outflow tract, the final segment of the embryonic heart. The values for the inner

---

[4] Note that increasing or decreasing the temperature also has an important effect through the viscosity of blood, see Section 13.6.

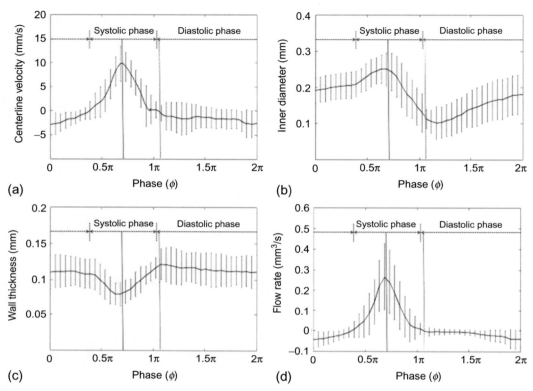

FIGURE 13.7 Cyclic variations of the centerline velocity (a), inner diameter (b), wall thickness of the OFT (c), and flow rate (d). The solid and dotted vertical lines represent the phase angle for the peak velocity and the appearance time of backflow, respectively. Each value is represented in mean SD ($n=8$). *(Reproduced from Lee et al. [50].)*

diameter (and wall thickness) have here been determined from high-speed videomicroscopy data. Using this figure, an important issue in hemodynamic studies can also be illustrated: when comparing embryos, variations exist that either occur naturally (e.g., variation within the control group) or as a result of an external agent (e.g., due to ligation of an artery). These variations can range from changes in heart rate, peak/mean velocity, duration of the systolic versus diastolic phases, and so on. Furthermore, the flow can change from unidirectional to bidirectional: measurements in hypothermic chicken embryos showed retrograde flow during the diastolic phase [38,50]. These changes can be inherent to the heart itself but can also be due to the resistance of the vascular system. Changes in local hemodynamic parameters, e.g., the shear stress levels in the vitelline circulation, thus have to be interpreted with great care, as they can easily be overshadowed by variations in the heart function. A possible solution can be to normalize all results so that relative changes can be observed [60]. Nevertheless, it can often be difficult to determine the exact underlying causes of changes in the local peak value of the WSS: these changes can be local (due to changes in local blood vessel diameter) or due to a change in the heart rate, stroke volume, distal topology, and so on.

13.4.2 Pressure

The local pressure can only be measured invasively, by puncturing a blood vessel with a thin probe (typically 5-7 μm in diameter). Due to the small observation volumes, such pressure measurements are generally done using servo-null micropipette systems [41,61]. In these systems, the pipette is filled with a conducting fluid. Due to the higher pressure inside the vessel that is investigated, blood attempts to enter the pipette and thus displace the conducting fluid. Changes in the conductance are registered and a counterpressure is applied to maintain a balance at the tip between blood and saline solution. This way, pressure can be measured with minimal fluid displacement. Measurements in the contracting heart itself are unfeasible, so usually the pressure in the heart is determined by measuring in a proximal artery [62]. Typical shear stresses in the cardiovascular system are of the order of a few Pa. This means that pressure *drop* measurements over individual vessel segments are very challenging but are possible: Lipowsky et al. [63] report pressure drop measurements in the cat mesentery vasculature using a dual servo-null system with an accuracy of approximately 2 Pa.

13.4.3 Cardiac Output

The cardiac output serves, as mentioned earlier, as the boundary condition for the flow in the remainder of the cardiovascular system. Unfortunately, it is currently very difficult to measure the complete time-dependent velocity field in the entire cross section (and thus cardiac outflow). The first reliable routine measurements have been done using Doppler ultrasound velocimetry, which can provide one velocity component along a profile. Velocities are determined from the Doppler shift in the sound scattered by moving red blood cells. For measurements in the embryonic circulation, relatively high ultrasound frequencies (20-55 MHz) are used [15,41]. This provides a measurement resolution of the order of tens of microns, sufficient for measurement of the heart and larger blood vessels. The limited penetration depth associated with these high frequencies (typically only a few mm) is not an issue given the dimensions of the embryo. The Doppler velocity measurements are generally combined with M-mode imaging (a method that reads out a single line of the image at high frame rate) to determine the variation in diameter. An example of both velocity and diameter measurement using M-mode is shown in Figure 13.8. This figure, adapted from Oosterbaan et al. [15], shows the velocity in the outflow tract of a chicken embryo at HH18; these velocities are the projected velocities along the insonification angle. The bottom figure shows the M-mode data, a single line oriented as shown in the right-hand figure (by approximation). The horizontal axis represents time (as with the velocity measurements in the top figure). The variability in the diameter of the outflow tract (denoted by "B" in the bottom half) and atrioventricular region (above the OFT) can be observed. The blue- and green-shaded areas indicate closure of the atrioventricular and outflow cushions, respectively.

From such data, the instantaneous flow rate has to be estimated using a number of approximations, e.g., that the OFT is axisymmetric and that the velocities are a good representation (e.g., centerline or mean) of the streamwise velocity profile. The latter also means that data are assumed to be obtained exactly at the mid-plane of the heart or OFT. Similar assumptions

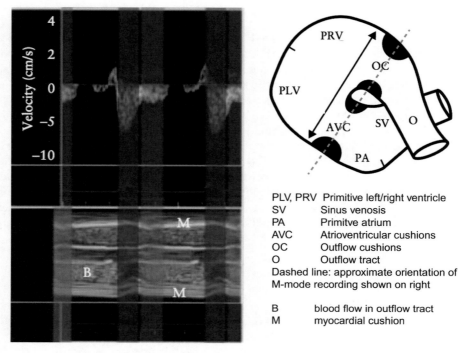

FIGURE 13.8   Doppler flow measurements in the outflow tract of an HH18 chicken embryo. (Top left) Velocity; (Bottom right) M-mode recording showing the variation of the OFT diameter. The blue and green shaded areas indicate closure of the AVC and outflow cushions, respectively; (Right) Overview of the embryonic heart and terminology and approximate location of the M-mode sample line. *(Adapted from Oosterbaan et al. [15].)*

have been made to determine the flow rate from velocities and diameter reported in Figure 13.7 (panel D).

13.4.4 Flow Rate and Velocities in Blood Vessels

Although Doppler ultrasound is viable for relatively large vessels, the resolution is inherently limited by the sound frequency and is not sufficient for measurement in the majority of the blood vessels in the embryonic circulation. By switching to visible light (in particular the monochromatic, coherent light of a laser), the resolution can be improved drastically. However, the Doppler shift will be difficult to detect, due to the orders of magnitude in difference between the moving red blood cells and the speed of light. This necessitates the use of a second signal (either unaffected by the flow, but generally at a different angle with respect to the flow direction), which is mixed with the first signal. From this heterodyned signal—orders of magnitude lower in frequency—the Doppler shift and thus local blood velocity can be determined. This technique is called laser-Doppler velocimetry (or anemometry; LDV, LDA). The intersection of the two laser beams forms a single measurement volume, in contrast to ultrasound methods. In the latter, a (near-)instantaneous *profile* can be reconstructed by making use of the temporal delay between transmitted and received pulses. To obtain a velocity profile using LDV, the measurement probe has to be translated accurately, collecting data point by point. This means that only time-averaged velocity profiles can be obtained. Riva et al. [64] reported the first LDV-based blood velocity and flow measurements, of flow in retinal arteries. The method is still used occasionally and extended [65], but it has largely been replaced by recent whole-field measurements.

A second important technique to measure mean velocities and flow rate in blood vessels is two-slit (or "dual slit") photometry [66]. This technique is based on the transmission of light through a blood vessel. Due to local variations in concentration and orientation of red blood cells, the light transmitted constantly fluctuates as blood passes. A second detector—placed slightly downstream—will observe approximately the same signal, but with a small time delay. By correlating the signals from the two detectors, this time delay can be obtained. The time delay can then be converted to a velocity assuming that the distance between the two detectors is known. While the technique is straightforward and relatively cheap, it has a number of drawbacks: it is invasive, as light source and detector have to be placed close to the blood vessel. The blood vessel diameters have to be determined separately before flow rates can be estimated. Furthermore, the velocity is an average over all radial positions, which means that a correction is required [44]. Despite these drawbacks, the method still finds a lot of use in physiological studies and its accuracy is constantly being improved [67].

13.4.5 Whole-Field Measurements: Micro-PIV

In the last decade, there has been a transition from single-point measurements to whole-field techniques. An extensive overview of these techniques is given in Vennemann et al. [36]. In particular, microscopic particle image velocimetry (micro-PIV) has proved to be a very powerful technique [68]. Unlike LDA or Doppler, it can provide instantaneous information of two velocity components in a plane. This means that it can provide all information that is required to capture the hemodynamics of a two-dimensional vascular network, such as the vitelline circulation. Furthermore, it removes the uncertainty in the angle between the sound propagation direction and the mean flow direction. The latter can be difficult to determine in *in vivo* experiments [15]. Due to the two-dimensional nature of the data, the WSS distribution can be determined by evaluating Equation (13.4). Both the flow gradient and wall location can be determined, the latter by finding locations where the flow approaches zero [38].

Micro-PIV can be seen as a natural extension of videomicroscopy: flows are investigated by imaging the displacement of elements (referred to as "tracers") carried by the flow. Initial applications relied on the tracking of individual objects, such as red blood cells [69]. This approach, generally, only works with high magnifications in capillaries, where individual cells can be identified and tracked. For lower magnifications (required to capture larger diameters and complete networks), it is no longer possible to unambiguously identify tracers. In this case, the common approach is to rely on a statistical method, in particular by using cross-correlation. In Figure 13.9, taken from Kloosterman et al. [21], an example is given of flow measurement using micro-PIV. Image pairs are recorded using specialized CCD cameras (top). Each image is then divided into smaller areas called interrogation areas, which are typically 16×16 to 64×64 pixels. These areas are cross-correlated with the equivalent areas in the

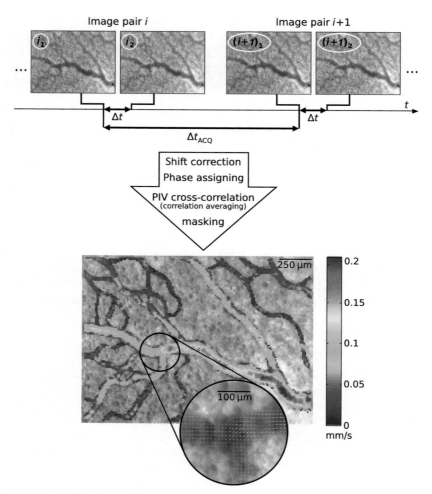

FIGURE 13.9   Micro-PIV in the vitelline circulation of an embryonic chicken using red blood cells as tracers. *(Reproduced from Kloosterman et al. [21].)*

second frame of the image pair. This yields estimates for the mean displacement in each interrogation area, which is converted to velocities making use of the temporal separation between the frames of each pair (Δt) and a scaling factor based on the magnification of the microscope. The bottom part of the figure shows a typical result: an instantaneous velocity vector field with a resolution determined by the size of the interrogation areas. The aforementioned use of special cameras to obtain image *pairs* is necessary to limit the displacement of tracers between subsequent frames. Very generally speaking, their displacement should be less than a quarter of the interrogation area size. This requires, depending on the magnification, temporal separations (Δt) of maximally hundreds of microseconds to milliseconds. This is beyond the frame rates of most conventional cameras ($1/\Delta t_{\mathrm{ACQ}}$ in Figure 13.9), which are generally limited to 10-30 frames per second (i.e., $\Delta t_{ACQ} \approx 0.033-0.1\,s$). A way to circumvent this is to use high-speed cameras [23]. This has the drawback that the noise levels are generally somewhat higher and that measurement sequences are limited to several seconds (i.e., a limited number of heart beats).

Before the velocities can be obtained from the data, a number of preprocessing steps are generally performed to optimize the performance of the PIV correlation algorithms. First, images are corrected for drift during the measurements. This process, "image registration," can be automated by evaluating the motion of tissue without blood motion: similar routines as used for PIV are applied to the region with distinctive features. This provides the shift with respect to a reference image (e.g., the first image in the series), which can be used to correct the drift. Once this process is completed, image processing can be used to remove imaging artifacts. In particular, debris or small air bubbles that might be present in between the microscope objective and measurement plane have to be removed. Without their removal, bright or dark spots that are persistently present in the data will lead to strong underestimation of the local blood velocity. This can be reduced significantly by subtracting the time-averaged image from the images in the data series. Note that this will only be effective if the mean intensity of all images is constant, something that can also readily be fixed by image preprocessing.

With the data preprocessed, one has to decide how the data are to be analyzed. If only the mean velocity fields are of interest, one can process the entire data series as a whole. Note that averaging of time-dependent data using PIV has to be done with care to avoid erroneous results [70]. If time-dependent velocities are required, either individual image pairs can be analyzed [46] or a form of phase-averaging can be applied. The latter results in considerably better signal-to-noise ratios [34] but requires information about the phase associated with each image pair. This information can be obtained by synchronizing the PIV data acquisition with an external signal, e.g., from an ultrasound probe aimed at the embryonic heart [34]. Alternatively, it can be obtained from the PIV data series itself by a technique called "self-gating," provided that the cardiac cycle is sampled with a sufficiently high resolution [48]. In this method, the data are first processed using an initial pass to determine the mean flow in the entire field of view. This provides a signal which can be used to assign a phase to the individual frame so that averaging of data within each group of appropriately sorted images can be performed in the actual PIV analysis.

PIV studies that are reported either use tracers that are naturally present (i.e., endogenous elements), in particular red blood cells, or use tracers that are artificial (exogenous). The endogenous approach is truly noninvasive, while the use of artificial tracers can provide superior resolution, well below the size of individual red blood cells [57,58]. Additionally, they are usually chosen to contain a fluorescent dye so that the signal-to-noise ratios can be improved significantly, as scattering by tissue can be removed by using appropriate optical filters. Some studies combine the benefits of both tracer types, for instance by using fluorescently labeled platelets [71], but this requires much more elaborate experimental protocols.

Artificial tracers have to be selected carefully in order to avoid influencing the model system. Fluorescently labeled liposomes have been used [34], but the majority of experiments use polymeric particles coated with polyethylene glycol. The latter drastically improves biocompatibility and avoids the particles sticking to the wall. Typically, particles in the range of 0.5-1 μm are used. Recent work has shown that for low-magnification experiments, the use of endogenous and exogenous tracer particles gives nearly identical results [47]. For higher-magnification experiments, the results will be different. This can largely be explained by the tracer size and the associated optical characteristics [47].

For most vascular networks, a single measurement plane is sufficient: the "thickness" of the measurement volume will be large compared to the diameters of the blood vessels in the two-dimensional network. The velocity fields that are obtained can be directly translated to mean or centerline velocities using a correction factor [72]. This correction factor is based on the nonlinear averaging along the optical axis and is dependent on the imaging characteristics of the microscope objective that is used. If three-dimensional flows are investigated, the entire volume can be reconstructed using a z-stack: data are recorded at a series of locations—usually using a computer-controlled translation stage—and the time-averaged three-dimensional velocity field can be reconstructed [38]. In these cases, the third velocity component can also be estimated, using the fact that the flow field is divergence free.

13.4.6 Scanning Imaging Methods

A drawback of the PIV measurements described in the previous section is that they rely on wide-field illumination. The entire sample is illuminated and light is scattered in the focal plane and everywhere outside. The use of fluorescent tracer particles improves this some-what, by limiting the signal to the flow domain. In some cases, the vascular system that is investigated may extend beyond the focal plane, so fluorescent light from out-of-focus objects will still reduce the image quality.

The contrast and resolution of microscopy images can significantly be increased by making use of confocal imaging [73]. In this method, the use of two pinholes in the optical path ensures that only the signal from one spot on the focal plane can reach the detector (or camera). While this eliminates background noise, it has two drawbacks: (1) The use of small apertures (pin-holes) reduces the amount of light that is collected, requiring long exposure times and (2) only a single point is sampled at a time. To reconstruct an entire image, the focal plane has to be scanned point by point, with a minimum exposure time. Scanning is generally done using either a spinning (Nipkow) disk or by actuated mirrors manipulating a laser beam (confocal laser scanning microscopy). Both methods require considerably more time to reconstruct an image than conventional wide-field imaging so that the effective imaging rate is much lower than using normal imaging. This can conflict with the limitations imposed by the maximum tracer displacement discussed before, as the interframe time Δt is relatively large. A compromise can be made by reducing the field-of-view. As an example of typical values, Patrick et al. [57] report frame rates of up to 100 Hz (i.e., $\Delta t = 10$ ms) for an image size of 512×128 pixels. The magnification that is used will restrict the maximum velocity that can be used with these frame rates. Nevertheless, an important advantage of this method is that it can determine *instantaneous* velocity fields, for instance around interacting red blood cells [57].

An alternative approach using confocal imaging is to forgo imaging of an entire image at once. Jones et al. [74] describes a method where the flow in blood vessels in mice is obtained by reading out a single line, chosen *perpendicular* to the flow direction. The data are collected and rearranged into a two-dimensional image, with one axis representing time and the other the position along the scan line. The velocity of tracers—in this case fluorescently labeled red blood cells—can then be determined by observing the shape of the tracers in the image: tracers with a high velocity in the mean flow direction will only appear in a few subsequent scanned lines. Tracers with a radial velocity component will appear as sheared objects. An alternative approach is based on scanning along the *centerline* of a blood vessel [56]. The displacement of tracers in subsequent scanned lines is proportional to the velocity—this method could be considered to be a one-dimensional PIV technique, with the interrogation area aligned with the blood vessel centerline. Both approaches, perpendicular or centerline scanning, have to be repeated for each vessel within a field-of-view, by manually positioning the optimal scan line. This means that generally only mean flow rates within a field-of-view can be determined, unlike whole-field methods. Recently, Landolt et al. [75] introduced a method where the image is not sampled along a line, but in a double-circular trajectory. This allowed them to determine velocity profiles simultaneously in the three branches of a bifurcation.

An important advantage of line-scanning is that more information can be extracted than just the velocity. As individual red blood cells are detected, it can provide local red blood cell density (i.e., hematocrit values) [56,74]. Being noninvasive, this approach is naturally preferable over conventional approaches that rely on taking blood samples [76]. In the study by Kamoun et al. [56], the authors were able to distinguish between vessels in normal and tumor regions in mice by analyzing flow, diameter, and hematocrit levels.

To be able to measure deep inside tissue, studies often utilized two-photon excitation microscopy [56,75]. In this technique, near-infrared is used to minimize scattering by tissue. The required absorption of two (or more) photons by fluorophores only occurs near the focal point of the laser, reducing out-of-focus contributions. This increases the image quality and penetration depth significantly. An alternative technique for visualizations within tissue that has recently been introduced is optical coherence tomography (OCT) [77]. This interferometric technique works similar to ultrasound imaging, but local regions of tissue are now sampled with a laser beam instead of an ultrasound pulse. Reflections of the laser light occur at all depths within the tissue, but through the use of a second, reference beam the system can isolate the reflections originating from a certain depth. One scan provides a single line measurement along the depth direction, so to construct a two- or three-dimensional image a point-by-point scan is again needed. Recent studies have introduced very fast OCT systems, which were, for instance, capable of obtaining four-dimensional reconstructions of the beating avian heart [78]. Recently, the technique has been extended to also obtain flow information by evaluating the Doppler shift of the reflected light (similar to LDA and Doppler ultrasound). Examples include flow measurements in the extraembryonic blood vessels [79], flow measurements in the outflow tract at HH18 [80,81], and even the complete WSS distribution in the heart [82]. While still in relatively early stages, OCT-based flow measurements can bridge the gap between conventional wide-field methods (micro-PIV) and ultrasound-based measurements. The former suffers from lack of optical access so that measurements are limited to early stages of development. Ultrasound, on the other hand, is inherently limited in spatial resolution so that detailed measurements are generally possible in later stages. OCT does not suffer from these limitations, but a drawback is the more complex equipment and processing.

As will be clear from this section, measurement techniques are still rapidly evolving. Most of the techniques presented here only focus on velocity fields (and thus also providing flow rates, topology, and wall shear stresses). While this is suitable to describe the hemodynamics of the circulation, it is not sufficient to study its physiological functioning. Recent developments have combined flow measurements with additional modalities, such as oxygen concentration measurements. An example is shown in Figure 13.10. This result was obtained by imaging the vasculature with two wavelengths. One of the wavelengths has different absorption coefficients for oxyhemoglobin and deoxyhemoglobin so that the local oxygen concentration in blood can be determined. This example of combined velocity and concentration measurements is just one of the many very promising possibilities that are now becoming feasible.

13.5 MECHANOTRANSDUCTION

13.5.1 Mechanotransduction in Cardiovascular Development

Mechanotransduction refers to the combined mechanisms by which cells convert mechanical stimuli to chemical signals. Biomechanical forces are translated into biological responses in a tissue-specific manner, so cells must have unique adaptations. Cellular responses can be as diverse as activation of receptor molecules, protein phosphorylation, transient changes in the localization or concentration of ions like Ca^{2+}, and activation of gene expression. In the cardiovascular system, pressure and flow forces are continuously present and are instrumental for proper cellular function as well as for transport throughout the body. For example, flow-induced vasodilation [83] is a mechanism by which increased blood flow is translated by the endothelial cells into production of nitric oxide which causes local relaxation of the underlying circular smooth muscle layer of that particular blood vessel.

The importance of mechanotransduction during cardiovascular development is illustrated by a number of experiments in the late 1990s in which experimentally altered blood flow through the embryonic chicken heart resulted in the development of cardiovascular anomalies [3,5,84,85]. This is especially interesting since these anomalies were reminiscent of prevalent congenital vascular and heart malformations in the human population, like ventricular septal defects and interruptions of the aorta. Hogers et al. [84] showed that permanent ligation of the right lateral vitelline vein (venous clip model), which carries the blood from the right part of the yolk sac vasculature into the embryo, temporarily changed the patterning of blood flow inside the heart and resulted in the development of structural cardiac and vascular malformations. Subsequent studies demonstrated that this transient change in flow patterning led to local changes in gene expression of flow-responsive genes in the embryonic endothelium/endocardium of the heart [86–88], and to changes in cardiac performance [89]. Interestingly, one of these genes was endothelial nitric oxide synthase (eNOS/NOS3), which encodes the enzyme that converts L-arginine into citrulline and nitric oxide [88]. eNOS expression in endothelial cells is partially controlled by the transcription factor Krüppel-like factor-2 (KLF2), which turned out to be mastering the larger part of the flow response in endothelial cells [90]. Recently, the mechanism by which KLF2 is activated by shear stress was further elucidated by Egorova et al. [54], who showed that activation of signaling through the transforming growth factor-β (TGFβ) pathway is an essential limiting step in this process.

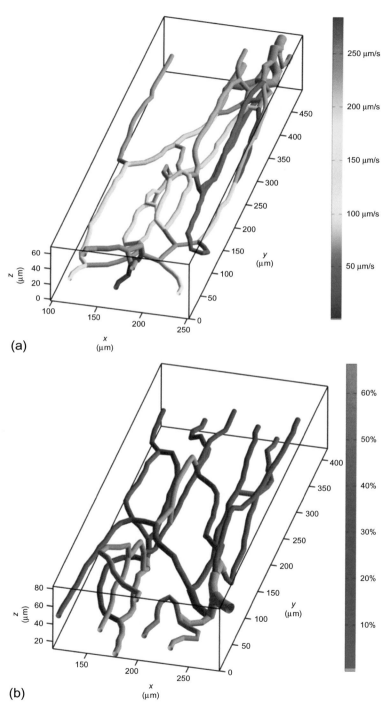

FIGURE 13.10 Color maps of mean cell velocity for network E (a) and mean red blood cell O_2 saturation for network F (b) measured *in vivo* for individual vessels (gray indicates that no measurement was made). *(Reproduced from Fraser et al. [55].)*

The mechanism by which TGFβ is activated by flow, however, is still unclear. Recent data by Clement et al. [91] suggest that endocytosis from specialized membrane regions, in this case the ciliary pocket region, is involved in the activation of TGFβ by shear forces.

13.5.2 Primary Cilia as Mechanosensors

In the last decades, various cellular components or compartments have been proposed as shear sensors. Much of this data are empirical and based on gene knockout or mutation studies which show a loss of functional response to shear stress. Many (membrane) molecules like cell-cell adhesion molecules (e.g., CD31, VE-Cadherin), cell-matrix adhesion molecules (e.g., various integrins), surface molecules (e.g., glycocalyx), receptor molecules (e.g., tyrosine kinase receptors), channels (e.g., TRP channel proteins, Connexins), and structural molecules (e.g., microfilaments, microtubules) have been identified this way (see, e.g., the reviews of Van der Heiden et al. [92] and Egorova et al. [93]). In addition, structural adaptations, like caveolae and primary cilia, have been suggested as shear sensors. With respect to the latter category, it is still unclear whether the geometry of these structures contributes to aid the mechanical strain or that these structures form "niches" in which the aforementioned proteins cluster to enhance functional impact.

Many cell types possess primary cilia which have been attributed various, often sensory, functions. Stress and stretch are among the mechanical sensory functions, but primary cilia have also been described to sense chemical properties (light, odor, growth hormones) of the cellular environment [94]. They serve a sensor function as well because their form resembles an antenna protruding from the cell surface. A primary cilium consists of a core of nine microtubule bundles which is covered by the membrane of the cell. Primary cilia differ from "regular" cilia (found in, for example, the airways and oviducts), in their lack of active moving or beating capacities. Cells can also only carry a single primary cilium [92,93]. The active moving capacity of "regular" cilia is based on the presence of an extra central pair of microtubules, radial spokes, and dynein molecules, which facilitate a rhythmic beating motion similar to that found in flagella in sperm cells. These moving cilia are, therefore, often referred to as $9+2$ cilia, in contrast to $9+0$ primary cilia. To date, endothelial cells have only been reported to carry primary cilia and it is, therefore, generally accepted that cilia do not play a role in the propulsion of the blood in the vasculature. Endothelial primary cilia are between 1 and 5 μm in length and 200 nm in diameter and are physically connected to the microtubules which comprise part of the skeleton of the cell, the cytoskeleton [92,95]. Other cytoskeletal elements are actin microfilaments and intermediate filaments. Microtubules are hollow cylinders made up of polymerized α- and β-tubulin dimers which grow from a nucleation center, the microtubule organizing center (MTOC), to the periphery of the cell where they bind to other proteins and cytoskeletal elements through linker proteins. In eukaryotic cells, the MTOC is called the centrosome and consists of two centrioles. One of the two centrioles is the mother centriole or basal body and forms the basis for the primary cilium in the G0 phase of the cell cycle. During cell division, both centrioles are involved in the mitotic spindle to pull apart the chromosomes. Therefore, dividing cells do not carry primary cilia. This is an interesting fact because we have shown that, although ciliated endothelial cells are more sensitive for shear stress, nonciliated cells do not completely lose their capacity to respond to this drag force (see more following). Endothelial primary cilia deflect in response to fluid flow [93,96], and this deflection results in a transient increase of intracellular calcium, and subsequent production of, e.g., nitric oxide [97–99]. Non-ciliated endothelial cells have lost this mechanism. One could argue that the primary cilium acts as a lever to increase (cytoskeletal) strain and thereby enhances flow sensitivity. In fact, increased strain in endothelial cells can lead to activation of cell-cell adhesion molecules (like CD31 and VE-Cadherin) and receptor molecules (like VEGFR2) and cellular "tensegrity" focuses strain to various hotspots in the cells, like the focal adhesions to which integrins are localized [92,93,100,101]. Endothelial cells in which the cytoskeletal microtubules were experimentally disrupted, while maintaining the primary cilium, showed a decrease in the response to fluid flow [3]. Modeling studies showed that there is a clear relation between the length of the cilium and the drag force and torque which is experienced by the cilium and the cell (reviewed in [102]). This is in line with the fact that cells can actively modify the length of the cilium by adjusting their cyclic adenosine monophosphate metabolism [103,104]. On the other hand, we have shown that even without the exposure to fluid flow, the endothelial cells without functional cilia are intrinsically different from their ciliated counterparts with respect to gene expression and, e.g., TGFβ activation status [53]. These studies were performed in cells derived from the Tg737—Oak Ridge Polycystic Kidney disease mouse model and changes were reversible upon reacquiring the primary cilium. This shows that modifying the force distribution inside the cells is not the only function of primary cilia in endothelial cells. The mechanism by which cells become inherently different upon losing their cilium is yet poorly understood. A clue can possibly be found in the properties of microtubules, including those in the ciliary axoneme, to function as substrates for cargo transport inside the cell. Another possibility is the fact that several intracellular signaling pathways rely on the presence of a primary cilium to function properly. Wnt, Hedgehog, and PDGF signaling pathways are good examples of that, reviewed in Egorova et al. [93] and van der Heiden et al. [92].

13.5.3 Loss of Cilia Is Instrumental for Heart Valve Development

Endothelial cells lose their cilia upon exposure to high and unidirectional blood flow [95]. On the other hand, cyclic flow reversals induce the protrusion of endothelial primary cilia. This is the reason for the distinct distribution of ciliated and nonciliated cells in regions of atherosclerosis and in the developing embryonic heart [105]. During the developmental stages shown in Figure 13.1, the heart is a looped tube in which so-called endocardial cushions indent the lumen at distinct locations [95,96]. These cushions consist of gelatinous matrix in between the myocardial and endothelial layers. They will eventually develop into the cardiac valves. From the atrioventricular cushions, which are located at the interface between the atrium and ventricle, the tricuspid and mitral valves will develop at the right and left side, respectively. In the outflow tract of the heart, a second pair of valves develops, the aortic and pulmonary semilunar valves in the systemic and pulmonary circulation, respectively. For proper development into valves, the cushions will have to be populated with cells which primarily derive from the endothelial cell layer in a process called endothelial-to-mesenchymal transition (EndoMT). Because the cushion indent into the lumen, local blood flow and shear stresses are high compared to the wider chambers of the heart (in this stage of development, the heart consists of only a common atrium and common ventricular chamber). Therefore, the cushion endothelial cells are nonciliated, in contrast to the cells of the atrial and ventricular chambers [95]. EndoMT is a process which relies on the activation of TGFβ signaling, especially through the activation of the activin-like kinase-5 (ALK5) receptor molecule. The presence of a primary cilium on endothelial cells inhibits the activation of ALK5 signaling in endothelial cells and thereby prevents EndoMT. On the other hand, nonciliated endothelial cells easily activate ALK5 signaling in response to fluid flow and acquire a mesenchymal (connective tissue-like) phenotype [54]. When considering this mechanism, it becomes easier to understand how the local increase in shear stress in the venous clip model affects heart development through abnormal activation of TGFβ and subsequent KLF2 signaling, with the endothelial cilia as key flow adaptations—or should we say sensors?

13.6 HEMORHEOLOGY

13.6.1 Macroscopic Rheological Behavior of Blood

Blood is a well-known example of a fluid with non-Newtonian behavior. In Newtonian fluids, a single parameter—the viscosity—is sufficient to link the deformation rate of fluid elements to the stress tensor. In other words, the internal friction that a fluid experiences is not dependent on how it is deformed. For non-Newtonian fluids, this is no longer the case. It is well known that higher deformation rates will lead to a lower effective viscosity of blood, usually referred to as "shear-thinning" behavior. The seminal work by Chien [106] explored this aspect of blood. For low shear values ($<1\,s^{-1}$), the viscosity exceeds 0.010 Pa s, while for large shear rate values ($>300\,s^{-1}$), it plateaus at values of around 0.0030-0.0035 Pa s. In the intermediate region, the effective viscosity exhibits power law behavior. The causes for this are the deformability of the red blood cells at higher shear rates and the fact that they form rouleaux (reversible stacks) at lower shear rates [106–108]. Based on experimental data from viscometry, several constitutive equations have been formulated that capture this behavior, such as the Carreau-Yasuda model [109,110]. As can be expected, the volume fraction of red blood cells present—the hematocrit—strongly influences the rheological behavior; higher hematocrit leads to more viscous blood [111]. Hematocrit can be determined by taking samples or *in vivo* [76]. Another important factor is temperature: decreasing the temperature of human blood from 36.5 to 22 °C leads to an increase in viscosity of 26%, while increasing the temperature to 39.5 °C reduces the viscosity by 11% [112]. These changes are due to a combination of the temperature-dependent viscosity of water and changes in the deformability of the red blood cells.

Virtually all experimental studies that investigate the role of WSS determine this quantity indirectly, by multiplying the local near-wall velocity gradient with the dynamics viscosity (Equation 13.4). In general, the non-Newtonian nature of blood is ignored and a single value for the viscosity is used that is close to plateau value observed at high shear rate (3-4.5 mPa s). This is motivated by the fact that generally the velocity gradients—and thus shear rates—are at a maximum near the wall. Unfortunately, the validity of this assumption is hardly ever investigated. Typical values of shear rates range from 200 to 600 s^{-1} [38] in the embryonic outflow tract, 50-70 s^{-1} in the larger vitelline arteries and veins [48] and even lower values for the remainder of the vitelline circulation [21].[5] Note that these values are for the systolic phase; they can fall to near zero during diastole. This means that using a single value for the viscosity will lead to significant errors in the estimation of the WSS, as the effective viscosity varies over an order of magnitude in this shear rate range.

13.6.2 Multiphase Aspects of Blood

For smaller blood vessels, the multiphase nature of blood will become more prominent [108,113]. From a biomechanical point of view, blood consists of plasma (mostly water with dissolved proteins, electrolytes, etc.) and red blood cells (erythrocytes). Other components, such as platelets and white blood cells (leukocytes) are much less abundant. The fact that blood is a suspension of red blood cells becomes apparent when blood is forced through geometries with a diameter that is of the same order or smaller than the dimensions of a red blood cell: the effective viscosity increases dramatically when blood is forced through capillaries smaller than 6 μm [114]. This is evident in Figure 13.11, which shows the effective viscosity as a function of diameter. As can be seen in the micrograph inset at the top, erythrocytes have to deform strongly and will only flow in single file through such blood vessels.

A second important effect is the red blood cells are not evenly distributed along the blood vessel: cells tend to migrate away from the wall, leaving a thin layer that is depleted of cells near the wall. This so-called marginal layer or cell-free layer has a thickness of the order of magnitude as the dimensions of a red blood cell [115], which is approximately 2-8 μm. The viscosity of this plasma layer is lower than the bulk viscosity of blood, leading to a significantly lower pressure drop. The thickness of the layer is relatively constant for various blood vessel diameters. For smaller blood vessels, the cell-free layer thickness is, therefore, larger in

FIGURE 13.11 Relative effective viscosity as a function of blood vessel diameter. The effective viscosity refers to the apparent viscosity of blood in a tube with a certain diameter, normalized with the viscosity of the continuous phase (medium/plasma). *(Reproduced from Pries and Secomb [114].)*

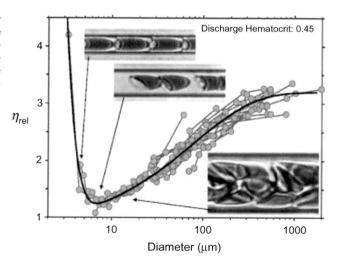

[5] An extensive listing of reported values is presented by Al-Roubaie et al. [2].

relative terms. The layer will occupy a larger fraction of the volume of the blood vessel and hence the effective viscosity will be lower. The "effective viscosity" here refers to the viscosity that is obtained when Poiseuille's law (Table 13.1) is used with a known blood flow rate, diameter, and pressure drop. The relative effective viscosity is obtained by normalization with the viscosity of the continuous phase (i.e., plasma). The phenomenon that the effective viscosity is a function of blood vessel diameter is usually referred to as the Fåhræus-Lindqvist effect [114,116]. The effect is considerable: the relative effective viscosity in blood vessels with a diameter of 6 μm is 1.25, while it increases to 3 for blood vessels of approximately 300 μm (see Figure 13.11). For diameters larger than 500 μm, the thickness of the cell-free layer is relatively small. The Fåhræus-Lindqvist effect thus becomes negligible and blood can safely be considered to be a continuous (i.e., one-phase) fluid. Note that the exact underlying mechanisms that cause the cell-free layer have not been clarified completely. Although lateral migration of single, isolated spherical particles due to rotation or inertia is understood relatively well [117], one may ask to what extent these concepts can be translated to blood flow. As shown in Figure 13.11, physiological values for the hematocrit of 40-50% mean that cell-cell interaction will be important and may even overshadow the forces that regulate the migration of dilute suspensions of solid particles.

13.6.3 Endothelial Surface Layer

The cell-free layer described above separates the bulk of the blood flow and the endothelial cells that line the blood vessel wall. However, an additional layer can be identified, which covers the endothelium: the endothelial surface layer (ESL) or glycocalyx [118,119]. Note that the two terms are often used interchangeably, but the ESL refers to the more general microenvironment that contains the glycocalyx and locally immobile plasma [118]. The thickness of the glycocalyx is variable, in the range of 0.1-1 μm, and the layer consists of a mixture of membrane-bound macromolecules [119].

While thin, the layer may have important consequences for mass and momentum transfer. Regarding mass transfer, it has been postulated that the dense network of macromolecules presents a barrier to molecules that attempt to migrate from blood into the endothelial cells. Disturbances in the layer have been linked with a wide variety of pathologies, such as diabetes, ischemia/reperfusion, and atherosclerosis [119].

The hydrodynamic relevance of the ESL was described by Smith et al. [58]. Using detailed velocity measurements, they were able to extrapolate the velocity field and found that the flow velocity reaches zero somewhat away from the wall. This indicates a reduction of the mobility of plasma due to the glycocalyx. Another implication of the ESL is that it may mitigate disturbances in the WSS. Secomb et al. [120] describe a model where the highly viscous layer forms a buffer between blood flow and endothelial cells, strongly attenuating rapid fluctuations. Furthermore, the layer plays an important role in mechanotransduction (see, e.g., the review by [121]).

In summary, it can be remarked that the exact details of the microenvironment near the blood vessel wall are unknown at this moment. The WSS experienced by endothelial cells will be influenced by the presence of a glycocalyx, the cell-free layer, and even the motion of individual blood cells [57]. The assumption that endothelial cells simply experience a force proportional to the product of a well-defined velocity gradient and viscosity may be overly simplistic. More research is needed to establish under which conditions this simplistic view will hold.

13.6.4 Human Versus Avian Blood

Due to the scarcity of experimental data available in the literature, data from various model systems are often directly compared. It is important to note that there can be large differences between species with regards to hemorheology. For instance, the size of red blood cells and the average hematocrit values will vary considerably [122,123]. Often, rheological parameters—in particular viscosity—obtained in human blood are directly used in the chicken embryo studies [23,34]. However, the hematocrit of chicken blood is significantly lower—it increases linearly during development from 19.4% at day 4 to a plateau of 33% at day 16 [124,125]. Note that determining these parameters during early development (HH12 to HH20) are especially cumbersome due to the minute volumes of blood.

Another fundamental difference between avian and mammalian blood is that the former are nucleated. The study by Gaehtgens et al. [126] showed that the nucleated avian cells deform less when forced through small capillaries; under similar conditions, the difference in deformation is approximately a factor of two. Human red blood cells are biconcave disc shaped and this leads to a large surface-to-volume ratio, facilitating rouleaux formation. As the nucleated avians cells are more spherical in shape, they are less prone to rouleaux formation [124,127].

The combined effects of the lower hematocrit and nucleated cells of avian blood lead to significantly different rheological behavior. The recent work by Al-Roubaie et al. [124] provided, for the first time, a thorough analysis of blood that confirms these changes. In Figure 13.12, the dynamic viscosity is shown as a function of shear rate, for three different

FIGURE 13.12   The dynamic viscosity of blood (human and avian) as a function of shear rate. The data for human blood are based on the Carreau-Yasuda model given by Gijsen et al. [110] based on original data by Thurston [111] and are for blood with a hematocrit of 44%. The data for avian blood are for hematocrits of 17%, 25%, and 52% at day 4 and was obtained by Al-Roubaie et al. [124].

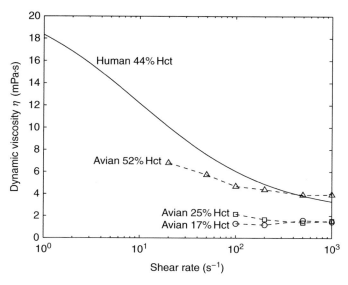

hematocrit values. Note that during early development (i.e., the first few days) the hematocrit is around 20%. Two main differences with human blood rheology are evident: (1) the viscosity is significantly lower and (2) the viscosity is relatively constant (i.e., the shear-thinning effects are diminished). The former effect can be attributed to the lower hematocrit, while the latter is linked with the lower deformability and inability to form rouleaux. A similar result was obtained earlier by Gaehtgens et al. [126]. This implies that using a constant viscosity in WSS measurements may be a reasonable approach after all, provided that the correct value for each particular developmental stage is used (as the stage determines the hematocrit and thus effectively the viscosity). Further evidence for this assumption is the observation mentioned in Section 13.3 that blood velocity profiles in the developing chicken embryo vasculature are parabolic (see also Figure 13.4). A strongly shear-thinning fluid would have lead to blunted profiles [110].

13.7 CONCLUSIONS AND OUTLOOK

The chicken embryo system is an accessible and versatile model for the human cardiovascular system. It allows state-of-the-art measurement techniques to study the role of hemodynamics during development, both normal and with interventions. The flow is dominated by viscous forces so that many phenomena that occur at later stages can be ignored: effects due to curvature, pulsatility, or turbulence are not likely to occur. Furthermore, the rheology of avian blood is much simpler than that of human blood; for most studies, a constant viscosity will give reasonable results. Due to this simplicity, having only two parameters for each blood vessel segment is sufficient to capture the local hemodynamics.

Experimental techniques have progressed very rapidly in the last decade, from quantitative visualizations to two-, three-, and even four-dimensional (i.e., time-resolved) flow measurements. With these techniques, virtually all hemodynamic parameters can be obtained (flow rates, WSS). The stage of development favors some modalities due to resolution and tissue penetration capabilities. An exciting new possibility is the combination of flow measurements with some form of functional imaging. Flow is important but is a means to an end. Dual-wavelength imaging systems will provide flow with concentration information so that the actual mass transport (the reason for circulation) can be studied. Such studies can lead to major breakthroughs with respect to the regulation and adaptation of the cardiovascular system.

Important open questions remain for the microscopic environment near the blood vessel wall. It is well established that endothelial cells respond to flow characteristics, and in particular to WSS. The exact mechanisms within the cell that are involved are still not completely resolved. Furthermore, the flow field at these scales is more complicated than most studies assume. Instead of a single-fluid with a constant viscosity, two-phase phenomena may start to play a role. The behavior of individual red blood cells will become relevant, for instance in the cell-free layer. The glycocalyx will also influence how forces are transmitted from flow to cells. A comprehensive model that encapsulates all these phenomena is currently not available.

These two major opportunities—combined flow/concentration measurements and microscale hemodynamics—ensure that the chicken embryo will be a common model system for many years to come.

References

[1] Chapman WB. The effect of the heart-beat upon the development of the vascular system in the chick. Am J Anat 1918;23(1):175–203.

[2] Clark ER, Clark EL. Microscopic observations on the growth of blood capillaries in the living mammal. Am J Anat 1939;64(2):251–301.

[3] Hierck BP, Van der Heiden K, Poelma C, Westerweel J, Poelmann RE. Fluid shear stress and inner curvature remodeling of the embryonic heart. Choosing the right lane! Sci World J 2008;8:212–22.

[4] Hove JR, Koster RW, Forouhar AS, Acevedo-Bolton G, Fraser SE, Gharib M. Intracardiac fluid forces are an essential epigenetic factor for embryonic cardiogenesis. Nature 2003;421(6919):172–7.

[5] Hogers B, De Ruiter MC, Grittenberger-de Groot AC, Poelmann RE. Unilateral vitelline vein ligation alters intracardiac blood flow patterns and morphogenesis in the chick embryo. Circ Res 1997;80:473–81.

[6] Le Noble F, Moyo D, Pardanaud L, Yuan L, Djonov V, Mathijssen R, et al. Flow regulates arterial-venous differentiation in the chick embryo yolk sac. Development 2004;132(2):361–75.

[7] Lucitti JL, Jones EAV, Huang C, Chen J, Fraser SE, Dickinson ME. Vascular remodeling of the mouse yolk sac requires hemodynamic force. Development 2007;134(18):3317–26.

[8] Stern CD. The chick: a great model system becomes even greater. Dev Cell 2005;8(1):9–17.

[9] Chapman SC, Collignon J, Schoenwolf GC, Lumsden A. Improved method for chick whole-embryo culture using a filter paper carrier. Dev Dyn 2001;220(3):284–9.

[10] Korn MJ, Cramer KS. Windowing chicken eggs for developmental studies. J Visual Exp 2007;8:306.

[11] Hamburger V, Hamilton HL. A series of normal stages in the development of the chick embryo. J Morphol 1951;88(1):49–92.

[12] Burggren WW, Warburton SJ, Slivkoff MD. Interruption of cardiac output does not affect short-term growth and metabolic rate in day 3 and 4 chick embryos. J Exp Biol 2000;203(24):3831–8.

[13] Männer J. Cardiac looping in the chick embryo: a morphological review with special reference to terminological and biomechanical aspects of the looping process. Anat Rec 2000;259(3):248–62.

[14] Manasek FJ, Monroe RG. Early cardiac morphogenesis is independent of function. Dev Biol 1972;27(4):584–8.

[15] Oosterbaan AM, Ursem NTC, Struijk PC, Bosch JG, Van Der Steen AFW, Steegers EAP. Doppler flow velocity waveforms in the embryonic chicken heart at developmental stages corresponding to 5–8 weeks of human gestation. Ultrasound Obstet Gynecol 2009;33(6):638–44.

[16] Patten BM. Early embryology of the chick. Rockville, MD. Wildside Press; 1951.

[17] Bellairs R, Osmond M. The atlas of chick development. London: Elsevier Academic Press; 2005.

[18] Risau W, Flamme I. Vasculogenesis. Annu Rev Cell Dev Biol 1995;11(1):73–91.

[19] Wilting J, Christ B. Embryonic angiogenesis: a review. Naturwissenschaften 1996;83(4):153–64.

[20] Jones EAV, le Noble F, Eichmann A. What determines blood vessel structure? Genetic prespecification vs. hemodynamics. Physiology 2006;21(6):388–95.

[21] Kloosterman A, Hierck BP, Westerweel J, Poelma C. Quantification of blood flow and topology in developing vascular networks. PLoS One 2014;9(5):e96856.

[22] Murray CD. The physiological principle of minimum work: I. the vascular system and the cost of blood volume. Proc Natl Acad Sci USA 1926;12(3):207.

[23] Lee JY, Lee SJ. Hemodynamics of the omphalo-mesenteric arteries in stage 18 chicken embryos and flow-structure relations for the microcirculation. Microvasc Res 2010;80(3):402–11.

[24] DeFouw DO, Rizzo VJ, Steinfeld R, Feinberg RN. Mapping of the microcirculation in the chick chorioallantoic membrane during normal angiogenesis. Microvasc Res 1989;38(2):136–47.

[25] White F. Viscous fluid flow. New York: Tata McGraw-Hill Education; 1974.

[26] McDonald DA. Blood flow in arteries. London: Edward Arnold; 1974.

[27] Pope SB. Turbulent flows. Cambridge, UK: Cambridge University Press; 2000.

[28] Berger SA, Talbot L, Yao LS. Flow in curved pipes. Annu Rev Fluid Mech 1983;15(1):461–512.

[29] Siggers JH, Waters SL. Steady flows in pipes with finite curvature. Phys Fluids 2005;17(7):077102.

[30] Dean WR. Note on the motion of fluid in a curved pipe. Phil Mag S 7 1927;4(20):208–23.

[31] Vincent PE, Plata AM, Hunt AAE, Weinberg PD, Sherwin SJ. Blood flow in the rabbit aortic arch and descending thoracic aorta. J R Soc Interface 2011;8(65):1708–19.

[32] Caro CG, Pedley TJ, Schroter RC, Seed WA, Parker KH. The mechanics of the circulation, vol. 192633236. Oxford: Oxford University Press; 1978.

[33] Ward-Smith AJ. Internal fluid flow: the fluid dynamics of flow in pipes and ducts. Oxford: Clarendon Press; 1980.

[34] Vennemann P, Kiger KT, Lindken R, Groenendijk BCW, Stekelenburg-de Vos S, Ten Hagen TLM, et al. In vivo micro particle image velocimetry measurements of blood-plasma in the embryonic avian heart. J Biomech 2006;39:1191–200.

[35] Wang CY, Bassingthwaighte JB. Blood flow in small curved tubes. J Biomech Eng 2003;125(6):910.

[36] Vennemann P, Lindken R, Westerweel J. In vivo whole-field blood velocity measurement techniques. Exp Fluids 2007;42:495–511.

[37] Alastruey J, Siggers JH, Peiffer V, Doorly D, Sherwin S. Reducing the data: analysis of the role of vascular geometry on blood flow patterns in curved vessels. Phys Fluids 2012;24:031902.

[38] Poelma C, Mari JM, Foin N, Tang M-X, Krams R, Caro CG, et al. 3D flow reconstruction using ultrasound PIV. Exp Fluids 2011;50(4):777–85.

[39] Wang Y, Dur O, Patrick MJ, Tinney JP, Tobita K, Keller BB, et al. Aortic arch morphogenesis and flow modeling in the chick embryo. Ann Biomed Eng 2009;37(6):1069–81.

[40] Clark EB, Hu N. developmental hemodynamic changes in the chick embryo from stage 18 to 27. Circ Res 1982;51(6):810–5.

[41] Hu N, Clark EB. Hemodynamics of the stage 12 to stage 29 chick embryo. Circ Res 1989;65(6):1665–70.

[42] Fung Y-C. Biomechanics: circulation. New York: Springer; 1997.

[43] Baker M, Wayland H. On-line volume flow rate and velocity profile measurement for blood in microvessels. Microvasc Res 1974;7(1):131–43.

[44] Pittman RN, Ellsworth ML. Estimation of red cell flow in microvessels: consequences of the baker-wayland spatial averaging model. Microvasc Res 1986;32(3):371–88.

[45] Groenendijk BCW, Stekelenburg-de Vos S, Vennemann P, Wladimiroff JW, Nieuwstadt FTM, Lindken R, et al. The endothelin-1 pathway and the development of cardiovascular defects in the haemodynamically challenged chicken embryo. J Vasc Res 2007;45(1):54–68.

[46] Lee JY, Ji HS, Lee SJ. Micro-PIV measurements of blood flow in extraembryonic blood vessels of chicken embryos. Physiol Meas 2007;28(10):1149–62.

[47] Poelma C, Kloosterman A, Hierck BP, Westerweel J. Accurate blood flow measurements: are artificial tracers necessary? PLoS One 2012;7(9):e45247.

[48] Poelma C, Vennemann P, Lindken R, Westerweel J. In vivo blood flow and wall shear stress measurements in the vitelline network. Exp Fluids 2008;45(4):703–13.

[49] Männer J, Thrane L, Norozi K, Yelbuz TM. In vivo imaging of the cyclic changes in cross-sectional shape of the ventricular segment of pulsating embryonic chick hearts at stages 14 to 17: a contribution to the understanding of the ontogenesis of cardiac pumping function. Dev Dyn 2009;238(12):3273–84.

[50] Lee S-J, Yeom E, Ha H, Nam K-H. Cardiac outflow and wall motion in hypothermic chick embryos. Microvasc Res 2011;82(3):296–303.

[51] Reichold J, Stampanoni M, Keller AL, Buck A, Jenny P, Weber B. Vascular graph model to simulate the cerebral blood flow in realistic vascular networks. J Cereb Blood Flow Metabol 2009;29(8):1429–43.

[52] Poelma C, Van der Heiden K, Hierck BP, Poelmann RE, Westerweel J. Measurements of the wall shear stress distribution in the outflow tract of an embryonic chicken heart. J R Soc Interface 2010;7(42):91–103.

[53] Egorova AD, Khedoe PPSJ, Goumans MJTH, Yoder BK, Nauli SM, ten Dijke P, et al. Lack of primary cilia primes shear-induced endothelial-to-mesenchymal transition novelty and significance. Circ Res 2011;108(9):1093–101.

[54] Egorova AD, Van der Heiden K, Van de Pas S, Vennemann P, Poelma C, DeRuiter MC, et al. Tgfβ/alk5 signaling is required for shear stress induced klf2 expression in embryonic endothelial cells. Dev Dyn 2011;240(7):1670–80.

[55] Fraser GM, Milkovich S, Goldman D, Ellis CG. Mapping 3-d functional capillary geometry in rat skeletal muscle in vivo. Am J Physiol Heart Circ Physiol 2012;302(3):H654–64.

[56] Kamoun WS, Chae S-S, Lacorre DA, Tyrrell JA, Mitre M, Gillissen MA, et al. Simultaneous measurement of rbc velocity, flux, hematocrit and shear rate in vascular networks. Nat Methods 2010;7(8):655–60.

[57] Patrick MJ, Chen C-Y, Frakes DH, Dur O, Pekkan K. Cellular-level near-wall unsteadiness of high-hematocrit erythrocyte flow using confocal μpiv. Exp Fluids 2011;50(4):887–904.

[58] Smith ML, Long DS, Damiano ER, Ley K. Near-wall μ-PIV reveals a hydrodynamically relevant endothelial surface layer in venules in vivo. Biophys J 2003;85(1):637–45.

[59] Lee JY, Lee SJ. Thermal effect on heart rate and hemodynamics in vitelline arteries of stage 18 chicken embryos. J Biomech 2010;43(16):3217–21.

[60] Oosterbaan AM, Poelma C. Bon E, Steegers-Theunissen RPM, Steegers EAP, Vennemann P, et al. Induction of shear stress in the homocysteine exposed chicken embryo J Med Biol Eng 2014;34(1):56–61.

[61] Bronzino JD. Biomedical engineering fundamentals. Boca Raton: CRC Press; 2006.

[62] Stekelenburg-de Vos S, Steendijk P, Ursem NTC, Wladimiroff JW, Delfos R, Poelmann RE. Systolic and diastolic ventricular function assessed by pressure-volume loops in the stage 21 venous clipped chick embryo. Pediatr Res 2005;57(1):16–21.

[63] Lipowsky HH, Kovalcheck S, Zweifach BW. The distribution of blood rheological parameters in the microvasculature of cat mesentery. Circ Res 1978;43(5):738–49.

[64] Riva C, Ross B, Benedek GB. Laser doppler measurements of blood flow in capillary tubes and retinal arteries. Invest Ophthalmol Vis Sci 1972;11(11):936–44.

[65] Nishihara K, Iwasaki W, Nakamura M, Higurashi E, Soh T, Itoh T, et al. Development of a wireless sensor for the measurement of chicken blood flow using the laser doppler blood flow meter technique. IEEE Trans Biomed Eng 2013;60:1645–53.

[66] Wayland H, Johnson PC. Erythrocyte velocity measurement in microvessels by a two-slit photometric method. J Appl Physiol 1967;22(2):333–7.

[67] Roman S, Lorthois S, Duru P, Risso F. Velocimetry of red blood cells in microvessels by the dual-slit method: Effect of velocity gradients. Microvasc Res 2012;84:249–61.

[68] Lindken R, Rossi M, Große S, Westerweel J. Micro-particle image velocimetry (μpiv): Recent developments, applications, and guidelines. Lab Chip 2009;9(17):2551–67.

[69] Klitzman B, Johnson PC. Capillary network geometry and red cell distribution in hamster cremaster muscle. Am J Physiol Heart Circ Physiol 1982;242(2):H211–9.

[70] Poelma C, Westerweel J. Generalized displacement estimation for averages of non-stationary flows. Exp Fluids 2011;50(5):1421–7.

[71] Tangelder GJ, Teirlinck HC, Slaaf DW, Reneman RS. Distribution of blood platelets flowing in arterioles. Am J Physiol Heart Circ Physiol 1985;248(3):H318–23.

[72] Kloosterman A, Poelma C, Westerweel J. Flow rate estimation in large depth-of-field micro-piv. Exp Fluids 2011;50(6):1587–99.

[73] Pawley JB. Handbook of biological confocal microscopy. New York: Springer; 1995.

[74] Jones EAV, Baron MH, Fraser SE, Dickinson ME. Measuring hemodynamic changes during mammalian development. Am J Physiol Heart Circ Physiol 2004;287(4):H1561–9.

[75] Landolt A, Obrist D, Wyss M, Barrett M, Langer D, Jolivet R, et al. Two-photon microscopy with double-circle trajectories for in vivo cerebral blood flow measurements. Exp Fluids 2013;54(5):1–8.

[76] Desjardins C, Duling BR. Microvessel hematocrit: measurement and implications for capillary oxygen transport. Am J Physiol Heart Circ Physiol 1987;252(3):H494–503.

[77] Huang D, Swanson EA, Lin CP, Schuman JS, Stinson WG, Chang W, et al. Optical coherence tomography. Science 1991;254(5035):1178–81.

[78] Jenkins MW, Adler DC, Gargesha M, Huber R, Rothenberg F, Belding J, et al. Ultrahigh-speed optical coherence tomography imaging and visualization of the embryonic avian heart using a buffered fourier domain mode locked laser. Opt Express 2007;15(10):6251–67.

[79] Davis A, Izatt J, Rothenberg F. Quantitative measurement of blood flow dynamics in embryonic vasculature using spectral doppler velocimetry. Anat Rec 2009;292(3):311–9.

[80] Liu A, Yin X, Shi L, Li P, Thornburg KL, Wang R, et al. Biomechanics of the chick embryonic heart outflow tract at hh18 using 4d optical coherence tomography imaging and computational modeling. PLoS One 2012;7(7):e40869.

[81] Ma Z, Liu A, Yin X, Troyer A, Thornburg K, Wang RK, et al. Measurement of absolute blood flow velocity in outflow tract of hh18 chicken embryo based on 4d reconstruction using spectral domain optical coherence tomography. Biomed Optics Express 2010;1(3):798.

[82] Peterson LM, Jenkins MW, Gu S, Barwick L, Watanabe M, Rollins AM. 4d shear stress maps of the developing heart using doppler optical coherence tomography. Biomed Optics Express 2012;3(11):3022.

[83] Gielen S, Sandri M, Erbs S, Adams V. Exercise-induced modulation of endothelial nitric oxide production. Curr Pharm Biotechnol 2011;12(9):1375–84.

[84] Hogers B, DeRuiter MC, Gittenberger-de Groot AC, Poelmann RE. Extraembryonic venous obstructions lead to cardiovascular malformations and can be embryolethal. Cardiovasc Res 1999;41(1):87–99.

[85] Saiki Y, Konig A, Waddell J, Rebeyka IM. Hemodynamic alteration by fetal surgery accelerates myocyte proliferation in fetal guinea pig hearts. Surgery 1997;122(2):412–9.

[86] Groenendijk BCW, Hierck BP, Gittenberger-de Groot AC, Poelmann RE. Development-related changes in the expression of shear stress responsive genes klf-2, et-1, and nos-3 in the developing cardiovascular system of chicken embryos. Dev Dyn 2004;230(1):57–68.

[87] Groenendijk BCW, Hierck BP, Vrolijk J, Baiker M, Pourquie MJBM, Gittenberger-de Groot AC, et al. Changes in shear stress-related gene expression after experimentally altered venous return in the chicken embryo. Circ Res 2005;96(12):1291–8.

[88] Groenendijk BCW, Van der Heiden K, Hierck BP, Poelmann RE. The role of shear stress on ET-1, KLF2, and NOS-3 expression in the developing cardiovascular system of chicken embryos in a venous ligation model. Physiology 2007;22(6):380.

[89] Broekhuizen MLA, Hogers B, DeRuiter MC, Poelmann RE, Gittenberger-De Groot AC, Wladimiroff JW, et al. Altered hemodynamics in chick embryos after extraembryonic venous obstruction. Ultrasound Obstet Gynecol 1999;13(6):437–45.

[90] Boon RA, Horrevoets AJ, et al. Key transcriptional regulators of the vasoprotective effects of shear stress. Hamostaseologie 2009;29(1):39–40.

[91] Clement CA, Ajbro KD, Koefoed K, Vestergaard ML, Veland IR, Henriques de Jesus MPR, et al. Tgf-β signaling is associated with endocytosis at the pocket region of the primary cilium. Cell Rep 2013;3(6):1806–14.

[92] Van der Heiden K, Egorova AD, Poelmann RE, Wentzel JJ, Hierck BP. Role for primary cilia as flow detectors in the cardiovascular system. Int Rev Cell Mol Biol 2011;290:87–119.

[93] Egorova AD, van der Heiden K, Poelmann RE, Hierck BP. Primary cilia as biomechanical sensors in regulating endothelial function. Differentiation 2012;83(2):S56–61.

[94] Singla V, Reiter JF. The primary cilium as the cell's antenna: signaling at a sensory organelle. Science 2006;313 (5787):629–33.

[95] Van der Heiden K, Groenendijk BP, Hierck BP, Hogers B, Koerten HK, Mommaas AM, et al. Monocilia on chicken embryonic endocardium in low shear stress areas. Dev Dyn 2006;235(1):19.

[96] ten Dijke P, Egorova AD, Goumans MJTH, Poelmann RE, Hierck BP. Tgf-β signaling in endothelial-to-mesenchymal transition: the role of shear stress and primary cilia. Sci. Signal 2012;5(212):pt2.

[97] Abdul-Majeed S, Moloney BC, Nauli SM. Mechanisms regulating cilia growth and cilia function in endothelial cells. Cell Mol Life Sci 2012;69(1):165–73.

[98] Nauli SM, Jin X, AbouAlaiwi WA, El-Jouni W, Su X, Zhou J, et al. Non-motile primary cilia as fluid shear stress mechanosensors. Methods Enzymol 2012;525:1–20.

[99] Nauli SM, Kawanabe Y, Kaminski JJ, Pearce WJ, Ingber DE, Zhou J. Endothelial cilia are fluid shear sensors that regulate calcium signaling and nitric oxide production through polycystin-1. Circulation 2008;117(9):1161–71.

[100] Ando J, Yamamoto K. Flow detection and calcium signaling in vascular endothelial cells. Cardiovasc Res 2013;99:260–8.

[101] Chrétien ML, Zhang M, Jackson MR, Kapus A, Langille BL. Mechanotransduction by endothelial cells is locally generated, direction-dependent, and ligand-specific. J Cell Physiol 2010;224(2):352–61.

[102] Hoey DA, Downs ME, Jacobs CR. The mechanics of the primary cilium: an intricate structure with complex function. J Biomech 2012;45(1):17–26.

[103] Besschetnova TY, Kolpakova-Hart E, Guan Y, Zhou J, Olsen BR, Shah JV. Identification of signaling pathways regulating primary cilium length and flow-mediated adaptation. Curr Biol 2010;20(2):182–7.

[104] Masyuk AI, Masyuk TV, Splinter PL, Huang BQ, Stroope AJ, LaRusso NF. Cholangiocyte cilia detect changes in luminal fluid flow and transmit them into intracellular Ca^{2+} and camp signaling. Gastroenterology 2006;131 (3):911–20.

[105] Van der Heiden K, Hierck BP, Krams R, de Crom R, Cheng C, Baiker M, et al. Endothelial primary cilia in areas of disturbed flow are at the base of atherosclerosis. Atherosclerosis 2008;196(2):542–50.

[106] Chien S. Shear dependence of effective cell volume as a determinant of blood viscosity. Science 1970;168:977–9.

[107] Baskurt OK, Meiselman HJ. Blood rheology and hemodynamics. In: Seminars in thrombosis and hemostasis, vol. 29. New York: Stratton Intercontinental Medical Book Corporation; 2003. p. 435–50.

[108] Popel AS, Johnson PC. Microcirculation and hemorheology. Annu Rev Fluid Mech 2005;37(1):43–69.

[109] Bird RB, Armstrong RC, Hassager O. Dynamics of polymer liquids. Fluid mechanics, vol. 1. New York: Wiley & Sons; 1987.

[110] Gijsen FJH, Van de Vosse FN, Janssen JD. The influence of the non-newtonian properties of blood on the flow in large arteries: steady flow in a carotid bifurcation model. J Biomech 1999;32(6):601–8.

[111] Thurston GB. Rheological parameters for the viscosity viscoelasticity and thixotropy of blood. Biorheology 1978;16(3):149–62.

[112] Çinar Y, Şenyol AM, Duman K. Blood viscosity and blood pressure: role of temperature and hyperglycemia. Am J Hypertens 2001;14(5):433–8.

[113] Lipowsky HH. Microvascular rheology and hemodynamics. Microcirculation 2005;12:5–15.

[114] Pries AR, Secomb TW. Rheology of the microcirculation. Clin Hemorheol Microcirc 2003;29(3):143–8.

[115] Kim S, Ong PK, Yalcin O, Intaglietta M, Johnson PC. The cell-free layer in microvascular blood flow. Biorheology 2009;46(3):181–9.

[116] Fåhræus R, Lindqvist T. The viscosity of the blood in narrow capillary tubes. Am J Physiol Legacy Content 1931;96(3):562–8.

[117] Matas JP, Morris JF, Guazzelli E. Lateral forces on a sphere. Oil Gas Sci Technol 2004;59(1):59–70.

[118] Pries AR, Secomb TW, Gaehtgens P. The endothelial surface layer. Pflügers Arch Eur J Physiol 2000;440 (5):653–66.

[119] Reitsma S, Slaaf DW, Vink H, van Zandvoort MAMJ, Oude Egbrink MGA. The endothelial glycocalyx: composition, functions, and visualization. Pflügers Arch Eur J Physiol 2007;454(3):345–59.

[120] Secomb TW, Hsu R, Pries AR. Effect of the endothelial surface layer on transmission of fluid shear stress to endothelial cells. Biorheology 2001;38(2–3):143–50.

[121] Weinbaum S, Zhang X, Han Y, Vink H, Cowin SC. Mechanotransduction and flow across the endothelial glycocalyx. Proc Natl Acad Sci USA 2003;100(13):7988–95.

[122] Duncan JR, Prasse KW, Mahaffey EA, et al. Veterinary laboratory medicine: clinical pathology. Ames: Iowa State University Press; 1994.

[123] Gulliver G. Observations on the sizes and shapes of the red corpuscles of the blood of vertebrates, with drawings of them to a uniform scale, and extended and revised tables of measurements. Proc Zool Soc Lond 1875;474–95.

[124] Al-Roubaie S, Jahnsen E, Mohammed M, Henderson-Toth C, Jones E. Rheology of embryonic avian blood. Am J Physiol Heart Circ Physiol 2011;301(6):H2473–81.

[125] Johnston PM. Hematocrit values for the chick embryo at various ages. Am J Physiol Legacy Content 1955;180 (2):361–2.

[126] Gaehtgens P, Will G, Schmidt F. Comparative rheology of nucleated and non-nucleated red blood cells. Pflugers Arch 1981;390(3):283–7.

[127] Mathur UB, Adaval SK. Laboratory studies on avian blood under simulated crash conditions. Ind J Aerospace Med 2007;Special Commemorative Volume:p. 52.

Index

Printed in the United States
By Bookmasters